Graduate Texts in Mathematics 236

T0186023

Graduate Texts in Mathematics

(continued after index)

John B. Garnett

Bounded Analytic Functions

Revised First Edition

 Springer

John B. Garnett
Department of Mathematics
6363 Mathematical Sciences
University of California, Los Angeles
Los Angeles, CA. 90095–1555
USA
jbg@math.ucla.edu

Previously published by Academic Press, Inc.
San Diego, CA 92101

Mathematics: Subject Classification (2000): 30D55 30–02 46315

ISBN 978-1-4419-2216-8   e-ISBN 978-0-387-49763-1

Printed on acid-free paper.

springer.com

To Dolores

Preface to Revised First Edition

This edition of *Bounded Analytic Functions* is the same as the first edition except for the corrections of several mathematical and typographical errors. I thank the many colleagues and students who have pointed out errors in the first edition. These include S. Axler, C. Bishop, A. Carbery, K. Dyakonov, J. Handy, V. Havin, H. Hunziker, P. Koosis, D. Lubinsky, D. Marshall, R. Mortini, A. Nicolau, M. O'Neill, W. Rudin, D. Sarason, D. Suárez, C. Sundberg, C. Thiele, S. Treil, I. Uriarte-Tuero, J. Väisälä, N. Varopoulos, and L. Ward.

I had planned to prepare a second edition with an updated bibliography and an appendix on results new in the field since 1981, but that work has been postponed for too long. In the meantime several excellent related books have appeared, including M. Andersson, *Topics in Complex Analysis*; G. David and S. Semmes, *Singular Integrals and Rectifiable Sets in $\mathbb{R}^n$* and *Analysis of and on Uniformly Rectifiable Sets*; S. Fischer, *Function theory on planar domains*; P. Koosis, *Introduction to H_p spaces, Second edition*; N. Nikolski, *Operators, Functions, and Systems*; K. Seip, *Interpolation and Sampling in Spaces of Analytic Functions*; and B. Simon, *Orthogonal Polynomials on the Unit Circle*.

Several problems posed in the first edition have been solved. I give references only to Mathematical Reviews. The question page 167 on when $\mathcal{E}_\infty$ contains a Blaschke product was settled by A. Stray in MR 0940287. M. Papadimitrakis, MR 0947674, gave a counterexample to the conjecture in Problem 5 page 170. The late T. Wolff, MR 1979771, had a counterexample to the Baernstein conjecture cited on page 260. S. Treil resolved the g^2 problem on page 319 in MR 1945294. A constructive Fefferman-Stein decomposition of functions in BMO($\mathbb{R}^n$) was given by the late A. Uchiyama in MR 1007515, and C. Sundberg, MR 0660188, found a constructive proof of the Chang-Marshall theorem. Problem 5.3 page 420 was resolved by Garnett and Nicolau, MR 1394402, using work of Marshall and Stray MR 1394401. Problem 5.4. on page 420 remains a puzzle, but Hjelle and Nicolau (Pacific Journal of Mathematics, 2006) have an interesting result on approximation of moduli. P. Jones, MR 0697611, gave a construction of the P. Beurling linear operator of interpolation.

I thank Springer and F. W. Gehring for publishing this edition.

John Garnett

Contents

List of Symbols

Symbol	Page	Symbol	Page
L^1_{loc}	215	$T(Q)$	290
$L^p_{\mathbb{R}}$	236	$u^+(t)$	116
$L^\infty_{\mathbb{R}}$	64	$u^*(t)$	27
$L_z(\zeta)$	392	$\tilde{u}(z)$	98
$\log^+|f(z)|$	33	UC	242
$\ell(Q)$	278	U_A	365
L	394	$v_r(z)$	35
$m(f) = \hat{f}(m) = f(m)$	180	VMO	242
$m(\lambda)$	20	VMOA	274
$M(T)$	128	VMO_A	375
$Mf(x)$	21	X	181
$M(d\mu)(t)$	28	X_ζ	207
$M_\delta(\varphi)$	242	$\beta\mathbb{N}$	180
$M_\mu f(x)$	45	$\delta(B)$	327
$\mathfrak{M}$	183	$\Delta(c, R)$	3
$\mathfrak{M}_A$	178	Δu	10
$\mathfrak{M}_\zeta$	183	$\varepsilon(\rho)$	250, 258
$\mathfrak{M}^D$	394	$\Gamma_\alpha(t)$	21
N	66	$\Gamma_\alpha(e^{i\theta})$	23
$N(\sigma)$	30	$\Gamma_\alpha^h(t)$	379
N^+	68	λ_φ	233
$N_\varepsilon(x)$	339	λ^*	433
norms: $\|f\|^p_{H^p}$	48, 49	Λ_α	102
$\|\varphi\|_*$	216	Λ^*	273
$\|\varphi\|'$	250	μ_φ	233
$\|u\|_{H_1}$	236	ν_f	371
$P_{z_0}(\theta)$	10	$\rho(z, w)$	2
$P_{z_0}(t)$	11	$\rho(m_1, m_2)$	392
$P(m)$	393	$\rho(S, T)$	305
$Q_r(\varphi)$	99	φ_I	216
$Q_z(\varphi)$	98	$\omega(\delta)$	101
QC	367	$\omega_f(\delta)$	101
Q_A	366	∇_g	228
$\mathbb{R}$ = real numbers	12	$|\nabla_g|^2$	228
$\text{R}(f, z_0)$	77	$*$	13
$S(\theta_0, h)$	371		
$\tilde{S}_j$	334		
S_h	153		
T	98		

$\bar{z}$ = complex conjugate of z

$\bar{f}$ = complex conjugate of f

$\bar{\mathfrak{F}} = \{\bar{f} : f \in \mathfrak{F}\}$, when $\mathfrak{F}$ is a set of functions

$\bar{E}$ = closure of E, when E is a point set

I

Preliminaries

As a preparation, we discuss three topics from elementary real or complex analysis which will be used throughout this book.

The first topic is the invariant form of Schwarz's lemma. It gives rise to the pseudohyperbolic metric, which is an appropriate metric for the study of bounded analytic functions. To illustrate the power of the Schwarz lemma, we prove Pick's theorem on the finite interpolation problem

$$f(z_j) = w_j, \qquad j = 1, 2, \dots, n,$$

with $|f(z)| \leq 1$.

The second topic is from real analysis. It is the circle of ideas relating Poisson integrals to maximal functions.

The chapter ends with a brief introduction to subharmonic functions and harmonic majorants, our third topic.

1. Schwarz's Lemma

Let D be the unit disc $\{z : |z| < 1\}$ in the complex plane and let B denote the set of analytic functions from D into $\overline{D}$. Thus $|f(z)| \leq 1$ if $f \in$ B . The simple but surprisingly powerful Schwarz lemma is this:

Lemma 1.1. *If $f(z) \in$ B , and if $f(0) = 0$, then*

(1.1)
$$|f(z)| \leq |z|, \quad z \neq 0,$$
$$|f'(0)| \leq 1 .$$

Equality holds in (1.1) at some point z if and only if $f(z) = e^{i\varphi}z$, φ a real constant.

The proof consists in observing that the analytic function $g(z) = f(z)/z$ satisfies $|g| \leq 1$ by virtue of the maximum principle.

We shall use the invariant form of Schwarz's lemma due to Pick. A *Möbius transformation* is a conformal self-map of the unit disc. Every Möbius transformation can be written as

$$\tau(z) = e^{i\varphi} \frac{z - z_0}{1 - \bar{z}_0 z}$$

with φ real and $|z_0| \le 1$. With this notation we have displayed $z_0 = \tau^{-1}(0)$.

Lemma 1.2. *If $f(z) \in B$, then*

(1.2) $$\frac{|f(z) - f(z_0)|}{|1 - \overline{f(z_0)} f(z)|} \le \left| \frac{z - z_0}{1 - \bar{z}_0 z} \right|, \quad z \ne z_0,$$

and

(1.3) $$\frac{|f'(z)|}{1 - |f(z)|^2} \le \frac{1}{1 - |z|^2}.$$

Equality holds at some point z if and only if $f(z)$ is a Möbius transformation.

The proof is the same as the proof of Schwarz's lemma if we regard $\tau(z)$ as the independent variable and

$$\frac{f(z) - f(z_0)}{1 - \overline{f(z_0)} f(z)}$$

as the analytic function. Letting z tend to z_0 in (1.2) gives (1.3) at $z = z_0$, an arbitrary point of D.

The *pseudohyperbolic distance* on D is defined by

$$\rho(z, w) = \left| \frac{z - w}{1 - \bar{w} z} \right|.$$

Lemma 1.2 says that analytic mappings from D to D are Lipschitz continuous in the pseudohyperbolic distance:

$$\rho(f(z), f(w)) \le \rho(z, w).$$

The lemma also says that the distance $\rho(z, w)$ is invariant under Möbius transformations:

$$\rho(z, w) = \rho(\tau(z), \tau(w)).$$

We write $K(z_0, r)$ for the noneuclidean disc

$$K(z_0, r) = \{z : \rho(z, z_0) < r\}, \quad 0 < r < 1.$$

Since the family B is invariant under the Möbius transformations, the study of the restrictions to $K(z_0, r)$ of functions in B is the same as the study of their restrictions to $K(0, r) = \{|w| < r\}$. In such a study, however, we must give $K(z_0, r)$ the coordinate function $w = \tau(z) = (z - z_0)/(1 - \bar{z}_0 z)$. For example, the set of derivatives of functions in B do not form a conformally invariant family, but the expression

(1.4) $$|f'(z)|(1 - |z|^2)$$

is conformally invariant. The proof of this fact uses the important identity

(1.5) $$1 - \left|\frac{z - z_0}{1 - \bar{z}_0 z}\right|^2 = \frac{(1 - |z|^2)(1 - |\bar{z}_0|^2)}{|1 - \bar{z}_0 z|^2} = (1 - |z|^2)|\tau'(z)|,$$

which is (1.3) with equality for $f(z) = \tau(z)$. Hence if $f(z) = g(\tau(z)) = g(w)$, then

$$|f'(z)|(1 - |z|^2) = |g'(w)||\tau'(z)|(1 - |z|^2) = |g'(w)|(1 - |w|^2)$$

and this is what is meant by the invariance of (1.4).

The noneuclidean disc $K(z_0, r)$, $0 < r < 1$, is the inverse image of the disc $|w| < r$ under

$$w = \tau(z) = \frac{z - z_0}{1 - \bar{z}_0 z}.$$

Consequently $K(z_0, r)$ is also a euclidean disc $\Delta(c, R) = \{z : |z - c| < R\}$, and as such it has center

(1.6) $$c = \frac{1 - r^2}{1 - r^2 |z_0|^2} z_0$$

and radius

(1.7) $$R = r \frac{1 - |z_0|^2}{1 - r^2 |z_0|^2}.$$

These can be found by direct calculation, but we shall derive them geometrically. The straight line through 0 and z_0 is invariant under τ, so that $\partial K(z_0, r) = \tau^{-1}(|w| = r)$ is a circle orthogonal to this line. A diameter of $K(z_0, r)$ is therefore the inverse image of the segment $[-r z_0/|z_0|, r z_0/|z_0|]$. Since $z = (w + z_0)/(1 + \bar{z}_0 w)$, this diameter is the segment

(1.8) $$[\alpha, \beta] = \left[\frac{|z_0| - r}{1 - r|z_0|} \frac{z_0}{|z_0|}, \frac{|z_0| + r}{1 + r|z_0|} \frac{z_0}{|z_0|}\right].$$

The endpoints of (1.8) are the points of $\partial K(z_0, r)$ of largest and smallest modulus. Thus $c = (\alpha + \beta)/2$ and $R = |\beta - \alpha|/2$ and (1.6) and (1.7) hold. Note that if r is fixed and if $|z_0| \to 1$, then the euclidean radius of $K(z_0, r)$ is asymptotic to $1 - |z_0|$.

Corollary 1.3. *If $f(z) \in B$, then*

(1.9) $$|f(z)| \le \frac{|f(0)| + |z|}{1 + |f(0)||z|}.$$

Proof. By Lemma 1.2, $\rho(f(z), f(0)) \leq |z|$, so that $f(z) \in \overline{K(f(0), |z|)}$. The bound on $|f(z)|$ then follows from (1.8). Equality can hold in (1.9) only if f is a Möbius transformation and $\arg z = \arg f(0)$ when $f(0) \neq 0$. $\square$

The pseudohyperbolic distance is a metric on D. The triangle inequality for ρ follows from

Lemma 1.4. *For any three points z_0, z_1, z_2 in D,*

$$(1.10) \qquad \frac{\rho(z_0, z_2) - \rho(z_2, z_1)}{1 - \rho(z_0, z_2)\rho(z_2, z_1)} \leq \rho(z_0, z_1) \leq \frac{\rho(z_0, z_2) + \rho(z_2, z_1)}{1 + \rho(z_0, z_2)\rho(z_2, z_1)}.$$

Proof. We can suppose $z_2 = 0$ because ρ is invariant. Then (1.10) becomes

$$(1.11) \qquad \frac{|z_0| - |z_1|}{1 - |z_0||z_1|} \leq \left| \frac{z_1 - z_0}{1 - \bar{z}_0 z_1} \right| \leq \frac{|z_0| + |z_1|}{1 + |z_0||z_1|}.$$

If $|z_1| = r$, then $z = (z_1 - z_0)/(1 - \bar{z}_0 z_1)$ lies on the boundary of the non-euclidean disc $K(-z_0, r)$, and hence $|z|$ lies between the moduli of the endpoints of the segment (1.8). That proves (1.11). Of course (1.10) and especially (1.11) are easy to verify directly. $\square$

Every Möbius transformation $w(z)$ sending z_0 to w_0 can be written

$$\frac{w - w_0}{1 - \bar{w}_0 w} = e^{i\varphi} \frac{z - z_0}{1 - \bar{z}_0 z}.$$

Differentiation then gives

$$(1.12) \qquad |w'(z_0)| = \frac{1 - |w_0|^2}{|z_0|^2}.$$

This identity we have already encountered as (1.3) with equality. By (1.12) the expression

$$(1.13) \qquad ds = \frac{2|dz|}{1 - |z|^2}$$

is a conformal invariant of the disc. We can use (1.13) to define the hyperbolic length of a rectifiable arc γ in D as

$$\int_\gamma \frac{2|dz|}{1 - |z|^2}.$$

We can then define the *Poincaré metric* $\psi(z_1, z_2)$ as the infimum of the hyperbolic lengths of the arcs in D joining z_1 to z_2. The distance $\psi(z_1, z_2)$ is then conformally invariant. If $z_1 = 0$, $z_2 = r > 0$, it is not difficult to see that

$$\psi(z_1, z_2) = 2 \int_0^r \frac{dx}{1 - |x|^2} = \log \frac{1 + r}{1 - r}.$$

Since any pair of points z_1 and z_2 can be mapped to 0 and $\rho(z_1, z_2) = |(z_2 - z_1)/(1 - \bar{z}_1 z_2)|$, respectively, by a Möbius transformation, we therefore have

$$\psi(z_1, z_2) = \log \frac{1 + \rho(z_1, z_2)}{1 - \rho(z_1, z_2)}.$$

A calculation then gives

$$\rho(z_1, z_2) = \tanh\left(\frac{\psi(z_1, z_2)}{2}\right)$$

Moreover, because the shortest path from 0 to r is the radius, the geodesics, or paths of shortest distance, in the Poincaré metric consist of the images of the diameter under all Möbius transformations. These are the diameters of D and the circular arcs in D orthogonal to ∂D. If these arcs are called lines, we have a model of the hyperbolic geometry of Lobachevsky.

In this book we shall work with the pseudohyperbolic metric ρ rather than with ψ, although the geodesics are often lurking in our intuition.

Hyperbolic geometry is somewhat simpler in the upper half plane $H = \{z = x + iy : y > 0\}$ In H

$$\rho(z_1, z_2) = \left| \frac{z_1 - z_2}{z_1 - \bar{z}_2} \right|$$

and the element of hyperbolic arc length is

$$ds = \frac{|dz|}{y}.$$

Geodesics are vertical lines and circles orthogonal to the real axis. The conformal self-maps of H that fix the point at ∞ have a very simple form:

$$\tau(z) = az + x_0, \qquad a > 0, \qquad x_0 \in \mathbb{R}.$$

Horizontal lines $\{y = y_0\}$ can be mapped to one another by these self-maps of H. This is not the case in D with the circles $\{|z| = r\}$. In H any two squares

$$\{x_0 < x < x_0 + h, h < y < 2h\}$$

are congruent in the noneuclidean geometry. The corresponding congruent figures in D are more complicated. For these and for other reasons, H is often the more convenient domain for many problems.

2. Pick's Theorem

A *finite Blaschke product* is a function of the form

$$B(z) = e^{i\varphi} \prod_{j=1}^{n} \frac{z - z_j}{1 - \bar{z}_j z}, \qquad |z_j| < 1.$$

The function B has the properties

    (i) B is continuous across ∂D,
    (ii) $|B| = 1$ on ∂D, and
    (iii) B has finitely many zeros in D.

These properties determine B up to a constant factor of modulus one. Indeed, if an analytic function $f(z)$ has (i)–(iii), and if $B(z)$ is a finite Blaschke product with the same zeros, then by the maximum principle, $|f/B| \leq 1$ and $|B/f| \leq 1$, on D, and so f/B is constant. The *degree* of B is its number of zeros. A Blaschke product of degree 0 is a constant function of absolute value 1.

Theorem 2.1 (Carathéodory).   *If $f(z) \in$ B , then there is a sequence $\{B_k\}$ of finite Blaschke products that converges to $f(z)$ pointwise on D.*

Proof.   Write

$$f(z) = c_0 + c_1 z + \cdots .$$

By induction, we shall find a Blaschke product of degree at most n whose first n coefficients match those of f;

$$B_n = c_0 + c_1 z + \cdots + c_{n-1} z^{n-1} + d_n z^n + \cdots .$$

That will prove the theorem. Since $|c_0| \leq 1$, we can take

$$B_0 = \frac{z + c_0}{1 + \bar{c}_0 z}.$$

If $|c_0| = 1$, then $B_0 = c_0$ is a Blaschke product of degree 0. Suppose that for each $g \in$ B we have constructed $B_{n-1}(z)$. Set

$$g = \frac{1}{z} \frac{f - f(0)}{1 - \overline{f(0)} f}$$

and let B_{n-1} be a Blaschke product of degree at most $n - 1$ such that $g - B_{n-1}$ has $n - 1$ zeros at 0. Then $zg - zB_{n-1}$ has n zeros at $z = 0$. Set

$$B_n(z) = \frac{z B_{n-1}(z) + f(0)}{1 + \overline{f(0)} z B_{n-1}(z)}.$$

Then B_n is a finite Blaschke product, degree$(B_n) = $ degree$(z B_{n-1}) \leq n$, and

$$f(z) - B_n(z) = \frac{zg(z) + f(0)}{1 + \overline{f(0)} zg(z)} - \frac{z B_{n-1}(z) + f(0)}{1 + \overline{f(0)} z B_{n-1}(z)}$$

$$= \frac{(1 - |f(0)|^2) z(g(z) - B_{n-1}(z))}{(1 + \overline{f(0)} zg(z))(1 + \overline{f(0)} z B_{n-1}(z))},$$

so that $f - B_n$ has a zero of order n at $z = 0$. $\square$

    The coefficient sequences $\{c_0, c_1, \ldots.\}$ of functions in B were characterized by Schur [1917]. Instead of giving Schur's theorem, we shall prove Pick's theorem (from Pick [1916]). For $\{z_1, \ldots, z_n\}$ a finite set of distinct points in D, Pick determined those $\{w_1, \ldots, w_n\}$ for which the interpolation

$$(2.1) \qquad f(z_j) = w_j, \quad j = 1, 2, \ldots, n,$$

has a solution $f(z) \in \mathscr{B}$.

Theorem 2.2. *There exists $f \in \mathscr{B}$ satisfying the interpolation (2.1) if and only if the quadratic form*

$$Q_n(t_1, \ldots, t_n) = \sum_{j,k=1}^{n} \frac{1 - w_j \bar{w}_k}{1 - z_j \bar{z}_k} t_j \bar{t}_k$$

is nonnegative, $Q_n \geq 0$. When $Q_n \geq 0$ there is a Blaschke product of degree at most n which solves (2.1).

Pick's theorem easily implies Carathéodory's theorem, but its proof is more difficult.

When $n = 2$ a necessary and sufficient condition for interpolation is given by (1.2) in Lemma 1.2. It follows that $Q_2 \geq 0$ if and only if $|w_1| \leq 1$ and (1.2) holds. This can of course be seen directly, since $Q_2 \geq 0$ if and only if $1 - |w_1| \geq 0$ and the determinant of Q_2 is nonnegative:

$$\frac{(1 - |w_1|^2)(1 - |w_2|^2)}{|1 - \bar{w}_1 w_2|^2} \geq \frac{(1 - |z_1|^2)(1 - |z_2|^2)}{|1 - \bar{z}_1 z_2|}.$$

By the useful identity (1.5), this last inequality can be rewritten

$$\left| \frac{w_1 - w_2}{1 - \bar{w}_1 w_2} \right| \leq \left| \frac{z_1 - z_2}{1 - \bar{z}_1 z_2} \right|,$$

which is (1.2).

Proof. We use induction on n. The case $n = 1$ holds because the Möbius transformations act transitively on D. Assume $n > 1$. Suppose (2.1) holds. Then clearly $|w_n| \leq 1$, and if $|w_n| = 1$, then the interpolating function is the constant w_n and $w_j = w_n, 1 \leq j \leq n - 1$. Suppose $Q_n \geq 0$. Setting $t_n = 1, t_j = 0, j < n$, we see $|w_n| \leq 1$; and if $|w_n| = 1$, then setting $t_j = 0, j \neq k, n$, we see by (1.2) as before that $w_k = w_n$. We can therefore take $B_n = w_n$ if $|w_n| = 1$. Thus the problem is trivial if $|w_n| = 1$, and in any event, $|w_n| \leq 1$.

Now assume $|w_n| < 1$. We move z_n and w_n to the origin. Let

$$z'_j = \frac{z_j - z_n}{1 - \bar{z}_n z_j}, \quad 1 \leq j \leq n; \qquad w'_j = \frac{w_j - w_n}{1 - \bar{w}_n w_j}, \quad 1 \leq j \leq n.$$

There is $f \in \mathscr{B}$ satisfying (2.1) if and only if

$$(2.2) \qquad g = \left(f\left(\frac{z + z_n}{1 + \bar{z}_n z} \right) - w_n \right) \bigg/ \left(1 - \bar{w}_n f\left(\frac{z + z_n}{1 + \bar{z}_n z} \right) \right)$$

is in B and solves

(2.3) $$g(z'_j) = w'_j, \quad 1 \le j \le n.$$

Also, f is a Blaschke product of degree at most n if and only if g is a Blaschke product of degree at most n.

On the other hand, the quadratic form Q'_n corresponding to the points $\{z'_1, \ldots, z'_{n-1}, 0\}$ and $\{w'_1, \ldots, w'_{n-1}, 0\}$ is closely related to Q_n. Since by a computation

$$\frac{1 - z'_j \bar{z}'_k}{1 - z_j \bar{z}_k} = \frac{1 - |z_n|^2}{(1 - \bar{z}_n z_j)(1 - z_n \bar{z}_k)} = \alpha_j \bar{\alpha}_k$$

and

$$\frac{1 - w'_j \bar{w}'_k}{1 - w_j \bar{w}_k} = \frac{1 - |w_n|^2}{(1 - \bar{w}_n w_j)(1 - w_n \bar{w}_k)} = \beta_j \bar{\beta}_k,$$

we have

$$\frac{1 - w'_j \bar{w}'_k}{1 - z'_j \bar{z}'_k} t_j \bar{t}_k = \frac{1 - w_j \bar{w}_k}{1 - z_j \bar{z}_k} \left(\frac{\beta_j}{\alpha_j} t_j \right) \overline{\left(\frac{\beta_k}{\alpha_k} t_k \right)}$$

and

(2.4) $$Q'_n(t_1, \ldots, t_n) = Q_n\left(\frac{\beta_1}{\alpha_1} t_1, \ldots, \frac{\beta_n}{\alpha_n} t_n \right).$$

Thus $Q'_n \ge 0$ if and only if $Q_n \ge 0$, and the problem has been reduced to the case $z_n = w_n = 0$.

Let us therefore assume $z_n = w_n = 0$. There is $f \in B$ such that $f(0) = 0$,

$$f(z_j) = w_j, \quad 1 \le j \le n - 1,$$

if and only if there is $g(z) = f(z)/z \in B$ such that

(2.5) $$g(z_j) = w_j/z_j, \quad 1 \le j \le n - 1.$$

Also, f is a Blaschke product of degree d if and only if g is a Blaschke product of degree $d - 1$. Now by induction, (2.5) has a solution if and only if the quadratic form

$$\tilde{Q}_{n-1}(s_1, \ldots, s_{n-1}) = \sum_{j,k=1}^{n-1} \frac{1 - (w_j/z_j)\overline{(w_k/z_k)}}{1 - z_j \bar{z}_k} s_j \bar{s}_k$$

is nonnegative. This means the theorem reduces to showing

$$Q_n \ge 0 \quad \Leftrightarrow \quad \tilde{Q}_{n-1} \ge 0$$

under the assumption $z_n = w_n = 0$.

Because $z_n = w_n = 0$, we have

$$Q_n(t_1, \ldots, t_n) = |t_n|^2 + 2 \operatorname{Re} \sum_{j=1}^{n-1} \bar{t}_j t_n + \sum_{j,k=1}^{n-1} \frac{1 - w_j \bar{w}_k}{1 - z_j \bar{z}_k} t_j \bar{t}_k.$$

Completing the square relative to t_n gives

$$Q_n(t_1, \ldots, t_n) = \left| t_n + \sum_{j=1}^{n-1} t_j \right|^2 + \sum_{j,k=1}^{n-1} \left(\frac{1 - w_j \bar{w}_k}{1 - z_j \bar{z}_k} - 1 \right) t_j \bar{t}_k.$$

Now

$$\frac{1 - w_j \bar{w}_k}{1 - z_j \bar{z}_k} - 1 = \frac{z_j \bar{z}_k - w_j \bar{w}_k}{1 - z_j \bar{z}_k} = \frac{1 - (w_j/z_j)(\overline{w_k/z_k})}{1 - z_j \bar{z}_k} z_j \bar{z}_k.$$

Hence

$$(2.6) \qquad Q_n(t_1, \ldots, t_n) = \left| \sum_{j=1}^{n} t_j \right|^2 + \tilde{Q}_{n-1}(z_1 t_1, \ldots, z_{n-1} t_{n-1}).$$

Thus $\tilde{Q}_{n-1} \geq 0$ implies $Q_n \geq 0$, and setting $t_n = - \sum_{1}^{n-1} t_j$, we see also that $Q_n \geq 0$ implies $\tilde{Q}_{n-1} \geq 0$. $\quad\square$

Corollary 2.3. *Suppose $Q_n \geq 0$. Then (2.1) has a unique solution $f(z) \in B$ if and only if $\det(Q_n) = 0$. If $\det(Q_n) = 0$ and $m < n$ is the rank of Q_n, then the interpolating function is a Blaschke product of degree m. Conversely, if a Blaschke product of degree $m < n$ satisfies (2.1), then Q_n has rank m.*

Proof. If $|w_n| = 1$ the whole thing is very trivial because then $Q_n = 0$, $m = 0$, and $B_n = w_n$. So we may assume $|w_n| < 1$. We may then suppose $z_n = w_n = 0$, because by (2.4), Q_n and Q'_n have the same rank, while by (2.2), the original problem has a unique solution if and only if the adjusted problem (2.3) has a unique solution. Also (2.3) can be solved with a Blaschke product of degree m if and only if (2.1) can be also.

So we assume $z_n = w_n = 0$. Then (2.1) has a unique solution if and only if (2.5) has a unique solution; and (2.1) can be solved with a Blaschke product of degree $m - 1$. Consequently, by induction, all assertions of the corollary will be proved when we show

$$(2.7) \qquad\qquad \operatorname{rank}(Q_n) = 1 + \operatorname{rank}(\tilde{Q}_{n-1}).$$

Writing $\tilde{Q}_{n-1} = (a_{j,k})$, we have

$$Q_n = \begin{bmatrix} 1 + z_j \bar{z}_k a_{j,k} & \begin{matrix} 1 \\ \vdots \\ 1 \end{matrix} \\ \hline 1 \quad \cdots \quad 1 & 1 \end{bmatrix},$$

which has the same rank as

$$
\begin{bmatrix}
& & & 0 \\
z_j \bar{z}_k a_{j,k} & & \vdots & \vdots \\
& & & 0 \\
\hline
1 & \cdots & & 1
\end{bmatrix},
$$

and the rank of this matrix is $1 + \operatorname{rank}(\tilde{Q}_{n-1})$.   □

Corollary 2.4. *Suppose $Q_n \geq 0$ and $\det(Q_n) > 0$. Let $z \in D, z \neq z_j, j = 1, 2, \ldots, n$. The set of values*

$$
W = \{ f(z) : f \in \mathrm{B} , f(z_j) = w_j, 1 \leq j \leq n \}
$$

is a nondegenerate closed disc contained in D. If $f \in \mathrm{B}$, and if f satisfies (2.1), then $f(z) \in \partial W$ if and only if f is a Blaschke product of degree n. Moreover, if $w \in \partial W$, there is a unique solution to (2.1) in B which also solves $f(z) = w$.

Proof. We may again suppose $z_n = w_n = 0$. Then $\det(\tilde{Q}_{n-1}) > 0$ by (2.7). By induction,

$$
\tilde{W} = \{ g(z) : g \in \mathrm{B} , g(z_j) = w_j/z_j, 1 \leq j \leq n - 1 \}
$$

is a closed disc contained in D. But then $W = \{ z\zeta : \zeta \in \tilde{W} \}$ is also a closed disc. Since $w \in \partial W$ if and only if $w/z \in \partial \tilde{W}$, the other assertions follow by induction.   □

We shall return to this topic in Chapter IV.

3. Poisson Integrals

Let $u(z)$ be a continuous function on the closed unit disc $\bar{D}$. If $u(z)$ is harmonic on the open disc D, that is, if

$$
\Delta u = \frac{\partial^2 u}{\partial x^2} + \frac{\partial^2 u}{\partial y^2} = 0,
$$

then $u(z)$ has the mean value property

$$
u(0) = \frac{1}{2\pi} \int_0^{2\pi} u(e^{i\theta}) \, d\theta.
$$

Let $z_0 = re^{i\theta_0}$ be a point in D. Then there is a similar representation formula for $u(z_0)$, obtained by changing variables through a Möbius transformation. Let $\tau(z) = (z - z_0)/(1 - \bar{z}_0 z)$. The unit circle ∂D is invariant under τ, and we may write $\tau(e^{i\theta}) = e^{i\varphi}$. Differentiation now gives

$$
(3.1) \qquad \frac{d\varphi}{d\theta} = \frac{1 - |z_0|^2}{|e^{i\theta} - z_0|^2} = \frac{1 - r^2}{1 - 2r\cos(\theta - \theta_0) + r^2} = P_{z_0}(\theta).
$$

This function $P_{z_0}(\theta)$ is called the *Poisson kernel* for the point $z_0 \in D$. Since $u(\tau^{-1}(z))$ is another function continuous on $\bar{D}$ and harmonic on D, the change of variables yields

$$u(z_0) = u(\tau^{-1}(0)) = \frac{1}{2\pi} \int_0^{2\pi} u(e^{i\theta}) P_{z_0}(\theta)\, d\theta.$$

This is the *Poisson integral formula*.

Notice that the Poisson kernel $P_z(\theta)$ also has the form

$$P_z(\theta) = \mathrm{Re}\, \frac{e^{i\theta} + z}{e^{i\theta} - z},$$

so that for $e^{i\theta}$ fixed, $P_z(\theta)$ is a harmonic function of $z \in D$. Hence the function defined by

(3.2) $$u(z) = \frac{1}{2\pi} \int P_z(\theta) f(\theta)\, d\theta$$

is harmonic on D whenever $f(\theta) \in L^1(\partial D)$. Since $P_z(\theta)$ is also a continuous function of θ, we get a harmonic function from (3.2) if we replace $f(\theta)\, d\theta$ by a finite measure $d\mu(\theta)$ on ∂D. The extreme right side of (3.1) shows that the Poisson integral formula may be interpreted as a convolution. If $z = re^{i\theta_0}$, then

$$P_z(\theta) = P_r(\theta_0 - \theta)$$

and (3.2) takes the form

$$u(z) = \frac{1}{2\pi} \int P_r(\theta_0 - \theta) f(\theta)\, d\theta = (P_r * f)(\theta_0).$$

This reflects the fact that the space of harmonic functions on D is invariant under rotations.

Map D to the upper half plane H by $w \to z(w) = i(1 - w)/(1 + w)$. Fix $w_0 \in D$ and let $z_0 = z(w_0)$ be its image in H. Our map sends ∂D to $\mathbb{R} \cup \{\infty\}$, so that if $w = e^{i\theta} \in \partial D$, and $w \neq -1$, then $z(w) = t \in \mathbb{R}$. Differentiation now gives

$$\frac{1}{2\pi} P_{w_0}(\theta) \frac{d\theta}{dt} = \frac{1}{\pi} \frac{y_0}{(x_0 - t)^2 + y_0^2} = P_{z_0}(t), \qquad z_0 = x_0 + iy_0.$$

The right side of this equation is the *Poisson kernel* for the upper half plane, $P_{z_0}(t) = P_{y_0}(x_0 - t)$. (The notation is unambiguous because $z_0 \in H$ but $y_0 \notin H$.) Pulling the Poisson integral formula for D over to H, we see that

(3.3) $$u(z) = \int P_z(t) u(t)\, dt = \int P_y(x - t) u(t)\, dt$$

whenever the function $u(z)$ is continuous on $\overline{H} \cup \{\infty\}$ and harmonic on H. When $t \in \mathbb{R}$ is fixed, the Poisson kernel for the upper half plane is a harmonic function of z, because

$$P_z(t) = \frac{1}{\pi} \operatorname{Im} \left(\frac{1}{t - z} \right).$$

From its defining formula we see that $P_z(t) \leq c_z/(1 + t^2)$, where c_z is a constant depending on z. Consequently, if $1 \leq q \leq \infty$, then $P_z(t) \in L^2(\mathbb{R})$, and the function

(3.4) $$u(z) = \int P_z(t) f(t) \, dt$$

is harmonic on H whenever $f(t) \in L^p(\mathbb{R})$, $1 \leq p \leq \infty$. Moreover, since $P_z(t)$ is a continuous function of t, (3.4) will still produce a harmonic function $u(z)$ if $f(t) \, dt$ is replaced by a finite measure $d\mu(t)$ or by a positive measure $d\mu(t)$ such that

$$\int \frac{1}{1 + t^2} \, d\mu(t) < \infty$$

(so that $\int P_z(t) \, d\mu(t)$ converges).

Now let $f(t)$ be the characteristic function of an interval (t_1, t_2). The resulting harmonic function

$$\omega(z) = \int_{t_1}^{t_2} P_y(x - t) \, dt,$$

called the *harmonic measure* of the interval, can be explicitly calculated. We get

$$\omega(z) = \frac{1}{\pi} \arg \left(\frac{z - t_2}{z - t_1} \right) = \frac{\alpha}{\pi},$$

where α is the angle at z formed by t_1 and t_2. See Figure I.1. This angle α is constant at points along the circular arc passing through t_1, z, and t_2, and α is the angle between the real axis and the tangent of that circular arc. A similar geometric interpretation of harmonic measure on the unit disc is given in Exercise 3.

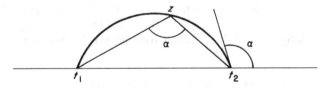

Figure I.1. A level curve of $\omega(z)$.

The Poisson integral formula for the upper half plane can be written as a convolution

$$u(z) = \int P_y(x - t)f(t)\, dt = (P_y * f)(t).$$

This follows from the formula defining the Poisson kernel, and reflects the fact that under the translations $z \to z + x_0$, x_0 real, the space of harmonic functions on H is invariant. The harmonic functions are also invariant under the dilations $z \to az$, $a > 0$, and accordingly we have

$$P_y(t) = (1/y)P_1(t/y),$$

which means P_y is *homogeneous of degree* -1 in y. The Poisson kernel has the following properties, illustrated in Figure I.2:

(i) $P_y(t) \geq 0$, $\int P_y(t)\, dt = 1$.
(ii) P_y is even, $P_y(-t) = P_y(t)$.
(iii) P_y is decreasing in $t > 0$.
(iv) $P_y(t) \leq 1/\pi y$.

For any $\delta > 0$,

(v) $\sup_{|t|>\delta} P_y(t) \to 0$ ($y \to 0$).
(vi) $\int_{|t|>\delta} P_y(t)\, dt \to 0$ ($y \to 0$).

Moreover, $\{P_y\}$ is a semigroup.

(vii) $P_{y_1} * P_{y_2} = P_{y_1+y_2}$.

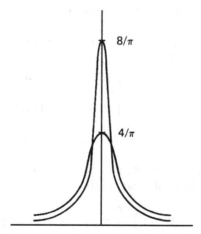

Figure I.2. The Poisson kernels $P_{1/4}$ and $P_{1/8}$.

The first six properties are obvious from the definition of $P_y(t)$, and properties (iv)–(vi) also follow from the homogeneity in y. Property (vii) means that if $u(z)$ is a harmonic function given by (3.4), then $u(z + iy_1)$ can be computed from $u(t + iy_1), t \in \mathbb{R}$, by convolution with P_y. To prove (vii), consider the harmonic function $u(x + iy) = P_{y_1+y}(x)$. This function extends continuously to $\overline{H} \cup \{\infty\}$. Consequently by (3.3),

$$P_{y_1+y_2}(x) = \int P_{y_2}(x - t)u(t)\,dt = (P_{y_1} * P_{y_2})(x).$$

An important tool for studying integrals like (3.4) is the *Minkowski inequality for integrals*:

If μ and v are σ-finite measures, if $1 \le p < \infty$, and if $F(x,t)$ is $v \times \mu$ measurable, then

$$\left\| \int F(x,t)dv(x) \right\|_{L^p(\mu)} \le \int \| F(x,t) \|_{L^p(\mu)}\,dv(x).$$

This is formally the same as Minkowski's inequality for sums of $L^p(\mu)$ functions and it has the same proof. The case $p = 1$ is just Fubini's theorem. For $p > 1$ we can suppose that $F(x, t) \ge 0$ and that $F(x, t)$ is a simple function, so that both integrals converge. Set

$$G(t) = \left(\int F(x,t)\,dv(x) \right)^{p-1}.$$

Then with $q = p/(p - 1)$,

$$\| G \|_{L^q(\mu)} = \left\| \int F(x,t)\,dv(x) \right\|_{L^p(\mu)}^{p-1},$$

and by Fubini's theorem and Hölder's inequality,

$$\begin{aligned} \left\| \int F(x,t)\,dv(x) \right\|_{L^p(\mu)}^p &= \int G(t) \int F(x,t)\,dv(x)\,d\mu(t) \\ &= \iint G(t)F(x,t)\,d\mu(t)\,dv(x) \\ &\le \int \| G \|_{L^q(\mu)} \| F(x,t) \|_{L^p(\mu)}\,dv(x) \\ &= \| G \|_{L^q(\mu)} \int \| F(x,t) \|_{L^p(\mu)}\,dv(x). \end{aligned}$$

Canceling $\| G \|_{L^q(\mu)}$ from each side now gives the Minkowski inequality.

Using Minkowski's inequality we obtain

$$(3.5) \qquad \left(\int |u(x, y)|^p\,dx \right)^{1/p} \le \| f \|_p, \quad 1 \le p < \infty,$$

if $u(x, y) = P_y * f(x)$, $f \in L^p$; and

$$(3.6) \qquad\qquad \int |u(x, y)|\,dx \le \int |d\mu|$$

if $u(x, y) = P_y * \mu = \int P_y(x - t) \, d\mu(t)$, where μ is a finite measure on $\mathbb{R}$. For $p = \infty$ the analog of (3.5), $\sup_x |u(x, y)| \leq \|f\|_\infty$, is trivial from property (i) of $P_y(t)$.

Theorem 3.1. (a)   *If $1 \leq p < \infty$ and if $f(x) \in L^p$, then*

$$\|P_y * f - f\|_p \to 0 \qquad (y \to 0).$$

(b)   *When $f(x) \in L^\infty$, $P_y * f$ converges weak-star to $f(x)$.*

(c)   *If $d\mu$ is a finite measure on $\mathbb{R}$, the measures $(P_y * \mu)(x) \, dx$ converge weak-star to $d\mu$.*

(d)   *When $f(x)$ is bounded and uniformly continuous on $\mathbb{R}$, $P_y * f(x)$ converges uniformly to $f(x)$.*

Statement (b) means that for all $g \in L^1$,

$$\int g(x)(P_y * f)(x) \, dx \to \int f(x)g(x) \, dx \quad (y \to 0).$$

Statement (c) has a similar meaning:

$$\int g(x)(P_y * \mu)(x) \, dx \to \int g(x) \, d\mu(x) \quad (y \to 0),$$

for all $g \in C_0(\mathbb{R})$, the continuous functions vanishing at ∞. It follows from Theorem 3.1 that $f \in L^p$ is uniquely determined by the harmonic function $u(z) = P_y * f(x)$ and that a measure μ is determined by its Poisson integral $P_y * \mu$. Note also that by (a) or (b)

$$\lim_{y \to 0} \|P_y * f\|_p = \|f\|_p, \quad 1 \leq p \leq \infty.$$

By (3.5) and property (vii), the function $\|P_y * f\|_p$ is monotone in y.

Besides Minkowski's inequality, the main ingredient of the proof of the theorem is the continuity of translations on L^p, $1 \leq p < \infty$: If $f_x(t) = f(t - x)$, then $\|f_x - f\|_p \to 0 (x \to 0)$. (To prove this approximate f in L^p norm by a function in $C_0(\mathbb{R})$.) The translations are not continuous on L^∞ nor are they on the space of finite measures; that is why we have weaker assertions in (b) and (c). The translations are of course continuous on the space of uniformly continuous functions, and for this reason (d) holds.

Proof. Let $f \in L^p$, $1 \leq p \leq \infty$. When $p = \infty$ we suppose in addition that f is uniformly continuous. Then

$$P_y * f(x) - f(x) = \int P_y(t)(f(x - t) - f(x)) \, dt.$$

Minkowski's inequality gives

$$\|P_y * f - f\|_p \leq \int P_y(t)\|f_t - f\|_p \, dt,$$

when $p < \infty$, because $P_y \geq 0$. The same inequality is trivial when $p = \infty$. For $\delta > 0$, we now have

$$\| P_y * f - f \|_p \leq \int_{|t| \leq \delta} P_y(t) \| f_t - f \|_p \, dt + \int_{|t| > \delta} P_y(t) \| f_t - f \|_p \, dt.$$

Since $\int P_y(t) \, dt = 1$, continuity of translations shows that $\int_{|t| \leq \delta}$ is small provided δ is small. With δ fixed,

$$\int_{|t| > \delta} \leq 2\| f \|_p \int_{|t| > \delta} P_y(t) \, dt \to 0 \quad (y \to 0)$$

by property (vi) of the Poisson kernel. That proves (a) and (d). By Fubini's theorem, parts (b) and (c) follow from (a) and (d), respectively. □

Corollary 3.2. *Assume $f(x)$ is bounded and uniformly continuous, and let*

$$u(x, y) = \begin{cases} (P_y * f)(x), & y > 0, \\ f(x), & y = 0. \end{cases}$$

Then $u(x, y)$ is harmonic on H and continuous on $\overline{\mathrm{H}}$.

This corollary follows from (d). We also need the local version of the corollary.

Lemma 3.3. *Assume $f(x) \in L^p, 1 \leq p \leq \infty$, and assume f is continuous at x_0. Let $u(x, y) = P_y * f(x)$. Then*

$$\lim_{(x, y) \to x_0} u(x, y) = f(x_0).$$

Proof. We have

$$|u(x, y) - f(x_0)| \leq \int_{|t| < \delta} P_y(t) |f(x - t) - f(x_0)| \, dt + \int_{|t| \geq \delta}.$$

With δ small and $|x - x_0|$ small, $\int_{|t| < \delta}$ is small. With δ fixed, $\int_{|t| \geq \delta}$ tends to zero with y. □

Notice that the convergence is uniform on a subset $E \subset \mathbb{R}$ provided the continuity of f is uniform over $x_0 \in E$ and provided $|f(x_0)|$ is bounded on E.

It is important that the Poisson integrals of L^p functions and measures are characterized by the norm inequalities like (3.5) and (3.6). The proof of this in the upper half plane requires the following lemma.

Lemma 3.4. *If $u(z)$ is harmonic on H and bounded and continuous on $\overline{\mathrm{H}}$ then*

$$u(z) = \int P_y(x - t)u(t) \, dt.$$

Proof. The lemma is not a trivial consequence of the definition of $P_z(t)$, because $u(z)$ may not be continuous at ∞. But let

$$U(z) = u(z) - \int P_y(x - t)u(t)\, dt.$$

Then $U(z)$ is harmonic on H , and bounded and continuous on $\overline{H}$, and $U \equiv 0$ on $\mathbb{R}$, by Lemma 3.3. Set

$$V(z) = \begin{cases} U(z), & y \geq 0, \\ -U(\bar{z}), & y < 0. \end{cases}$$

Then V is a bounded harmonic function on the complex plane, because V has the mean value property over small discs. By Liouville's theorem, V is constant; $V(z) = V(0) = 0$. Hence $U(z) = 0$ and the lemma is proved. □

Theorem 3.5. *Let $u(z)$ be a harmonic function on the upper half plane H . Then*

(a) *If $1 < p \leq \infty$, u is the Poisson integral of a function in L^p if and only if*

(3.7)
$$\sup_y \int \|u(x + y)\|_{L^p(dx)} < \infty.$$

(b) *$u(z)$ is the Poisson integral of a finite measure on $\mathbb{R}$ and only if*

(3.8)
$$\sup_y \int |u(x + iy)|\, dx < \infty.$$

(c) *$u(z)$ is positive if and only if*

$$u(z) = cy + \int P_y(x - t)\, d\mu(t),$$

where

$$c \geq 0, \quad \mu \geq 0, \quad and \quad \int \frac{d\mu(t)}{1 + t^2} < \infty.$$

Proof. We have already noted that (3.7) and (3.8) are necessary conditions because of Minkowski's inequality. Suppose $u(z)$ satisfies (3.7) or (3.8). Then we have the estimate

(3.9)
$$|u(z)| \leq \left(\frac{2}{\pi y}\right)^{1/p} \sup_{\eta > 0} \|u(x, \eta)\|_{L^p(dx)},$$

which we now prove: Write $\zeta = \xi + i\eta$. Then by Hölder's inequality,

$$
\begin{aligned}
|u(z)| &= \frac{1}{\pi y^2} \left| \iint_{\Delta(z,y)} u(\zeta)\, d\xi\, d\eta \right| \\
&\leq \left(\frac{1}{\pi y^2} \iint_{\Delta(z,y)} |u(\zeta)|^p\, d\xi\, d\eta \right)^{1/p} \\
&\leq \left(\frac{1}{\pi y^2} \int_0^{2y} \int_{-\infty}^{\infty} |u(\xi + i\eta)|^p\, d\xi\, d\eta \right)^{1/p} \\
&\leq \left(\frac{2}{\pi y} \right)^{1/p} \sup_{\eta > 0} \left(\int |u(\xi + i\eta)|^p\, d\xi \right)^{1/p}.
\end{aligned}
$$

The estimate (3.9) tells us $u(z)$ is bounded on $y > y_n > 0$, and Lemma 3.4 then gives

$$
u(z + iy_n) = \int P_y(x - t) u(t + iy_n)\, dt.
$$

Let y_n decrease to 0. If $1 < p \leq \infty$, the sequence $f_n(t) = u(t + iy_n)$ is bounded in L^p. By the Banach–Alaoglu theorem, which says the closed unit ball of the dual of a Banach space is compact in the weak-star topology, $\{f_n\}$ has a weak-star accumulation point $f \in L^p$. Since Poisson kernels are in L^q, $q = p/(p-1)$, we have

$$
u(z) = \lim_n u(z + iy_n) = \lim_n \int P_y(x - t) f_n(t)\, dt = \int P_y(x - t) f(t)\, dt.
$$

The proof of (b) is the same except that now the measures $u(t + iy_n)\, dt$, which have bounded norms, converge weak-star to a finite measure on $\mathbb{R}$.

The easiest proof of (c) involves mapping H back onto D, using the analog of (b) for harmonic functions on the disc, and then returning to H. A harmonic function $u(z)$ on D is the Poisson integral of a finite measure v on ∂D if and only if $\sup_r \int |u(re^{i\theta})|\, d\theta < \infty$. The measure v is then a limit of the measures $u(re^{i\theta})d\theta/2\pi$ in the weak-star topology on measures on ∂D. If $u(z) \geq 0$, then the measures $u(re^{i\theta})\, d\theta$ are positive and bounded since

$$
\frac{1}{2\pi} \int u(re^{i\theta})\, d\theta = u(0),
$$

and so the limit v exists and v is a positive measure. That proves the disc version of (c). Now map D to H by $w \to z(w) = i(1 - w)/(1 + w)$. The harmonic function u on H is positive if and only if the harmonic function $u(z(w))$, which is positive, is the Poisson integral of a positive measure v on ∂D. Consider first the case when v is supported on the point $w = -1$, which corresponds to $z = \infty$. Then

$$u(z(w)) = v(\{-1\})P_w(-1) = v(\{-1\})\frac{1 - |w|^2}{|1 + w|^2}$$
$$= v(\{-1\}) \operatorname{Im} z = v(\{-1\})y.$$

Now assume $v(\{-1\}) = 0$. The map $z(w)$ moves v onto a finite positive measure $\tilde{v}$ on $\mathbb{R}$, and for $t = z(e^{i\theta})$

$$P_w(\theta) = \pi(1 + t^2)P_z(t).$$

In this case we have

$$u(z) = \int P_y(x - t)\, d\mu(t),$$

where

$$\mu = \pi(1 + t^2)\tilde{v}.$$

The general case is the sum of the two special cases already discussed. $\quad\square$

Part c) of Theorem 3.5 is known as Herglotz's theorem. The results in this section also hold in D, where they are easier to prove, when we write

$$u(re^{i\theta}) = \frac{1}{2\pi} \int P_r(\theta - \varphi)f(\varphi)\, d\varphi,$$
$$P_r(\theta - \varphi) = \frac{1 - r^2}{1 - 2r\cos(\theta - \varphi) + r^2}, \quad z = re^{i\theta}.$$

Most of these results also hold if $\{P_y(t)\}$ is replaced by some other approximate identity. Suppose $\{\varphi_y(t)\}_{y>0}$ is a family of integrable functions on $\mathbb{R}$ such that

(a) $\int \varphi_y(t)\, dt = 1$,
(b) $\|\varphi_y\|_1 \leq M$,

and such that for any $\delta > 0$,

(c) $\lim_{y \to 0} \sup_{|t| > \delta} |\varphi_y(t)| = 0$,
(d) $\lim_{y \to 0} \int_{|t| > \delta} |\varphi_y(t)|\, dt = 0$.

Then the reader can easily verify that Theorem 3.1 and its corollary hold for $\varphi_y * f$ in place of $P_y * f$.

4. Hardy–Littlewood Maximal Function

To each function f on $\mathbb{R}$ we associate two auxiliary functions that respectively measure the size of f and the behavior of the Poisson integral of f. The first auxiliary function can be defined whenever f is a measurable function on any measure space (X, μ). This is the *distribution function*

$$m(\lambda) = \mu(\{x \in X : |f(x)| > \lambda\}),$$

defined for $\lambda > 0$. The distribution function $m(\lambda)$ is a decreasing function of λ, and it determines the L^p norms of f. If $f \in L^\infty$, then $m(\lambda) = 0$ for $\lambda \geq \|f\|_\infty$, and $m(\lambda) > 0$ for $\lambda < \|f\|_\infty$; and so we have

$$\|f\|_\infty = \sup\{\lambda : m(\lambda) > 0\}.$$

Lemma 4.1. *If (X, μ) is a measure space, if $f(x)$ is measurable, and if $0 < p < \infty$, then*

$$(4.1) \qquad \int |f|^p \, d\mu = \int_0^\infty p\lambda^{p-1} m(\lambda) \, d\lambda.$$

Proof. We may assume f vanishes except on a set of σ-finite measure, because otherwise both sides of (4.1) are infinite. Then Fubini's theorem shows that both sides of (4.1) equal the product measure of the ordinate set $\{(x, \lambda) : 0 < \lambda < |f(x)|^p\}$. That is,

$$\int |f|^p \, d\mu = \iint_0^{|f|} p\lambda^{p-1} \, d\lambda \, d\mu = \int_0^\infty p\lambda^{p-1} \mu(|f| > \lambda) \, d\lambda$$

$$= \int_0^\infty p\lambda^{p-1} m(\lambda) \, d\lambda. \quad \square$$

We shall also need a simple estimate of $m(\lambda)$ known as *Chebychev's inequality.* Let $f \in L^p, 0 < p < \infty$ and let

$$E_\lambda = \{x \in X : |f(x)| > \lambda\},$$

so that $\mu(E_\lambda) = m(\lambda)$. Chebychev's inequality is

$$m(\lambda) \leq \|f\|_p^p / \lambda^p.$$

It follows from the observation that

$$\lambda^p \mu(E_\lambda) \leq \int_{E_\lambda} |f|^p \, d\mu \leq \|f\|_p^p.$$

A function f that satisfies

$$m(\lambda) \leq A/\lambda^p$$

is called a *weak L^p function.* Thus Chebychev's inequality states that every L^p function is a weak L^p function. The function $|x \log x|^{-1}$ on $[0, 1]$ is not in L^1, but it satisfies $m(\lambda) = o(1/\lambda) (\lambda \to \infty)$, and so it is weak L^1.

The other auxiliary function we shall define only for functions on $\mathbb{R}$. Recall Lebesgue's theorem that if $f(x)$ is locally integrable on $\mathbb{R}$, then

$$(4.2) \qquad \lim_{\substack{h \to 0 \\ k \to 0}} \frac{1}{h+k} \int_{x-h}^{x+k} f(t)\, dt = f(x)$$

for almost every $x \in \mathbb{R}$. To make Lebesgue's theorem quantitative we replace the limit in (4.2) by the supremum, and we put the absolute value inside the integral. Write $|I|$ for the length of an interval I. The *Hardy–Littlewood maximal function* of f is

$$Mf(x) = \sup_{x \in I} \frac{1}{|I|} \int_I |f(t)|\, dt$$

for f locally integrable on $\mathbb{R}$. Now if $f \in L^p,\ p \geq 1$, then $Mf(x) < \infty$ almost everywhere. This follows from Lebesgue's theorem, but we shall soon see a different proof in Theorem 4.3 below. The important thing about Mf is that it majorizes many other functions associated with f.

Theorem 4.2. *For $\alpha > 0$ and $t \in \mathbb{R}$, let $\Gamma_\alpha(t)$ be the cone in H with vertex t and angle $2 \arctan \alpha$, as shown in Figure I.3,*

$$\Gamma_\alpha(t) = \{(x, y) : |x - t| < \alpha y,\ 0 < y < \infty\}.$$

Let $f \in L^1(dt/(1 + t^2))$ and let $u(x, y)$ be the Poisson integral of $f(t)$,

$$u(x, y) = \int P_y(s) f(x - s)\, ds.$$

Then

$$(4.3) \qquad \sup_{\Gamma_\alpha(t)} |u(x, y)| \leq A_\alpha Mf(t), \quad t \in \mathbb{R},$$

where A_α is a constant depending only on α.

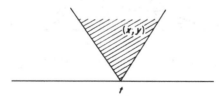

Figure I.3. The cone $\Gamma_\alpha(t)$, $\alpha = \frac{2}{3}$.

The condition $f \in L^1(dt/(1 + t^2))$ merely guarantees that $\int P_y(s) f(x - s)\, ds$ converges.

Proof. We may assume $t = 0$. Let us first consider the points $(0, y)$ on the axis of the cone $\Gamma_\alpha(0)$. Then

$$u(0, y) = \int P_y(s) f(s)\, ds,$$

and the kernel $P_y(s)$ is a positive even function which is decreasing for positive s. That means $P_y(s)$ is a convex combination of the box kernels $(1/2h)\chi_{(-h,h)}(s)$ that arise in the definition of Mf. Take step functions $h_n(s)$, which are also nonnegative, even, and decreasing on $s > 0$, such that $h_n(s)$ increases with n to $P_y(s)$. Then $h_n(s)$ has the form

$$\sum_{j=1}^{N} a_j \chi_{(-x_j, x_j)}(s)$$

with $a_j \geq 0$, and $\int h_n\, ds = \sum_j 2x_j a_j \leq 1$. See Figure I.4. Hence

$$\left| \int h_n(s) f(s)\, ds \right| \leq \int h_n(s) |f(s)|\, ds \leq \sum_{j=1}^{N} 2x_j a_j \frac{1}{2x_j} \int_{-x_j}^{x_j} |f(s)|\, ds \leq Mf(0).$$

Then by monotone convergence

$$|u(0, y)| \leq \int P_y(s) |f(s)|\, ds \leq Mf(0).$$

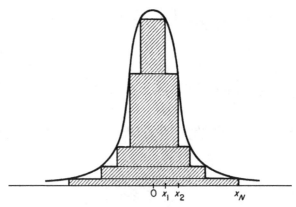

Figure I.4. $P_y(s)$ and its approximation $h_n(s)$, which is a positive combination of box kernels $(1/2x_j)\chi_{(-x_j, x_j)}(s)$.

Now fix $(x, y) \in \Gamma_\alpha(0)$. Then $|x| < \alpha y$, and $P_y(x - s)$ is majorized by a positive even function $\psi(s)$, which is decreasing on $s > 0$, such that

$$\int \psi(s)\, ds \leq A_\alpha = 1 + \frac{2\alpha}{\pi}.$$

The function is $\psi(s) = \sup\{P_y(x - t) : |t| > s\}$. Approximating $\psi(s)$ from below by step functions $h_n(s)$ just as before, we have

$$\int \psi(s)|f(s)| \, ds \le A_\alpha M f(0)$$

and

$$|u(x, y)| \le \int \psi(s)|f(s)| \, ds \le A_\alpha M f(0),$$

which is (4.3).   □

Theorem 4.2 is, with the same proof, true for Poisson integrals of functions on ∂D, where the cone is replaced by the region

$$\Gamma_\alpha(e^{i\varphi}) = \left\{ z : \frac{|e^{i\varphi} - z|}{1 - |z|} < \alpha, |z| < 1 \right\}, \quad \alpha > 1,$$

which is asymptotic, as $z \to e^{i\varphi}$, to an angle with vertex $e^{i\varphi}$. The theorem is quite general. The proof shows it is true with $P_y(x - s)$ replaced by any kernel $\varphi_y(x - s)$ which can be dominated by a positive, even function $\psi(s)$, depending on (x, y), provided that ψ is decreasing on $s > 0$ and that $\int \psi(s) \, ds \le A_\alpha$ whenever $(x, y) \in \Gamma_\alpha(t)$ (see Stein, [1970]).

The Hardy–Littlewood maximal theorem is this:

Theorem 4.3. *If $f \in L^p(\mathbb{R})$, $1 \le p \le \infty$, then $M f(t)$ is finite almost everywhere.*

(a)   *If $f \in L^1(\mathbb{R})$, then Mf is weak L^1,*

$$|\{t \in \mathbb{R} : M f(t) > \lambda\}| \le (2/\lambda)\|f\|_1, \quad \lambda > 0.$$

(b)   *If $f \in L^p(\mathbb{R})$, with $1 < p \le \infty$, then $M f \in L^p(\mathbb{R})$ and*

$$\|M f\|_p \le A_p \|f\|_p,$$

where A_p depends only on p.

In (a) we have used $|E|$ to denote the Lebesgue measure of $E \subset \mathbb{R}$. That $Mf < \infty$ almost everywhere follows from (a) or (b). Condition (a) says the operator Mf is *weak-type* 1–1. The weak-type inequality in (a) is the best possible result on Mf when $f \in L^1$. Notice that if $f \not\equiv 0$, then Mf cannot possibly be in L^1, because for large x

$$M f(x) \ge \frac{1}{|4x|} \int_{-x}^{3x} |f(t)| \, dt \ge \frac{c}{|x|}$$

if $\|f\|_1 \ne 0$. If f is supported on a finite interval I, then $\int_I M f(t) dt < \infty$ if and only if $\int_I |f| \log^+ |f| \, dt < \infty$; we leave the proof as an exercise. By letting $f(t) = (1/h)\chi_{(0,h)}(t)$, and sending $h \to 0$, one can see that the constant in (a) cannot be improved upon.

The proof of Theorem 4.3 will use two additional theorems: a covering lemma of Vitali type for part (a) and the Marcinkiewicz interpolation theorem for part (b).

Lemma 4.4. *Let μ be a positive Borel measure on $\mathbb{R}$ and let $\{I_1, \ldots, I_n\}$ be a finite family of open intervals in $\mathbb{R}$. There is a subfamily $\{J_1, \ldots, J_m\}$ such that the J_i are pairwise disjoint and such that*

$$\sum_{i=1}^{m} \mu(J_i) \geq \frac{1}{2}\mu\left(\bigcup_{j=1}^{n} I_j\right).$$

Proof. By induction $\{I_1, \ldots, I_n\}$ can be replaced by a subfamily of intervals such that no interval I_j is contained in the union of the others and such that the refined family has the same union as the original family. Write the I_j in the refined family as (α_j, β_j) and index them so that

$$\alpha_1 \leq \alpha_2 \leq \cdots \alpha_n.$$

Then $\beta_{j+1} > \beta_j$ since otherwise $I_{j+1} \subset I_j$, and $\alpha_{j+1} > \beta_{j-1}$ since otherwise $I_j \subset I_{j-1} \cup I_{j+1}$. Therefore the even-numbered intervals and the odd-numbered intervals comprise pairwise disjoint subfamilies. Then

$$\sum_{j \text{ even}} \mu(I_j) + \sum_{j \text{ odd}} \mu(I_j) \geq \mu\left(\bigcup_{j=1}^{n} I_j\right),$$

and for $\{J_i\}$ we take either the even-numbered intervals or the odd-numbered intervals, which ever gives the larger sum. $\square$

Proof of Theorem 4.3(a). Assume $f \in L^1$ and let $\lambda > 0$. Then the set $E_\lambda = \{t : Mf(t) > \lambda\}$ is open, and therefore measurable. For each $t \in E_\lambda$ we have an open interval I containing t such that

$$\frac{1}{|I|} \int_I |f| \, ds > \lambda,$$

which is the same as

(4.4) $$|I| < \frac{1}{\lambda} \int_t |f| \, ds.$$

Let K be a compact subset of E_λ and cover K by finitely many intervals $I_1, \ldots, I_n$ that satisfy (4.4). Applying the lemma to $\{I_1, \ldots, I_n\}$ gives us pairwise disjoint intervals $J_1, J_2, \ldots, J_m$, that satisfy (4.4) such that

$$\left|\bigcup_{j=1}^{n} I_j\right| \leq 2 \sum_{j=1}^{m} |J_j|.$$

Then

$$|K| \le \left| \bigcup_{i=1}^{n} I_j \right| \le 2 \sum_j \frac{1}{\lambda} \int_{J_j} |f| \, ds \le \frac{2}{\lambda} \int |f| \, ds.$$

Letting $|K|$ increase to $|E_\lambda|$ gives us part (a). $\quad\square$

The proof of part (b) depends on the interpolation theorem of Marcinkiewicz.

Theorem 4.5. *Let (X, μ) and (Y, ν) be measure spaces, and let $1 < p_1 \le \infty$. Suppose T is a mapping from $L^1(X, \mu) + L^{p_1}(X, \mu)$ to ν-measurable functions such that*

(i) $\qquad |T(f + g)(y)| \le |Tf(y)| + |Tg(y)|$;

(ii) $\quad \nu(\{y : |Tf(y)| > \lambda\}) \le (A_0/\lambda)\|f\|_1, \quad f \in L^1$;

(iii) $\quad \nu(\{y : |Tf(y)| > \lambda\}) \le ((A_1/\lambda)\|f\|_{p_1})^{p_1}, \quad f \in L^{p_1}$;

(when $p_1 = \infty$ we assume instead that

$$\|Tf\|_\infty \le A_1 \|f\|_\infty).$$

Then for $1 < p < p_1$,

$$\|Tf\|_p \le A_p \|f\|_p, \quad f \in L^p,$$

where A_p depends only on A_0, A_1, p, and p_1.

The hypothesis that the domain of T is $L^1(X, \mu) + L^{p_1}(X, \mu)$ is just a device to make sure Tf is defined when $f \in L^p, 1 \le p \le p_1$. For $f \in L^p$, write $f = f\chi_{|f|>1} + f\chi_{|f|\le1} = f_0 + f_1$. Then $|f_0| \le |f|^p \in L^1$ and $|f_1| \le |f|^{p/p_1} \in L^{p_1}$. Before proving the theorem, let us use it to prove the remaining part (b) of the Hardy–Littlewood maximal theorem.

Proof of Theorem 4.3(b). In this case the measure spaces are both $(\mathbb{R}, dx)$. The operator M clearly satisfies the subadditivity condition (i). Condition (ii) we proved as part (a) of the Hardy–Littlewood theorem. We take $p_1 = \infty$ and condition (iii) holds with $A_1 = 1$. The Marcinkiewicz theorem then tells us

$$\|Mf\|_p \le A_p \|f\|_p, \quad 1 < p \le \infty$$

which is the assertion in part (b) of Theorem 4.3. It of course follows that $Mf < \infty$ almost everywhere. $\quad\square$

Proof of Theorem 4.5. Fix $f \in L^p, 1 < p < p_1$, and, $\lambda > 0$. Let

$$E_\lambda = \{y : |Tf(y)| > \lambda\}.$$

We are going to get a tight grip on $\nu(E_\lambda)$ and then use Lemma 4.1 to estimate $\|Tf\|_p$. The clever Marcinkiewicz idea is to split f at $\lambda/2A_1$. Write

$$f_0 = f\chi_{\{x:|f(x)|>\lambda/2A_1\}}, \quad f_1 = f\chi_{\{x:|f(x)|\le\lambda/2A_1\}}.$$

Then $|Tf(y)| \le |Tf_0(y)| + |Tf_1(y)|$, and $E_\lambda \subset B_\lambda \cup C_\lambda$, where

$$B_\lambda = \{y : |Tf_0(y)| > \lambda/2\}, \quad C_\lambda = \{y : |Tf_1(y)| > \lambda/2\}.$$

Now by (ii) we have

$$v(B_\lambda) \le 2\frac{A_0}{\lambda}\|f_0\|_1 \le 2\frac{A_0}{\lambda}\int_{|f|>\lambda/2A_1} |f|\,d\mu.$$

To estimate $v(C_\lambda)$ we consider two cases. If $p_1 = \infty$, then $\|f_1\|_\infty < \lambda/2A_1$ and by (iii) $C_\lambda = \varnothing$. (This explains the presence of A_1 in the definitions of f_0 and f_1.) If $p_1 < \infty$, we have by (iii)

$$v(C_\lambda) \le \left(\frac{2}{\lambda}A_1\|f_1\|_{p_1}\right)^{p_1} \le \frac{(2A_1)^{p_1}}{\lambda^{p_1}}\int_{|f|\le\lambda/2A_1} |f|^{p_1}\,d\mu.$$

We bound $v(E_\lambda)$ by $v(B_\lambda) + v(C_\lambda)$ and use Lemma 4.1. The case $p_1 = \infty$ is easier:

$$\|Tf\|_p^p = \int_0^\infty p\lambda^{p-1}v(E_\lambda)d\lambda \le \int_0^\infty p\lambda^{p-1}\left(\frac{2A_0}{\lambda}\int_{|f|>\lambda/2A_1} |f|\,d\mu\right)d\lambda$$

$$\le 2A_0p\int |f|\int_0^{2A_1|f|}\lambda^{p-2}d\lambda\,d\mu = 2^p\frac{A_0A_1^{p-1}p}{p-1}\int |f|^p\,d\mu,$$

because $p - 2 > -1$. Hence $Tf \in L^p$. If $p_1 < \infty$, we have the same thing plus an additional term bounding $v(C_\lambda)$:

$$\|Tf\|_p^p \le \int_0^\infty p\lambda^{p-1}\left(\frac{2A_0}{\lambda}\int_{|f|>\lambda/2A_1} |f|\,d\mu\right)d\lambda$$

$$+ \int_0^\infty p\lambda^{p-1}\frac{(2A_1)^{p_1}}{\lambda^{p_1}}\int_{|f|\le\lambda/2A_1} |f|^{p_1}\,d\mu\,d\lambda.$$

The first integral we just estimated in the proof for $p_1 = \infty$. The second integral is

$$(2A_1)^{p_1}p\int |f|^{p_1}\int_{2A_1|f|}^\infty \lambda^{p-p_1-1}\,d\lambda\,d\mu = \frac{(2A_1)^p p}{p_1-p}\int |f|^p\,d\mu$$

because $p - p_1 - 1 < -1$. Altogether this gives

$$\|Tf\|_p \le A_p\|f\|_p$$

with

$$A_p^p \le 2^p A_1^{p-1}\left(\frac{A_0 p}{p-1} + \frac{A_1 p}{p_1 - p}\right),$$

which proves the theorem. $\square$

It is interesting that as $p \to 1$, $A_p \leq A/(p-1)$ for some constant A, and if $p_1 = \infty$, then $\lim_{p \to \infty} A_p \leq A_1$. For the maximal function we obtain

$$A_p^p = p2^{p+1}/(p-1).$$

Other splittings of $f = f_0 + f_1$ give more accurate estimates of the dependencies of A_p on A_0 and A_1 (see Zygmund [1968, Chapter XII]).

5. Nontangential Maximal Function and Fatou's Theorem

Fix $\alpha > 0$, and consider the cones

$$\Gamma_\alpha(t) = \{z \in H \; : |x - t| < \alpha y\}, \quad t \in \mathbb{R}.$$

If $u(z)$ is a harmonic function on H , the *nontangential maximal function* of u at $t \in \mathbb{R}$ is

$$u^*(t) = \sup_{\Gamma_\alpha(t)} |u(z)|.$$

The value of u^* depends on the parameter α, but since α has been fixed we will ignore that distinction.

Theorem 5.1. *Let $u(z)$ be harmonic on H and let $1 \leq p < \infty$. Assume*

$$\sup_y \int |u(x + iy)|^p \, dx < \infty.$$

If $p > 1$, then $u^*(t) \in L^p$, and

(5.1) $$\|u^*\|_p^p \leq B_p \sup_y \int |u(x + iy|^p \, dx.$$

If $p = 1$, then u^* is weak $I.^1$, and

(5.2) $$|\{t : u^*(t) > \lambda\} \leq \frac{B_1}{\lambda} \sup_y \int |u(x + iy)| \, dx.$$

The constants B_p depend only on p and α.

Proof. Let $p > 1$. Then $u(z)$ is the Poisson integral of a function $f(t) \in L^p(\mathbb{R})$, and

$$\|f\|_p \leq \sup_y \left(\int |u(x + iy)|^p \, dx \right)^{1/p}.$$

Theorem 4.2 says that $u^*(t) \leq A_\alpha M f(t)$, and the Hardy–Littlewood theorem then yields (5.1).

If $p = 1$, we know only that $u(z)$ is the Poisson integral of a finite measure μ on $\mathbb{R}$ and

$$\int |d\mu| \leq \sup_y \int |u(x + iy)|\, dx,$$

because μ is a weak-star limit of the measures $u(x + iy)\, dx$, $y \to 0$. Define

$$M(d\mu)(t) = \sup_{t \in I}(|\mu|(I)/|I|).$$

The proof of Theorem 4.2 shows that $u^*(t) \leq A_\alpha M(d\mu)(t)$. And the proof of part (a) of Theorem 4.3 shows that $M(d\mu)(t)$ is weak L^1 and

$$|\{t : M(d\mu)(t) > \lambda\}| \leq \frac{2}{\lambda} \int d|\mu|.$$

Therefore (5.2) holds in the case $p = 1$. $\square$

The nontangential maximal function u^* will be more important to us than the Hardy–Littlewood maximal function Mf. The next corollary is stated to emphasize the strength of Theorem 5.1.

Corollary 5.2. *If $u(z)$ is harmonic on H and if $p > 1$, then*

$$\int \sup_y |u(x + iy)|^p\, dx \leq B_p \sup_y \int |u(x + iy)|^p\, dx.$$

Note that Corollary 5.2 is false at $p = 1$. Take $u(x, y) = P_y * f(x)$, $f \in L^1$, $f > 0$. Then $\sup_y |u(x, y)| \geq Mf(x)$ and $Mf \notin L^1$.

Theorem 5.3 (Fatou).   *Let $u(z)$ be harmonic on H , let $1 \leq p \leq \infty$ and assume*

$$\sup_y \|u(x + iy)\|_{L^p(dx)} < \infty.$$

Then for almost all t the nontangential limit

$$\lim_{\Gamma_\alpha(t) \ni z \to t} u(z) = f(t)$$

exists.

   If $p > 1$, $u(z)$ is the Poisson integral of the boundary value function $f(t)$, and if $1 < p < \infty$,

$$\|u(x + iy) - f(x)\|_p \to 0 \quad (y \to 0).$$

   If $p = 1$, then $u(z)$ is the Poisson integral of a finite measure μ on $\mathbb{R}$, and μ is related to the boundary value function $f(t)$ by

$$d\mu = f(t)\, dt + dv,$$

where dv is singular to Lebesgue measure.

Proof. First, let $1 \leq p < \infty$ and assume $u(z)$ is the Poisson integral of a function $f(t) \in L^p$. We will show $u(z)$ has nontangential limit $f(t)$ for almost all t. We can assume f is real valued. Let

$$\Omega_f(t) = \varlimsup_{z \to t} u(z) - \varliminf_{z \to t} u(z),$$

where z is constrained to $\Gamma_\alpha(t)$. Then by the maximal theorem $\Omega_f(t) \leq 2u^*(t) \leq 2A_\alpha M f(t)$, so that Ω_f, as well as the limes superior and the limes inferior, is finite almost everywhere. The function $\Omega_f(t)$ represents the nontangential oscillation of u at t, and u has a nontangential limit at t if and only if $\Omega_f(t) = 0$.

By Theorem 5.1, and by Chebychev's inequality if $p > 1$, we have

(5.3) $$|\{t : \Omega_f(t) > \varepsilon\}| \leq B_p \left(\frac{2}{\varepsilon} \|f\|_p\right)^p.$$

Now if $g \in L^p$ and if in addition $g \in C_0(\mathbb{R})$, then by Theorem 3.1, $\Omega_g = 0$ for all t, and so $\Omega_f = \Omega_{f+g}$. Take $g \in C_0(\mathbb{R})$ so that $\|f + g\|_p \leq \varepsilon^2$. Then

$$|\{t : \Omega_f(t) > \varepsilon\}| = |\{t : \Omega_{f+g}(t) > \varepsilon\}| \leq B_p \left(\frac{2}{\varepsilon} \|f + g\|_p\right)^p \leq c_p \varepsilon^p.$$

Consequently $\Omega_f(t) = 0$ almost everywhere and u has a nontangential limit almost everywhere. The limit coincides with $f(t)$ almost everywhere because $u(x, y)$ converges in L^p norm to $f(x)$. That proves the theorem in the case $1 < p < \infty$ and, provided that $u(z)$ is the Poisson integral of an L^1 function, in the case $p = 1$.

Let $p = \infty$, and let $u(z) = (P_y * f)(x)$, with $f(t) \in L^\infty$. Let $A > 0$ and write $f(t) = f_1(t) + f_2(t)$ where $f_2 = 0$ on $(-A, A)$ and $f_1 \in L^1$. Then $u(z) = u_1(z) + u_2(z)$, where $u_j(z) = (P_y * f_j)(x)$, $j = 1, 2$. It was proved above that $u_1(z)$ has nontangential limit $f_1(t)$ almost everywhere, and by Lemma 3.3 $u_z(z)$ has limit $f_2(t) = 0$ everywhere on $(-A, A)$. Hence $u(z)$ converges to $f(t)$ nontangentially almost everywhere on $(-A, A)$. Letting $A \to \infty$ we have the result for $p = \infty$.

Now let $p = 1$ and assume

$$\sup_y \|u(x + iy)\|_{L^1(dx)} < \infty.$$

Then $u(z)$ is the Poisson integral of a finite measure μ on $\mathbb{R}$. Write $d\mu = f(t)\,dt + dv$, where dv is singular to dx, and let $u_1(z) = (P_y * f)(x)$, $u_2(z) = (P_y * v)(x)$. Then $u(z) = u_1(z) + u_2(z)$. It was shown above that $u_1(z)$ has nontangential limit $f(t)$ almost everywhere. Because v is singular, the next lemma shows $u_2(z)$ has nontangential limit zero almost everywhere, and that concludes the proof. $\square$

Lemma 5.4. *If v is a finite singular measure on $\mathbb{R}$, then $(P_y * v)(x)$ converges nontangentially to zero almost everywhere.*

Proof. We may assume $v \geq 0$. Because v is singular, we have

(5.4) $$\lim_{h \to 0} \frac{v((t - h, t + h))}{2h} = 0$$

for Lebesgue almost all t. Indeed, if (5.4) were not true, there would be a compact set K such that $|K| > 0$, $v(K) = 0$, and

$$\overline{\lim_{h \to 0}} \frac{v((t - h, t + h))}{2h} > a > 0, \quad \text{all} \quad t \in K.$$

Cover K by finitely many intervals I_j such that $v(\cup I_j) < \varepsilon$ and such that $v(I_j) > a|I_j|$. By the covering lemma 4.4, pairwise disjoint intervals $\{J_i\}$ can be chosen from the $\{I_j\}$ such that

$$|K| \leq 2 \sum |J_i| < \frac{2}{a} \sum v(J_i) < \frac{2\varepsilon}{a},$$

a contradiction for ε sufficiently small.

Suppose (5.4) holds at $t \in \mathbb{R}$. Let $z \in \Gamma_\alpha(t)$ and suppose for simplicity that $\text{Re } z = t$. Since $v \geq 0$, we have

$$(P_y * v)(t) = \int_{|s-t|<Ay} P_y(t - s) \, dv(s) + \int_{|s-t| \geq Ay} P_y(t - s) dv(s).$$

The second integral does not exceed $(\pi A^2 y)^{-1} \int dv$. If we approximate $P_y(s)\chi_{|s|<Ay}(s)$ from below by even step functions, as in the proof of Theorem 4.2, we see that

$$\int_{|s-t|<Ay} P_y(t - s) dv(s) \leq \sup_{h<Ay} \frac{v((t - h, t + h))}{2h}.$$

Choosing $A = A(y)$ so that $Ay \to 0$ $(y \to 0)$ but $A^2 y \to \infty$ $(y \to 0)$, we obtain $P_y * v(t) \to 0 (y \to 0)$ if (5.4) holds at t. The estimates when $|x - t| < \alpha y$ are quite similar and we leave them to the reader. $\quad \square$

A positive measure σ on $\mathbb{H}$ is called a *Carleson measure* if there is a constant $N(\sigma)$ such that

(5.5) $$\sigma(Q) \leq N(\sigma)h$$

for all squares

$$Q = \{x_0 < x < x_0 + h, 0 < y < h\}.$$

The smallest such constant $N(\sigma)$ is the *Carleson norm* of σ.

Lemma 5.5. *Let σ be a positive measure on $\mathbb{H}$, and let $\alpha > 0$. Then σ is a Carleson measure if and only if there exists $A = A(\alpha)$ such that*

(5.6) $$\sigma(\{|u(z)| > \lambda\}) \leq A|\{t : u^*(t) > \lambda\}|, \quad \lambda > 0,$$

for every harmonic function $u(z)$ on H *, where $u^*(t)$ is the nontangential maximal function of $u(z)$ over the cone $\{|x - t| < \alpha y\}$. If A is the least constant such that (5.6) holds, then*

$$c_1(\alpha)A \leq N(\sigma) \leq c_2(\alpha)A.$$

Proof.   We take $\alpha = 1$. The proof for a different α is similar. Assume σ is a Carleson measure. The open set $\{t : u^*(t) > \lambda\}$ is the union of a disjoint sequence of open intervals $\{I_j\}$, with centers $c(I_j)$. Let T_j be the tent

$$T_j = \{z : |x - c(I_j)| + y < |I_j|/2\},$$

an isosceles right triangle with hypotenuse I_j. If $|u(z)| > \lambda$, then $u^*(t) > \lambda$ on the interval $\{|t - x| < y\}$ and this interval is contained in some I_j. See Figure I.5. Consequently,

$$\{z : |u(z)| > \lambda\} \subset \bigcup_{j=1}^{\infty} T_j.$$

By (5.5) we therefore have

$$\sigma(\{z : |u(z)| > \lambda\}) \leq \sum_{j} \sigma(T_j) \leq N(\sigma) \sum_{j} |I_j| = N(\sigma)|\{t : u^*(t) > \lambda\}|,$$

and (5.6) holds.

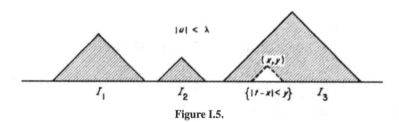

Figure I.5.

Conversely, let I be an interval $\{x_0 < t < x_0 + h\}$ and let $u(z) = P_y * f(x)$ with $f(x) = 4\lambda \chi_I(x)$. Then $u(z) > \lambda$ on the square Q with base I, so that by (5.6) and the maximal theorem,

$$\sigma(Q) \leq A|\{t : u^*(t) > \lambda\}| \leq (AC/\lambda)\|f\|_1 \leq ACh,$$

and σ is a Carleson measure. $\square$

Theorem 5.6.   (Carleson).   *Let $f \in L^p(\mathbb{R})$ and let $u(z)$ denote the Poisson integral of f. If σ is a positive measure on the upper half plane, then the following are equivalent.*

(a) σ *is a Carleson measure.*
(b)   *For $1 < p < \infty$, and for all $f \in L^p(\mathbb{R})$, $u(z) \in L^p(\sigma)$.*

(c) *For* $1 < p < \infty$,

$$\int |u(z)|^p d\sigma \le C_p \int |f|^p \, dt, \quad f \in L^p.$$

(d) *For all* $f \in L^1(\mathbb{R})$, *we have the distribution function inequality*

$$\sigma(\{z : |u(z)| > \lambda\}) \le (c_1/\lambda) \int |f(t)| \, dt, \quad \lambda > 0.$$

If (b) *or* (c) *holds for one value of* p, $1 < p < \infty$, *then* (a) *holds.*

The constants C_p *depend only on* p *and the constant* $N(\sigma)$ *in* (5.5). *In fact, if* (a) *holds, we can take* $C_p = N(\sigma)B_p$ *where* B_p *is the constant in Theorem* 5.1 *with* $\alpha = 1$. *If* (c) *or* (d) *holds then* (5.5) *holds with* $N(\sigma) \le 4^p C_p$.

Proof. If (a) holds, then by (5.6) and Theorem 5.1, (c) and (d) hold. Clearly, (c) implies (b), and if (b) holds for some p, then the closed graph theorem for Banach spaces shows that (c) holds for the same value of p.

Now suppose that (d) holds or that (c) holds for some p, $1 < p < \infty$. As in the proof of Lemma 5.5, take $I = \{x_0 < t < x_0 + h\}$, and set $u(z) = P_y * f(x)$, $f(t) = 4\chi_I(t)$. Then $\|f\|_p = 4h^{1/p}$, and $u(z) > 1$ on $Q = I \times (0, h)$. Hence

$$\sigma(Q) \le \sigma(\{|u(z)| > 1\}) \le C_p \int |f|^p \, dt = 4^p C_p h,$$

and (5.5) holds. $\square$

6. Subharmonic Functions

Let Ω be an open set in the plane. A *subharmonic function* on Ω is a function $v : \Omega \to [-\infty, \infty]$ such that
(a) v is *upper semicontinous*:

$$v(z_0) \ge \lim_{z \to z_0} v(z), \quad z_0 \in \Omega,$$

(b) for each $z_0 \in \Omega$ there is $r(z_0) > 0$ such that the disc $\Delta(z_0, r(z_0)) = \{z : |z - z_0| < r(z_0)\}$ is contained in Ω and such that for every $r < r(z_0)$,

(6.1)
$$v(z_0) \le \frac{1}{\pi r^2} \iint_{|z - z_0| < r} v(z) \, dx \, dy.$$

The semicontinuity guarantees that v is measurable and bounded above on any compact subset of Ω. Therefore the integral in (6.1) either converges or diverges to $-\infty$.

Every harmonic function is subharmonic, but our primary example of a subharmonic function is $v(z) = \log|f(z)|$, where $f(z)$ is an analytic function on Ω. It is clear that $v(z)$ is upper semicontinuous. Condition (6.1) is trivial at a point z_0 for which $f(z_0) = 0$. If $f(z_0) \neq 0$, then $\log f(z)$ has a single-valued determination on some neighborhood of z_0, and $v(z) = \mathrm{Re}(\log f(z))$ is harmonic on this neighborhood. Hence (6.1) holds with equality if $f(z_0) \neq 0$.

Lemma 6.1. (Jensen's Inequality). *Let (X, μ) be a measure space such that μ is a probability measure, $\mu(X) = 1$. Let $v \in L^1(\mu)$ be a real function, and let $\varphi(t)$ be a convex function on $\mathbb{R}$. Then*

$$\varphi\left(\int v \, d\mu\right) \leq \int \varphi(v) \, d\mu.$$

Proof. The convexity of φ means that $\varphi(t)$ is the supremum of the linear functions lying below φ:

$$\varphi(t_0) = \sup\{at_0 + b : at + b \leq \varphi(t), t \in \mathbb{R}\}.$$

Whenever $at + b \leq \varphi(t)$, we have

$$a\left(\int v \, d\mu\right) + b = \int (av + b) \, d\mu \leq \int \varphi(v) \, d\mu,$$

and the supremum of the left sides of these inequalities is $\varphi(\int v \, d\mu)$. □

Jensen's inequality is also true if $\int v \, d\mu = -\infty$, provided that φ is defined at $t = -\infty$ and increasing on $[-\infty, \infty)$. The proof is trivial in that case.

Theorem 6.2. *Let $v(z)$ be a subharmonic function on Ω, and let $\varphi(t)$ be an increasing convex function on $[-\infty, \infty)$, continuous at $t = -\infty$. Then $\varphi \circ v$ is a subharmonic function on Ω.*

Proof. Since every convex function is continuous on $\mathbb{R}$, φ is continuous on $[-\infty, \infty)$. It follows immediately that $\varphi \circ v$ is upper semicontinuous. If $z_0 \in \Omega$ and if $r < r(z_0)$, then because φ is increasing

$$\varphi(v(z_0)) \leq \varphi\left(\frac{1}{\pi r^2} \iint\limits_{\Delta(z_0, r)} v(z) \, dx \, dy\right).$$

By Jensen's inequality, then

$$\varphi(v(z_0)) \leq \frac{1}{\pi r^2} \iint\limits_{\Delta(z_0, r)} \varphi(v) \, dx \, dy,$$

which is (6.1) for $\varphi(v)$. □

For example, if $f(z)$ is analytic on Ω, then $|f(z)|^p = \exp(p \log |f(z)|)$ is a subharmonic function on Ω if $0 < p < \infty$, and

$$\log^+ |f(z)| = \max(\log |f(z)|, 0)$$

is also a subharmonic function on Ω. Notice the contrast with the situation for harmonic functions, where we have $|u|^p$ subharmonic only for $p \geq 1$ (by Hölder's inequality).

Theorem 6.3. *Let $v : \Omega \to [-\infty, \infty]$ be an upper semicontinuous function. Then v is subharmonic on Ω if and only if the following condition holds: If $u(z)$ is a harmonic function on a bounded open subset W of Ω and if*

$$\overline{\lim_{W \ni z \to \zeta}} (v(z) - u(z)) \leq 0$$

for all $\zeta \in \partial W$, then

$$v(z) \leq u(z), \quad z \in W.$$

Proof. Assume $v(z)$ is subharmonic on Ω. Let $u(z)$ and W be as in the above statement. Then $V(z) = v(z) - u(z)$ is subharmonic on W, and

$$\overline{\lim_{W \ni z \to \zeta}} V(z) \leq 0$$

for all $\zeta \in \partial W$.

Using a standard maximum principle argument we now show $V \leq 0$ in W. We can assume W is connected. Let $a = \sup_W V(z)$ and suppose $a > 0$. Let $\{z_n\}$ be a sequence in W such that $V(z_n) \to a$. Since $a > 0$, the z_n cannot accumulate on ∂W, and there is a limit point $z \in W$. By the semicontinuity, $V(z) = a$, and the set

$$E = \{z \in W : V(z) = a\}$$

is not empty. The set E is closed because V is upper semicontinuous and has maximum value a.

If $z_0 \in E$, then because $V(z) \leq a$ on W, the mean value inequality (6.1) shows $V(z) = a$ almost everywhere on $\Delta(z_0, r)$, for some $r > 0$. Hence E is dense in $\Delta(z_0, r)$. Because E is closed this means $\Delta(z_0, r) \subset E$, and E is open. Since W was assumed to be connected, we have a contradiction and we conclude that $a \leq 0$.

Conversely, let $z_0 \in \Omega$ and let $\overline{\Delta(z_0, r)} \subset \Omega$. Since v is upper semicontinuous there are continuous functions $u_n(z)$ decreasing to $v(z)$ on $\partial \Delta(z_0, r)$ as $n \to \infty$. Let $U_n(z)$ be the harmonic function on $\Delta(z_0, r)$ with boundary values $u_n(z)$. After a suitable change of scale, U_n is obtained from u_n by the Poisson integral formula for the unit disc. From Section 3 we know that U_n is continuous on $\overline{\Delta(z_0, r)}$. By hypothesis we have $v(z_0) \leq U_n(z_0)$, and hence

$$v(z_0) \leq \lim_n \frac{1}{2\pi} \int u_n(z_0 + re^{i\theta}) \, d\theta = \frac{1}{2\pi} \int v(z_0 + re^{i\theta}) \, d\theta$$

by monotone convergence. Averaging these inequalities against $r\,dr$ then gives (6.1), and so $v(z)$ is subharmonic. $\square$

The proof just given shows that if $v(z)$ is subharmonic on Ω, then (6.1) holds for any $r > 0$ such that $\overline{\Delta(z_0, r)} \subset \Omega$. It also shows that we can replace area means by circular means in (6.1). The condition

$$v(z_0) \le \frac{1}{2\pi} \int v(z_0 + re^{i\theta})\, d\theta,$$

$0 < r < r(z_0)$, is therefore equivalent to (6.1).

Corollary 6.4. *If Ω is a connected open set and if $v(z)$ is a subharmonic function on Ω such that $v(z) \not\equiv -\infty$, then whenever $\overline{\Delta(z_0, r)} \subset \Omega$,*

$$\frac{1}{2\pi} \int v(z_0 + re^{i\theta})\, d\theta = -\infty.$$

Proof. Let $u_n(z)$ be continuous functions decreasing to $v(z)$ on $\partial\Delta(z_0, r)$, and let $U_n(z)$ denote the harmonic extension of u_n to $\Delta(z_0, r)$. If

$$\frac{1}{2\pi} \int v(z_0 + re^{i\theta})\, d\theta = -\infty,$$

then since v is bounded above and since Poisson kernels are bounded and positive, we have

$$\frac{1}{2\pi} \int P_z(\theta) v(z_0 + re^{i\theta})\, d\theta = -\infty, \quad |z| < 1.$$

Consequently $U_n(z) \to -\infty$ for each $z \in \Delta(z_0, r)$, and by Theorem 6.3 $v = -\infty$ on $\Delta(z_0, r)$. The nonempty set

$$\{z \in \Omega : v(z) \equiv -\infty \text{ on a neighborhood of } z\}$$

is then open and closed, and we again have a contradiction. $\square$

Theorem 6.5. *Let $v(z)$ be a subharmonic function in the unit disc D. Assume $v(z) \not\equiv -\infty$. For $0 < r < 1$, let*

$$v_r(z) = \begin{cases} v(z), & |z| \le r, \\ \dfrac{1}{2\pi} \displaystyle\int P_{z/r}(\theta) v(re^{i\theta})\, d\theta, & |z| < r. \end{cases}$$

Then $v_r(z)$ is a subharmonic function in D, $v_r(z)$ is harmonic on $|z| < r$, $v_r(z) \ge v(z)$, $z \in D$, and $v_r(z)$ is an increasing function of r.

Proof. By Corollary 6.4 and by Section 3 we know $v_r(z)$ is finite and harmonic on $\Delta(0, r) = \{|z| < r\}$. To see that $v_r(z)$ is upper semicontinuous at a point $z_0 \in \partial\Delta(0, r)$ we must show

$$v(z_0) \ge \varlimsup_{\substack{z \to z_0 \\ |z| < r}} v_r(z).$$

This follows from the approximate identity properties of the Poisson kernel and from the semicontinuity of v. Write $z_0 = re^{i\theta_0}$. For $\varepsilon > 0$ there is $\delta > 0$

such that $v(re^{i\theta}) < v(z_0) + \varepsilon$ if $|\theta - \theta_0| < \delta$. Then if $|z| < r$ and if $|z - z_0|$ is small,

$$v_r(z) \le \frac{1}{2\pi} \int_{|\theta - \theta_0| \le \delta} P_{z/r}(\theta)(v(z_0) + \varepsilon) \, d\theta$$

$$+ \frac{1}{2\pi} \left(\sup_\theta v(re^{i\theta}) \right) \int_{|\theta - \theta_0| > \delta} P_{z/r}(\theta) \, d\theta$$

$$\le v(z_0) + 2\varepsilon.$$

Hence v_r is upper semicontinuous.

If we again take continuous functions $u_n(z)$ decreasing to $v(z)$ on $\partial \Delta(0, r)$, then as in the proof of Corollary 6.4 we have

$$v(z) \le v_r(z).$$

Because v is subharmonic, this inequality shows that $v_r(z)$ satisfies the mean value inequality (6.1) at each point z_0 with $|z_0| = r$. Consequently $v_r(z)$ is a subharmonic function on D.

If $r > s$, then $v_r = (v_s)_r$, and since for any subharmonic function v, $v_r(z) \ge v(z)$, the functions $v_r(z)$ increase with r. $\quad\square$

Corollary 6.6. *If $v(z)$ is a subharmonic function on D, then*

$$m(r) = \frac{1}{2\pi} \int v(re^{i\theta}) \, d\theta$$

is an increasing function of r.

The subharmonic function $v(z)$ on Ω has a *harmonic majorant* if there is a harmonic function $U(z)$ such that $v(z) \le U(z)$ throughout Ω. If Ω is connected, if $v(z) \not\equiv -\infty$ in Ω, and if $v(z)$ has a harmonic majorant, then the Perron process for solving the Dirichlet problem produces the *least harmonic majorant* $u(z)$, which is a harmonic function majorizing $v(z)$ and satisfying $u(z) \le U(z)$ for every other harmonic majorant $U(z)$ of $v(z)$ (see Ahlfors [1966] or Tsuji [1959]). Since we are interested only in simply connected domains, we shall not need the beautiful Perron process to obtain harmonic majorants. We can use the Poisson kernel instead.

Theorem 6.7. *Let $v(z)$ be a subharmonic function in the unit disc D. Then v has a harmonic majorant if and only if*

$$\sup_r \frac{1}{2\pi} \int v(re^{i\theta}) \, d\theta = \sup_r v_r(0) < \infty.$$

The least harmonic majorant of $v(z)$ is then

$$u(z) = \lim_{r \to 1} \int P_{z/r}(\theta) v(re^{i\theta}) \, d\theta / 2\pi = \lim_{r \to 1} v_r(z).$$

Proof. If $\sup_r v_r(0)$ is finite, then by Harnack's theorem the functions $v_r(z)$ increase to a finite harmonic function $u(z)$ on D. Since $v(z) \leq v_r(z)$, $u(z)$ is a harmonic majorant of $v(z)$. Conversely, if $U(z)$ is harmonic on D, and if $U(z) \geq v(z)$ on D, then by Theorem 6.3, $U(z) \geq v_r(z)$ for each r. Consequently, $\sup_r v_r(0) < \infty$, and again $u(z) = \lim_r v_r(z)$ is finite and harmonic. Since $v_r(z) \leq U(z)$, we have $u(z) \leq U(z)$, and so $u(z)$ is the least harmonic majorant. □

Since by continuity $u(z) = \lim_{r \to 1} u(rz)$, the least harmonic majorant of $v(z)$ can also be written

$$u(z) = \lim_{r \to 1} \int P_z(\theta) v(re^{i\theta}) \, d\theta / 2\pi.$$

In particular, if $v(z) \geq 0$ and if $v(z)$ has a harmonic majorant, then its least harmonic majorant is the Poisson integral of the weak-star limit of the bounded positive measures $v(re^{i\theta}) \, d\theta / 2\pi$.

Theorem 6.8. *Let $v(z)$ be a subharmonic function in the upper half plane* H *. If*

$$\sup_y \int |v(x + iy)| \, dx = M < \infty,$$

then $v(z)$ has a harmonic majorant in H *of the form*

$$u(z) = \int P_y(x - t) \, d\mu(t),$$

where μ is a finite signed measure on $\mathbb{R}$*.*

Proof. The inequality

$$(6.2) \qquad v(z) \leq \frac{2}{\pi y} \sup_\eta \int |v(\xi + i\eta)| \, d\xi, \qquad z = x + iy, \quad y > 0,$$

is proved in the same way that the similar inequality (3.9) was proved to begin the proof of Theorem 3.5.

Fix $y_0 > 0$ and consider the harmonic function

$$u(z) = u_{y_0}(z) = \int P_{y - y_0}(x - t) v(t, y_0) \, dt,$$

defined on the half plane $\{y > y_0\}$. We claim $v(z) \leq u(z)$ on $y > y_0$. To see this, let $\varepsilon > 0$ and let $A > 0$ be large. Let $u_n(t)$ be continuous functions decreasing to $v(t + iy_0)$ on $[-A, A]$, and let

$$U_n(z) = \int_{-A}^{A} P_{y - y_0}(x - t) u_n(t) \, dt, \qquad y > y_0,$$

be the Poisson integral of u_n. The function

$$V(z) = v(z) - \varepsilon \log|z + i| - U_n(z)$$

is subharmonic on $y > y_0$. With ε fixed we have $\lim_{z \to \infty} V(z) = -\infty$, by (6.2), and if A is large we have

$$\overline{\lim_{z \to (t, y_0)}} V(z) \le 0$$

for $|t| \ge A$, again by (6.2). If $|t| < A$, then $\overline{\lim}_{z \to (t, y_0)} V(z) \le v(t, y_0)$ $-u_n(t, y_0) \le 0$. It follows from Theorem 6.3 and a conformal mapping that $V(z) \le 0$ on $y > y_0$. Sending $n \to \infty$, then $A \to \infty$, and then $\varepsilon \to 0$, we obtain $v(z) \le u(z)$ on $y > y_0$. The measures $v(t, y_0)\, dt$ remain bounded as $y_0 \to 0$, and if $d\mu(t)$ is a weak-star cluster point, then

$$\lim_{y_0 \to 0} u_{y_0}(z) = \int P_y(x - t)\, d\mu(t)$$

is a harmonic majorant of $v(z)$. $\quad \square$

The function $u(z)$ is actually the least harmonic majorant of $v(z)$, but we shall not use this fact.

Notes

See the books of Ahlfors [1973] and Carathéodory [1954, Vol. II], and Nevanlinna's paper [1929] for further applications of Schwarz's lemma.

Pick [1916], studied the finite interpolation problem (2.1) for functions mapping the upper half plane to itself. Theorem 2.2 follows easily from Pick's work via conformal mappings (see Nevanlinna [1919]). The proof in the text is from Marshall [1976c], who had earlier [1974] published a slightly different proof. See Sarason [1967], Sz.-Nagy and Korányi [1956] and Donoghue [1974] for operator-theoretic approaches to that theorem.

The coefficient sequences for functions having positive real parts on D were characterized by Carathéodory [1911] and by Toeplitz [1911]. Schur's theorem [1917], is that there exists $f \in B$ with expansion

$$f(z) = c_0 + c_1 z + \cdots + c_n z^n + O(z^{n+1})$$

if and only if the matrix $I_n - A_n^* A_n$ is nonnegative definite, where I_n is the $(n + 1) \times (n + 1)$ identity matrix and

$$A_n = \begin{bmatrix} c_0 & c_1 & \cdots & & c_n \\ 0 & c_0 & & & c_{n-1} \\ \vdots & & & & \vdots \\ 0 & & & 0 & c_0 \end{bmatrix}.$$

A proof is outlined in Exercise 21 of Chapter IV. See Tsuji [1959], for a derivation of Schur's theorem from the Carathéodory-Toeplitz result. Pick's theorem and Schur's theorem are both contained in a recent result by Cantor [1981] who found the matrix condition necessary and sufficient for interpolation by finitely many derivatives of a function in B at finitely many points in D.

The maximal function was introduced by Hardy and Littlewood [1930], but its importance was not widely recognized until much later. In their proof Hardy and Littlewood used rearrangements of functions. Lemma 4.4 is from Garsia's book [1970], where it is credited to W. H. Young. Also see Stein [1970] for another covering lemma, which is valid in $\mathbb{R}^n$, and for a more general discussion of maximal functions and approximate identities.

The books by Zygmund [1968] and by Stein and Weiss [1971] contain more information on the Marcinkiewicz theorem and other theorems on interpolation of operators.

Fatou's theorem is from his classic paper [1906], which was written not long after the introduction of the Lebesgue integral. Theorem 5.6 is from Carleson [1958, 1962a]. This proof of Lemma 5.5 is due to E. M. Stein.

We have barely touched the vast theory of subharmonic functions. Among the numerous important references we mention Tsuji [1959] and Hayman and Kennedy [1976] as guides to the literature.

Some authors call the inequality

$$\log |f(0)| \leq \frac{1}{2\pi} \int \log |f(e^{i\theta})| \, d\theta$$

and its relatives Jensen's inequality, but we reserve the name for the inequality of Lemma 6.1. Because the logarithm is concave, the two candidates for the name "Jensen's inequality" actually go in opposite directions.

Exercises and Further Results

1. Let $f(z) \in B$ satisfy $f(0) = 0, |f'(0)| = \delta > 0$. If $|z| < \eta < \delta$, then

$$|f(z)| \geq \left(\frac{\delta - \eta}{1 + \eta \delta} \right) |z|,$$

and in the disc $\{|z| < \eta\}$, $f(z)$ takes on each value w,

$$|w| < \left(\frac{\delta - \eta}{1 - \eta \delta} \right) \eta$$

exactly one time. (Hint: If $g(z) = f(z)/z$, then $\rho(g(z), f'(0)) \leq |z|$.)

2. Suppose $f(z) \in B$.

(a) If $f(z)$ has two distinct fixed points in D, then $f(z) = z$.

(b) Let $\varepsilon > 0$, and suppose $\rho(f(z_1), z_1) < \varepsilon$, and $\rho(f(z_2), z_2) < \varepsilon$ for z_1 and z_2 distinct points of D. If z lies on the hyperbolic geodesic arc joining z_1

to z_2, then

$$\rho(f(z), z) \leq C\varepsilon^{1/2}$$

with C an absolute constant. (Hint: We can assume that $z = 0$, $z_1 = -r < 0$, $z_2 = s > 0$, and that $f(z_1) = z_1$. Then $f(0) \in K(z_1, r) \cap K(z_2, s + \varepsilon)$ and the euclidean description of the discs $K(z_1, r)$ and $K(z_2, s + \varepsilon)$ yield an upper bound for $|f(0)| = \rho(f(0), 0)$.)

(c) Let $e^{i\theta}$ and $e^{i\varphi}$ be distinct points of ∂D. Suppose $\{z_n\}$ and $\{w_n\}$ are sequences in D such that $z_n \to e^{i\theta}$, $w_n \to e^{i\varphi}$ and such that $\rho(f(z_n), z_n) \to 0$, $\rho(f(w_n), w_n) \to 0$. Then f fixes each point of the geodesic from $e^{i\theta}$ to $e^{i\varphi}$, and hence $f(z) = z$.

This result will reappear in Chapter X, Exercise 9.

3. Let $I = (\theta_1, \theta_2)$ be an arc on the unit circle ∂D. Then

$$\omega(z) = \int_{\theta_1}^{\theta_2} P_z(\theta) \, d\theta/2\pi, \qquad z \in D,$$

is the harmonic measure of I. Show

$$\omega(z) = \alpha/\pi - (\theta_2 - \theta_1)/2\pi,$$

where $\alpha = \arg((e^{i\theta_2} - z)/(e^{i\theta_1} - z))$, as shown in Figure I.6 (see Carathéodory [1954] or Nevanlinna [1953]).

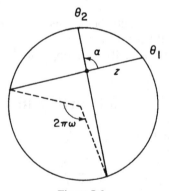

Figure I.6.

4. For $z, w \in D$ define the Gleason distance as

$$d(z, w) = \sup\{|f(z) - f(w)| : f \in B\}$$

and define the Harnack distance as

$$H(z, w) = \sup \log \left\{ \frac{u(z)}{u(w)} : u \text{ harmonic on } D, u > 0 \right\}.$$

Then

(i) $H(z, w) = \log \left(\dfrac{1 + \rho(z, w)}{1 - \rho(z, w)} \right),$

(ii) $\dfrac{4d(z, w)}{4 + d^2(z, w)} = \rho(z, w),$

where $\rho(z, w) = |(z - w)/(1 - \bar{w}z)|$. To prove (i) compare Poisson kernels; to prove (ii) consider $(f(z) - f(w))/(1 - \overline{f(w)}f(z))$, $f \in B$.

A related identity,

$$\|P_z - P_w\|_1 = \frac{1}{2\pi} \int |P_z(\theta) - P_w(\theta)| \, d\theta = 2 - \frac{2}{\pi} \cos^{-1} \rho(z, w),$$

follows from Exercise 3 and conformal invariance. For (i) and (ii) see König [1969] or Bear [1970].

5. If $g(z)$ is bounded and analytic in the angle

$$\Gamma = \left\{ z \in D : \frac{|1 - z|}{1 - |z|} < \alpha \right\}, \quad \alpha > 1,$$

and if $\lim_{x \to 1} g(x) = a$, then in any smaller angle Γ',

$$\lim_{\Gamma' \ni z \to 1} g(z) = a.$$

(Hint: Normal families.)

6. (Julia's lemma). The region

$$W_k = \left\{ \frac{|1 - z|^2}{1 - |z|^2} < k \right\}, \quad k > 0,$$

is a disc in D tangent to ∂D at $z = 1$. It is called an *orocycle*. Let $f \in B$ and suppose there is a sequence $\{z_n\}$ in D such that $z_n \to 1$, $f(z_n) \to 1$, and

$$\frac{1 - |f(z_n)|}{1 - |z_n|} \to A < \infty.$$

Then $f(W_k) \subset W_{Ak}$ for all $k > 0$. Equivalently,

$$\frac{|1 - f(z)|^2}{1 - |f(z)|^2} \le A \frac{|1 - z|^2}{1 - |z|^2}, \quad z \in D.$$

In particular, $A > 0$. (Hint: Choose r_n so that $(1 - |z_n|)/(1 - r_n) \to k$. Then the noneuclidean discs $K(z_n, r_n)$ converge to W_k, while the noneuclidean discs $(Kf(z_n), r_n)$ converge to W_{Ak}. But by Schwarz's lemma, $(Kf(z_n, r_n)) \subset K(f(z_n), r_n)$. See Julia [1920] and Carathéodory [1929].)

★**7.** (The angular derivative). For $f \in B$ set

$$B = \sup_{z \in D} \frac{|1 - f(z)|^2}{1 - |f(z)|^2} \Big/ \frac{|1 - z|^2}{1 - |z|^2}.$$

(a) Suppose $B = \infty$. If $z_n \to 1$, then by Julia's lemma

$$\frac{1 - |f(z_n)|}{1 - |z_n|} \to \infty.$$

Consequently

$$\lim_{\Gamma \ni z \to 1} \frac{1 - f(z)}{1 - z} = \infty$$

within any angle

$$\Gamma = \left\{ z \in D : \frac{|1 - z|}{1 - |z|} < \alpha \right\}, \qquad \alpha > 1.$$

(b) Suppose $B < \infty$. Then $B > 0$ and by Julia's lemma

$$\varliminf_{x \to 1} \frac{1 - |f(z)|}{1 - |z|} \geq B.$$

Take $z_n = x_n$ real, $x_n \to 1$. Then

(E.1) $$\qquad |1 - f(x_n)|^2 \leq B(1 - |f(x_n)|^2) \frac{1 - x_n}{1 + x_n}$$

and $f(x_n) \to 1$. Moreover, the inequalities

$$\frac{1 - |f(x_n)|}{1 - x_n} \frac{1 + x_n}{1 + |f(x_n)|} \leq \frac{|1 - f(x_n)|}{1 + |f(x_n)|} \frac{1 - |f(x_n)|}{1 - |f(x_n)|} \frac{1 - x_n^2}{(1 - x_n)^2}$$

$$\leq \frac{|1 - f(x_n)|^2}{1 - |f(x_n)|^2} \frac{1 - x_n^2}{(1 - x_n)^2}$$

show that

$$\varlimsup \frac{1 - |f(x_n)|}{1 - x_n} \leq B.$$

So when $z_n = x_n$, the hypotheses of Julia's lemma hold with $A = B$. Furthermore,

$$B = \lim \frac{1 - |f(x_n)|}{1 - x_n} \leq \varliminf \frac{|1 - f(x_n)|}{1 - x_n}$$

$$\leq \varlimsup \frac{|1 - f(x_n)|}{1 - x_n} \leq B^{1/2} \lim \left\{ \frac{1 - |f(x_n)|^2}{1 - x_n^2} \right\}^{1/2} = B,$$

by (E.1), so that

$$\left| \frac{1 - f(x_n)}{1 - x_n} \right| \to B, \qquad \frac{|1 - f(x_n)|}{1 - |f(x_n)|} \to 1.$$

It follows that $\arg(1 - f(x)) \to 0$ as $x \to 1$ and consequently that

$$\lim_{x \uparrow 1} \frac{1 - |f(x)|}{1 - x} = \lim_{x \uparrow 1} \frac{|1 - f(x)|}{1 - x} = \lim_{x \uparrow 1} \frac{1 - f(x)}{1 - x} = B.$$

(c) Again suppose $B < \infty$. Then in any angle $\Gamma = \{|1 - z|/(1 - |z|) < \alpha\}$,

$$\left| \frac{1 - f(z)}{1 - z} \right| \le 2B\alpha.$$

Using Exercise 5, conclude that

$$\lim_{\Gamma \ni z \to 1} \frac{1 - f(z)}{1 - z} = B \qquad \text{and} \qquad \lim_{\Gamma \ni z \to 1} f'(z) = B,$$

for any angle Γ.

(d) There is another proof using the Poisson integral representation of positive harmonic functions. Write $F(z) = (1 + f(z))/(1 - f(z))$. Then

$$\text{Re } F(z) = \frac{1 - |f(z)|^2}{|1 - f(z)|^2} > 0,$$

and there is $\mu > 0$ such that

$$\text{Re } F(z) = \int_{-\pi}^{\pi} \frac{1 - |z|^2}{|e^{i\theta} - z|^2} \, d\mu(\theta)$$

$$= \mu(\{0\}) \frac{1 - |z|^2}{|1 - z|^2} + \int_{-\pi}^{\pi} \frac{1 - |z|^2}{|e^{i\theta} - z|^2} d\mu_0(\theta),$$

where μ_0 has no mass at 0. Clearly $B \le 1/(\mu/\{0\})$. Also

$$\frac{1 - |f(z)|^2}{|1 - f(z)|^2} \bigg/ \frac{1 - |z|^2}{|1 - z|^2} = \mu(\{0\}) + \int_{-\pi}^{\pi} \frac{|1 - z|^2}{|e^{i\theta} - z|^2} d\mu_0(\theta),$$

and when $z \to 1$ within an angle Γ, the integral has limit 0. Consequently $B = 1/\mu(\{0\})$. Since

$$\frac{1 + f(z)}{1 - f(z)} = \frac{1 + z}{1 - z} \mu(\{0\}) + \int_{-\pi}^{\pi} \frac{e^{i\theta} + z}{e^{i\theta} - z} \, d\mu_0(\theta) + ic,$$

similar reasoning shows that

$$\lim_{\Gamma \ni z \to 1} \frac{1 - z}{1 - f(z)} = \frac{1}{2} \lim_{\Gamma \ni z \to 1} \frac{(1 - z)(1 + f(z))}{1 - f(z)} = \mu(\{0\}).$$

8. Suppose $f(z)$ is a function from D to D such that whenever z_1, z_2, z_3 are distinct points of D there exists $g \in B$ (depending on z_1, z_2, z_3) such that

$$g(z_j) = f(z_j), \qquad j = 1, 2, 3.$$

Then f is analytic. (Either use Pick's theorem or show directly that $f(z)$ satisfies the Cauchy–Riemann equations.)

9. Let $u(z)$ be a real-valued harmonic function on the unit disc D. Show $u(z)$ is the difference of two positive harmonic functions if and only if

$$\sup_r \int |u(re^{i\theta})| \, d\theta < \infty,$$

and if and only if $u(z)$ is the Poisson integral of a finite signed measure on ∂D. Give an example of a harmonic function not the difference of two positive harmonic functions.

10. Let $E \subset \mathbb{R}$ be compact, $|E| = 0$. Then there are harmonic functions $V_n(z)$ on H , $n = 1, 2, \ldots$, such that

   (i) $V_n(z) \geq 0$,
   (ii) $\lim_{H \ni z \to t} V_n(z) = +\infty$, all $t \in E$,
   (iii) $\lim_{n \to \infty} V_n(z) = 0$, all $z \in H$.

11. (a) Write $f \in L^1_{\text{loc}}$ if f is measurable and if $|f|$ is integrable over any compact set. If $f \in L^1_{\text{loc}}(\mathbb{R})$, then almost everywhere

$$\lim_{h \to 0} \frac{1}{2h} \int_{x-h}^{x+h} |f(t) - f(x)| \, dt = 0.$$

The set of x for which this holds is called the *Lebesgue set* of $f(x)$. (Hint: The proof of Theorem 5.3 shows that

$$f(x) = \lim_{h \to 0} \frac{1}{2h} \int_{x-h}^{x+h} f(t) \, dt$$

almost everywhere. Replacing f by $|f - c|$, c rational, then yields the result.)
   (b) Let $\varphi(x) \in L^1$, $\int \varphi(x)dx = 1$. Let $\Psi(x)$ be the least even decreasing majorant of $|\varphi(x)|$,

$$\Psi(x) = \sup_{|t| \geq |x|} |\varphi(t)|.$$

If $\psi(x) \in L^1$, then the operator

$$M_\varphi f(x) = \sup_{y>0} \frac{1}{y} \int \varphi\left(\frac{x-t}{y}\right) f(t) \, dt$$

is weak-type 1–1,

$$|\{x : M_\varphi f(x) > \lambda\}| \le \frac{C}{\lambda}\|f\|_1,$$

and M_φ is bounded on L^p, $1 < P \le \infty$,

$$\|M_\varphi f\|_p \le C_p\|f\|_p, \qquad 1 < p \le \infty.$$

If $f \in L^p$, $1 \le p \le \infty$, then

(E.2) $$f(x) = \lim_{y \to 0} \frac{1}{y} \int \varphi\left(\frac{x-t}{y}\right) f(t)\, dt$$

almost everywhere. More precisely, (E.2) holds at every point of the Lebesgue set of f, which is independent of φ.

(c) Formulate and prove a similar result about nontangential convergence and nontangential maximal functions.

12. If $f(x)$ has support a bounded interval I, then $\int_I Mf\, dx < \infty$ if and only if $\int_I |f| \log^+ |f|\, dx < \infty$ (Stein [1969]).

13. Let μ be a positive Borel measure on $\mathbb{R}$, finite on compact sets, and define

$$M_\mu f(x) = \sup_{I \ni x} \frac{1}{\mu(I)} \int_I |f|\, d\mu.$$

Show

$$\mu(\{M_\mu f(x) > \lambda\}) \le \frac{C}{\lambda} |f|\, d\mu$$

and

$$\int |M_\mu f|^p d\mu \le C_p \int |f|^p d\mu, \qquad 1 < p < \infty.$$

Conclude that for μ-almost all x,

$$\lim_{h \to 0} \frac{1}{\mu(x-h, x+h)} \int_{x-h}^{x+h} |f(t) - f(x)|\, d\mu(t) = 0, \qquad f \in L^1(\mu).$$

14. If μ is a positive singular measure on $\mathbb{R}$, then almost everywhere with respect to μ,

$$\lim_{h \to 0} \frac{\mu((x-h, x+h))}{2h} = \infty.$$

15. Let $v(z)$ be a positive subharmonic function on the unit disc and define

$$m(r) = \frac{1}{2\pi} \int v(re^{i\theta})\, d\theta.$$

Then $m(r)$ is an increasing convex function of $\log r$. That is, if $\log r = t \log r_1 + (1-t)\log r_2$, $0 < t < 1$, then

$$m(r) \le t\, m(r_1) + (1-t)\, m(r_2).$$

16. If $\varphi(z)$ is of class C^2 on a neighborhood of the closed disc $\overline{\Delta(z_0, R)} = \{z : |z - z_0| \le R\}$, then

$$\varphi(z_0) = \frac{1}{2\pi} \int_0^{2\pi} \varphi(z_0 + Re^{i\theta})\, d\theta - \frac{1}{2\pi} \int\!\!\int_{\Delta(z_0, R)} \log\left|\frac{R}{z - z_0}\right| \Delta\varphi(z)\, dx\, dy$$

by Green's theorem.

★**17.** Let $v(z) \ne -\infty$ be upper semicontinuous on a plane domain Ω.
  (a) If $v \in C^2$ and if $\Delta v = \partial^2 v/\partial x^2 + \partial^2 v/\partial y^2 \ge 0$, then v is subharmonic on Ω. (Use Exercise 16.)
  (b) Let $\chi(z) \in C^\infty(\mathbb{R}^2)$ satisfy

$$\chi \ge 0, \qquad \chi(z) = 0, \quad |z| > 1, \qquad \int \chi(z)dx\, dy = 1.$$

Assume χ is radial, $\chi(z) = \chi(|z|)$. Set $\chi_\varepsilon(z) = \varepsilon^{-2}\chi(z/\varepsilon)$. Then

$$v * \chi_\varepsilon(z) = \int v(w)\chi_\varepsilon(z - w)dx\, dy$$

is C^∞ on $\Omega_\varepsilon = \{z \in \Omega : \text{dist}\,(z, \partial\Omega) > \varepsilon\}$ Prove v is subharmonic if and only if

$$\Delta(v * \chi_\varepsilon) \ge 0$$

on Ω_ε, for all $\varepsilon > 0$. In that case $v * \chi_\varepsilon$ decreases to $v(z)$ everywhere on Ω.
  (c) Let $C_0^\infty(\Omega)$ denote the space of infinitely differentiable functions having compact support contained in Ω. If $\varphi \in C_0^\infty(\Omega)$ and $f \in C^2(\Omega)$, then

$$\int\!\!\int_\Omega f \Delta\varphi\, dx\, dy = \int\!\!\int_\Omega \varphi\, \Delta f\, dx\, dy$$

(Green's theorem again). If

$$\int\!\!\int_\Omega f \Delta\, \varphi\, dx\, dy \ge 0$$

when $\varphi \ge 0, \varphi \in C_0^\infty(\Omega)$, then f is subharmonic on Ω (take $\varphi(z) = \chi_\varepsilon(z_0 - z)$). The *weak Laplacian* of $v \in L^1_{\text{loc}}(\Omega)$ is the functional

$$C_0^\infty(\Omega) \ni \varphi \to \int\!\!\int_\Omega v\Delta\varphi\, dx\, dy.$$

If v is upper semicontinuous and if its weak Laplacian is nonnegative, that is, if

$$\iint_{\Omega} v \Delta \varphi \, dx \, dy \geq 0$$

whenever $\varphi \geq 0$, $\varphi \in C_0^{\infty}(\Omega)$, then $v * \chi_{\varepsilon}$ is subharmonic for every $\varepsilon > 0$, so that v is subharmonic. Therefore $v(z)$ is subharmonic if and only if its weak Laplacian is nonnegative.

(d)   The weak Laplacian of $\log|z - z_0|$, $z_0 \in \Omega$, is a familiar measure. Find the measure.

(e)   If $v(z)$ is subharmonic on Ω, then by the Riesz representation theorem, the weak Laplacian is a positive Borel measure finite on compact subsets of Ω. Denote this measure by Δv. Then on Ω_{ε}

$$v(z) = \frac{-1}{2\pi} \int_{\Omega_{\varepsilon}} \log \frac{1}{|z - w|} d \Delta v(w) + h_{\varepsilon}(z)$$

where h_{ε} is harmonic on Ω_{ε}. This is called the *Riesz decomposition* of the subharmonic function $v(z)$. (see F. Riesz [1926]).

II

H^p Spaces

The classical theory of the Hardy spaces H^p is a mixture of real and complex analysis. This chapter is a short introduction to this theory, with special emphasis put on the results and techniques we will need later.

The theory has three cornerstones:

   (i)   nontangential maximal functions;
   (ii) the subharmonicity of $|f|^p$ and $\log |f|$ for an analytic function $f(z)$;
   (iii) the use of Blaschke products to reduce problems to the case of a non-vanishing analytic function.

There are two H^p theories, one for the disc and another for the upper half plane. We introduce these twin theories simultaneously.

1. Definitions

Let $0 < p < \infty$ and let $f(z)$ be an analytic function on D. We say $f \in H^p = H^p(D)$ if

$$\sup_r \frac{1}{2\pi} \int |f(re^{i\theta})|^p \, d\theta = \|f\|_{H^p}^p < \infty.$$

If $p = \infty$, we say $f \in H^\infty$ if $f(z)$ is a bounded analytic function on D and we write

$$\|f\|_\infty = \sup_{z \in D} |f(z)|.$$

Thus the unit ball of H^∞, $\{f \in H^\infty : \|f\|_\infty \le 1\}$ is the class B considered in the first two sections of Chapter I. The remarks on subharmonic functions in Chapter I show that $f(z) \in H^p$ *if and only if the subharmonic function* $|f(z)|^p$ *has a harmonic majorant*, and that for $p < \infty$, $\|f\|_{H^p}^p$ is the value of the least harmonic majorant at $z = 0$. This second definition of H^p in terms of harmonic majorants is conformally invariant. It is used to define H^p functions on any plane domain or Riemann surface.

However, the H^p spaces on the upper half plane H have a special definition that is not conformally invariant. Let $f(z)$ be an analytic function on H. For $0 < p < \infty$, we say $f(z) \in H^p = H^p(dt)$ if

$$\sup_y \int |f(x + iy)|^p \, dx = \|f\|^p_{H^p} < \infty.$$

When $p = \infty$ we write H^∞ for the bounded analytic functions on H, and we give H^∞ the norm $\|f\|_\infty = \sup_H |f(z)|$. Note that the definition of $H^p(dt)$ involves all $y, 0 < y < \infty$, instead of only small values of y, like, say, $0 < y < 1$. For example, if

$$g(z) = \frac{e^{-iz/p}}{(i + z)^{2/p}},$$

then $\int |g(x + iy)|^p \, dx = \pi e^y (1 + y)^{-1}$, but $g(z) \notin H^p(dt)$.

Let $z = \tau(w) = i(1 - w)/(1 + w)$ be the conformal mapping of D onto H. Clearly $f \circ \tau \in H^\infty(D)$ if and only if $f \in H^\infty(dt)$. However, for $p < \infty$, $H^p(D)$ and $H^p(dt)$ are unfortunately not transformed into each other. For example, $H^p(D)$ contains nonzero constants, but $H^p(dt)$ does not. In order to treat $H^p(D)$ and $H^p(dt)$ together, we prove two simple lemmas.

Lemma 1.1. *If $0 < p < \infty$ and if $f(z) \in H^p(dt)$, then the subharmonic function $|f(z)|^p$ has a harmonic majorant $u(z)$ in H and*

$$u(i) \leq (1/\pi)\|f\|^p_{H^p}.$$

Proof. This follows from Theorem 6.8 of Chapter I. □

Lemma 1.1 shows that if $f(z) \in H^p(dt)$, then $g(w) = f(\tau(w)) \in H^p(D)$ and $\|g\|_{H^p} \leq (1/\pi)\|f\|_{H^p}$.

Lemma 1.2. *If $0 < p < \infty$ and if $f(z)$ is an analytic function in the upper half plane such that the subharmonic function $|f(z)|^p$ has a harmonic majorant, then*

$$F(z) = \frac{\pi^{-1/p}}{(z + i)^{2/p}} f(z)$$

is in $H^p(dt)$ and

(1.1) $$\|F\|^p_{H^p} \leq u(i)$$

where $u(z)$ is the least harmonic majorant of $|f(z)|^p$.

Proof. Let $u(z)$ be the least harmonic majorant of $|f(z)|^p$. The positive harmonic function $u(z)$ has the form

(1.2) $$u(z) = cy + \int P_y(x - t) \, d\mu(t),$$

where $c \geq 0$ and where μ is a positive measure on $\mathbb{R}$ such that

$$\int (1 + t^2)^{-1} d\mu(t) < \infty.$$

Consequently,

$$|F(z)|^p = \frac{1}{\pi(x^2 + (1+y)^2)} |f(z)|^p,$$

so that by (1.2) and Fubini's theorem,

$$\int |F(x+iy)|^p dx \leq c + \frac{1}{\pi} \int \frac{1}{1+x^2} \int P_y(x-t) \, d\mu(t) dx$$

$$= c + \int P_y(t) \int P_1(x-t) d_\mu(x) dt = c + (u(i) - c),$$

Therefore, $F(z) \in H^p(dt)$. The proof of Theorem I.6.8 shows that

$$\int |F(x+iy)|^p dx$$

is a decreasing function of y. Hence,

$$\|F\|_{H^p}^p = \lim_{y \to 0} \int |F(x+iy)|^p dx$$

and

$$\|F\|_{H^p}^p = -c + \lim_{y \to 0} u((1+y)i) = \frac{1}{\pi} \int \frac{d\mu(t)}{1+t^2} \leq u(i). \quad \square$$

Lemma 1.2 shows that if $g(w) \in H^p(D)$, then

$$F(z) = \frac{\pi^{-1/p}}{(z+i)^{2/p}} g(w(z)) \in H^p(dt),$$

where $w(z) = \tau^{-1}(z)$, and $\|F\|_{H^p} \leq \|g\|_{H^p}$. We shall see in Section 3 that the converse of Lemma 1.2 is true and that $\|F\|_{H^p} = \|g\|_{H^p}$. Notice that because $(z+i)^{-2/p}$ has no zeros on H , the family of zero sets of H^p functions is invariant under $z = \tau(w)$.

For $p \geq 1$, H^p is a normed linear space. For $p < 1$, the inequality

$$|z_1 + z_2|^p \leq |z_1|^p + |z_2|^p$$

shows that H^p is a metric space with metric

$$d(f, g) = \|f - g\|_{H^p}^p.$$

Theorem 1.3. *For $0 < p \leq \infty$, H^p is complete.*

Proof. We can assume $p < \infty$. We give the proof in the upper half plane; the reasoning for the disc is very similar. The key inequality

(1.3) $$|f(x + iy)| \leq (2/\pi y)^{1/p} \|f\|_{H^p}, \quad y > 0,$$

follows from (6.2) of Chapter I. It shows that any H^p Cauchy sequence $\{f_n\}$ converges pointwise on $\mathcal{H}$ to an analytic function $f(z)$. Fatou's lemma then shows

$$\int |f(x + iy) - f_n(x + iy)|^p dx \leq \varliminf_{m \to \infty} \int |f_m(x + iy) - f_n(x + iy)|^p dx$$
$$\leq \varliminf_{m \to \infty} \|f_m - f_n\|_{H^p}^p.$$

Hence $\|f - f_n\|_{H^p}^p \leq \varliminf_{m \to \infty} \|f_m - f_n\|_{H^p}^p$, and H^p is complete. $\quad\square$

2. Blaschke Products

We show that the zeros $\{z_n\}$ of a nonzero H^p function on the disc satisfy Blaschke's condition

(2.1) $$\sum (1 - |z_n|) < \infty.$$

It is noteworthy that (2.1) does not depend on p. When (2.1) holds, special H^∞ functions called Blaschke products will be constructed to have $\{z_n\}$ as zeros. Blaschke products will play an expanding role in the later chapters of this book.

Theorem 2.1. *Let $f(z)$ be an analytic function on the disc, $f \not\equiv 0$, and let $\{z_n\}$ be the zeros of $f(z)$. If $\log|f(z)|$ has a harmonic majorant, then*

$$\sum (1 - |z_n|) < \infty.$$

If $f(0) \neq 0$ and if $u(z)$ is the least harmonic majorant of $\log|f(z)|$, then

$$\sum (1 - |z_n|) \leq u(0) - \log|f(0)|.$$

Proof. Replacing $f(z)$ by $f(z)/z^N$ if necessary, we can assume $f(0) \neq 0$. Then by Theorem I.6.7,

$$\sup_r \int \log|f(re^{i\theta})| d\theta/2\pi = u(0),$$

where u is the least harmonic majorant of $\log|f(z)|$. Fix $r < 1$ so that $|z_n| \neq r$ for all n, and let $z_1, \ldots, z_n$ be those zeros with $|z_j| < r$. Then $f(rz)$ is analytic on the closed disc and $f(rz)$ has zeros $z_1/r, z_2/r, \ldots, z_n/r$. Let $B_r(z) = \Pi_{j=1}^n (z - z_j/r)/(1 - \bar{z}_j z/r)$, a finite Blaschke product with the same zeros as $f(rz)$, and let $g(z) = f(rz)/B_r(z)$. Then g is analytic and zero free on $\bar{D}$, so that

$$\log|g(0)| = \frac{1}{2\pi} \int \log|g(e^{i\theta})| d\theta.$$

Since $|g(e^{i\theta})| = |f(re^{i\theta})|$, this gives the familiar Jensen formula

$$\log|f(0)| + \sum_{|z_j|<r} \log \frac{r}{|z_j|} = \frac{1}{2\pi} \int \log|f(re^{i\theta})| d\theta.$$

Letting r tend to 1 then yields

$$\sum \log \frac{1}{|z_j|} = \lim_{r\to 1} \frac{1}{2\pi} \int \log|f(re^{i\theta})| d\theta - \log|f(0)| = u(0) - \log|f(0)|.$$

Since $1 - |z_j| \le \log(1/|z_j|)$, the theorem is proved. $\square$

If $f \in H^p(D)$, then $\log|f| \le (1/p)|f|^p$, and $\log|f|$ has a harmonic majorant. Hence, if $f \in H^p(D)$, or by Lemma 1.1 if $f(w) = F(z(w))$ where $F \in H^p(dt)$, then $\sum(1 - |z_n|) < \infty$.

Theorem 2.2.   *Let $\{z_n\}$ be a sequence of points in D such that*

$$\sum(1 - |z_n|) < \infty.$$

Let m be the number of z_n equal to 0. Then the Blaschke product

(2.2)
$$B(z) = z^m \prod_{|z_n|\ne 0} \frac{-\bar{z}_n}{|z_n|} \frac{z - z_n}{1 - \bar{z}_n z}$$

converges on D. The function $B(z)$ is in $H^\infty(D)$ and the zeros of $B(z)$ are precisely the points z_n, each zero having multiplicity equal to the number of times it occurs in the sequence $\{z_n\}$. Moreover $|B(z)| \le 1$ and

$$|B(e^{i\theta})| = 1$$

almost everywhere.

By definition, a *Blaschke product* on D is a function of the form (2.2).

Proof.   We can suppose $|z_n| > 0$ for all n. Let

$$b_n(z) = \frac{-\bar{z}_n}{|z_n|} \frac{z - z_n}{1 - \bar{z}_n z}.$$

Then the product $\prod b_n$ converges on D to an analytic function having $\{z_n\}$ for zeros if and only if $\sum|1 - b_n(z)|$ converges uniformly on each compact subset of D. But by a calculation,

$$|1 - b_n(z)| = \frac{|z_n + z|z_n||}{|z_n||1 - \bar{z}_n z|}(1 - |z_n|) \le \frac{1 + |z|}{1 - |z|}(1 - |z_n|)$$

and the convergence follows from (2.1).

Since $|b_n(z)| \le 1$, it is clear that $B(z) \in H^\infty$ and $|B(z)| \le 1$. The bounded harmonic function $B(z)$ has nontangential limits $|B(e^{i\theta})| \le 1$ almost everywhere. To see that $|B(e^{i\theta})| = 1$ almost everywhere, set $B_n(z) = \prod_1^n b_k(z)$. Then B/B_n is another Blaschke product and

$$\left| \frac{B(0)}{B_n(0)} \right| \le \frac{1}{2\pi} \int \frac{|B(e^{i\theta})|}{|B_n(e^{i\theta})|} d\theta = \frac{1}{2\pi} \int |B(e^{i\theta})| d\theta.$$

Letting $n \to \infty$ now yields

$$\frac{1}{2\pi} \int |B(e^{i\theta})| d\theta = 1.$$

so that $|B(e^{i\theta})| = 1$ almost everywhere. □

The purpose of the convergence factors $-\bar{z}_n/|z_n|$ is to make $\sum \arg b_n(z)$ converge. To remember the convergence factors, note that they are chosen so that $b_n(0) > 0$.

By Theorem 2.1 and Theorem 2.2, the analytic function $f(z)$ has a factorization

$$f(z) = B(z)g(z), \quad z \in D,$$

where $B(z)$ is a Blaschke product and where $g(z)$ has no zeros on D, if and only if the subharmonic function $\log |f(z)|$ has a harmonic majorant.

In the upper half plane condition (2.1) is replaced by

(2.3) $$\sum \frac{y_n}{1 + |z_n|^2} < \infty, \quad z_n = x_n + iy_n,$$

and the Blaschke product with zeros $\{z_n\}$ is

$$B(z) = \left(\frac{z-i}{z+i} \right)^m \prod_{z_n \ne i} \frac{|z_n^2 + 1|}{z_n^2 + 1} \frac{z - z_n}{z - \bar{z}_n}.$$

If the moduli $|z_n|$ are bounded, (2.3) becomes $\sum' y_n < \infty$, and the convergence factors are not needed because $\prod((z - z_n)/(z - \bar{z}_n))$ already converges.

Theorem 2.3 (F. Riesz). *Let $0 < p < \infty$. Let $f(z) \in H^p(D)$, $f \not\equiv 0$, let $\{z_n\}$ be the zeros of $f(z)$, and let $B(z)$ be the Blaschke product with zeros $\{z_n\}$. Then $g(z) = f(z)/B(z)$ is in $H^p(D)$ and*

$$\|g\|_{H^p} = \|f\|_{H^p}.$$

Proof. It was noted above that $B(z)$ converges when $f \in H^p$. Let B_n be the finite Blaschke product with zeros $z_1, z_2, \ldots, z_n$, and let $g_n = f/B_n$. Fix $r < 1$. Then by Theorem I.6.6,

$$\int |g_n(re^{i\theta})|^p \frac{d\theta}{2\pi} \le \lim_{R \to 1} \int \frac{|f(Re^{i\theta})|^p}{|B_n(Re^{i\theta})|^p} \frac{d\theta}{2\pi}.$$

If $1 - R$ is small, then $|B_n(Re^{i\theta})| > 1 - \varepsilon$, so that

$$\int |g_n(re^{i\theta})|^p \frac{d\theta}{2\pi} \leq \lim_{R \to 1} \int |f(Re^{i\theta})|^p \frac{d\theta}{2\pi} = \|f\|_{H^p}^p.$$

Since $|g_n|$ increases to $|g|$, and since $|g| \geq |f|$, this gives $\|g\|_{H^p}^p = \|F\|_{H^p}^p$. □

Theorem 2.3 is also true for $H^p(dt)$, because the proof of Theorem I.6.8 shows that

$$\sup_y \int |f(x+iy)|^p dx = \lim_{y \to 0} \int |f(x+iy)|^p dx.$$

Blaschke products have a simple characterization in terms of harmonic majorants.

Theorem 2.4. *Let $f(z) \in H^\infty(D)\|f\|_\infty \leq 1$. Then the following are equivalent.*

(a) $f(z) = \lambda B(z)$, *where λ is constant, $|\lambda| = 1$, and $B(z)$ is a Blaschke product.*
(b) $\lim_{r \to 1} \int \log|f(re^{i\theta})| d\theta/2\pi = 0$.
(c) *The least harmonic majorant of $\log|f(z)|$ is 0.*

Proof.   Theorem I.6.7 shows that (b) and (c) are equivalent.
Suppose $f(z)$ is the Blaschke product with zeros $\{z_n\}$, and let $\varepsilon > 0$. We may divide $f(z)$ by a finite Blaschke product $B_n(z)$ so that $|(f/B_n)(0)| > 1 - \varepsilon$. Since B_n is continuous on $\bar{D}$ and $|B_n(e^{i\theta})| = 1$,

$$\lim_{r \to 1} \int \log|f(re^{i\theta})| d\theta = \lim_{r \to 1} \int \log\left|\frac{f}{B_n}(re^{i\theta})\right| d\theta.$$

But since $\log|f/B_n|$ is subharmonic and negative,

$$\log(1 - \varepsilon) \leq \int \log\left|\frac{f}{B_n}(re^{i\theta})\right| \frac{d\theta}{2\pi} \leq 0.$$

Therefore (b) holds.
Suppose (c) holds. Let $g(z) = f(z)/B(z)$, where $B(z)$ is the Blaschke product formed from the zeros of $f(z)$. Then

$$\log|f(z)| \leq \log|g(z)| \leq 0$$

because $\|f\|_\infty \leq 1$. Since $\log|g(z)|$ is a harmonic majorant of $\log|f(z)|$, (c) implies that $\log|g(z)| = 0$. Hence $g(z) = \lambda$, where λ is a constant and $|\lambda| = 1$, and so (a) holds. □

3. Maximal Functions and Boundary Values

Let $f(z) \in H^p(dt)$. If $p > 1$, we know from Chapter I that the nontangential maximal function $f^*(t)$ is in L^p, and that $f(z)$ converges nontangentially to an L^p function $f(t)$ almost everywhere. An important feature of the H^p spaces is that these results remain true for all p, $0 < p \leq \infty$.

Theorem 3.1. *Let $0 < p < \infty$ and let $f(z)$ be a function in $H^p(dt)$. Then for any $\alpha > 0$, the nontangential maximal function*

$$f^*(t) = \sup_{z \in \Gamma_\alpha(t)} |f(z)|$$

is in $L^p(\mathbb{R})$ and

(3.1)
$$\|f^*\|_p^p \leq A_\alpha \|f\|_{H^p}^p,$$

where the constant A_α depends only on α. Moreover, for almost all $t \in \mathbb{R}$, $f(z)$ has a nontangential limit $f(t) \in L^p(\mathbb{R})$ satisfying

(3.2)
$$\int |f(t)|^p dt = \|f\|_p^p = \|f\|_{H^p}^p$$

and

(3.3)
$$\lim_{y \to 0} \|f(t + iy) - f(t)\|_p^p = 0.$$

Proof. Except for the fact that the constant A_α in (3.1) does not depend on p, the case $p > 1$ of the theorem is proved in Section 5 of Chapter I. To stretch p below 1 we use Theorem 2.3 above. Suppose $f \in H^p$, $f \not\equiv 0$. Let $B(z)$ be the Blaschke product formed from the zeros of $f(z)$ and let $g(z) = f(z)/B(z)$. Then $\|g\|_{H^p} = \|f\|_{H^p}$, and since $|f(z)| \leq |g(z)|$, $|f^*(t)| \leq |g^*(t)|$. Let $p_1 > 1$. Since g has no zeros and $g \in H^p$, the analytic function g^{p/p_1} is in H^{p_1}. Consequently $(g^*)^{p/p_1} = (g^{p/p_1})^*$ is in L^{p_1}, and hence by Theorem I.5.1,

$$\|g^*\|_p^p = \|(g^{p/p_1})^*\|_{p_1}^{p_1} \leq B_{p_1} \|g\|_{H^p}^p.$$

Taking $p_1 = 2$ we see that (3.1) holds with a constant independent of p, because $f^* \leq g^*$.

Now $G(z) = (g(z))^{p/p_1}$ has a nontangential limit $G(t)$ almost everywhere. Taking p_1/p to be a positive integer m, we see that $g(z) = G(z)^m$ also has non-tangential limits. Since $B(z)$ has boundary values almost everywhere we conclude that $f(z)$ has a nontangential limit almost everywhere. It now follows from (3.1) and the dominated convergence theorem that $f(t) \in L^p(\mathbb{R})$ and that (3.2) and (3.3) hold. □

The equality (3.2) establishes an isometry between $H^p(dt)$ and a closed subspace of $L^p(\mathbb{R})$. For $p < 1$, although L^p is not a Banach space, it is, like H^p, a complete metric space under the metric $d(f, g) = \|f - g\|_p^p$. When $p \geq 1$, this space of boundary values of H^p functions has a simple characterization.

Corollary 3.2. *Let $1 \leq p \leq \infty$ and let $f(t) \in L^p(\mathbb{R})$. Then $f(t)$ is almost everywhere the nontangential limit of an $H^p(dt)$ function if and only if its Poisson integral*

$$f(z) = P_y * f(x)$$

is analytic on H *. The Poisson integral $f(z)$ is then the corresponding H^p function.*

Proof. If $f \in L^p(\mathbb{R})$ and if $f(z) = P_y * f(x)$ is analytic, then by Chapter I, $f(z) \in H^p(dt)$ and $f(z)$ converges nontangentially to $f(t)$.

Conversely, suppose $f(z)$ is some H^p function. If $p > 1$, then by Theorem 5.3, Chapter I, $f(z)$ has nontangential limit $f(t)$ and $f(z) = P_y * f(x)$. The case $p = 1$ requires Theorem 3.1. When $f(z) \in H^1$, (1.3) and Lemma I.3.4 yield

$$f(z + i\varepsilon) = \int P_z(t) f(t + i\varepsilon) dt, \qquad \varepsilon > 0.$$

By (3.3), this means $f(z)$ is the Poisson integral of its boundary function $f(t)$. $\square$

See Exercise 2 for some other characterizations of the boundary values of H^p functions. Because of (3.2), we often identify $f(z) \in H^p$ with its boundary function $f(t)$.[†] However, we prescribe no method of regaining $f(z)$ from $f(t)$ when $p < 1$.

Theorem 3.1 is of course also true on the unit disc. Equality (3.2) for the correspondence $f(z) \to f(e^{i\theta})$ then shows that $H^p(D)$ is isometric to a closed subspace of the Lebesgue space $L^p(d\theta/2\pi)$. And if $p \geq 1$, $f(z)$ is the Poisson integral of $f(e^{i\theta})$. The analog of (3.3),

$$\lim_{r \to 1} \int |f(re^{i\theta}) - f(e^{i\theta})|^p \frac{d\theta}{2\pi} = 0,$$

coupled with the fact that $f(rz), r < 1$, is trivially the uniform sum of its Taylor series, tells us this: *For $p < \infty$, the space of boundary functions of $H^p(D)$ coincides with the closure in $L^p(d\theta/2\pi)$ of the analytic polynomials.* Moreover, if a sequence $p_n(e^{i\theta})$ of polynomials converges to $f(e^{i\theta})$ in L^p, then by (3.2) for the disc, $\|p_n - f\|_{H^p}^p \to 0$. Since $H^p(D) \subset H^1(D), p \geq 1$, we see that for $p \geq 1$, the boundary function of an $H^p(D)$ function has Fourier series

$$f(e^{i\theta}) \sim \sum_{n=0}^{\infty} a_n e^{in\theta}$$

[†] And we often write $\|f\|_p$ for the equal $\|f\|_{H^p}$.

supported on the nonnegative integers, and its Fourier coefficients

$$a_n = \frac{1}{2\pi} \int f(e^{i\theta})e^{-in\theta}\,d\theta = \lim_{r \to 1} \frac{1}{2\pi i} \int_{|z|=r} \frac{f(z)\,dz}{z^{n+1}}$$

are the Taylor coefficients of the H^p function $f(z) = \sum a_n z^n$. Thus H^p theory is a natural bridge between Fourier analysis and complex analysis.

In the upper half plane we see from (3.3) that the uniformly continuous functions in $H^\infty \cap H^p$ are dense in H^p, because for $f \in H^p$,

$$|f(x + iy)| \le \left(\frac{2}{\pi y}\right)^{1/p} \|f\|_{H^p}, \quad y > 0,$$

and

$$|f'(x + iy)| \le \frac{2}{y}\left(\frac{4}{\pi y}\right)^{1/p} \|f\|_{H^p},$$

by the preceding inequality and by Schwarz's lemma, scaled to the disc $\Delta(z, y/2)$. We will need some very smooth classes of analytic functions that are dense in $H^p(dt)$ and that will play the role of the polynomials in the disc case. Let N be a positive integer, and let $\mathfrak{A}_N$ be the family of $H^\infty(dt)$ functions satisfying

(i)　$f(z)$ is continuous on $\overline{H}$ and $f(t)$ is infinitely differentiable, $f \in C^\infty$.
(ii)　$\lim_{|z| \to \infty} |z|^N |f(z)| = 0$, $z \in \overline{H}$.

Corollary 3.3. *Let N be a positive integer. For $0 < p < \infty$, the class $\mathfrak{A}_N$ is dense in $H^p(dt)$. For $f(z) \in H^\infty$, there are functions $f_n(z)$ in $\mathfrak{A}_N$ such that $\|f_n\|_\infty \le \|f\|_\infty$ and such that $|f_n(t)| \le |f(t)|$ almost everywhere.*

Proof. Were it not for the decay condition (ii), we could approximate f by the smooth functions $f(z + i/n)$, which converge in H^p norm to $f(t)$ if $p < \infty$ and which converge boundedly pointwise almost everywhere to $f(t)$ if $p = \infty$. Now there are some special functions $g_k(z)$ such that

(a)　$g_k(z) \in \mathfrak{A}_N$,
(b)　$|g_k(z)| \le 1, z \in \overline{H}$,
(c)　$g_k(z) \to 1, z \in \overline{H}$.

Before we construct the functions g_k, we note that the functions

$$f_n(z) = g_n(z)f(z + i/n)$$

in $\mathfrak{A}_N$ then give the desired approximation.

The heart of the proof, constructing the functions $g_k(z)$, will be done in the unit disc, where $w = -1$ corresponds to $z = \infty$. Let $\alpha_k < 1, \alpha_k \to 1$. The function

$$h_k(w) = \left(\frac{w + \alpha_k}{1 + \alpha_k w}\right)^{N+1}$$

has an $(N + 1)$-fold zero at $-\alpha_k$. These $h_k(w)$ are bounded by 1, and with N fixed they converge to 1 uniformly on compact subsets of $\bar{D}\backslash\{-1\}$. Then the functions

$$g_k(z) = h_k(\alpha_k, w), \quad w = \frac{i - z}{i + z},$$

satisfy (a)–(c). □

Now we can clarify the relation between $H^p(D)$ and $H^p(dt)$.

Corollary 3.4. *Let* $0 < p < \infty$, *let* $f(z)$ *be an analytic function in the upper half plane and let*

$$F(z) = \frac{\pi^{-1/p}}{(z + i)^{2/p}} f(z).$$

Then $|f(z)|^p$ *has a harmonic majorant if and only if* $F(z) \in H^p$. *In that case*

$$(3.4) \qquad \qquad \|F\|_{H^p}^p = u(i),$$

where $u(z)$ *is the least harmonic majorant of* $|f(z)|^p$.

Proof. Let $g(w) = f(z), z = i(1 - w)/(1 + w)$. The corollary asserts that $g \in H^p(D)$ if and only if $F \in H^p(dt)$, and that $\|g\|_p = \|F\|_p$. If $N > 2/p$ and if $F \in \mathfrak{A}_N$, then the corresponding function $g(w)$ is bounded on D and, because $d\theta/2\pi$ corresponds to $dt/\pi(1 + t^2)$,

$$\int |g(\theta)|^p \frac{d\theta}{2\pi} = \int |F(t)|^p dt.$$

By the density of $\mathfrak{A}_N$ in H^p, it follows from (3.2) that $g \in H^p(D)$ whenever $F \in H^p(dt)$ and that

$$\|g\|_{H_p} = \|F\|_{H_p}.$$

Since Lemma 1.2 shows $F \in H^p(dt)$ if $g(w) \in H^p(D)$, that concludes the proof. □

Incidentally, the fact that we have the equality (3.4) instead of the inequality (1.1) means that in the formula (1.2) for the least harmonic majorant of $|f(z)|^p$, the constant term is $c = 0$. We shall see in Section 4 that the least harmonic majorant is the Poisson integral of the L^1 function $|f(t)|^p$.

The next corollary is noteworthy because of the recent discovery, for $p \leq 1$, of its converse, which will be proved in Chapter III.

Corollary 3.5. *Let* $0 < p < \infty$ *and let* $u(z)$ *be a real-valued harmonic function on the upper half plane* H . *If* $u(z)$ *is the real part of a function* $f(z) \in H^p$, *then*

$$u^*(t) = \sup_{\Gamma_\alpha(t)} |u(z)|$$

is in $L^p(\mathbb{R})$.

The proof from Theorem 3.1 is trivial.

Let us reexamine the boundary values in the case $p = 1$. If the harmonic function $u(z)$ satisfies

$$\sup_y \int |u(x + iy)|dx < \infty,$$

then $u(z)$ need not be the Poisson integral of its nontangential limit $u(t)$. All we can say is that $u(z)$ is the Poisson integral of a finite measure. However, if $u(z)$ is also an analytic function, then the measure is absolutely continuous, and its density is the boundary value $u(t)$. The reason is that the maximal function $u^*(t)$ is integrable.

Theorem 3.6.   *If $f(z) \in H^1(dt)$, then $f(z)$ is the Poisson integral of its boundary values:*

$$(3.5) \qquad\qquad f(z) = \int P_y(x - t)f(t)dt.$$

Conversely, if μ is a finite complex measure on $\mathbb{R}$ such that the Poisson integral $f(z) = P_y * \mu(x)$ is an analytic function on H , then μ is absolutely continuous and

$$d\mu = f(t) = dt,$$

where $f(t)$ is the boundary function of the Poisson integral $f(z)$ of μ.

Proof.   If $f(z) \in H^1$, then (3.5) was already obtained in the proof of Corollary 3.2. Conversely, if $f(z) = P_y, *\mu(x)$ is an analytic function, then by Minkowski's integral inequality it is an H^1 function and hence it is the Poisson integral of its boundary value $f(t)$. The difference measure $dv(t) = d\mu(t) - f(t)dt$ has Poisson integral zero, and so $v = 0$ by Theorem I.3.1.   □

Lemma 3.7.   *Let $f(z) \in H^1$. Then the Fourier transform*

$$\hat{f}(s) = \int_{-\infty}^{\infty} f(t)e^{-2\pi i st}\, dt = 0$$

for all $s \le 0$.

Proof.   By the continuity of $f \to \hat{f}$, we may suppose $f \in \mathfrak{A}_N$. Then for $s \le 0$, $F(z) = f(z)e^{-2\pi i sz}$ is also in $\mathfrak{A}_N$. The result now follows from Cauchy's theorem because

$$\int_0^\pi |F(Re^{i\theta})| R\, d\theta \to 0 \qquad (R \to \infty).\ \ \square$$

Notice that

$$(3.6) \qquad\qquad P_z(t) = \frac{1}{2\pi i}\left(\frac{1}{t - z} - \frac{1}{t - \bar{z}}\right).$$

Also notice that for $f \in H^1$, Lemma 3.7 applied to $(t - \bar{z})^{-1} f(t)$ yields

$$\int \frac{f(t)}{t - \bar{z}} dt = 0, \quad \text{Im } z > 0.$$

Theorem 3.8.   *Let $d\mu(t)$ be a finite complex measure on $\mathbb{R}$ such that either*

(a) $$\int \frac{d\mu(t)}{t - z} = 0 \qquad on \quad \text{Im } z < 0,$$

or

(b) $$\hat{\mu}(s) = \int e^{-2\pi i s t} d\mu(t) = 0 \qquad on \quad s < 0.$$

Then $d\mu$ is absolutely continuous and $d\mu = f(t)\,dt$, where $f(t) \in H^1$.

Proof.   If (a) holds, then by (3.6) $f(z) = P_y * \mu(x)$ is analytic and the result follows from Theorem 3.6.

Assume (b) holds. The Poisson kernel $P_y(t)$ has Fourier transform

$$\int e^{-2\pi i s t} P_y(t)\,dt = e^{-2\pi |s| y},$$

because $\hat{P}_y(-s) = \overline{\hat{P}_y(s)}$ since P_y is real, and because if $s \leq 0$, $e^{-2\pi i s z}$ is the bounded harmonic function with boundary values $e^{-2\pi i s t}$. Let $f_y(x) = P_y * \mu(x)$. By Fubini's theorem, f_y has Fourier transform

$$\hat{f}_y(s) = \begin{cases} e^{-2\pi x y} \hat{\mu}(s), & s \geq 0, \\ 0, & s < 0. \end{cases}$$

Since $\hat{f}_y \in L^1$, Fourier inversion implies

$$f_y(x) = \int e^{2\pi i x s} \hat{f}_y(s)\,ds = \int_0^\infty e^{2\pi i (x+iy)s} \hat{\mu}(s)\,ds.$$

Differentiating under the integral sign then shows that $f(z) = f_y(x)$ is analytic, and Theorem 3.6 now implies that $f(z) \in H^1$. $\square$

The disc version of Theorem 3.6, or equivalently Theorem 3.8, is one half of the famous F. and M. Riesz theorem. The other half asserts that if $f(z) \in H^1$, $f \not\equiv 0$, then $|f(t)| > 0$ almost everywhere. This is a consequence of a stronger result proved in the next section.

The theorem on Carleson measures, Theorem I.5.6, also extends to the H^p spaces, $0 < p \leq 1$, because the key estimate in its proof was the maximal theorem.

Theorem 3.9 (Carleson). *Let σ be a positive measure in the upper half plane. Then the following are equivalent:*

(a) σ *is a Carleson measure: for some constant $N(\sigma)$,*

$$\sigma(Q) \le N(\sigma)h$$

for all squares

$$Q = \{x_0 < x < x_0 + h, 0 < y < h\}.$$

(b) *For $0 < p < \infty$,*

$$\int |f|^p d\sigma \le A\|f\|_{H^p}^p, \quad f \in H^p.$$

(c) *For some p, $0 < p < \infty$, $f \in L^p(\sigma)$ for all $f \in H^p$.*

Proof. That (a) implies (b) follows from (3.1) and Lemma I.5.5 just as in the proof of Theorem I.5.6.

Trivially, (b) implies (c). On the other hand, if (c) holds for some fixed $p < \infty$, then (b) holds for the same value p. This follows from the closed graph theorem, which is valid here even when $p < 1$ (see Dunford and Schwartz [1958, p. 57]). One can also see directly that if there are $\{f_n\}$ in H^p with $\|f_n\|_p = 1$ but $\int |f_n|^p d\sigma \to \infty$, then the sum $\Sigma\alpha_n f_n$, when the α_n, are chosen adroitly, will give an H^p function for which (c) fails.

Now suppose (b) holds for some $p > 0$. Let Q be the square $\{x_0 < x < x_0 + y_0, 0 < y < y_0\}$ and let

$$f(z) = \left(\frac{1}{\pi}\frac{y_0}{(z - \bar{z}_0)^2}\right)^{1/p},$$

where $z_0 = x_0 + iy_0$. Then $f \in H^p$ and $\|f\|_p^p = \int P_{z_0}(t)dt = 1$. Since $|f(z)|^p \ge (5\pi y_0)^{-1}, z \in Q$, we have

$$\sigma(Q) \le \sigma(\{z : |f(z)| > (5\pi y_0)^{-1/p}\}) \le 5\pi Ay_0,$$

so that (a) holds. $\square$

4. $(1/\pi)\int(\log|f(t)|/(1+t^2))\,dt > -\infty$

A fundamental result of H^p theory is that the condition of this section's title characterizes the moduli of H^p functions $|f(t)|$ among the positive L^p functions. In the disc, this result is due to Szegö for $p = 2$ and to F. Riesz for the other p. For functions analytic across ∂D, the inequality (4.1) below was first noticed by Jensen [1899] and for this reason the inequality is sometimes called Jensen's inequality. We prefer to use that name for the inequality about averages and convex functions given in Theorem I.6.1.

In this section the important thing about an H^p function will be the fact that the subharmonic function

$$\log|f(re^{i\theta})| \le (1/p)|f(re^{i\theta})|^p$$

is majorized by a positive L^1 function of θ. It will be simpler to work at first on the disc.

Theorem 4.1. *If* $0 < p \leq \infty$ *and if* $f(z) \in H^p(D)$, $f \not\equiv 0$, *then*

$$\frac{1}{2\pi} \int \log |f(e^{i\theta})| d\theta > -\infty.$$

If $f(0) \neq 0$, *then*

(4.1) $$\log |f(0)| \leq \frac{1}{2\pi} \int \log |f(e^{i\theta})| d\theta,$$

and more generally, if $f(z_0) \neq 0$

(4.2) $$\log |f(z_0)| \leq \frac{1}{2\pi} \int \log |f(e^{i\theta})| P_{z_0}(\theta) d\theta.$$

Proof. By Theorem I.6.7 and by the subharmonicity of $\log |f|$,

$$\log |f(z)| \leq \lim_{r \to 1} \frac{1}{2\pi} \int \log |f(re^{i\theta})| P_z(\theta) d\theta.$$

Since $\log |f(re^{i\theta})| \to \log |f(e^{i\theta})|$ almost everywhere, and since these functions are bounded above by the integrable function $(1/p)|f^*(\theta)|^p$, where f^* is the maximal function, we have

$$\int \log^+ |f(re^{i\theta})| P_{z_0}(\theta) \, d\theta \to \int \log^+ |f(e^{i\theta})| P_{z_0}(\theta) \, d\theta.$$

Fatou's lemma can now be applied to the negative parts to give us

$$\lim_{r \to 1} \frac{1}{2\pi} \int \log |f(re^{i\theta})| P_{z_0}(\theta) \, d\theta \leq \frac{1}{2\pi} \int \log |f(e^{i\theta})| P_{z_0}(\theta) \, d\theta.$$

This proves (4.2) and (4.1). The remaining inequality follows by removing any zero at the origin. $\square$

Note that the same result in the upper half plane

(4.3) $$\log |f(z_0)| \leq \int \log |f(t)| P_{z_0}(t) dt, \quad f \in H^p,$$

follows from Theorem 4.1 and from Lemma 1.1 upon a change of variables.

Corollary 4.2. *If* $f(z) \in H^p$ *and if* $f(t) = 0$ *on a set of positive measure, then* $f = 0$.

Corollary 4.2 gives the other half of the F. and M. Riesz theorem. If $d\mu(t)$ is a finite measure such that $p_y * \mu(x)$ is analytic, then not only is $d\mu$ absolutely continuous to dt, but also dt is absolutely continuous to $d\mu$.

Corollary 4.3. *Let* $0 < p, r \leq \infty$. *If* $f(z) \in H^p$ *and if the boundary function* $f(t) \in L^r$, *then* $f(z) \in H^r$.

This corollary is often written

$$H^p \cap L^r \subset H^r.$$

Proof. Applying Jensen's inequality, with the convex function $\varphi(s) = \exp(rs)$ and with the probability measure $P_y(x-t)dt$, to (4.3) gives

$$|f(z)|^r \le \int |f(t)|^r P_y(x-t)\,dt.$$

Integration in x then yields $f \in H^r$. $\square$

Theorem 4.4. *Let $h(t)$ be a nonzero nonnegative function in $L^p(\mathbb{R})$. Then there is $f(z) \in H^p(dt)$ such that $|f(t)| = h(t)$ almost everywhere if and only if*

(4.4)
$$\int \frac{\log h(t)}{1+t^2}\,dt > -\infty.$$

Proof. It has already been proved that (4.4) is a necessary condition. To show (4.4) is sufficient, note that since $\log h \le (1/p)|h|^p$, (4.4) holds if and only if $\log h \in L^1(dt/(1+t^2))$. Let $u(z)$ be the Poisson integral of $\log h(t)$ and let $v(z)$ be any harmonic conjugate of $u(z)$. (Since H is simply connected, there exists a harmonic conjugate function $v(z)$ such that $u + iv$ is analytic. The conjugate function $v(z)$ is unique except for an additive constant.) The function we are after is

$$f(z) = e^{u(z)+iv(z)},$$

which is an analytic function on H . When $p < \infty$ Jensen's inequality again gives

$$|f(z)|^p \le \int P_y(x-t)|h(t)|^p\,dt$$

and therefore $f \subset H^p$. If $p = \infty$ then u is bounded above and so $f \in H^\infty$. $\square$

When $p = \infty$, (4.4) is especially important. Let $h \ge 0$, $h \in L^\infty$, and suppose $1/h \in L^\infty$. Then $f = e^{u+iv} \in H^\infty$ and also

$$1/f = e^{-(u+iv)} \in H^\infty.$$

In other words, f is an *invertible function* in H^∞. We write $f \in (H^\infty)^{-1}$. For emphasis, we state this fact separately, writing $g = \log h$, so that $g \in L^\infty$ if $h \in L^\infty$ and $1/h \in L^\infty$.

Theorem 4.5. *Every real-valued function $g(t)$ in L^∞ has the form $\log|f(t)|$, where $f \in (H^\infty)^{-1}$ is an invertible function in H^∞.*

In the language of uniform algebra theory, Theorem 4.5 asserts that H^∞ is a *strongly logmodular subalgebra* of L^∞. It is also a *logmodular subalgebra* of L^∞, which means that the set

$$\log |(H^\infty)^{-1}| = \{\log |f(t)| : f \in H^\infty, 1/f \in H^\infty\}$$

is dense in $L^\infty_{\mathbb{R}}$, the space of real L^∞ functions. The Banach algebra aspects of H^∞ will be discussed in some detail later; for the present we only want to say that Theorem 4.4 is a powerful tool for constructing H^p functions.

Let $h(t) \geq 0$ satisfy

$$\int \frac{|\log h(t)|dt}{1+t^2} < \infty.$$

The function

$$f(z) = e^{u(z)+iv(z)},$$

where

(4.5) $u(z) = P_y * (\log h)(x)$

and where $v(z)$ is a harmonic conjugate function of $u(z)$, is called an *outer function*. The outer function $f(z)$ is determined by $h(t)$ except for the unimodular constant factor arising from the choice of the conjugate function $v(z)$. The function $|f(z)|$ has boundary values $h(t)$ almost everywhere, and Jensen's inequality with (4.5) shows that $f(z) \in H^p$ if and only if $h(t) \in L^p$. Outer functions in H^p can be characterized in several ways.

Theorem 4.6. *Let $0 < p \leq \infty$ and let $f(z) \in H^p$, $f \not\equiv 0$. Then the following are equivalent.*

(a)   *$f(z)$ is an outer function.*
(b)   *For each $z \in H$, equality holds in (4.3); that is,*

(4.6) $\log |f(z)| = \int \log |f(t)| P_y(x - t)dt.$

(c)   *For some point $z_0 \in H$, (4.6) holds.*
(d)   *If $g(z) \in H^p$ and if $|g(t)| = |f(t)|$ almost everywhere, then*

$$|g(z)| \leq |f(z)|, \quad z \in H.$$

(e)   *$f(z)$ has no zeros in H and the harmonic function $\log |f(z)|$ is the Poisson integral of a function $k(t)$ such that*

$$\int \frac{|k(t)|dt}{1+t^2} < \infty, \quad e^{k(t)} \in L^p.$$

Proof. First, (e) is merely a rewording of the definition of an outer function in H^p, because any function $f(z)$ without zeros is an exponential, $f = e^{u+iv}$, $u = \log |f|$. Thus (a) and (e) are equivalent.

By definition, (a) implies (b). If (b) holds, then we see (d) holds by applying (4.3) to the function $g(z)$ in (d). Moreover, if (d) holds, and if $g(z)$ is an outer function determined by $\log|f(z)|$, then the analytic function $f(z)/g(z)$ satisfies

$$|f(z)/g(z)| = 1,$$

so that $f = \lambda g$, $|\lambda| = 1$, and f is an outer function. Hence (a), (b), (d), and (e) are equivalent.

Trivially, (b) implies (c). Now assume (c) and again let $g(z)$ be an outer function determined by $\log|f(t)|$. Then $|f(z)/g(z)| \leq 1$, and if (c) holds, the maximum principle shows that $|f/g| = 1$, and so $f(z)$ is an outer function. $\quad\square$

The function $S(z) = e^{iz}$ has no zeros in the upper half plane, and $S(z) \in H^\infty$, but $S(z)$ is not an outer function, because $\log|S(z)| = -y$ is not a Poisson integral.

Corollary 4.7. *If $f(z) \in H^p$ and if for some $r > 0$, $1/f(z) \in H^r$, then $f(z)$ is an outer function.*

This holds because f satisfies (4.6).

Corollary 4.8. *Let $f(z) \in H^p$. Either of the following two conditions imply that $f(z)$ is an outer function.*

(a) $\operatorname{Re} f(z) \geq 0, z \in H$.
(b) *There exists a C^1 are Γ terminating at 0 such that*

$$f(H) \subset \mathbb{C}\backslash\Gamma.$$

Proof. If (a) holds then $f + \varepsilon$ is an outer function for any $\varepsilon > 0$, because $(f + \varepsilon)^{-1} \in H^\infty$. Now, since $\operatorname{Re} f \leq 0$,

$$\int \log|f(t) + \varepsilon| P_y(x - t)dt \to \int \log|f(t)| P_y(x - t)dt$$

as $\varepsilon \to 0$, by dominated convergence on $\{t : |f(t)| \geq \frac{1}{2}\}$ and by monotone convergence on $\{t : |f(t)| < \frac{1}{2}\}$. Hence (4.6) holds for f and f is an outer function.

The above argument shows that $f(z)$ is an outer function if $\operatorname{Re} f(z) > 0$ on the set $\{|f(z)| < 1\}$. Now assume (b). Replacing $f(z)$ by $\lambda f(z)$, $|\lambda| = 1$, we can also assume that Γ has tangent vector $(1, 0)$ at $z = 0$. This means that if δ is sufficiently small, the analytic function

$$g(z) = (f(z)/\delta)^{1/5}$$

satisfies $\operatorname{Re} g(z) > 0$ if $|g(z)| < 1$. Hence g is an outer function, and $f = \delta^5 g^5$ is an outer function. $\quad\square$

5. The Nevanlinna Class

In this section we continue to use the fact that $\log |f(z)|$ has a harmonic majorant if $f(z) \in H^p$, but now the important thing will be that the least harmonic majorant is a Poisson integral.

An analytic function $f(z)$ on D or $\mathbb{H}$ is in the *Nevanlinna class*, $f \in N$, if the subharmonic function $\log^+ |f(z)|$ has a harmonic majorant. This definition is conformally invariant and the Nevanlinna classes on D and $\mathbb{H}$ therefore coincide, but it is easier to discuss N on the disc, where it is characterized by

$$(5.1) \qquad\qquad \sup_r \int \log^+ |f(re^{i\theta})| d\theta/2\pi < \infty.$$

It is clear that $H^p \subset N$ for all $p > 0$, because $\log^+ |f(z)| \le (1/p)|f(z)|^p$.

Theorem 5.1. *Let $f(z)$ be an analytic function on D, $f \not\equiv 0$. Then $f \in N$ if and only if $\log |f(z)|$ has least harmonic majorant the Poisson integral of a finite measure on ∂D.*

Proof. If

$$\log |f(z)| \le \int P_z(\theta) d\mu(\theta)$$

for some finite measure, then

$$\log^+ |f(z)| \le \int P_z(\theta) d\mu^+(\theta),$$

where μ^+ is the positive part of μ, and hence $\log^+ |f(z)|$ has a harmonic majorant.

Conversely, if $f \in N$, then $\log^+ |f(z)|$ is majorized by some positive harmonic function $U(z)$. Since $\log |f(z)| \le \log^+ |f(z)|$, $\log |f(z)|$ has a least harmonic majorant $u(z)$, and clearly $u(z) \le U(z)$. Thus

$$u(z) = U(z) - (U(z) - u(z))$$

is the difference of two positive harmonic functions, and consequently $u(z)$ is the Poisson integral of a finite measure. $\quad\square$

The proof of Theorem 5.1 really shows that a subharmonic function $v(z)$ is majorized by a Poisson integral if and only if $v^+ = \max(v, 0)$ has a harmonic majorant, and that when this is the case, the least harmonic majorant of $v(z)$ is a Poisson integral.

Lemma 5.2. *Let $f(z) \in N$, $f \not\equiv 0$. Let $B(z)$ be the Blaschke product formed from the zeros of $f(z)$. Then $B(z)$ converges, and $g(z) = f(z)/B(z)$ is in N. Moreover, $\log |g(z)|$ is the least harmonic majorant of $\log |f(z)|$.*

Proof. Since $\log |f(z)|$ has a harmonic majorant, we know from Section 2 that $B(z)$ converges. Let $u(z)$ be the least harmonic majorant of $\log |f(z)|$.

Then since $|B(z)| \leq 1$, it is clear that

$$u(z) \leq \log |g(z)|.$$

On the other hand

$$\log |B(z)| = \log |f(z)| - \log |g(z)| \leq u(z) - \log |g(z)|.$$

By Theorem 2.4, this means that

$$0 \leq u(z) - \log |g(z)|$$

and hence $u(z) = \log |g(z)|$. It of course follows that $g(z) \in N$. $\quad\square$

Theorem 5.3. *Let $f(z) \in N$, $f \not\equiv 0$. Then $f(z)$ has a nontangential limit $f(e^{i\theta})$ almost everywhere, and*

$$(5.2) \qquad\qquad \log |f(e^{i\theta})| \in L^1(d\theta).$$

The least harmonic majorant of $\log |f(z)|$ has the form $\int P_z(\theta) d\mu(\theta)$, where

$$(5.3) \qquad\qquad d\mu(\theta) = \log |f(\theta)| d\theta / 2\pi + 2\mu_s,$$

with $d\mu$, singular to $d\theta$.

Proof. Let $g(z) = f(z)/B(z)$, where $B(z)$ is the Blaschke product with the same zeros as $f(z)$. We knew from Theorem 5.1 and Lemma 5.2 that

$$(5.4) \qquad\qquad \log |g(z)| = \int P_z(\theta) \, d\mu(\theta).$$

Write $d\mu = k(\theta)(d\theta/2\pi) + d\mu_s$, where $d\mu_s$, is singular to $d\theta$. By (5.4) and by Theorem I.5.3, $\log |g(z)|$ has nontangential limit $k(\theta)$ almost everywhere. Since $|B(e^{i\theta})| = 1$ almost everywhere, it follows that $\log |f(z)|$ has nontangential limit $k(\theta)$ almost everywhere. Therefore (5.2) and (5.3) will be proved once we show there exist nontangential limits for $f(z)$.

By (5.4), $\log |g(z)|$ is the difference of two positive harmonic functions: $\log |g| = u_1 - u_2$, $u_j \geq 0$. Let $v_j(z)$ be a harmonic conjugate function of $u_j(z)$. Because D is simply connected, $v_j(z)$ is well defined, and $v_j(z)$ is unique if we set $v_j(0) = 0$. Then

$$\log g(z) = (u_1 + iv_1) - (u_2 + iv_2) + ic$$

with c a real constant, and hence

$$g(z) = \frac{e^{ic} e^{(-u_2 + iv_2)}}{e^{-(u_1 + iv_1)}}.$$

The bounded functions $e^{-(u_j + iv_j)}$ have nontangential limits almost everywhere, and these limits cannot vanish on a set of positive measure. Consequently g and $f = Bg$ have nontangential limits. $\quad\square$

If $f(z) \in N$, $f \not\equiv 0$, then by (5.2)

$$\int \log |f(\theta)| d\theta > -\infty.$$

However, the sharper inequality

(5.5) $$\log |f(z)| \leq \int \log |f(\theta)| P_z(\theta) \frac{d\theta}{2\pi},$$

which was proved for H^p functions in Section 4, can fail for $f(z) \in N$. Consider the function

$$g(z) = \exp \frac{1+z}{1-z}.$$

Then $\log |g(z)| = (1 - |z|^2)/(|1 - z|^2) = P_z(1)$. The function g is in N, and the measure determined by $\log |g(z)|$ is the unit charge at $e^{i\theta} = 1$. Since

$$\log |g(0)| = 1 > 0 = \int \log |g(\theta)| \frac{d\theta}{2\pi},$$

(5.5) fails for $g(z)$.

This counterexample contains the only thing that can go wrong with (5.5) for a function in N. The right side of (5.5) is a harmonic function; it majorizes $\log |f(z)|$ if and only if it is bigger than the least harmonic majorant of $\log |f(z)|$, which is the Poisson integral of the measure $d\mu$, in (5.3). Comparing (5.3) with (5.5), we therefore see that (5.5) is true for $f(z) \in N$ if and only if the singular term $d\mu$, in (5.3) is nonpositive.

The functions in N for which $d\mu_s, \leq 0$ form a subclass of N called N^+. We give the classical definition: *Let $f(z) \in N$. We say $f(z) \in N^+$ if*

$$\lim_{r \to 1} \int \log^+ |f(re^{i\theta})| d\theta = \int \log^+ |f(e^{i\theta})| d\theta.$$

Theorem 5.4. *Let $f(z) \in N$, $f \not\equiv 0$. Then the following are equivalent.*

(a) $f(z) \in N^+$

(b) *The least harmonic majorant of $\log^+ |f(\theta)|$ is*

$$\int \log^+ |f(\theta)| P_z(\theta) \frac{d\theta}{2\pi}.$$

(c) *For all $z \in D$,*

$$\log |f(z)| \leq \int \log |f(\theta)| P_z(\theta) \frac{d\theta}{2\pi}.$$

(d) *The least harmonic majorant of $\log |f(z)|$ is the Poisson integral of*

$$d\mu = \log |f(\theta)| \frac{d\theta}{2\pi} + d\mu_s,$$

where $d\mu_s \perp d\theta$ and

$$d\mu_s \leq 0.$$

Proof. We have already proved that (c) and (d) are equivalent. If $f(z) \in N$, then $\log^+ |f(z)|$ has least harmonic majorant

$$U(z) = \int P_z(\theta)dv(\theta),$$

where the positive measure v is the weak-star limit of the measures

$$\log^+ |f(re^{i\theta})|d\theta/2\pi,$$

by Theorem I.6.7 and the remark thereafter. By Fatou's lemma,

$$\int \log^+ |f(\theta)| P_z(\theta)\frac{d\theta}{2\pi} \leq \lim_{r \to 1} \int \log^+ |f(re^{i\theta})| P_z(\theta)\frac{d\theta}{2\pi} = \int P_z(\theta)dv(\theta),$$

and hence

(5.6) $$\log^+ |f(\theta)|\frac{d\theta}{2\pi} \leq dv,$$

because a measure is determined by its Poisson integral. By definition $f(z) \in N^+$ if and only if the two sides of (5.6), which are positive, have the same integral. Thus $f \in N^+$ if and only if

$$\log^+ |f(\theta)|\frac{d\theta}{2\pi} = dv$$

and (a) and (b) are equivalent.

Finally, a comparison of the least harmonic majorants of $\log |f(z)|$ and of $\log^+ |f(z)|$, as in the proof of Theorem 5.1, shows that (b) and (d) are equivalent. $\square$

It follows from (b) or (c) that

$$N \supset N^+ \supset H^p, \qquad p > 0.$$

It also follows that $f(z) \in H^p$ if and only $f(z) \in N^+$ and $f(e^{i\theta}) \in L^p$. This fact generalizes Corollary 4.3 and it has the same proof, using Jensen's inequality with (c). This fact can be written

$$N^+ \cap L^p = H^p, \qquad p > 0.$$

The example given before Theorem 5.4 shows that $N \cap L^p \neq H^p$.

We return to Theorem 5.3 and use formula (5.3) to obtain an important factorization theorem for functions in N. Let $f(z) \in N$, $f \not\equiv 0$. Let $B(z)$ be the Blaschke product formed from the zeros of $f(z)$, and let $g(z) = f(z)/B(z)$. Then $g(z) \in N$, and $\log |g(z)|$ is the Poisson integral of the measure μ in (5.3). By Theorem 5.4, $f \in N^+$ if and only if $g \in N^+$.

It is not hard to recover g, and therefore also $f = Bg$, from the measure μ. We actually did that in the proof of Theorem 5.3, but let us now do it again more carefully. Write

(5.7) $d\mu = \log|f(\theta)|d\theta/2\pi - (d\mu_1 - d\mu_2),$

where $d\mu_j \geq 0$ and $d\mu_j \perp d\theta$. The function

$$F(z) = \exp\left(\int \frac{e^{i\theta} + z}{e^{i\theta} - z} \log|f(e^{i\theta})|\frac{d\theta}{2\pi}\right)$$

is an outer function on the disc, because

$$F(z) = e^{u(z)+iv(z)},$$

where $u(z)$ is the Poisson integral of $\log|f(e^{i\theta})|$ and $v(z)$ is a conjugate function of $u(z)$ normalized by $v(0) = 0$. Among the outer functions associated with $\log|f(\theta)|$, $F(z)$ is determined by the condition $F(0) > 0$.

Similarly, let

(5.8) $S_j(z) = \exp\left(-\int \frac{e^{i\theta} + z}{e^{i\theta} - z}d\mu_j(\theta)\right), \quad j = 1, 2.$

Then $S_j(z)$ is analytic on D, and $S_j(z)$ has the following properties:

(i) $S_j(z)$ has no zeros in D.
(ii) $|S_j(z)| \leq 1,$
(iii) $|S_j(e^{i\theta})| = 1$ almost everywhere, and
(iv) $S_j(0) > 0.$

Properties (i) and (iv) are immediate from (5.8). Since

(5.9) $\log|S_j(z)| = -\int P_z(\theta)d\mu_j(\theta),$

property (ii) holds because $\mu_j \geq 0$ and property (iii) follows from Lemma I.5.4 because $d\mu_j \perp d\theta$. A function with properties (i)–(iv) is called a *singular function*. Every singular function $S(t)$ has the form (5.8) for some positive singular measure. This measure is determined by (5.9).

We now have

$$\log|g(z)| = \log|F(z)| + \log|S_1(z)| - \log|S_2(z)|$$

by the decomposition (5.7) of μ. Since $g(z)$ has no zeros on D, $\log g(z)$ is single valued and hence

$$\log g(z) = ic + \log F(z) + \log S_1(z) - \log S_2(z),$$

c a real constant, so that

$$g(z) = e^{ic}F(z)S_1(z)/S_2(z).$$

We have now proved most of the Canonical Factorization theorem:

Theorem 5.5. *Let $f(z) \in N$, $f \not\equiv 0$. Then*

$$(5.10) \qquad f(z) = CB(z)F(z)S_1(z)/S_2(z), \quad |C| = 1,$$

where $B(z)$ is a Blaschke product, $F(z)$ is an outer function, and $S_1(z)$ and $S_2(z)$ are singular functions. Except for the choice of the constant C, $|C| = 1$, the factorization (5.10) is unique. Every function of the form (5.10) is in N.

Proof. We have already derived the factorization (5.10). There can be no difficulties about the uniqueness of the factors because $B(z)$ is determined by the zeros of $f(z)$, and as $|B(e^{i\theta})| = |S_1(e^{i\theta})| = |S_2(e^{i\theta})| = 1$ almost everywhere, $F(z)$ is determined by $\log |f(e^{i\theta})|$. S_1 and S_2 are then determined by the least harmonic majorant of $\log |f|$. If $f(z)$ is a function of the form (5.10), then

$$\log |f(z)| \leq \log |F(z)| + \log |S_1(z)| - \log |S_2(z)|,$$

so that $\log |f(z)|$ is majorized by a Poisson integral. It now follows from Theorem 5.1 that $f(z) \in N$. □

Corollary 5.6. *Let $f(z) \in N$, $f \not\equiv 0$. Then in (5.10) the singular factor $S_2 \equiv 1$ if and only if $f(z) \in N^+$.*

Proof. $S_2(z) \equiv 1$ if and only if

$$d\mu = \log |f(\theta)|(d\theta/2\pi) - d\mu_1,$$

with $d\mu_1 \geq 0$, and this holds if and only if $f(z) \in N^+$. □

For emphasis we state

Corollary 5.7. *If $f(z) \in H^p$, $p > 0$, then $f(z)$ has a unique decomposition*

$$f(z) = CB(z)S(z)F(z),$$

where $|C| = 1$, $B(z)$ is a Blaschke product, $S(z)$ is a singular function, and $F(z)$ is an outer function in H^p.

Corollary 5.8. *Let $f(z) \in N^+$. Then $f(z) \in H^p$ if and only if the outer factor $F(z)$ is in H^p.*

The proofs of these corollaries are left to the reader.

6. Inner Functions

An *inner function* is a function $f(z) \in H^\infty$ such that $|f(e^{i\theta})| = 1$ almost everywhere. Every Blaschke product is an inner function, and so is every singular function

$$S(z) = \exp\left(-\int \frac{e^{i\theta} + z}{e^{i\theta} - z} d\mu(\theta)\right),$$

where the measure $d\mu$ is positive and singular to $d\theta$. By the Factorization Theorem 5.5, every inner function has the form

$$f(z) = e^{ic} B(z) S(z),$$

where c is a real constant, B is a Blaschke product, and S is a singular function. If $f(z) \in N^+$ and if $|f(e^{i\theta})| = 1$ almost everywhere, then f is an inner function, because $N^+ \cap L^\infty = H^\infty$. However, if $f(z) = 1/S(z)$, where S is a nonconstant singular function, then $f \in N$ and $|f(e^{i\theta})| = 1$ almost everywhere, but $f(z)$ is not an inner function. If it were, then for $z \in D$ we would have $|f(z)| \leq 1$ and $|1/f(z)| \leq 1$, but this is impossible.

Theorem 6.1. *Let $B(z)$ be a Blaschke product with zeros $\{z_n\}$, and let $E \subset \partial D$ be the set of accumulation points of $\{z_n\}$. Then $B(z)$ extends to be analytic on the complement of*

$$E \cup \{1/\bar{z}_n : n = 1, 2, \ldots\}$$

in the complex plane: In particular $B(z)$ is analytic across each arc of $(\partial D)\backslash E$. On the other hand, the function $|B(z)|$ does not extend continuously from D to any point of E.

Proof. Let $\{B_n(z)\}$ be the finite Blaschke products converging to $B(z)$. Then

$$B_n(1/\bar{z}) = 1/\overline{B_n(z)}$$

by reflection, and $\lim_{n \to \infty} B_n(z)$ exists on $\{z : |z| > 1\}\backslash\{1/\bar{z}_n : n = 1, 2, \ldots\}$ and the limit is an analytic function on that region. If $z_0 \in \partial D\backslash E$ and if $\delta > 0$ is small, then each B_n is analytic on $\Delta = \Delta(z_0, \delta)$ and $\{B_n(z)\}$ converges boundedly on $\partial\Delta\backslash\partial D$. By the Poisson integral formula for Δ, for example, this means $\{B_n(z)\}$ converges on Δ. Hence $B(z)$ is analytic except on

$$E \cup \{1/\bar{z}_n : n = 1, 2, \ldots\}.$$

If $z_0 \in E$, then

$$\lim_{D \ni z \to z_0} |B(z)| = 0,$$

while since $|B(e^{i\theta})| = 1$ almost everywhere,

$$\overline{\lim_{D \ni z \to z_0}} |B(z)| = 1.$$

Thus $|B|$ does not extend continuously to z_0. □

Aside from the Blaschke products, the simplest inner function is the singular function

$$S(z) = \exp\left(\frac{z+1}{z-1}\right),$$

generated by the point mass at 1. A calculation shows that S and all its derivatives have nontangential limit 0 at $e^{i\theta} = 1$. Under $w = (1+z)/(1-z)$, the disc D is conformally mapped to the right half plane $\{\mathrm{Re}\ w > 0\}$, so that $z = 1$ corresponds to $w = \infty$, and so that $\partial D \setminus \{1\}$ corresponds to the imaginary axis. Thus $S(z) = e^{-w}$ is analytic across $\partial D \setminus \{1\}$ and $S(z)$ wraps $\partial D \setminus \{1\}$ around ∂D infinitely often. The vertical line $\mathrm{Re}\ w = \alpha$, $\alpha > 0$, comes from the *orocycle*

$$C_\alpha = \left\{ \frac{1 - |z|^2}{|1 - z|^2} = \alpha \right\}.$$

This is a circle, with center $\alpha/(1+\alpha)$ and radius $1/(1+\alpha)$, which is tangent to ∂D at 1. On this orocycle $|S(z)| = e^{-\alpha}$, and S wraps $C_\alpha \setminus \{1\}$ around $|\zeta| = e^{-\alpha}$ infinitely often. The function $S(z)$ has no zeros in D, but for every ζ, $0, < |\zeta| < 1$, $S(z) = \zeta$ infinitely often in every neighborhood of $z = 1$.

Recall the notation

$$\Gamma_\alpha(e^{i\theta}) = \left\{ z \in D : \frac{|e^{i\theta} - z|}{1 - |z|} < \alpha \right\}, \qquad \alpha > 1$$

for the conelike region in D with vertex $e^{i\theta}$.

Theorem 6.2. *Let $S(z)$ be the singular function determined by the measure μ on ∂D, and let $E \subset \partial D$ be the closed support of μ. Then $S(z)$ extends analytically to $\mathbb{C} \setminus E$. In particular $S(z)$ is analytic across each arc of $(\partial D) \setminus E$. On the other hand, $|S(z)|$ does not extend continuously from D to any point of E. For any $\alpha > 1$ and for μ-almost all θ*

$$\lim_{z \to e^{i\theta}} S(z) = 0, \qquad z \in \Gamma_\alpha(e^{i\theta}).$$

If

(6.1)
$$\lim_{h \to 0} \frac{\mu((\theta - h, \theta + h))}{h \log 1/h} = \infty,$$

then every derivative $S^{(n)}(z)$ of $S(z)$ satisfies

(6.2)
$$\lim_{z \to e^{i\theta}} S^{(n)}(z) = 0, \qquad z \in \Gamma_\alpha(e^{i\theta}).$$

Proof. For any measure μ on ∂D, the function

$$\int \frac{e^{i\theta} + z}{e^{i\theta} - z} d\mu(\theta)$$

is analytic at all points not in the closed support E of μ. Hence $S(z)$ is analytic on $\mathbb{C} \setminus E$.

If μ is singular, then

(6.3)
$$\lim_{h \to 0} \frac{\mu((\theta - h, \theta + h))}{2h} = \infty$$

for μ-almost all θ. This follows from Lemma I.4.4 by repeating the proof of Lemma I.5.4, with dx and $d\mu$ interchanged. For $z \in \Gamma_\alpha(e^{i\theta})$ and for $|\varphi - \theta| < 1 - |z|^2$ we have

$$P_z(\varphi) \geq \frac{c_2}{1 - |z|^2}.$$

Setting $h = 1 - |z|^2$, we therefore obtain

$$-\log|S(z)| \geq \frac{c_2}{h}\mu((\theta - h, \theta + h)) \to \infty$$

as $z \to e^{i\theta}$, $z \in \Gamma_\alpha(e^{i\theta})$, whenever (6.3) holds at θ. This shows $|S|$ does not extend continuously at any point of E.

If (6.1) holds at θ, then by similar reasoning

$$\lim_{\Gamma_\alpha(\theta) \ni z \to e^{i\theta}} (-\log|S(z)| + n\log(1 - |z|^2)) = \infty$$

for every $n = 1, 2, \ldots$. Hence we have

(6.4)
$$\lim_{\Gamma_\alpha(\theta) \ni z \to e^{i\theta}} \frac{|S(z)|}{(1 - |z|^2)^n} = 0.$$

Now fix z in $\Gamma_\alpha(\theta)$ and consider two discs

$$\Delta_1 = \Delta(z, a(1 - |z|^2)), \quad \Delta_2 = \Delta(z, \tfrac{1}{2}a(1 - |z|^2)).$$

If $f(\zeta)$ is analytic on Δ_1, and if

$$\sup_{\Delta_1} \frac{|f(\zeta)|}{(1 - |\zeta|^2)^n} < \varepsilon,$$

then

$$\sup_{\Delta_2} |f^{(n)}(\zeta)| < C(a, n)\varepsilon,$$

where $C(a, n)$ depends only on a and n. This is an easy consequence of Schwarz's lemma on Δ_1, or of the Poisson integral formula for Δ_1. By (6.4) we therefore have

$$S^{(n)}(z) \to 0$$

as $z \to e^{i\theta}$, $z \in \Gamma_\alpha(\theta)$, whenever (6.1) holds at $e^{i\theta}$. $\quad\square$

Theorem 6.3. *Let $f \in H^p$, $p > 0$, and let Γ be an open arc on ∂D. If $f(z)$ is analytic across Γ, then its inner factor and its outer factor are analytic across Γ. If $f(z)$ is continuous across Γ, then its outer factor is continuous across Γ.*

Proof. Write $f = BSF$, where B is a Blaschke product, S is a singular function, and F is an outer function. We may suppose $f \not\equiv 0$. If f is analytic or continuous across Γ, then F is bounded on any compact subset of Γ, because $|F| = |f|$ on Γ.

Suppose f is analytic across Γ. If the zeros of B had an accumulation point on Γ, then f would have a zero of infinite order at some point of Γ. This is impossible and so B is analytic across Γ. Let μ be the measure determining S. If μ had a point charge at $e^{i\theta} \in \Gamma$, then S, and hence also f, would have a zero of infinite order at $e^{i\theta}$, by Theorem 6.2. Thus $\mu(\{e^{i\theta}\}) = 0$ for all $\theta \in \Gamma$. Now if $\mu(K) > 0$ for some compact subset K of Γ, then K is uncountable and by Theorem 6.2, $f(z)$ has infinitely many zeros on K. Therefore $\mu(K) = 0$ and S is analytic across Γ. Hence $F = f/BS$ is analytic across Γ.

Suppose f is continuous across Γ. Let $K = \{\theta \in \Gamma : f(e^{i\theta}) = 0\}$. Then, as a function on Γ, F is continuous at each point of K, because $|F| = |f|$ on Γ and $|F| = 0$ on K. On $\Gamma \backslash K$, $|f| > 0$, so that B and S cannot tend to zero at any point of $\Gamma \backslash K$. Then B and S are analytic across $\Gamma \backslash K$ and F is continuous on Γ. The Poisson integral representation now implies that F is continuous on $D \cup \Gamma$. $\square$

A compact set K in the plane has positive *logarithmic capacity* if there is a positive measure σ on K with $\sigma \neq 0$ such that the *logarithmic potential*

$$U_\sigma(z) = \int_K \log \frac{1}{|\zeta - z|} d\sigma(\zeta)$$

is bounded on some neighborhood of K. If $K \subset D$, then K has positive capacity if and only if K supports a positive mass σ for which *Green's potential*

(6.5) $$G_\sigma(z) = \int_K \log \left| \frac{1 - \bar{\zeta}z}{\zeta - z} \right| d\sigma(\zeta)$$

is bounded on D, because the term $\int_K \log |1 - \bar{\zeta}z| d\sigma(\zeta)$ is always bounded on D. An arbitrary set E is said to have positive capacity if some compact subset of E has positive capacity. Since $\log 1/|\zeta|$ is locally integrable with respect to area, any set of positive area has positive capacity. There are perfect sets of capacity zero, but these sets are very thin. For example, the Cantor ternary set on $[0, 1]$ has positive capacity (see Tsuji [1959]).

The Green's potential $G_\sigma(z)$ in (6.5) clearly satisfies $G_\sigma(z) \geq 0, z \in D$. Since σ is finite and supported at a positive distance from ∂D, we have

(6.6) G_σ *is continuous and zero at each point of* ∂D.

Further information about logarithmic capacity and potentials can be found in Tsuji's book [1959], but we shall need only the facts cited above.

If $B(z)$ is a Blaschke product, let us agree to also call

$$e^{ic} B(z)$$

a *Blaschke product* when c is a real constant.

Theorem 6.4 (Frostman). *Let* $f(z)$ *be a nonconstant inner function on the unit disc. Then for all* $\zeta, |\zeta| < 1$, *except possibly for a set of capacity zero,* *the function*

$$f_\zeta(z) = \frac{f(z) - \zeta}{1 - \bar\zeta f(z)}$$

is a Blaschke product.

Proof. Let K be a compact set of positive capacity and let σ be a positive mass on K such that $G_\sigma(z)$ is bounded on D. We shall show

$$\sigma(\{\zeta \in K : f_\zeta \text{ is not a Blaschke product}\}) = 0$$

and that will prove the theorem. Let

$$F(z) = G_\sigma(f(z)) = \int_K \log\left|\frac{1 - \bar\zeta f(z)}{\zeta - f(z)}\right| d\sigma(\zeta).$$

Then $V \geq 0$ and V is bounded. Because $f(z)$ is an inner function, (6.6) and dominated convergence imply that

$$\lim_{r \to 1} \int V(re^{i\theta}) d\theta = 0.$$

Hence, by Fatou's lemma,

$$\int_K \lim_{r \to 1} \int \log|f_\zeta(r^{i\theta})| \frac{d\theta}{2\pi} d\sigma(\zeta) = 0.$$

Because $\sigma \geq 0$ and $\log|f_\zeta| \leq 0$, this means

$$\lim_{r \to 1} \int \log|f_\zeta(re^{i\theta})| \frac{d\theta}{2\pi} = 0$$

for σ-almost every ζ. Theorem 2.4 then shows that f_ζ is a Blaschke product for σ-almost all ζ. $\square$

Corollary 6.5. *The set of Blaschke products is uniformly dense in the set of inner functions.*

Proof. If $f(z)$ is an inner function and if $|\zeta|$ is small, then

$$\|f - f_\zeta\|_\infty < \varepsilon.$$

By Frostman's theorem f_ζ is a Blaschke product for many small ζ. $\square$

Corollary 6.5 should be compared to Carathéodory's theorem I.2.1, in which a weaker form of convergence, namely, pointwise bounded convergence, is obtained, but in which it is assumed only that $\|f\|_\infty \leq 1$.

Let $f \in H^\infty(D)$ and let $z_0 \in \partial D$. The *cluster set* of f at z_0 is

$$\text{Cl}(f, z_0) = \bigcap_{r > 0} \overline{f(D \cap \Delta(z_0, r))}.$$

Thus $\zeta \in \mathrm{Cl}(f, z_0)$ if and only if there are points z_n in D tending to z_0 such that $f(z_n) \to \zeta$. The cluster set is a compact, nonempty, connected plane set. It is a singleton if and only if f is continuous on $D \cup \{z_0\}$. The *range set* of f at z_0 is

$$\mathrm{R}\,(f, z_0) = \bigcap_{r>0} f(D \cap \Delta(z_0, r)),$$

so $\zeta \in \mathrm{R}\,(f, z_0)$ if and only if there are points z_n in D tending to z_0 such that $f(z_n) = \zeta, n = 1, 2, \ldots$ In other words, the range set is the set of values assumed infinitely often in each neighborhood of z_0. The range set $\mathrm{R}\,(f, z_0)$ is a G_δ set. Clearly $\mathrm{R}\,(f, z_0) \subset \mathrm{Cl}(f, z_0)$. If $f(z)$ is analytic across z_0, and not constant, then $\mathrm{Cl}(f, z_0) = f(z_0)$, and $\mathrm{R}\,(f, z_0) = \varnothing$.

Theorem 6.6. *Let $f(z)$ be an inner function on D, and let $z_0 \in \partial D$ be a singularity of $f(z)$ (that is, a point at which $f(z)$ does not extend analytically). Then*

$$\mathrm{Cl}(f, z_0) = \bar{D}$$

and

$$\mathrm{R}\,(f, z_0) = D \backslash L,$$

where L is a set of logarithmic capacity zero.

Theorem 6.6 shows that, despite Fatou's theorem, the boundary behavior of an H^∞ function can be rather wild. For example, if $f(z)$ is a Blaschke product whose zeros are dense on ∂D, or if $f(z)$ is the singular function determined by a singular measure with closed support ∂D, then the conclusion of Theorem 6.6 holds at every $z_0 \in \partial D$, even though $f(z)$ has nontangential limits almost everywhere.

Proof. Since sets of capacity zero have no interior, the assertion about the range sets implies the assertion about the cluster set.

We are assuming $f(z)$ is not analytic across any arc containing z_0. If $f(z)$ is a Blaschke product, then z_0 is an accumulation point of the zeros of $f(z)$. Thus $0 \in \mathrm{R}\,(f, z_0)$ if f is a Blaschke product. In general

$$f_\zeta(z) = \frac{f(z) - \zeta}{1 - \bar{\zeta} f(z)}$$

is a Blaschke product when $\zeta \notin L$, a set of capacity zero. Since f_ζ also has a singularity at z_0, we see that for $\zeta \notin L$, z_0 is an accumulation point of the zeros of f_ζ, therefore z_0 is an accumulation point of the ζ-points of $f(z)$. $\square$

That proves Theorem 6.6, but by using Theorem 6.2 we can get more precise information. Suppose $f(z)$ is an inner function, suppose z_0 is a singularity of $f(z)$ and suppose $\zeta \in D \backslash \mathrm{R}\,(f, z_0)$. Then the inner function f_ζ is not a Blaschke product. Moreover, the proof above shows its Blaschke factor cannot have zeros accumulating at z_0. Write $f_\zeta = B_\zeta S_\zeta$. We have just showed that B_ζ is analytic across z_0. Hence S_ζ has a singularity at z_0. Two cases now arise.

If z_0 is an isolated singularity of S_ζ on ∂D, then the singular measure μ giving rise to S_ζ contains an atom at z_0. In this case S_ζ and all its derivatives tend nontangentially to 0 at z_0. It follows readily that $f_\zeta(z)$ and all its derivatives tend nontangentially to 0, so that nontangentially $f(z)$ tends to ζ, while all the derivatives of $f(z)$ tend to 0. These conclusions also hold if μ satisfies (6.1) at z_0. It is quite clear that for fixed z_0, there can be at most one point $\zeta \in D \backslash R(f, z_0)$ at which these conclusions can hold.

The alternative case is that $\mu(\{z_0\}) = 0$. Then by Theorem 6.2, z_0 is the limit of a sequence of points $\{e^{i\theta_n}\}$ at each of which S_ζ has nontangential limit 0, and therefore at each of which $f(z)$ has nontangential limit ζ. In this case Theorem 6.2 tells us even more. Either μ is continuous on some neighborhood of z_0, or μ assigns positive mass to each point in a sequence $e^{i\theta_n}$ tending to z_0. If μ is not continuous, there are $e^{i\theta_n} \to z_0$ such that at each point $e^{i\theta_n}$, $f(z)$ tends nontangentially to ζ, and each derivative $f^{(k)}(z)$ tends nontangentially to 0. If μ is continuous, then by Theorem 6.2 each neighborhood of z_0 contains uncountably many $e^{i\theta}$ at which $f(z)$ tends nontangentially to ζ.

The above reasoning can be summarized as follows:

Theorem 6.7. *Let $f(z)$ be an inner function on D and let $z_0 \in \partial D$ be a singularity of $f(z)$. For $|\zeta| < 1$ at least one of the following holds.*

(a)   *ζ is in the range set of f at z_0.*

(b)   *$f(z)$ has nontangential limit ζ at z_0, and each derivative $f^{(n)}$ has nontangential limit 0 at z_0.*

(c)   *z_0 is the limit of a sequence of points $e^{i\theta_n}$ on ∂D, and (b) holds at each $e^{i\theta_n}$.*

(d)   *Each neighborhood of z_0 on ∂D contains uncountably many points at which $f(z)$ has nontangential limit ζ.*

Considerably more about cluster theory can be found in the interesting books of Noshiro [1960] and of Collingwood and Lohwater [1966].

7. Beurling's Theorem

Let H be a separable Hilbert space with basis $\{\xi_0, \xi_1, \xi_2, \dots\}$. The *shift operator S* on H is defined by

$$S(\xi_j) = \xi_{j+1}$$

or, equivalently,

$$S\left(\sum a_j \xi_j\right) = \sum a_j \xi_{j+1}.$$

Beurling used inner functions to characterize the (closed) invariant subspaces for S. If we identify H with H^2 by taking $\xi_k = e^{ik\theta}$, then the operator S becomes multiplication by z,

$$S(f) = zf(z).$$

A subspace M of H^2 is *invariant* under S if $zM \subset M$, that is, if $zf(z) \in M$ whenever $f \in M$. Equivalently, M is invariant if $p(z)M \subset M$ for every polynomial $p(z)$. Since M is closed, this is the same as saying $H^\infty M \subset M$ where $H^\infty M = \{gf : g \in H^\infty, f \in M\}$, because by Exercise 4 each $g \in H^\infty$ is a pointwise bounded limit of a sequence of polynomials.

Theorem 7.1 (Beurling). *Let M be a subspace of H^2 invariant under S. If $M \neq \{0\}$, then there is an inner function $G(z)$ such that*

(7.1) $$M = GH^2 = \{G(z)f(z) : f \in H^2\}.$$

The inner function G is unique except for a constant factor. Every subspace of the form (7.1) is an invariant subspace for S.

Proof. Every subspace of the form (7.1) is closed in L^2, because $|G| = 1$. Every such subspace is clearly invariant under multiplication by z. Moreover if G_1 and G_2 are inner functions such that

$$G_1 H^2 = G_2 H^2,$$

then $G_1 = G_2 h, G_2 = G_1 k, h, k \in H^2$. This clearly means $G_1 = \lambda G_2$, $|\lambda| = 1$, because G_1/G_2 and G_2/G_1 are both inner functions in H^∞.

Now let M be an invariant subspace, $M \neq \{0\}$. Then there is $f \in M$, $f = a_k z^k + a_{k+1} z^{k+1} + \ldots$ with $a_k \neq 0$. Choose $f \in M$ with the least such k and write $M = z^k M_1$. Then $M_1 \subset H^2$ is also invariant (and closed), and we might as well assume $M = M_1$. Thus we assume M contains a function f_0 with $f_0(0) \neq 0$.

Let g_0 be the orthogonal projection of $1 \in H^2$ onto M. Then $g_0 \in M$ and $1 - g_0$ is orthogonal to M. Consequently

$$\frac{1}{2\pi} \int e^{in\theta} g_0(\theta)(1 - \overline{g_0(\theta)})d\theta = 0, \quad n = 0, 1, 2, \ldots$$

because $z^n g_0(z) \in M$. Since for $n \geq 1$, $z^n g_0(z)$ vanishes at $z = 0$, this gives

$$\frac{1}{2\pi} \int e^{in\theta} |g_0(\theta)|^2 d\theta = \frac{1}{2\pi} \int e^{in\theta} g_0(\theta) \, d\theta = 0,$$

$n = 1, 2, \ldots$. Hence the Fourier coefficients of $|g_0|^2$ vanish except at $n = 0$ and $|g_0|^2$ is constant,

$$|g_0|^2 = c = \frac{1}{2\pi} \int |g_0|^2 d\theta = \|g_0\|_2^2.$$

If $g_0 = 0$, then $1 = 1 - g_0$ is orthogonal to M and all functions in M vanish at $z = 0$, contrary to our assumption. Thus $g_0 \neq 0$, and

$$G = g_0/\|g_0\|_2$$

is an inner function in M.

Clearly $GH^2 \subset M$, because M is invariant. Now suppose $h \in M$ is orthogonal to GH^2. Then as $g_0 = \|g_0\|G \in GH^2$,

$$\int he^{-in\theta}\overline{g_0(\theta)}\frac{d\theta}{2\pi} = 0, \quad n = 0, 1, 2, \ldots.$$

But $1 - g_0$ is orthogonal to M, and $z^n h \in M, n = 1, 2, \ldots$, so that

$$0 = \int he^{in\theta}(1 - \overline{g_0(\theta)})\frac{d\theta}{2\pi} = -\int he^{in\theta}\overline{g_0(\theta)}\frac{d\theta}{2\pi}, \quad n = 1, 2, \ldots,$$

since $z^n h(z)$ vanishes at $z = 0$. Hence the L^1 function $h\overline{g_0}$ has zero Fourier series, and so $h\overline{g_0} = 0$. As $|g_0| > 0$, we see that $h = 0$ and hence $M = GH^2$. $\square$

Let G_1 and G_2 be two inner functions and let $M_1 = G_1H^2, M_2 = G_2H^2$ be their invariant subspaces. Then $M_1 \subset M_2$ if and only if $G_1 \in M_2$. This happens if and only if G_2 *divides* $G_1, G_1 = G_2h, h \in H^2$. When G_2 divides G_1 the quotient G_1/G_2 is another inner function. Now let G be a nonempty family of inner functions. There is a smallest invariant subspace M containing G. It is simply the intersection of all invariant subspaces containing G. This subspace M has the form $M = G_0H^2$ for some inner function G_0, so that G_0 divides every function in G. If G_1 is another inner function which divides every function in G, then $M_1 = G_1H^2$ contains G, so that $M \subset M_1$ and G_1 divides G_0. Thus G has a *greatest common divisor* G_0, which is unique except for a constant factor. We have proved the following:

Corollary 7.2. *Every nonempty family* G *of inner functions has a greatest common divisor.*

There is another way to prove Corollary 7.2 that gives us an idea of what G_0 looks like. Write $G_0 = B_0S_0$ with B_0 a Blaschke product and S_0 a singular function determined by the measure μ_0. If $G = BS$ is in G, then B_0 divides B and S_0 divides S. This means on the one hand that the zeros of B include the zeros of B_0, and on the other hand that $\mu_0 \leq \mu$, where μ is the measure associated with S. Hence B_0 is the Blaschke product with zeros $\bigcap\{G^{-1}(0) : G \in G\}$. And μ_0 is the greatest lower bound of the set L of measures attached to the functions in G. Any nonempty set L of positive Borel measures has a greatest lower bound μ_0. For any Borel set E, $\mu_0(E)$ is defined by

$$\mu_0(E) = \inf \sum_{j=1}^{N} \mu_j(E_j),$$

where the μ_j run through L and where $\{E_1, \ldots, E_N\}$ runs through all partitions of E into Borel sets.

Let P denote the set of polynomials in z.

Corollary 7.3. *Let $f(z) \in H^2$. Then $f(z)$ is an outer function if and only if* P $f = \{p(z)f(z) : p \in$ P $\}$ *is dense in H^2.*

Proof. Let M be the closure of P f in H^2. Then M is invariant under the shift operator, so that $M = GH^2$ for some inner function $G(z)$. Since $f \in M$, we have $f = Gh, h \in H^2$. So if f is an outer function, then G is constant and $M = H^2$.

Now write $f = Fh$, with F an inner function and h an outer function. If F is not constant, then FH^2 is a proper closed invariant subspace containing P f and so P f is not dense in H^2. □

Corollary 7.3 says that for any outer function $f(z) \in H^2$, there are polynomials $p_n(z)$ such that

$$(7.2) \qquad \int |1 - p_n f|^2 \frac{d\theta}{2\pi} \to 0.$$

A sharper version of (7.2) can be proved directly.

Theorem 7.4. *Let $f(z)$ be an outer function. Then there are functions $\{f_n\}$ in H^∞ such that*

$$(7.3) \qquad |f_n(z)f(z)| \le 1,$$

$$(7.4) \qquad f_n(\theta)f(\theta) \to 1 \quad \text{almost everywhere.}$$

Proof. Let

$$u_n(\theta) = \min(A_n, -\log|f(\theta)|),$$

where A_n is a large number to be determined later. Let f_n be the outer function with $\log|f_n| = u_n$ and with $f_n(0)f(0) > 0$. Then $|f_n| < e^{A_n}$, and since $u_n(\theta) + \log|f(\theta)| \le 0$, (7.3) holds. If $A_n \to \infty$, then $\|u_n + \log|f|\|_1 \to 0$ and $f_n(0)f(0) \to 1$. Let $A_n \to \infty$ so fast that

$$\sum(1 - f_n(0)f(0)) < \infty.$$

Then $\sum \|1 - f_n f\|_2^2 = \sum\{1 + \|f_n f\|_2^2 - 2Re f_n(0)f(0)\} \le 2\sum\{1 - f_n(0)f(0)\} < \infty$, which implies (7.4). □

The invariant subspaces of $H^p, 0 < p < \infty$, are described in Exercise 18. An invariant subspace of H^∞ is an ideal in the ring H^∞. The *weak-star* topology on H^∞ is defined by the basic open sets

$$\bigcap_{j=1}^n \left\{ f \in H^\infty : \left| \int f F_j d\theta - \int f_0 F_j d\theta \right| < 1 \right\},$$

where $F_1, \ldots, F_n \in L^1$ and $f_0 \in H^\infty$. It is the weak-star topology of L^∞ restricted to the subspace H^∞.

Theorem 7.5. *Let I be a nonzero ideal in H^∞. If I is weak-star closed, then there is an inner function G such that*

$$(7.5) \qquad\qquad I = GH^\infty$$

The inner function G is unique except for a constant factor, and every set of the form (7.5) is a weak-star closed ideal in H^∞.

Proof. It is clear that G is essentially unique and that (7.5) defines a weak-star closed ideal.

Now let $I \neq \{0\}$ be a weak-star closed ideal. Let M be the closure of I in H^2. We claim

$$(7.6) \qquad\qquad M \cap H^\infty = I.$$

Since $M = GH^2$ for an inner function G and $M \cap H^\infty = GH^\infty$, (7.6) implies (7.5).

Clearly $I \subset M \cap H^\infty$. Let $g \in M \cap H^\infty$. Then there are g_n in I such that

$$\|g_n - g\|_2 \to 0.$$

We shall modify (a subsequence of) the g_n by taking functions $h_n \in H^\infty$ so that

$$\|h_n g_n\|_\infty \leq \|g\|_\infty$$

and

$$h_n g_n \to g \qquad \text{almost everywhere.}$$

This implies $h_n g_n \to g$ weak-star, and since $g_n h_n \in I$ and I is weak-star closed, it follows that $g \in I$. The modification resembles the proof of Theorem 7.4. We may assume that $\|g\|_\infty = 1$ and that

$$\|g_n - g\|_2 \leq 1/n^2,$$

so that $g_n \to g$ almost everywhere. Since $\log |x| \leq |x| - 1$, it follows that

$$\int \log^+ |g_n| \frac{d\theta}{2\pi} \leq \int_{|g_n|>1} (|g_n| - 1) \frac{d\theta}{2\pi} \leq \int |g_n - g| \frac{d\theta}{2\pi} \leq \frac{1}{n^2}.$$

Let h_n be the outer function with

$$\log |h_n| = -\log^+ |g_n|, \qquad h_n(0) > 0.$$

Then $|h_n g_n| \leq 1$ and

$$1 - h_n(0) = 1 - \exp \frac{1}{2\pi} \int (-\log^+ |g_n|) d\theta \leq 1 - e^{-1/n^2} \leq \frac{1}{n^2}.$$

Hence $\sum \|1 - h_n\|_2^2 = \sum(1 + \|h_n\|_2^2 - 2Reh_n(0)) \leq 2\sum(1 - h_n(0)) < \infty$
and $h_n \to 1$ almost everywhere. $\quad\square$

The most interesting ideals of H^∞ are the maximal (proper) ideals. Since H^∞ is a Banach algebra with unit, the maximal ideals are the kernels of the homomorphisms $m : H^\infty \to \mathbb{C}$ (see Chapter V). If a weak-star closed ideal GH^∞ is maximal, then the inner function G has no proper divisors. Any singular function $S(z)$ has infinitely many divisors; for example, any power $S^t(z), 0 < t < 1$, of S is a divisor of S. Hence G has no divisors if and only if

$$G(z) = \lambda \frac{z - z_0}{1 - \bar{z}_0 z}, \quad |z_0| < 1, \quad |\lambda| = 1.$$

In that case GH^∞ is the maximal ideal

(7.7) $\qquad\qquad\qquad \{f \in H^\infty : f(z_0) = 0\}$

and the complex homomorphism is

$$m(f) = f(z_0).$$

These are the only weak-star closed maximal ideals.

H^∞ has many other maximal ideals that are not weak-star closed. For example, let $S(z)$ be a nonconstant singular function. Then $S(z)$ is not invertible in the ring H^∞, so that by Zorn's lemma $S(z)$ lies in some maximal ideal. But because $S(z)$ has no zero in D, this maximal ideal does not have the form (7.7).

The maximal ideals of H^∞ will be studied in Chapters V and X below. Here we only say that (7.7) does describe all the maximal ideals of H^∞ which can be obtained constructively. (The term constructive will remain undefined.)

Notes

See Parreau [1951] and Rudin [1955a] for discussions of the Hardy spaces on general domains, where it is necessary to define H^p in terms of harmonic majorants.

Theorem 2.3 is due to F. Riesz [1923]. Blaschke products were introduced by Blaschke [1915]. Theorem 2.4 was published by Frostman [1935].

Theorem 3.1 is from Hardy and Littlewood [1930]. The analogs of (3.2) and (3.3) for the disc had already been proved earlier by F. Riesz [1923]. Theorem 3.6 is from the famous paper [1916] of F. and M. Riesz; the proof in the text is apparently due to Bochner. Some applications of that fundamental result are included among the exercises for this chapter. Theorem 3.9 will be very important to us later.

On the circle, the integrability of $\log|f|$, $f \in H^p$, was first noticed by Szegö [1921] for $p = 2$ and by F. Riesz [1923] for the other p. The Canonical Factorization theorem is due to Smirnov [1929]. See F. Riesz [1930] for a parallel result on subharmonic functions outlined in Exercise 20.

Theorem 6.4 is from Frostman's thesis [1935], an important paper linking function theory to potential theory. See Seidel [1934] and the books of Collingwood and Lohwater [1966], Noshiro [1960], and Tsuji [1959] for further results on the boundary behavior of inner functions.

Beurling's theorem is from his famous paper [1948]. The books of Helson [1964] and Hoffman [1962a] give more thorough discussions of invariant subspaces.

Different approaches to Hardy space theory are presented in the books of Duren [1970], Hoffman [1962a], Privalov [1956], and Zygmund [1968].

Exercises and Further Results

1. Suppose $f \in H^p$. Then

$$f = gh$$

with $g, h \in H^{2p}$ and $\|g\|_{2p} = \|h\|_{2p} = \|f\|_p^{1/2}$. Also,

$$f = f_1 + f_2$$

with $f_j \in H^p$ an outer function and $\|f_j\|_p \leq \|f\|_p$. (Factor out the Blaschke product. If f is an inner function take $f_1 = (f + 1)/2$, $f_2 = (f - 1)/2$.)

2. (a) Let $f(t) \in L^p(\mathbb{R})$, $1 \leq p \leq \infty$. Then f is the nontangential limit of an $H^p(dt)$ function if and only if one (and hence all) of the following conditions hold:

(i)   The Poisson integral of $f(t)$ is analytic on H .
(ii) When $p < \infty$,

$$\int \frac{f(t)}{t - z} dt = 0, \quad \operatorname{Im} z < 0,$$

and when $p = \infty$,

$$\int f(t) \left(\frac{1}{t - z} - \frac{1}{t - z_0} \right) dt = 0, \quad \operatorname{Im} z < 0,$$

where z_0 is any fixed point in the lower half plane.
(iii) For all $g \in H^q$, $q = p/(p - 1)$,

$$\int fg \, dt = 0.$$

(iv) For $1 \leq p \leq 2$ (so that the Fourier transform is defined on L^p by Plancherel's theorem)

$$\hat{f}(s) = \lim_{N \to \infty} \int_{-N}^{N} f(t) e^{-2\pi i s t} dt = 0$$

almost everywhere on $s < 0$.

Part (iv) is one form of the Paley–Wiener theorem.

(b)   Now let $f \in L^p(\partial D)$, $1 \le p \le \infty$. Then f is the nontangential limit of an H^p function if and only if one of the following holds:

1.   The Poisson integral of f is analytic on D.
2. $\int e^{in\theta} f(e^{i\theta})d\theta = 0$, $n = 1, 2, \dots$.
3.   If

$$H_0^q = \left\{ g \in H^q : g(0) = \frac{1}{2\pi} \int g(e^{i\theta})d\theta = 0 \right\},$$

with $q = p/(p - 1)$, then

$$\int fg \, d\theta = 0$$

for all $g \in H_0^q$.
4.   On $|z| > 1$,

$$\frac{1}{2\pi i} \int \frac{f(\zeta)d\zeta}{\zeta - z} = 0.$$

If $f \in H^p$, $p \ge 1$, then

$$f(z) = \frac{1}{2\pi i} \int_{|\zeta|=1} \frac{f(\zeta)d\zeta}{\zeta - z}, \quad |z| < 1.$$

3.   (The jump theorem).   Let $f \in L^1(\partial D)$. Then on $|z| < 1$

$$\frac{1}{2\pi i} \int \frac{f(\zeta)d\zeta}{\zeta - z} - \frac{1}{2\pi i} \int \frac{f(\zeta)d\zeta}{\zeta - 1/\bar{z}} = \frac{1}{2\pi} \int f(e^{i\theta}) P_z(\theta)d\theta.$$

Consequently,

$$\lim_{r \uparrow 1} \left(\frac{1}{2\pi i} \int \frac{f(\zeta)d\zeta}{\zeta - re^{i\theta}} - \frac{1}{2\pi i} \int \frac{f(\zeta)d\zeta}{\zeta - e^{i\theta}/r} \right) = f(e^{i\theta})$$

almost everywhere. This result is more transparent on the upper half plane. See Exercise III.10 below for further information.

4.   Let $f(e^{i\theta}) \in L^\infty(\partial D)$. Then f is the nontangential limit of an H^∞ function if and only if there exists a uniformly bounded sequence of analytic polynomials $p_n(z)$ such that $p_n(e^{i\theta}) \to f(e^{i\theta})$ almost everywhere. If $f \in H^\infty$ then $p_n(z) \to f(z)$, $z \in D$, and $P_n(z)$ may be chosen so that $\| p_n \|_\infty \le \| f \|_\infty$. (Hint: Use Cesáro means, or approximate $f(rz)$, $r < 1$, by $\lambda p(z)$ where $p(z)$ is a Taylor polynomial for $f(rz)$ and $\lambda = \| f \|_\infty / \| p \|_\infty$.)

5.   (a)   If $f(z) \in H^p(D)$, then

$$|f(z)| \le \left(\frac{1}{1 - |z|^2} \right)^{1/p} \| f \|_{H^p}$$

and the derivatives $f^{(n)}(z)$ satisfy

$$|f^{(n)}(z)| \le C_{n,p} \frac{1}{(1 - |z|)^{n+(1/p)}} \|f\|_{H^p}.$$

These estimates are sharp for every p (except for constant factors); for example, take $f(z) = ((1 - |z_0|^2)/(1 - \bar{z}_0 z)^2)^{1/p}$.

(b) If $f \in H^2$, then by Fourier series or by Green's theorem,

$$J_2 = \int (1 - |z|) |f'(z)|^2 dx\, dy \le c \|f\|_2^2.$$

More generally, if $q \ge 2$, then

$$J_q = \int_0^1 (1 - r)^{2-2/q} \left(\int |f'(re^{i\theta})|^q d\theta \right)^{2/q} dr \le C_q \|f\|_2^2.$$

Hint: Let $\rho = (1 + r)/2$. Then

$$|f'(re^{i\theta})|^{q-2} \le \frac{C}{(1 - r)^{(q-2)/2}} \left(\int |f'(\rho e^{i\varphi})|^2 d\varphi \right)^{(q-2)/2}.$$

(c) If $f \in H^2$ and if $q > 2$, then

$$I_q = \int_0^1 (1 - r)^{-2/q} \left[\int |f(re^{i\theta})|^q d\theta \right]^{2/q} dr$$

satisfies

$$I_q \le C_q'(|f(0)|^2 + J_q).$$

(Hint: Integrate by parts, using

$$\left| \frac{\partial}{\partial r} |f(re^{i\theta})|^q \right| \le c_q |f(re^{i\theta})|^{q-1} |f'(re^{i\theta})|.$$

Then apply Hölder's inequality and the Cauchy–Schwarz inequality.)

(d) If $0 < p < 1$ and if $f \in H^p$, then by part (a),

$$\int |f(re^{i\theta})| d\theta \le c_p \|f\|_{H^p}^p \left\{ \frac{1}{(1 - r)^{1/p}} \|f\|_{H^p} \right\}^{1-p},$$

so that

$$\int |f(re^{i\theta})| d\theta \le c_p' \|f\|_{H^p}^{1-p} (1 - r)^{-\frac{(p-1)^2}{p}} \left(\int |f(re^{i\theta})| d\theta \right)^p.$$

Consequently

$$\int_0^1 (1 - r)^{(1/p)-2} \int |f(re^{i\theta})| d\theta\, dr \le C_p \|f\|_{H^p}^{1-p} \int_0^1 (1 - r)^{-p} \left(\int |f(re^{i\theta})| d\theta \right)^p dr.$$

If f has the form $g^{2/p}$, $g \in H^2$, then the integral on the right is $I_{2/p}$, and hence

$$\int_0^1 (1-r)^{(1/p)-2} \int |f(re^{i\theta})| d\theta \, dr \leq C'_p \|f\|_{H^p}$$

for all $f \in H^p$, $0 < p < 1$.

See Hardy and Littlewood [1932a] or Duren [1970] for further details.

6. When $1 \leq p < \infty$ it follows from Exercise 2 that the dual space of $H^p(D)$ is L^q/H_0^q, $q = p/(p-1)$. When $0 < p < 1$, H^p has a dual space which can be identified with a space of Lipschitz continuous analytic functions. We outline the proof in the case $\frac{1}{2} < p < 1$. For $0 < \alpha < 1$, let A_α be the space of analytic functions $\varphi(z)$ on D satisfying the Lipschitz condition

$$|\varphi(z_1) - \varphi(z_2)| \leq K|z_1 - z_2|^\alpha.$$

(a) If $\varphi(z)$ is analytic on D, then $\varphi \in A_\alpha$ if and only if

$$|\varphi'(z)| \leq c(1 - |z|)^{\alpha-1}.$$

If $\varphi \in A_\alpha$, the estimate above follows from Cauchy's theorem for the circle $|\zeta - z| = 1 - |z|$. For the converse, integrate φ' along the contour pictured in Figure II.1.

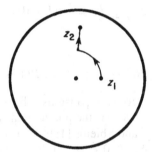

Figure II.1.

(b) Let $\frac{1}{2} < p < 1$ and set $\alpha = 1/p - 1$. If $f \in H^p$ and $\varphi \in A_\alpha$, then by Green's Theorem

$$\frac{1}{2\pi i} \int_{|z|=r} f(z)\overline{\varphi(z)}\frac{dz}{z} = f(0)\overline{\varphi(0)} + \frac{1}{\pi} \iint_{|z|<r} f(z)\overline{\varphi'(z)} \, dx \, dy.$$

By part (a) and by 5(d),

$$\iint_{|z|<1} |f(z)| \, |\varphi'(z)| \, dx \, dy \leq C\|f\|_{H^p},$$

Hence the limit

$$L_\varphi(f) = \lim_{r\to 1} \frac{1}{2\pi i} \int_{|z|=r} f(z)\overline{\varphi(z)}\frac{dz}{z}$$

exists and defines a bounded linear functional on H^p. If $\varphi(z) = \sum a_n z^n$, then

$$L_\varphi(z^n) = \bar{a}_n.$$

(c) Conversely, suppose L is a linear functional on H^p, $\frac{1}{2} < p < 1$, such that $|L(f)| \le B\|f\|_{H^p}$, $f \in H^p$. For $|w| < 1$, set

$$f_w(z) = \frac{1}{1 - \bar{w}_z} = \sum_{n=0}^{\infty} \bar{w}^n z^n.$$

The series converges in H^p norm (it converges uniformly) and $f_w \in H^p$. Define

$$\varphi(w) = \overline{L(f_w)} = \sum_{n=0}^{\infty} \overline{L(z^n)} w^n.$$

Then

$$|\varphi'(w)| \le B\|z/(1 - \bar{w}_z)^2\|_{H^p},$$

and the easy estimate

$$\int \frac{1}{|e^{i\theta} - w|^{2p}} d\theta \le C_p (1 - |w|)^{1-2p},$$

valid for $2p > 1$, yields

$$|\varphi'(w)| \le cB(1 - |w|)^{\frac{1}{p}-2} = cB(1 - |w|)^{\alpha-1}.$$

Hence $\varphi \in A_\alpha$. By the series expansions above, $L_\varphi(z^n) = L(z^n), n = 0, 1, 2, \dots$, and so $L = L_\varphi$ because the polynomials are dense in H^p.

See Duren, Romberg, and Shields [1969] for the full story, $0 < p < 1$. On the other hand, the spaces $L^p, 0 < p < 1$, have no nonzero bounded linear functionals.

7. (Lindelöf[1915]).   Let $f(z) \in H^\infty$.
  (a) If

$$m = \text{ess} \varlimsup_{\theta \to 0} |f(e^{i\theta})| = \lim_{\delta \to 0} \|f(e^{i\theta})\chi_{(-\delta,\delta)}(\theta)\|_\infty,$$

then

$$\varlimsup_{D \ni z \to 1} |f(z)| \le m.$$

(Use the subharmonicity of $\log|f(z)|$.)
  (b) If

$$\lim_{\theta \downarrow 0} f(e^{i\theta}) = \alpha,$$

then

$$\lim_{\Omega_\delta^+ \ni z \to 1} f(z) = \alpha,$$

where $\Omega_\delta^+ = \{z \in D : \arg(1 - z) < \pi/2 - \delta\}, \delta > 0$. (Consider $\log|f(z) - \alpha|$.)

(c) If

$$\lim_{\theta \downarrow 0} f(e^{i\theta}) = \alpha, \quad \lim_{\theta \uparrow 0} f(e^{i\theta}) = \beta,$$

then $\alpha = \beta$ and

$$\lim_{D \ni z \to 1} f(z) = \alpha.$$

(d) Boundedness is not essential for these results. It is enough to assume that for some $\eta > 0$, $f \in H^p (D \cap \{|z - 1| < \eta\})$, that is, that $|f|^p$ has a harmonic majorant on $D \cap \{|z - 1| < \eta\}$.

8. (a) (Hardy's inequality). If $f = \sum_{n=0}^\infty a_n z^n \in H^1$, then

$$\sum_{n=0}^\infty \frac{|a_n|}{n + 1} \leq \pi \|f\|_1.$$

(Write $f = gh, \|g\|_2^2 = \|h\|_2^2 = \|f\|_1, g, h \in H^2$. If $g = \sum b_n z^n, h = \sum c_n z^n$, set $G = \sum |b_n| z^n, H = \sum |c_n| z^n$, and $F = GH$. Then

$$\sum_{n=0}^\infty \frac{|a_n|}{n + 1} \leq \sum_{n=0}^\infty \frac{1}{n + 1} \sum_{k=0}^n |b_k| |C_{n-k}| = \frac{1}{2\pi} \int F(e^{i\theta}) \overline{\varphi(e^{i\theta})} d\theta,$$

where $\varphi \in L^\infty, \|\varphi\|_\infty = \pi$.)

(b) Let $f = \sum a_n z^n \in H^1$. Then $f(e^{i\theta})$ is equal almost everywhere to a function of bounded variation if and only if $f' \in H^1$. In that case $f(e^{i\theta})$ is absolutely continuous, $f(z)$ is continuous on $|z| \leq 1$, and

$$\sum |a_n| < \infty.$$

When $f' \in H^1$,

$$i e^{i\theta} \lim_{r \to 1} f'(re^{i\theta}) = \frac{d}{d\theta} f(e^{i\theta})$$

almost everywhere (Hardy and Littlewood [1927], and Smirnov [1933]).

★**9.** Let Ω be a simply connected open set bounded by a Jordan curve Γ, and let $f : D \to \Omega$ be a conformal mapping. By a famous theorem of Carathéodory (see Ahlfors [1973] or Tsuji [1959]) the mapping $f(z)$ has a one-to-one continuous extension from $\bar{D}$ onto $\bar{\Omega}$.

(a) The curve Γ is rectifiable if and only if $f' \in H^1$.

(b)   If Γ is rectifiable and if $E \subset \Gamma$ is a closed set, then

$$\text{length}(E) = \int_{f^{-1}(E)} |f'(e^{i\theta})| d\theta.$$

Thus E has length zero if and only if $f^{-1}(E)$ has length zero.

(c)   Moreover, if Γ is rectifiable, then the mapping $f(z)$ is conformal (angle preserving) at almost every $e^{i\theta}$. More precisely, if f' has nontangential limit at $e^{i\theta}$, and if γ is a curve in D terminating at $e^{i\theta}$ such that the angle

$$\alpha = \lim_{\gamma \ni z \to e^{i\theta}} \arg(1 - e^{-i\theta}z)$$

exists and $\alpha \neq \pm \pi/2$, then the curve $f(y)$ meets the normal line to Γ at $f(e^{i\theta})$ in the same angle α.

Parts (a) and (b) are due to F. and M. Riesz [1916]; see also Smirnov [1933].

★10.   (The local Fatou theorem). Let $u(z)$ be a harmonic function on D. We say that $u(z)$ is nontangentially bounded at $e^{i\theta} \in T = \partial D$ if $u(z)$ is bounded on some cone

$$\Gamma_\alpha(e^{i\theta}) = \left\{ z \in D : \frac{|z - e^{i\theta}|}{1 - |z|} < \alpha \right\}, \quad \alpha > 1.$$

If E is a measurable subset of T and if $u(z)$ is nontangentially bounded at each point of E, then $u(z)$ has a nontangential limit at almost every point of E. By elementary measure theory there is a compact set $F \subset E, |E \backslash F| < \varepsilon$, and there are $M > 0$ and $\alpha > 1$ such that $|u(z)| \leq M$ on

$$\Omega = \bigcup_{e^{i\theta} \in F} \Gamma_\alpha(e^{i\theta}).$$

The domain Ω is simply connected (it is a union of rays from the origin) and $\partial\Omega$ is a rectifiable curve ($\partial\Omega$ consists of F and a union of tentlike curves over the components of $T \backslash F$, and the tent over an arc $I \subset T$ has length not exceeding $c(\alpha \ni |I|)$. By Exercise 9, $u(z)$ has a nontangential limit, from within Ω, at almost every point of F. Nontangential convergence from within all of D now follows easily: If $e^{i\theta}$ is a point of density of F, that is, if

$$\lim_{\delta \to 0} \frac{|F \cap [\theta - \delta, \theta + \delta]|}{2\delta} = 1$$

and if $\beta > \alpha$, then for some $r < 1$,

$$\{|z| > r\} \cap \Gamma_\beta(e^{i\theta}) \subset \Omega.$$

Consequently,

$$\lim_{\Gamma_\beta(e^{i\theta}) \ni z \to e^{i\theta}} u(z)$$

exists for all $\beta > 1$ for almost all $e^{i\theta} \in F$.

The same conclusion holds if it is merely assumed that $u(z)$ is bounded below on some $\Gamma_\alpha(e^{i\theta})$ for each $e^{i\theta} \in E$. It also holds when the harmonic function is replaced by a meromorphic function. It follows from the above reasoning that a meromorphic function having zero nontangential limit on a set of positive measure must vanish identically on D. The corresponding assertion for radial limits is false (see Privalov [1956] or Bagemihl and Seidel [1954]).

The original source for this theorem is Privalov [1919]; see also Zygmund [1968, Vol. II]. A different elementary proof will be given in Chapter IX.

11. (Plessner's theorem). Let $f(z)$ be meromorphic in D. A point $e^{i\theta} \in T$ is a *Plessner point* for $f(z)$ if the angular cluster set of $f(z)$ at $e^{i\theta}$ is the full Riemann sphere S^2. In other words, for all $\alpha > 1$ and for all $r < 1$,

$$f(\Gamma_\alpha(e^{i\theta}) \cap \{|z| > r\})$$

is dense on S^2. The circle T splits into three disjoint Borel sets, $T = N \cup P \cup G$, such that

  (i) $|N| = 0$,
  (ii) Every point of P is a Plessner point for $f(z)$, and
  (iii) $f(z)$ has finite nontangential limit at each point of G.

(For rational $w \in \mathbb{C}$, let $E_w = \{e^{i\theta} : (f - w)^{-1}$ is nontangentially bounded at $e^{i\theta}\}$. Then $P = T \setminus \cup E_w$ is the set of Plessner points for $f(z)$, and $(f - w)^{-1}$ has nonzero nontangential limit at almost every $e^{i\theta} \in E_w$. See Plessner [1927] or Tsuji [1959].)

12. (Morera's theorem). Suppose that $g(z) \in H^1(|z| > 1)$ that is, that $g(1/z) \in H^1(D)$. Also suppose that $f(z) \in H^1(D)$ and that $f(e^{i\theta}) = g(e^{i\theta})$ almost everywhere on an arc $I \subset T$. Set

$$F(z) = \begin{cases} f(z), & |z| < 1, \\ g(z), & |z| > 1. \end{cases}$$

Then for $z_0 \in I$ and for $\delta > 0$ small,

$$\frac{1}{2\pi i} \int_{|\zeta - z_0| = \delta} \frac{F(\zeta)}{\zeta - z} d\zeta = F(z), \quad z \in \Delta(z_0, \delta) \setminus T,$$

so that $F(z)$ has an analytic extension across $I \cap \{|z - z_0| < \delta\}$. Thus $F(z)$ extends analytically across I.

13. Let $f(z)$ be analytic on D.
   (a) If $\mathrm{Re}\, f(z) \geq 0$, then $f \in H^p$ for all $p < 1$. (Write $f = e^\varphi$, $|\mathrm{Im}\, \varphi| \leq \pi/2$. Then $|f(z)|^p = e^{p\mathrm{Re}\varphi(z)} \leq (\cos p(\pi/2))^{-1} \mathrm{Re}\,(f^p(z))$, so that $|f(z)|^p$ has a harmonic majorant.)
   (b) If $f \in H^1$ and if $f(e^{i\theta})$ is real, then f is constant. This result is sharp, because $[(1 + z)/(1 - z)] \in H^p$ for all $p < 1$, by part (a).

(c) If $f \in H^{1/2}$ and if $f(e^{i\theta})$ is real and positive almost everywhere, then f is constant.

(d)   There are local versions of (b) and (c). Let I be an arc on T. If $f \in H^1$ and if $f(e^{i\theta})$ is real almost everywhere on I, then f has an analytic extension across I. This follows from Morera's theorem. If $h(z) = \overline{f(1/\bar{z})}$ then $h \in H^1(|z| > 1)$ and $f = h$ almost everywhere on I.

Similarly, if $f \in H^{1/2}$ and if $f(e^{i\theta}) \geq 0$ almost everywhere on I, then f extends analytically across I. Write $f = Bg^2$, where $g \in H^1$ and B is a Blaschke product. Then $\overline{g(1/\bar{z})} \in H^1(|z| > 1)$ and on I

$$B(z)g^2(z) = g(z)\overline{g(1/\bar{z})},$$

so that by Morera's theorem and reflection both $B(z)g(z)$ and $g(z)$ extend across I.

The $H^{1/2}$ results have another proof using maximal functions and subharmonicity instead of Riesz factorization. Let $v(z) = \mathrm{Im}(f^{1/2}(z))$, with the root chosen so that $v(z) \geq 0$. Then $v(z)$ is well-defined and subharmonic on D and $0 \leq v(z) \leq |f(z)|^{1/2}$, so that by the maximal theorem and Theorem I.6.7, v has least harmonic majorant

$$u(z) = \int P_z(\theta)v(e^{i\theta})d\theta/2\pi.$$

If $f(e^{i\theta}) \geq 0$ almost everywhere, then $v(z) = 0$ and $f(z)$ is constant. If $f(e^{i\theta}) \geq 0$ almost everywhere on an arc I, then for each $z_0 \in I$ there is $\delta > 0$ such that $0 \leq u(z) \leq 1$ on $V = D \cap \{|z - z_0| < \delta\}$. Then $0 \leq v(z) \leq 1$ on V, so that $|\mathrm{Im}(1 + f(z))| \leq 2\,\mathrm{Re}(1 + f(z)), z \in V$. Hence $1 + f \in H^1(V)$. Using a conformal map from V onto D, we see that $1 + f$ is analytic across $T \cap \partial V$. This proof is due to Lennart Carleson.

Part (c) is due to Helson and Sarason [1967] and, independently, Neuwirth and Newman [1967]. The $H^{1/2}$ result in part (d) is from Koosis [1973].

14.   (a)   The local version of Corollary 4.8(a) is valid when $p \geq 1$. Suppose $f \in H^1(D)$. If $\mathrm{Re}\, f \geq 0$ almost everywhere on an arc $I \subset T$, then the inner factor of $f(z)$ is analytic across I.

The case $p > 1$, which we shall use in Chapter IV, is easier. Let $u(z)$ be the Poisson integral of $\chi_1(\theta) \arg f(e^{i\theta})$ and let $v(z)$ be the harmonic conjugate of $u(z)$. By Exercise 13(a) $e^{-i(u+iv)} \in H^p$ for all $p < 1$, so that $F = fe^{-i(u+iv)} \in H^{1/2}$. Since $F \geq 0$ on I, the inner factor is analytic across I.

The following proof for $p = 1$ is due to P. Koosis. Replacing I by a subarc we can suppose that f has nontangential limits at the endpoints θ_1, θ_2 of I and that $\mathrm{Re}\, f(e^{i\theta_j}) > 0, j = 1, 2$. Let Γ be a circular arc in D joining $e^{i\theta_1}$ to $e^{i\theta_2}$. Varying Γ, we may assume $\inf_\Gamma |f(z)| > 0$. Let U be the domain bounded by $\Gamma \cup I$, and let τ be a conformal mapping from D onto U. We are going to show that $F = f \circ \tau$ is an outer function. Because τ can be computed explicitly, it will follow easily that

$$\lim_{r \to 1} \int_J \log |f(re^{i\theta})| d\theta = \int_J \log |f(e^{i\theta})| d\theta$$

for any compact subarc J of I, which means that the inner factor of $f(z)$ has no singularities on I.

Let Γ_1 be a compact subarc of Γ such that Re $f \geq 0$ on $\Gamma \backslash \Gamma_1$ and let $\gamma = \tau^{-1}(\Gamma_1)$. Then Re $F \geq 0$ almost everywhere on $T \backslash \gamma$, while $F(e^{i\theta})$ is C^∞ and nowhere zero on γ. Let $u \in C^\infty(T)$ satisfy $u = 0$ on $T \backslash \gamma$, and $|u - \arg F| < \pi/2$ on γ, and let $g(z) = u(z) + iv(z)$ where $u(z)$ is the Poisson integral of $u(e^{i\theta})$ and $v(z)$ is its conjugate function. In the next chapter we shall see that $g(z) \in H^\infty$. Then since Re $Fe^{-ig} \geq 0$ almost everywhere and $Fe^{-ig} \in H^1$, Corollary 4.8 shows that Fe^{-ig} is an outer function.

(b) When $p < 1$ the result of part (a) above fails. Let $\{z_k\}$ be a Blaschke sequence and let $B_n(z)$ be the finite Blaschke product with zeros z_k, $1 \leq k \leq n$, normalized so that $B_n(0) > 0$. Let $v_n \in L^\infty$, $v_n(e^{i\theta}) = \pm \pi/2$, be such that

$$|v_n(e^{i\theta}) - \arg B_n(e^{i\theta})| \leq \pi/2$$

modulo 2π. Let $v_n(z)$ be the Poisson integral of $v_n(e^{i\theta})$ and let $u_n(z)$ be its conjugate function, $u_n(0) = 0$. Then Re $(e^{u_n - iv_n}) \geq 0$, so that

$$G_n = B_n e^{u_n - iv_n} \in H^p$$

for any $p < 1$, and Re $G_n(e^{i\theta}) \geq 0$ almost everywhere.

Now for fixed $p < 1$ and for $\{z_1, z_2, \ldots, z_n\}$ already selected, we can choose z_{n+1} with $1 - |z_{n+1}|$ so small that we have

$$\|G_{n+1} - G_n\|_p^p < 2^{-n}.$$

Then G_n converges almost everywhere and in H^p to a limit G and Re $G(e^{i\theta}) \geq 0$. Now $G(z_n) = 0, n = 1, 2, \ldots$, but $G \not\equiv 0$, because $|G(0)| = \lim_n B_n(0)$. No constraints have been made on $\arg(z_n)$ and we can arrange that $\{z_n\}$ is dense on T. Then the inner factor of G does not extend across any arc of T.

15. Suppose $f(z) \in N$, the Nevanlinna class. For any $\alpha > 1$, the non-tangential maximal function

$$f^*(\theta) = \sup_{\Gamma_\alpha(\theta)} |F(z)|$$

satisfies

$$|\{\theta : f^*(\theta) > \lambda\}| \leq \frac{A_\alpha}{\log \lambda} \sup_{r < 1} \int \log^+ |f(re^{i\theta})| \frac{d\theta}{2\pi}, \quad \lambda > 1,$$

where A_α depends only on α. From this estimate it follows that $f(z)$ has nontangential limit $f(e^{i\theta})$ almost everywhere and that

$$\int \log^+ |f(e^{i\theta})|d\theta \leq \sup_{r<1} \int \log^+ |f(re^{i\theta})|d\theta/2\pi.$$

16. Let $f(z)$ be analytic on D. Then $f(z) \in N$ if and only if $f = f_1/f_2$, where $f_j \in H^\infty$, $j = 1, 2$, and f_2 is nowhere zero. The denominator f_2 can be taken to be an outer function if and only if $f \in N^+$ (F. and R. Nevanlinna [1922]).

17. Let P be the set of polynomials in z, and let $f \in H^2(D)$. Then $\overline{P \ f} = GH^2$, where G is the inner factor of f.

18. (a) Let $0 < p < \infty$. If M is a closed subspace of H^p invariant under multiplication by z, then $M = GH^p$ for some inner function G. (See the proof of Theorem 7.5.)

(b) Now let M be a closed subspace of L^p, $0 < p \leq \infty$ (if $p = \infty$ assume M is weak-star closed). Suppose $zM \subset M$. If $zM = M$, so that $\bar{z}M = M$, then $M = \chi_E L^p$, for some Borel set E. If $zM \subsetneq M$, then $M = UH^p$ for some $U \in L^\infty$ such that $|U| = 1$ almost everywhere.

★★**19.** (Littlewood's theorem on subharmonic functions). Let μ be a positive measure on D satisfying

(E.1) $$\int \log \frac{1}{|\zeta|} d\mu(\zeta) < \infty.$$

The *Green's potential*

$$U_\mu(z) = \int \log \left| \frac{1 - \bar{\zeta}z}{\zeta - z} \right| d\mu(\zeta)$$

is superharmonic on D ($-U_\mu$ is subharmonic). If μ is discrete with unit masses, $\mu = \sum \delta_{z_n}$, $z \neq 0$, then the hypothesis (E.1) is equivalent to the Blaschke condition $\sum (1 - |z_n|) < \infty$, and $U_\mu(z) = -\log |B(z)|$, where $B(z)$ is the Blaschke product with zeros $\{z_n\}$. Littlewood proved [1929]

(i) $\lim_{r \to 1} \int U_\mu(re^{i\theta})d\theta = 0$

and

(ii) $\lim_{x \to 1} U_\mu(re^{i\theta}) = 0$

almost everywhere. These generalize the corresponding results about Blaschke products from Section 2.

After a calculation, the identity

$$\frac{1}{2\pi} \int \log \left| \frac{re^{i\theta} - z}{1 - re^{-i\theta}z} \right| d\theta = \begin{cases} \log |z|, & r < |z| \leq 1, \\ \log r, & |z| \leq r, \end{cases}$$

follows from the mean value property of harmonic functions. Then Fubini's Theorem yields (i).

We prove (ii) in the upper half plane for vertical convergence. Because the problem is local, we assume μ has support the unit cube $Q = [0, 1] \times (0, 1]$.

The hypothesis (E.1) transforms into

$$\int_Q y \, d\mu(x, y) < \infty,$$

and the potential now has the form

$$U_\mu(z) = \int_Q \log \left| \frac{z - \bar{\zeta}}{z - \zeta} \right| d\mu(\zeta).$$

We must show $\lim_{y \to 0} U_\mu(x + iy) = 0$ almost everywhere on $[0,1]$. Let μ_n be the restriction of μ to the strip

$$A_n = Q \cap \{2^{-n} < \operatorname{Im}\zeta < 2^{-n-1}\},$$

and let v_n be the vertical projection of $2^{-n}\mu_n$ onto $[0, 1]$, $v_n(E) = 2^{-n}\mu(E \times (2^{-n}, 2^{-n+1}])$. By hypothesis, we have

$$\sum_{n=1}^{\infty} \|v_n\| < \infty.$$

Write

$$U_\mu(z) = U_1(z) + U_2(z),$$

where

$$U_1(z) = \int_{|\zeta - z| \geq y/4} \log \left| \frac{z - \bar{z}}{z - \zeta} \right| d\mu(\zeta),$$

and

$$U_2(z) = \int_{|\zeta - z| < y/4} \log \left| \frac{z - \bar{\zeta}}{z - \zeta} \right| d\mu(\zeta),$$

$y = \operatorname{Im} z$. We discuss U_1 and U_2 separately.

When $|\zeta - z| > y/4$, we have

$$\log \left| \frac{z - \bar{\zeta}}{z - \zeta} \right| \leq c \frac{y\eta}{(x - \xi)^2 + (y + \eta)^2}, \qquad \zeta = \xi + i\eta.$$

Hence

$$U_1(z) \leq c \sum_{n=1}^{\infty} \int \frac{y\eta}{(x - \xi)^2 + (y + \eta)^2} d\mu_n(\xi)$$

$$\leq c \sum_{n=1}^{\infty} \int \frac{y\eta}{(x - \xi)^2 + (y + \eta)^2} d\mu_n(\xi) + cM \left(\sum_{n=N+1}^{\infty} dv_n \right)(x),$$

where $M(dv)$ denotes the Hardy–Littlewood maximal function of the measure v. For any fixed N,

$$\lim_{y \to 0} \sum_{1}^{N} \frac{y\eta}{(x-\xi)^2 + (y+\eta)^2} d\mu_n(\xi) = 0,$$

so that by the weak type estimate for $M(dv)$,

$$\left| \left\{ x : \lim_{y \to 0} U_1(x+iy) > \varepsilon \right\} \right| \leq \lim_{N \to \infty} \frac{c}{\varepsilon} \sum_{n=N+1}^{\infty} \|v_n\| = 0.$$

To treat $U_2(z)$ we suppose $z \in A_n$. Then

$$U_2(z) \leq c2^n \sum_{k=n-1}^{n+1} \int_{|\xi - x| < 2^{-n}} \log \left| \frac{2^{-n}}{x - \xi} \right| dv_k(\xi),$$

because $\{|\zeta - z| < y/4\}$ meets only A_{n-1}, A_n, and A_{n+1}. Since the integrand is positive and even, we have

$$\int_{|\xi - x| < 2^{-n}} \log \left| \frac{2^{-n}}{x - \xi} \right| dv_k(\xi) \leq CM(dv_k)(x) \int_{0}^{2^{-n}} \log \left| \frac{2^{-n}}{t} \right| dt$$

$$\leq C2^{-n} M(dv_k)(x).$$

Hence

$$U_1(x+iy) \leq \sum_{k=n-1}^{n+1} M(dv_k)(x)$$

and $U_1(x+iy) \to 0$ $(y \to 0)$ almost everywhere.

The analog of (ii) is false for nontangential convergence. Let μ be a finite discrete measure with point masses on a set nontangentially sense or ∂D. (W. Rudin).

★20. The Canonical Factorization theorem has a generalization to subharmonic functions. The proof uses the Riesz decomposition theorem (Chapter I, Exercise 17) and Littlewood's theorem (Exercise 19).

(a) Let $v(z)$ be a subharmonic function on D, $v \not\equiv -\infty$, and let $\mu \geq 0$ be its weak Laplacian. Then for $|z| < r < 1$,

$$v(z) = u_r(z) - \frac{1}{2\pi} \int_{|\zeta| < r} \log \left| \frac{1 - \bar{\zeta}z}{\zeta - z} \right| d\mu(\zeta)$$

where $u_r(z)$ is harmonic on $|z| < r$.

(b) If $v(z)$ has a harmonic majorant then its least harmonic majorant is $v(z) = \lim_{r \to 1} u_r(z)$, and we have

$$v(z) = u(z) - (1/2\pi)U_\mu(z).$$

In particular, the potential converges wherever $v(z) > -\infty$, and Littlewood's theorem applies to $U_\mu(z)$

(c)   If

$$\sup_r \int v^+(re^{i\theta})d\theta/2\pi < \infty$$

then the least harmonic majorant $u(z)$ is the Poisson integral of a finite measure $k\,d\theta/2\pi + d\theta$, where σ is singular, so that by part (ii) of Littlewood's theorem, $v(z)$ has notangential limit $k(e^{i\theta})$ almost everywhere.

(d)   In the special case $v(z) = \log|f(z)|$, $f \in N$, $f(0) \neq 0$, we obtain Theorem 5.5 as a corollary of the results in (b) and (c).

See F. Riesz [1930].

III

Conjugate Functions

After some preliminaries, which identify the conjugate operator with the Hilbert transform, we prove the famous Marcel Riesz theorem and some of its variants. Then we discuss the more recent, but also most basic, theorem that the conjugate function and the nontangential maximal function belong to the same L^p classes.

1. Preliminaries

Let $f(\theta) \in L^1(T)$, T denoting the unit circle ∂D. Supposing for a moment that $f(\theta)$ is real valued, we let $u(z)$ be the Poisson integral of $f(\theta)$ and let $\tilde{u}(z)$ denote the harmonic conjugate function of $u(z)$, normalized so that $\tilde{u}(0) = 0$. Since

$$P_z(\varphi) = \operatorname{Re} \frac{e^{i\varphi} + z}{e^{i\varphi} - z},$$

we have

$$(u + i\tilde{u})(z) = \frac{1}{2\pi} \int \frac{e^{i\varphi} + z}{e^{i\varphi} - z} f(\varphi) \, d\varphi$$

and

$$\tilde{u}(z) = \frac{-1}{2\pi} \int Q_z(\varphi) f(\varphi) \, d\varphi,$$

where

$$-Q_z(\varphi) = \operatorname{Im}\left(\frac{e^{i\varphi} + z}{e^{i\varphi} - z}\right) = \frac{2r \sin(\theta - \varphi)}{1 - 2r \cos(\theta - \varphi) + r^2}, \quad z = re^{i\theta}.$$

These formulae define the *conjugate function* $\tilde{u}(z)$ even when $f(\theta)$ is complex valued.

The kernel $-Q_z(\varphi) = Q_r(\theta - \varphi)$ is the *conjugate Poisson kernel*.[†] The kernels Q_r do not behave at all like an approximate identity, because $Q_r(\theta)$ is an odd function and because $\|Q_r\|_1 \sim \log 1(1 - r)$. Nevertheless, the conjugate function $\tilde{u}(z)$ has nontangential limits almost everywhere on T.

Lemma 1.1. *If $f \in L^1(T)$, then $\tilde{u}(z)$ has nontangential limit $\tilde{f}(\theta)$ almost everywhere on T.*

Proof. We may suppose $f(\theta) \geq 0$. The analytic function $g(z) = u(z) + i\tilde{u}(z)$ then has nonnegative real part, and $G(z) = g(z)/(1 + g(z))$ is bounded. The function $G(z)$ has nontangential limit $G(\theta)$ almost everywhere, and $G(\theta) = 1$ on at most a set of measure zero. Consequently $g = G/(1 - G)$ has a finite nontangential limit almost everywhere, and so does its imaginary part $\tilde{u}(z)$. $\quad\square$

The linear mapping

$$f(\theta) \to \tilde{f}(\theta)$$

is called the *conjugation operator*. The conjugate function $\tilde{f}$ can also be calculated using a principal value integral.

Lemma 1.2. *Let $f(\theta) \in L^1(T)$. For almost every θ*

$$(1.1) \qquad \tilde{f}(\theta) = \lim_{\varepsilon \to 0} \frac{1}{2\pi} \int_{|\theta - \varphi| > \varepsilon} \cot\left(\frac{\theta - \varphi}{2}\right) f(\varphi)\, d\varphi.$$

In particular, the limit in (1.1) exists almost everywhere. Moreover,

$$(1.2) \qquad \left| \tilde{u}(re^{i\theta}) - \frac{1}{2\pi} \int_{|\theta - \varphi| > 1 - r} \cot\left(\frac{\theta - \varphi}{2}\right) f(\varphi)\, d\varphi \right| \leq CMf(\theta),$$

where C is an absolute constant and where Mf is the Hardy–Littlewood maximal function of f.

Proof. Notice that for $\theta \neq 0$,

$$\lim_{r \to 1} Q_r(\theta) = \lim_{r \to 1} \frac{2r \sin\theta}{1 - 2r\cos\theta + r^2} = \frac{\sin\theta}{1 - \cos\theta} = \cot\frac{\theta}{2} = Q_1(\theta),$$

which is the kernel in (1.1). Set $\varepsilon = 1 - r$. For $\varepsilon < \theta < \pi$, we have

$$Q_1(\theta) - Q_r(\theta) = \frac{(1 - r)^2 \sin\theta}{(1 - \cos\theta)(1 - 2r\cos\theta + r^2)}$$

$$= \frac{1 - r}{1 + r} Q_1(\theta) P_r(\theta) \leq \frac{1 - r}{1 + r} Q_1(1 - r) P_r(\theta).$$

[†] The minus sign in $-Q_z(\varphi)$ is to ensure that $Q_z(\varphi) = Q_r(\varphi)$ when $z = r$. With this notation, $P_z - iQ_z$ is analytic in $|z| < 1$ and $e^{i\theta} \to P_r(\theta) + iQ_r(\theta)$ extends to be analytic in $|z| < 1$.

Since $(1 - r)Q_1(1 - r) \leq 2$, this gives

(1.3) $$|Q_1(\theta) - Q_r(\theta)| \leq \frac{2}{1 + r} P_r(\theta)$$

for $\varepsilon < |\theta| < \pi$, On the other hand, for $|\theta| \leq \varepsilon = 1 - r$, we have

(1.4) $$|Q_r(\theta)| \leq 2/\varepsilon.$$

Now

$$\left| \tilde{u}(re^{i\theta}) - \frac{1}{2\pi} \int_{|\theta - \varphi| > \varepsilon} \cot\left(\frac{\theta - \varphi}{2}\right) f(\varphi) \, d\varphi \right|$$

$$\leq \frac{1}{2\pi} \int_{|\varphi| \leq \varepsilon} |Q_r(\varphi)| |f(\theta - \varphi)| \, d\varphi$$

$$+ \frac{1}{2\pi} \int_{|\varphi| > \varepsilon} |Q_1(\varphi) - Q_r(\varphi)| |f(\theta - \varphi)| \, d\varphi.$$

By (1.3), (1.4), and Theorem I.4.2, the last two integrals are dominated by $CMf(\theta)$ and (1.2) is proved.

To prove (1.1) we use the fact that the odd functions $Q_r(\theta)$ and $Q_1(\theta)\chi_{|\theta| > \varepsilon}$ are orthogonal to constants. Thus

$$\tilde{u}(re^{i\theta}) - \frac{1}{2\pi} \int_{\varepsilon < |\varphi| < \pi} \cot\left(\frac{\varphi}{2}\right) f(\theta - \varphi) \, d\varphi$$

$$= \frac{1}{2\pi} \int_{|\varphi| \leq \varepsilon} Q_r(\varphi)(f(\theta - \varphi) - f(\theta)) \, d\varphi$$

$$+ \frac{1}{2\pi} \int_{\varepsilon < |\varphi| < \pi} \frac{1 - r}{1 + r} Q_1(\varphi) P_r(\varphi)(f(\theta - \varphi) - f(\theta)) \, d\varphi,$$

and so by (1.3) and (1.4),

$$\left| \tilde{u}(re^{i\theta}) - \frac{1}{2\pi} \int_{\varepsilon < |\varphi| < \pi} \cot\left(\frac{\varphi}{2}\right) f(\theta - \varphi) \, d\varphi \right|$$

$$\leq \frac{1}{\pi \varepsilon} \int_{|\varphi| \leq \varepsilon} |f(\theta - \varphi) - f(\theta)| \, d\varphi$$

$$+ \frac{1}{2\pi} \int \frac{2}{1 + r} P_r(\varphi) |f(\theta - \varphi) - f(\theta)| \, d\varphi.$$

By Chapter I, Exercise 11, the last two integrals tend to zero on the Lebesgue set of f. □

It is not difficult to prove the nontangential analog of (1.2),

$$\left| \tilde{u}(z) - \frac{1}{2\pi} \int_{|\theta - \varphi| > \varepsilon} \cot\left(\frac{\theta - \varphi}{2}\right) f(\varphi) \, d\varphi \right| \leq C_\alpha Mf(\theta)$$

when $z \in \Gamma_\alpha(\theta)$ and $\varepsilon = 1 - |z|$. The details are left to the reader.

The inequality

$$\left| \frac{2}{\theta} - \cot \frac{\theta}{2} \right| \le \frac{2}{\pi}$$

shows that the principal value in (1.1) has the same behavior as the *Hilbert transform*,

$$(1.5) \qquad Hf(\theta) = \lim_{\varepsilon \to 0} \frac{1}{\pi} \int_{\varepsilon < |\varphi - \theta| < \pi} \frac{f(\varphi)}{\theta - \varphi} d\varphi = \lim_{\varepsilon \to 0} H_\varepsilon f(\theta).$$

Although $Hf(\theta) \ne \tilde{f}(\theta)$, the difference arises by convolving f with the bounded function $(2/\theta - \cot \theta/2)$, so that $Hf(\theta)$ exists almost everywhere and

$$(1.6) \qquad |\tilde{f}(\theta) - Hf(\theta)| \le (2/\pi)\|f\|_1.$$

There exist continuous $f(\theta)$ for which $\tilde{f}(\theta)$ is not even bounded. For example, let $F = u + i\tilde{u}$ be the conformal mapping of D onto $\{z : |x| < 1/(1 + y^2)\}$ with $F(0) = 0$, $F'(0) > 0$. Then Carathéodory's theorem on the continuity of conformal mappings implies that $u(\theta) = \lim_{r \to 1} u(re^{i\theta})$ is continous and that $\tilde{u}(\theta)$ is unbounded. A more elementary example can be given using (1.5). Let $f(\theta)$ be an odd function, so that no cancellation can occur in (1.5) at $\theta = 0$. Then $\lim_{r \to 1} \tilde{u}(r)$ behaves like

$$\lim_{\varepsilon \to 0} \frac{2}{\pi} \int_\varepsilon^\pi \frac{f(\theta)}{\theta} d\theta,$$

and this integral can diverge even though f is continuous.

When $f(\theta)$ is a continuous function on T, we write

$$\omega(\delta) = \omega_f(\delta) = \sup_{|\theta - \varphi| < \delta} |f(\theta) - f(\varphi)|$$

for the *modulus of continuity* of f. The modulus of continuity is a nondecreasing function satisfying

$$\lim_{\delta \to 0} \omega(\delta) = 0 \quad \text{and} \quad \omega(\delta_1 + \delta_2) \le \omega(\delta_1) + \omega(\delta_2).$$

A function is called *Dini continuous* if

$$\int_0^a \frac{\omega(t)}{t} dt < \infty$$

for some $a > 0$.

Theorem 1.3. *If $f(\theta)$ is a Dini continuous function on T, then $\tilde{f}$ exists at every point of T, $\tilde{f}$ is a continuous function, and*

$$(1.7) \qquad \omega_{\tilde{f}}(\delta) \le C \left(\int_0^\delta \frac{\omega(t)}{t} dt + \delta \int_\delta^\pi \frac{\omega(t)}{t^2} dt \right),$$

where C is a constant not depending on f or ω.

Observe that if δ_0 is small and if $0 < \delta < \delta_0$, then

$$\delta \int_\delta^\pi \frac{\omega(t)}{t^2} \, dt \leq \int_\delta^{\delta_0} \frac{\omega(t)}{t} \, dt + \frac{\delta}{\delta_0} \int_{\delta_0}^\pi \frac{\omega(t)}{t} \, dt,$$

so that when f is Dini continuous the continuity of $\tilde{f}$ follows from (1.7). For $0 < \alpha < 1$, we say $f \in \Lambda_\alpha$ if $\omega(\delta) = O(\delta^\alpha)$. Then (1.7) shows that conjugation preserves the Lipschitz classes Λ_α.

Proof. If $b(\theta)$ is a bounded function, then the convolution

$$b * f(\theta) = \frac{1}{2\pi} \int b(\varphi) f(\theta - \varphi) \, d\varphi$$

satisfies

$$\omega_{b*f}(\delta) \leq \|b\|_\infty \omega_f(\delta) \leq C \|b\|_\infty \delta \int_\delta^\pi \frac{\omega(t)}{t^2} \, dt.$$

It is therefore enough to show that $Hf(\theta)$ exists almost everywhere and that Hf has continuity (1.7).

Since

$$Hf(\theta) = \lim_{\varepsilon \to 0} \frac{1}{\pi} \int_{|\varphi - \theta| > \varepsilon} \frac{f(\varphi) - f(\theta)}{\theta - \varphi} \, d\varphi,$$

the Dini continuity ensures that Hf exists at every point.

Let $|\theta_1 - \theta_2| = \delta$ and take $\theta_3 = (\theta_1 + \theta_2)/2$. Because a constant function has Hilbert transform zero we can assume $f(\theta_3) = 0$. We assume $\theta_1 < \theta_2$. Then

$$\begin{aligned}
|Hf(\theta_1) - Hf(\theta_2)| \leq{} & \frac{1}{\pi} \int_{\theta_1 - \delta}^{\theta_1 + \delta} \frac{|f(\varphi) - f(\theta_1)|}{|\varphi - \theta_1|} \, d\varphi \\
&+ \frac{1}{\pi} \int_{\theta_2 - \delta}^{\theta_2 + \delta} \frac{|f(\varphi) - f(\theta_2)|}{|\varphi - \theta_2|} \, d\varphi \\
&+ \frac{1}{\pi} \int_{\theta_1 - \delta}^{\theta_1} \frac{|f(\varphi)|}{|\varphi - \theta_2|} \, d\varphi + \frac{1}{\pi} \int_{\theta_2}^{\theta_2 + \delta} \frac{|f(\varphi)|}{|\varphi - \theta_1|} \, d\varphi \\
&+ \frac{1}{\pi} \int_{3\delta/2 < |\varphi - \theta_3| < \pi} |f(\varphi)| \left| \frac{1}{\theta_1 - \varphi} - \frac{1}{\theta_2 - \varphi} \right| \, d\varphi.
\end{aligned}$$

The first two integrals are dominated by $C \int_0^\delta (\omega(t)/t) \, dt$. Since $f(\theta_3) = 0$, the second two integrals are each bounded by

$$\omega\left(\frac{3\delta}{2}\right) \int_\delta^{2\delta} \frac{dt}{t} \leq C\omega(\delta) \leq c \int_{\delta/2}^\delta \frac{\omega(t)}{t} \, dt.$$

Again using $f(\theta_3) = 0$, we can bound the fifth integral by

$$C|\theta_2 - \theta_1| \int_{3\delta/2}^{\pi} \frac{\omega(t)}{t^2} dt \leq C\delta \int_{\delta}^{\pi} \frac{\omega(t)}{t^2} dt.$$

Together, these estimates give us (1.7). □

Corollary 1.4. *Let I be an open arc on T. Let $f(\theta) \in L^1$ and assume f is Dini continuous on the arc I. Then $\tilde{f}(\theta)$ is continuous at each point of I.*

Proof. First note that if $f(\theta) = 0$ on I, then by (1.1), $\tilde{f}$ is real analytic on I. If J is any compact subarc of I, there is a function $g(\theta)$ Dini continuous on T such that $g = f$ on a neighborhood of J. By Theorem 1.3 and the preceding remark,

$$\tilde{f} = \tilde{g} + (f - g)^{\sim}$$

is continuous on J. □

There is a close connection between conjugate functions and partial sums of Fourier series. Let $f(\theta) \in L^1$ have Fourier series

$$\sum_{-\infty}^{\infty} a_n e^{in\theta}.$$

Suppose for convenience that $f(\theta)$ is real, so that $a_{-n} = \overline{a_n}$. Since

$$P_r(\theta) + i Q_r(\theta) = \frac{1 + re^{i\theta}}{1 - re^{i\theta}} = 1 + 2 \sum_{n=1}^{\infty} r^n e^{in\theta}$$

and since P_r and Q_r are real, we have

$$P_r(\theta) = \sum_{-\infty}^{\infty} r^{|n|} e^{in\theta},$$

$$Q_r(\theta) = \sum_{n \neq 0} (-i) \operatorname{sgn}(n) r^{|n|} e^{in\theta}.$$

Hence

$$u(re^{i\theta}) = P_r * f(\theta) = \sum_{-\infty}^{\infty} a_n r^{|n|} e^{in\theta}$$

and

$$\tilde{u}(re^{i\theta}) = Q_r * f(\theta) = -i \sum_{n \neq 0} \operatorname{sgn}(n) a_n r^{|n|} e^{in\theta}.$$

In particular, if $f(\theta)$ is a trigonometric polynomial $\sum_{-N}^{N} a_n e^{in\theta}$, then $\tilde{f}(\theta)$ is a trigonometric polynomial of the same degree

$$(1.8) \qquad \tilde{f}(\theta) = \sum_{-N}^{N} m(n) a_n e^{in\theta},$$

where

$$m(n) = \begin{cases} -i, & n > 0, \\ 0, & n = 0, \\ i, & n < 0, \end{cases}$$

is the *Fourier multiplier* associated with the conjugation operator. Parseval's theorem now gives

Theorem 1.5. If $f \in L^2$, then $\tilde{f} \in L^2$ and

$$\|\tilde{f}\|_2^2 = \|f\|_2^2 - |a_0|^2,$$

where $a_0 = (1/2\pi) \int f(\theta) \, d\theta$.

Now consider the operator

$$P(f) = \tfrac{1}{2}(f + i\tilde{f}) + \tfrac{1}{2}a_0,$$

which sends $\sum_{-\infty}^{\infty} a_n e^{in\theta}$ into $\sum_0^{\infty} a_n e^{in\theta}$. The operator P discards the a_n for $n < 0$, and so P is the orthogonal projection of L^2 onto H^2. In any norm under which the linear functional $f \to a_0(f)$ is continuous, the operator P is bounded if and only if the conjugation operator is bounded.

The operator $f \to e^{-in\theta} P(e^{in\theta} f)$ discards the coefficients a_k for $k < -n$ and leaves the other coefficients unchanged. The operator $f \to e^{i(n+1)\theta} P(e^{-i(n+1)\theta} f)$ similarly removes a_k for $k < n + 1$. Consequently

$$e^{-in\theta} P(e^{in\theta} f) - e^{i(n+1)\theta} P(e^{-i(n+1)\theta} f) = S_n(f),$$

the nth partial sum $\sum_{-n}^{n} a_k e^{ik\theta}$ of the Fourier series. Similar reasoning shows that

$$P(f) = \lim_{n \to \infty} e^{in\theta} S_n(e^{-in\theta} f)$$

for $f \in L^2$. This means that the famous Marcel Riesz theorem

$$\|\tilde{f}\|_p \le A_p \|f\|_p, \qquad 1 < p < \infty,$$

is equivalent to either of the inequalities

$$\|Pf\|_p \le B_p \|f\|_p, \qquad 1 < p < \infty$$

or

$$\sup_n \|S_n f\|_p \le C_p \|f\|_p, \qquad 1 < p < \infty.$$

It is not hard to see that the last inequality holds if and only if

$$\|S_n f - f\|_p \to 0 \, (n \to \infty), \quad 1 < p < \infty.$$

(For one implication use the uniform boundedness principle; for the other use the L^p density of the trigonometric polynomials.)

In the upper half plane let $u(z)$ be the Poisson integral of $f(t) \in L^p, 1 \leq p < \infty$. The conjugate function $\tilde{u}(z)$ is now defined by

$$\tilde{u}(z) = \frac{1}{\pi} \int \frac{x-t}{(x-t)^2 + y^2} f(t) \, dt = Q_y * f(x), \quad z = x + iy,$$

where

$$Q_y(t) = \frac{1}{\pi} \frac{t}{t^2 + y^2}$$

is the *conjugate kernel* for the upper half plane. The integral defining $\tilde{u}(z)$ converges because $Q_y \in L^q$ for all $q > 1$. Since

$$P_y(x-t) + iQ_y(x-t) = \frac{1}{\pi i} \frac{1}{t-z},$$

the function $u + i\tilde{u}$ is analytic in the upper half plane. This choice of $\tilde{u}$ involves a normalization different from the one used in the disc. Instead of $\tilde{u}(i) = 0$ we require $\lim_{y \to \infty} \tilde{u}(x + iy) = 0$, because only with this normalization is it possible for $\tilde{u}(z)$ to be the Poisson integral of an L^p function, $p < \infty$. Because $Q_y \notin L^1$, for $p = \infty$ we revert to the normalization used on the disc and write

$$\tilde{u}(z) = \int \left(Q_y(x-t) + \frac{1}{\pi} \frac{t}{1+t^2} \right) f(t) \, dt.$$

Then $\tilde{u}(i) = 0$ and the integral is absolutely convergent when $f \in L^\infty$.

The results obtained above for conjugate functions on the disc can be proved in a similar way for the upper half plane, and we shall not carry out the detailed arguments. We shall, however, point out some minor differences between the two cases.

When $p < \infty$, the limit of the conjugate kernels $Q_y(t)$, as $y \to 0$, coincides with the Hilbert transform kernel $1/\pi t$. Thus

$$\left| \tilde{u}(x+iy) - \frac{1}{\pi} \int_{|x-t|>y} \frac{f(t)}{x-t} \, dt \right| \leq CMf(x),$$

and as $y \to 0$ both quantities in this expression converge almost everywhere to the same function $\tilde{f}(x) = Hf(x)$.

The Hilbert transform of $f \in L^\infty$ is defined almost everywhere by

$$Hf(x) = \lim_{y \to 0} \tilde{u}(x+iy) = \lim_{\varepsilon \to 0} \frac{1}{\pi} \int_{|x-t|>\varepsilon} \left(\frac{1}{x-t} + \frac{t}{1+t^2} \right) f(t) \, dt.$$

The normalization $\tilde{u}(i) = 0$ conveniently makes these integrals converge for large t.

When $y > 0$ is fixed, the function $K_y = P_y + iQ_y = -1/\pi i(t + iy)$ is in L^2, and its Fourier transform

$$\hat{K}_y(s) = \lim_{N \to \infty} \frac{-1}{\pi i} \int_{-N}^{N} \frac{e^{-2\pi ist}}{t + iy} \, dt$$

can be evaluated by Cauchy's theorem,

$$\hat{K}_y(s) = \begin{cases} 2e^{-2\pi sy}, & s > 0, \\ 0, & s < 0. \end{cases}$$

Since $\hat{P}_y(s) = e^{-2\pi |s|y}$, this gives

$$\hat{Q}_y(s) = \begin{cases} -ie^{-2\pi |s|y}, & s > 0, \\ ie^{-2\pi |s|y}, & s < 0. \end{cases}$$

It now follows from Plancherel's theorem that $Q_y * f$ converges in L^2 norm as $y \to 0$. Since $Q_y * f \to Hf$ almost everywhere, we see that $Hf \in L^2$ and $\|Q_y * f - Hf\|_2 \to 0$, and we have the identities

$$(1.9) \qquad \widehat{Hf}(s) = (\tilde{f})\hat{\,}(s) = -i\frac{s}{|s|} \hat{f}(s),$$

$$(1.10) \qquad \|Hf\|_2 = \|\tilde{f}\|_2 = \|f\|_2.$$

2. The L^p Theorems

Fix $\alpha > 1$. The maximal conjugate function of $f \in L^1(T)$ is

$$(\tilde{f})^*(\theta) = \sup_{\Gamma_\alpha(\theta)} |\tilde{u}(z)|,$$

where $\tilde{u}$ is the conjugate Poisson integral of $f(\theta)$, and where $\Gamma_\alpha(\theta)$ is the cone

$$\left\{ \frac{|e^{i\theta} - z|}{1 - |z|} < \alpha, |z| < 1 \right\}.$$

Theorem 2.1. *There is a constant A_α depending only on α such that*

$$(2.1) \qquad |\{\theta : (\tilde{f})^*(\theta) > \lambda\}| \le (A_\alpha/\lambda)\|f\|_1$$

if $f \in L^1(T)$.

Proof. If we can prove (2.1) for all positive $f \in L^1$ with a constant C_α, then (2.1) holds in general with $A_\alpha = 8C_\alpha$. So we assume $f \ge 0$. Then $F(z) = (P_r + iQ_r) * f(\theta)$, $z = re^{i\theta}$, is an analytic function on D with

$$F(0) = \int f(\theta) \frac{d\theta}{2\pi} = \|f\|_1,$$

and Re $F(z) > 0$. With respect to the right half plane, the subset of the imaginary axis $\{is : |s| > \lambda\}$ has harmonic measure

$$h(w) = \frac{1}{\pi} \int_{|s|>\lambda} \frac{u}{(v-s)^2 + u^2}\, ds, \quad w = u + iv.$$

Clearly $h(w) \geq 0$ and $h(w) \geq \frac{1}{2}$ if Im $w \geq \lambda$. On the positive real axis

$$h(u) = \frac{2}{\pi} \int_\lambda^\infty \frac{u}{s^2 + u^2}\, ds \leq \frac{2u}{\pi\lambda}.$$

The composition $g(z) = h(F(z))$ is a positive harmonic function on D. It is the Poisson integral of a positive measure with mass

$$g(0) = h\left(\int f(\theta)\frac{d\theta}{2\pi}\right) = h(\|f\|_1).$$

If $|\text{Im } F(z)| > \lambda$, then $g(z) > \frac{1}{2}$, so that

$$\{\theta : (\tilde{f})^*(\theta) > \lambda\} \subset \{\theta : g^*(\theta) > \tfrac{1}{2}\}.$$

By the weak-type estimate for the nontangential maximal function, valid for Poisson integrals of positive measures (Theorem I.5.1), we have

$$|\{\theta : (\tilde{f})^*(\theta) > \lambda\}| \leq 2B_\alpha g(0) = 2B_\alpha h(\|f\|_1) \leq (4B_\alpha/\pi\lambda)\|f\|_1,$$

where the constant B_α depends only on α. Thus (2.1) holds for any $f \in L^1$ with $A_\alpha = 32B_\alpha/\pi$. $\quad\square$

The proof of (2.1) for the upper half plane differs from the above argument in one detail.

Lemma 2.2. *If μ is a finite measure on $\mathbb{R}$ with Poisson integral $u(z)$, then*

$$\int d\mu = \lim_{y\to\infty} \int \frac{y^2}{t^2 + y^2}d\mu(t) = \lim_{y\to\infty} \pi y u(iy).$$

The lemma is elementary.

To prove (2.1) for the upper half plane, let $f \in L^1(\mathbb{R})$, $f \geq 0$. Then $F(z) = (u + i\tilde{u})(z) = (P_y + iQ_y) * f(x)$, $z = x + iy$, is analytic on H and Re $f(z) > 0$. Then $g(z) = h(F(z))$ is a positive harmonic function on H . Moreover, since $0 \leq g(z) \leq 1$, $g(z)$ is the Poisson integral of its boundary values $g(t)$. Again we have

$$\{t : (\tilde{f})^*(t) > \lambda\} \subset \{t : g^*(t) > \tfrac{1}{2}\},$$

where $(\tilde{f})^*(t) = \sup_{\Gamma_\alpha(t)} |\tilde{u}(z)|$ is the half-plane maximal conjugate function. Consequently

$$|\{t : (\tilde{f})^*(t) > \lambda\}| \leq 2B_\alpha \int_R g(t)\, dt.$$

But by the lemma,

$$\int g(t)dt = \lim_{y\to\infty} \pi y h(F(iy)) = \lim_{y\to\infty} y \int_{|s|>\lambda} \frac{u(iy)\,ds}{(u(iy))^2 + (\tilde{u}(iy) - s)^2}.$$

Since $\lim_{y\to\infty} \tilde{u}(iy) = 0$, another use of the lemma yields

$$\int g(t)dt = \frac{1}{\pi}\int f(t)\,dt \int_{|s|>\lambda} \frac{ds}{s^2} = \frac{2\|f\|_1}{\pi\lambda}.$$

Therefore we have

$$|\{t : (\tilde{f})^*(t) > \lambda\}| \le (4B_\alpha/\pi\lambda)\|f\|_1$$

when $f \in L^1(\mathbb{R})$ and $f \ge 0$.

Theorem 2.1 shows that the conjugation operator is weak type 1–1 on $L^1(T)$ or on $L^1(\mathbb{R})$. We saw in Section 1 that it is also a bounded operator on L^2, and so the Marcinkiewicz theorem gives

$$(2.2) \qquad \|\tilde{f}\|_p \le A_p\|f\|_p, \qquad 1 < p \le 2.$$

For $2 < p < \infty$, (2.2) now follows by a duality argument. From (1.8) or (1.10) and polarization we have

$$\int \tilde{f}g\,dt = -\int f\tilde{g}\,dt$$

when $f,\ g \in L^2(\mathbb{R})$ or $L^2(T)$. Let us consider $L^p(\mathbb{R})$ only. If $p > 2$ and if $f \in L^2(\mathbb{R}) \cap L^p(\mathbb{R})$, then with $q = p/(p-1)$,

$$\|\tilde{f}\|_p = \sup\left\{\left|\int \tilde{f}g\,dt\right| : g \in L^2(\mathbb{R}) \cap L^q(\mathbb{R}),\ \|g\|_q \le 1\right\},$$

whether $\|\tilde{f}\|_p$ is finite or not. But then since $q < 2$,

$$\|\tilde{f}\|_p = \sup\left\{\left|\int f\tilde{g}\,dt\right| : g \in L^2(\mathbb{R}) \cap L^q(\mathbb{R}),\ \|g\|_q \le 1\right\} \le A_q\|f\|_p.$$

Since $L^2 \cap L^p$ is dense in L^p, (2.2) therefore holds for $p > 2$ with constant $A_p = A_q$. We have proved the following theorem, due to Marcel Riesz.

Theorem 2.3. *If $1 < p < \infty$, there is a constant A_p such that*

$$\|\tilde{f}\|_p \le A_p\|f\|_p$$

if $f \in L^p(\mathbb{R})$ or $f \in L^p(T)$.

From the interpolation we see that

$$(2.3) \qquad\qquad A_p \sim A/(p-1), \qquad p \to 1,$$
$$(2.4) \qquad\qquad A_p \sim Ap, \qquad p \to \infty.$$

These estimates are sharp, except for the choice of the constant A. By duality, (2.3) is sharp if (2.4) is sharp. Let $f(t) = \chi_{(0,1)}(t)$ Then $\|f\|_p = 1$ for all p, and

$$\tilde{f}(t) = \frac{1}{\pi}(\log|t| - \log|t - 1|),$$

so that

$$\|\tilde{f}\|_p \geq \frac{1}{\pi}\left(\int_0^1 |\log t|^p dt\right)^{1/p} = \frac{1}{\pi}\left(\int_0^\infty x^p e^{-x} dx\right)^{1/p} = \frac{1}{\pi}(\Gamma(p+1))^{1/p}.$$

Stirling's formula then shows

$$\lim_{p\to\infty} \frac{1}{p}\|\tilde{f}\|_p \geq e/\pi.$$

Together, Riesz's theorem and the maximal theorem give

$$\|(\tilde{f})^*\|_p \leq B_p\|f\|_p, \quad 1 < p < \infty,$$

where

$$B_p \sim C/(p-1)^2, \quad p \to 1.$$

This estimate on B_p can be improved to

$$B_p \sim B/(p-1), \quad p \to 1,$$

by interpolating directly between (2.1) and the L^2 estimate. At the other end we have $B_p \sim Bp$ $(p \to \infty)$, because the constants in the maximal theorem are bounded as $p \to \infty$.

Notice the fundamental difference between Theorem 2.3 and Theorem 1.3. In Theorem 1.3 the essential ingredient is the smoothness of $f(\theta)$; while Theorem 2.3 depends on the cancellation of the kernel.

Theorem 2.4. *Let $F(z)$ be an analytic function on the disc D. If $\operatorname{Re} F(z) > 0$, then $F \in H^p$ for all $p < 1$, and $\|F\|_{H^p} \leq C_p|F(0)|$.*

This result can be proved from Theorem 2.1 because $\operatorname{Re} F$ is the Poisson integral of a finite measure and because any weak L^1 function on T is in L^p for all $p < 1$. However, Theorem 2.4 also has a simple direct proof.

Proof. Write $F = |F|e^{i\varphi}$ where $|\varphi| < \pi/2$. Then F^p is analytic on D and

$$F^p = |F|^p(\cos p\varphi + i \sin p\varphi).$$

Since $p < 1$, this means

$$|F|^p \leq C_p\operatorname{Re}(F^p),$$

where $C_p = (\cos p\pi/2)^{-1}$. Hence

$$\frac{1}{2\pi} \int |F(re^{i\theta})|^p \, d\theta \leq C_p \frac{1}{2\pi} \int \mathrm{Re}(F^p(re^{i\theta})) d\theta = C_p \mathrm{Re}(F^p(0))$$
$$\leq C_p |F(0)|^p,$$

as desired. $\quad\square$

Corollary 2.5. *If $F(z)$ is analytic on D and if $|\arg F(z)| \leq \lambda \leq \pi$, then $F \in H^p$ for all $p < \pi/2\lambda$*

Proof. Use Theorem 2.4 on $F^{\pi/2\lambda}$ $\quad\square$

Corollary 2.6. *If $f(\theta) \in L^\infty(T)$ and if $\|f\|_\infty \leq 1$, then for $p < \pi/2$*

$$\frac{1}{2\pi} \int e^{p|\tilde{f}|} \, d\theta < C_p.$$

If $f(\theta)$ is continuous on T, then

$$\frac{1}{2\pi} \int e^{p|\tilde{f}|} \, d\theta < \infty$$

for all $p < \infty$.

Proof. If $f \in L^\infty$, $\|f\|_\infty \leq 1$, then

$$F = \exp(\pm\tfrac{1}{2}\pi i(f + i\tilde{f}))$$

maps the disc into the right half plane. Hence $F \in H^p$ for all $p < 1$. If $f(\theta)$ is continuous, there is a trigonometric polynomial g such that $\|f - g\|_\infty < \varepsilon$. Then $\tilde{g}$ is bounded because $\tilde{g}$ is another trigonometric polynomial. Hence

$$\exp p|\tilde{f}| \leq \exp(p|\tilde{f} - \tilde{g}|) \exp p|\tilde{g}|$$

and the last function is integrable if $p < \pi/2\varepsilon$. $\quad\square$

Theorem 2.7. *Let $E \subset T$ be a measurable set with measure $|E|$ and let $f = \chi_E$. Then the distribution function*

$$|\{\theta : |\tilde{f}(\theta)| > \lambda\}|$$

depends only on $|E|$.

Proof. Let $F(z) = (P_r + iQ_r) * f(\theta)$, $z = re^{i\theta}$. Then $0 < \mathrm{Re}\ F(z) < 1$, $\mathrm{Re}\ F(e^{i\theta}) = \chi_E(\theta)$ almost everywhere, and $F(0) = |E|/2\pi$. Let $h(w)$ be the harmonic function on the strip $0 < \mathrm{Re}\ w < 1$ with boundary values $\chi_{\{w:|\mathrm{Im}\ w|>\lambda\}}$. Then the bounded harmonic function $h(F(z))$ has nontangential limit almost everywhere equal to the characteristic function of $\{\theta : |\tilde{f}(\theta)| > \lambda\}$. Hence

$$|\{\theta : |\tilde{f}(\theta)| > \lambda\}| = 2\pi h(F(0)) = 2\pi h(|E|/2\pi).$$

which proves Theorem 2.7. $\quad\square$

3. Conjugate Functions and Maximal Functions

The Riesz theorem fails when $p = 1$ and when $p = \infty$. A counterexample for the case $p = \infty$ was given in Section 1, and a duality argument then shows that the theorem fails for $p = 1$.

However, there is a related inequality valid for all finite p that implies the Riesz theorem for $1 < p < \infty$. Let $u(z)$ be harmonic on the upper half plane H . Using the cones

$$\Gamma(t) = \{z \in \text{H} \ : \ |x - t| < y\}, \quad t \in \mathbb{R},$$

define the nontangential maximal function

$$u^*(t) = \sup_{\Gamma(t)} |u(z)|.$$

Theorem 3.1. *If $0 < p < \infty$ and if $u(z)$ is a real-valued harmonic function on* H *such that $u^* \in L^p$, then there is a harmonic conjugate function $v(z)$ for $u(z)$ such that*

$$(3.1) \qquad \sup_{y>0} \int |v(x + iy)|^p dx \leq C_p \int |u^*(t)|^p \, dt.$$

On the unit disc we have the same inequality:

$$(3.2) \qquad \sup_{y<1} \int |\tilde{u}(re^{i\theta})|^p \, d\theta \leq c_p \int |u^*(\theta)|^p d\theta, \quad 0 < p < \infty,$$

where $\tilde{u}$ is the usual conjugate function normalized by $\tilde{u}(0) = 0$ and where

$$(3.3) \qquad u^*(\theta) = \sup \left\{ |u(z)| : \frac{|e^{i\theta} - z|}{1 - |z|} < 2 \right\}.$$

Of course it is not crucial for Theorem 3.1 that the cones $\Gamma(t)$ have aperture $\pi/2$. From Theorem 3.6 below it will follow that (3.1) remains true for cones of any angle.

For $p > 1$, Theorem 3.1 follows from the Riesz theorem, so we are saying something new only for $p \leq 1$. Aided by the Hardy–Littlewood maximal theorem, Theorem 3.1 itself implies the Riesz theorem. This theorem was first proved by Burkholder, Gundy, and Silverstein [1971] using Brownian motion. The elementary proof given below was recently discovered by Paul Koosis. We restrict ourselves to the case $p < 2$, and we first give the proof on the line. After concluding the proof there we indicate how to adapt the argument to the circle.

Lemma 3.2. *If $F(z) = u(z) + iv(z)$ is of class H^2, if*

$$m(\lambda) = |\{t : u^*(t) > \lambda\}|,$$

and if

$$\mu(\lambda) = |\{t : |v(t)| > \lambda\}|,$$

then

(3.4) $$\mu(\lambda) \le 2m(\lambda) + \frac{2}{\lambda^2} \int_0^\lambda sm(s)\, ds.$$

Proof. Let

$$U_\lambda = \{t : u^*(t) > \lambda\},$$

so that $m(\lambda) = |U_\lambda|$, and let $E_\lambda = \mathbb{R} \backslash U_\lambda$. Form the region $R = \bigcup_{t \in E_\lambda} \Gamma(t)$. The boundary ∂R consists of two subsets,

$$E_\lambda = \mathbb{R} \cap \partial R \quad \text{and} \quad \Gamma = \{y > 0\} \cap \partial R.$$

The set Γ is the union of some tents whose bases are the component intervals of the open set U_λ, as shown in Figure III.1.

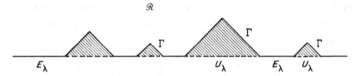

Figure III.1. On R , $|u(z)| \le \lambda$.

Since $F(z) \in H^2$, Cauchy's theorem and the density of $\mathfrak{A}_N$ in H^1 yields

$$\int_{\partial R} F^2(z)\, dz = 0.$$

Expanding out the real part of the integral gives

$$\int_{E_\lambda} (u^2 - v^2)\, dx + \int_\Gamma (u^2 - v^2)\, dx - \int_\Gamma 2uv\, dy = 0.$$

On Γ, $|dy| = dx$ and $-2uv \le u^2 + v^2$, so that

$$-\int_\Gamma 2uv\, dy \le \int_\Gamma (u^2 + v^2)\, dx.$$

Hence we have

$$0 \le \int_{E_\lambda} (u^2 - v^2)\, dx + 2 \int_\Gamma u^2\, dx$$

and

(3.5) $$\int_{E_\lambda} v^2\, dx \le \int_{E_\lambda} u^2\, dx + 2 \int_\Gamma u^2\, dx.$$

Along Γ, we have $|u(z)| \le \lambda$, so that

$$\int_\Gamma u^2\, dx \le \lambda^2 \int_\Gamma dx = \lambda^2 |U_\lambda| = \lambda^2 m(\lambda).$$

To estimate the other term on the right side of (3.5), write

$$\int_{E_\lambda} u^2 dx \le \int_{E_\lambda} (u^*)^2 dx = \iint_{\substack{u^*\le\lambda \\ 0<s<u^*}} 2s\, ds\, dx$$

$$= \int_0^\lambda \int_{\{s<u^*<\lambda\}} dx\, 2s\, ds = \int_0^\lambda (m(s) - m(\lambda))2s\, ds$$

$$= \int_0^\lambda 2sm(s)ds - \lambda^2 m(\lambda).$$

With Chebychev's inequality, (3.5) now yields

$$|\{t \in E_\lambda : |v(t)| > \lambda\}| \le (1/\lambda^2) \int_0^\lambda 2sm(s)ds - m(\lambda) + 2m(\lambda).$$

Then since

$$\mu(\lambda) \le |U_\lambda| + |\{t \in E_\lambda : |v(t)| > \lambda\}|,$$

(3.4) is proved. □

Inequality (3.4) is the main step in the proof of Theorem 3.1. The rest of the proof is more routine.

Lemma 3.3. *If $0 < p \le 2$, and if $u^* \in L^p$, then*

$$\left(\int |u(x+iy)|^2 dx\right)^{1/2} \le 2y^{1/2-1/p} \left(\int |u^*|^p\, dt\right)^{1/p}, \quad y > 0.$$

Proof. Fix $y > 0$. Then

$$|u(x+iy)|^p \le \inf_{|t-x|<y} |u^*(t)|^p \le (1/2y) \int_{x-y}^{x+y} |u^*(t)|^p\, dt,$$

and so

$$\sup_x |u(x+iy)| \le (2y)^{-1/p} \|u^*\|_p.$$

Since $p \le 2$, this yields

$$\left(\int |u(x+iy)|^2 dx\right)^{1/2} \le \left\{(2y)^{(p-2)/p}\|u^*\|_p^{2-p} \int |u(x+iy)|^p dx\right\}^{1/2}$$

$$\le (2y)^{1/2-1/p} \left(\int |u^*|^p dt\right)^{1/p},$$

as asserted. □

To prove Theorem 3.1 for the line and for $0 < p < 2$, assume $u^* \in L^p$ and fix $y_0 > 0$. By Lemma 3.3 there is a conjugate function $v(z)$ defined on $y > y_0$

such that $f = u + iv$ satisfies

$$\sup_{y > y_0} \int |f(x + iy)|^2 dx < \infty.$$

For every $y_0 > 0$ there is only one such v, and hence v does not depend on y_0. Let

$$\mu(\lambda) = |\{x : |v(x + iy_0)| > \lambda\}|.$$

Since $u_0(z) = u(z + iy_0)$ has $u_0^* \le u^*$, Lemma 3.2 implies that

$$\mu(\lambda) \le 2m(\lambda) + \frac{2}{\lambda^2} \int_0^\lambda sm(s)\, ds,$$

where $m(\lambda) = |\{t : u^*(t) > \lambda\}|$. Integrating this inequality against $p\lambda^{p-1}\, d\lambda$ now gives

$$
\begin{aligned}
\int |v(x + iy_0)|^p dx &= \int_0^\infty p\lambda^{p-1}\mu(\lambda)d\lambda \\
&\le 2\int_0^\infty p\lambda^{p-1}m(\lambda)d\lambda + 2\int_0^\infty psm(s)\int_s^\infty \lambda^{p-3}\, d\lambda\, ds \\
&= 2\|u^*\|_p^p + \frac{2}{2-p}\int_0^\infty ps^{p-1}m(s)\, ds \\
&= 2\left(1 + \frac{1}{2-p}\right)\|u^*\|_p^p.
\end{aligned}
$$

The right side is independent of y_0 and we have (3.1) with constant

$$c_p = 2\left(1 + \frac{1}{2-p}\right), \qquad p < 2.$$

It does not matter that this constant blows up as $p \to 2$ because the Riesz theorem gives another proof of (3.1) for $p > 1$. $\quad\square$

The case $p \ge 2$ of Theorem 3.1 can of course be proved without recourse to the Riesz theorem. The case $p > 2$ follows from the case $p < 2$ by a duality, and an interpolation then gives the remaining case $p = 2$.

On the unit disc the same reasoning can be used once we establish an inequality like (3.4). Let $f = u + i\tilde{u} \in H^2$, where $\tilde{u}(0) = 0$. Let $U_\lambda = \{\theta : u^*(\theta) > \lambda\}$, where u^* is defined by (3.3), let $E_\lambda = T \setminus U_\lambda$, and let $m(\lambda) = |U_\lambda|$. Also let $\mu(\lambda) = |\{\theta : |\tilde{u}(\theta)| > \lambda\}|$. If $m(\lambda)$ is not small we automatically have

$$(3.6) \qquad \mu(\lambda) \le Cm(\lambda) + \frac{1}{\lambda^2}\int_0^\lambda sm(s)\, ds,$$

which is the inequality we are after, and so we assume $m(\lambda)$ is small. As before, form

$$R = \bigcup_{\theta \in E_\lambda} \left\{ z : \frac{|e^{i\theta} - z|}{1 - |z|} < 2 \right\}.$$

Now $\partial R = E_\lambda \cup \Gamma$, where Γ is a union of tentlike curves, one over each component of U_λ. By Cauchy's theorem

$$(u(0))^2 = \frac{1}{2\pi i} \int_{\partial R} f^2(z) dz/z.$$

On E_λ,

$$\frac{1}{2\pi i} \frac{dz}{z} = \frac{d\theta}{2\pi},$$

while on Γ,

$$\frac{1}{2\pi i} \frac{dz}{z} = \frac{d\theta}{2\pi} \pm \frac{dr}{2\pi i r}.$$

If $m(\lambda) = |U_\lambda|$ is small enough, then $1 - r$ is small on Γ, and a calculation then shows that

$$\left| \frac{dr}{rd\theta} \right| < 1$$

almost everywhere on Γ. Taking the real part of the integral above then gives

$$0 \le \frac{1}{2\pi} \int_{E_\lambda} (u^2 - (\tilde{u})^2) \, d\theta + \frac{1}{2\pi} \int_{\Gamma} (u^2 - (\tilde{u})^2) \, d\theta + \frac{1}{2\pi} \int_{\Gamma} 2u\tilde{u} \frac{dr}{r}.$$

We have the estimate

$$\int_{\Gamma} 2|u\tilde{v}| \left| \frac{dr}{r} \right| \le \int_{\Gamma} (u^2 + (\tilde{u})^2) \, d\theta,$$

so that we obtain

$$\int_{E_\lambda} (\tilde{u})^2 \, d\theta \le \int_{E_\lambda} u^2 \, d\theta + 2 \int_{\Gamma} u^2 \, d\theta$$

when $m(\lambda)$ is small. This inequality implies (3.6) just as in the proof of Lemma 3.2 and consequently (3.6) is true for all values of $m(\lambda)$. The remainder of the proof of Theorem 3.1 now shows that (3.2) holds for $0 < p < 2$, and the Riesz theorem gives (3.2) for $2 \le p < \infty$.

Corollary 3.4. *If $0 < p < \infty$ and if $u(z)$ is a real-valued harmonic function, then $u(z) = \mathrm{Re}\ f(z)$, $f \in H^p$, if and only if $u^* \in L^p$. There are constants c_1 and c_2, depending only on p, such that*

$$c_1 \|u^*\|_p \le \|f\|_p \le c_2 \|u^*\|_p.$$

Proof. The inequality $c_1\|u^*\|_p \le \|f\|_p$ was proved in Chapter II. The other inequality is immediate from the theorem. $\square$

Corollary 3.4, which is of course equivalent to the theorem, is very important because it enables the H^P spaces, $p < \infty$, to be defined without any reference to analytic functions. The H^P spaces can further be characterized without recourse to harmonic functions (see Fefferman and Stein [1972]).

Corollary 3.5. *If $0 < p < \infty$ and if $u(z)$ is a harmonic function such that $u^* \in L^p$, then there is a conjugate function v such that $v^* \in L^P$ and*

$$\|v^*\|_p^p \le C_p \|u^*\|_p^p.$$

Proof. By Corollary 3.4, $f = u + iv$ is in H^p with $\|f\|_p^p \le c_2^p \|u^*\|_p^p$. The maximal theorem then shows that $f^* \in L^p$ with $\|f^*\|_p^p \le Ac_2^p \|u^*\|_p^p$. Since trivially $v^* \le f^*$, we conclude that $\|v^*\|_p^p \le C_p \|u^*\|_p^p$. $\square$

There is another inequality even stronger than (3.1) in which the nontangential maximal function is replaced by the vertical maximal function. Again let $u(z)$ be a harmonic function on H , and write

$$u^+(t) = \sup_{y>0} |u(t + iy)|, \quad t \in \mathbb{R}.$$

Obviously $u^+(t) \le u^*(t)$. Conversely, we have

Theorem 3.6. *If $0 < p < \infty$ and if $u(z)$ is harmonic on H , then*

$$\int |u^*|^p \, dt \le c_p \int |u^+|^p \, dt,$$

where c_p is a constant depending only on p.

As a corollary, we see that a harmonic function $u(z)$ is the real part of an H^P function if and only if $u^+ \in L^p$, and consequently that (3.1) holds for cones of any angle.

The proof of Theorem 3.6 rests on a remarkable inequality due to Hardy and Littlewood.

Lemma 3.7. *If $u(z)$ is harmonic on the disc $\Delta(z_0, R)$ and if $0 < p < \infty$, then*

(3.7) $$|u(z_0)| \le K_p \left(\frac{1}{\pi R^2} \iint_{\Delta(z_0, R)} |u(z)|^p dx \, dy \right)^{1/p},$$

where K_P depends only on p.

When $p \ge 1$, this lemma is a trivial application of Hölder's inequality and the mean value property. When $p < 1$ the inequality is rather surprising, because $|u(z)|^p$ is not always subharmonic.

Proof. We only have to treat the case $p < 1$. We may change variables and assume $z_0 = 0$, $R = 1$. Write

$$m_q(r) = \left(\frac{1}{2\pi} \int_0^{2\pi} |u(re^{i\theta})|^q \, d\theta \right)^{1/q}.$$

We can assume that

$$2 \int_0^1 m_p(r)^p r \, dr = \frac{1}{\pi} \iint_{\Delta(0,1)} |u(z)|^p dx \, dy = 1$$

and then that $m_\infty(r) = \sup\{|u(z)| : |z| = r\} > 1$ for $0 < r < 1$, because otherwise (3.7) is true with $K_P = 1$.

Since $p < 1$ we have

$$m_1(r) \le \frac{1}{2\pi} \int |u(re^{i\theta})|^p d\theta m_\infty(r)^{1-p} = m_p(r)^p m_\infty(r)^{1-p}.$$

Also, estimating the supremum of the Poisson kernel gives

$$m_\infty(\rho) \le \frac{2m_1(r)}{1 - \rho/r}, \quad 0 < \rho < r < 1.$$

Set $\rho = r^\alpha$, where $\alpha > 1$ will be determined later. Then

$$\int_{1/2}^1 \log m_\infty(r^\alpha) \frac{dr}{r} \le \int_{1/2}^1 \log \left(\frac{2}{1 - \rho/r} \right) \frac{dr}{r} + p \int_{1/2}^1 \log m_p(r) \frac{dr}{r}$$

$$+ (1 - \rho) \int_{1/2}^1 \log m_\infty(r) \frac{dr}{r}.$$

The first integral converges with value C_α, and the second integral is bounded by

$$\int_{1/2}^1 m_p^p(r) \frac{dr}{r} \le 4 \int_0^1 m_p^p(r) r \, dr \le 2$$

by our assumption, and so we have

$$\int_{1/2}^1 \log m_\infty(r^\alpha) \frac{dr}{r} \le (C_\alpha + 2) + (1 - p) \int_{1/2}^1 \log m_\infty(r) \frac{dr}{r}.$$

Since we have assumed $\log m_\infty(r) \ge 0$, a change of variables gives

$$\int_{1/2}^1 \log m_\infty(r) \frac{dr}{r} \le \int_{(1/2)^\alpha}^1 \log m_\infty(r) \frac{dr}{r} = \alpha \int_{1/2}^1 \log m_\infty(r^\alpha) \frac{dr}{r}$$

$$\le \alpha(C_\alpha + 2) + \alpha(1 - p) \int_{1/2}^1 \log m_\infty(r) \frac{dr}{r}.$$

Taking $\alpha(1 - p) < 1$ now yields

$$\inf_{1/2 < r < 1} \log m_\infty(r) < \frac{\alpha(C_\alpha + 2)}{1 - \alpha(1 - p)} \frac{1}{\log 2} = \log K_p,$$

and (3.7) follows from the maximum principle. □

Lemma 3.7 provides an alternate proof of Lemma 3.3 but with a different constant.

Proof of Theorem 3.6. Fix q with $0 < q < p$. Let $z = x + iy \in \Gamma(t) = \{|x - t| < y\}$. Then u is harmonic on $\Delta(z, y/2)$, and Lemma 3.7 gives

$$|u(z)|^q \leq \frac{4K_q^q}{\pi y^2} \iint_{\Delta(z,y/2)} |u(\xi + i\eta)|^q d\xi \, d\eta \leq \frac{C_q}{y} \int_{|\xi - x| < y/2} |u^+(\xi)|^q d\xi$$

$$\leq \frac{C_q}{y} \int_{|\xi - t| < 3y/2} |u^+(\xi)|^q d\xi,$$

since $|x - t| < y$. The last integral is dominated by $C_q' Mg(t)$ where $g = (u^+)^q$ and where Mg is the Hardy–Littlewood maximal function of g. We have proved that

$$|u^*(t)|^p \leq C_q''(Mg(t))^{p/q}.$$

Since $p/q > 1$ and $g \in L^{p/q}$, the maximal theorem gives us

$$\int |u^*(t)|^p \, dt \leq C \int |Mg(t)|^{p/q} \, dt \leq C \int |u^+(t)|^p \, dt,$$

and Theorem 3.6 is proved. □

Notes

For the classes Λ_α, Theorem 1.3 dates back to Privalov [1916]. The weak-type estimate

(N.1) $$|\{\theta : |\tilde{f}(\theta)| > \lambda\}| \leq (A/\lambda)\|f\|,$$

and its corollary

(N.2) $$\|\tilde{f}\|_p \leq A_p \|f\|_1, \quad 0 < p < 1,$$

which of course follow from Theorem 2.1, are due to Kolmogoroff [1925]. The weak-type estimate for the Hilbert transform had been published earlier by Besicovitch [1923]. M. Riesz announced his theorem in [1924], but he delayed publishing the proof until [1927]. Corollary 2.6 is due to Zygmund [1929] and Theorem 2.7 is from Stein and Weiss [1959].

There are numerous proofs of the Riesz theorem. The one in the text was chosen because it is short and because it stresses harmonic estimates. The proof of Theorem 2.1 copies the Carleson proof of Kolmogoroff's inequality in Katznelson [1968]. I learned this proof of Theorem 2.7 from Brian Cole, while driving in Los Angeles.

The real-variables proofs of the Riesz theorem, which are more amenable to higher-dimensional generalizations, lead to the theory of singular integrals (see Calderón and Zygmund [1952], and Stein [1970]). The Calderón-Zygmund approach to Theorem 2.1 is outlined in Exercise 11, and some of its ideas will appear at the end of Chapter VI. Another real variables proof, relying on a beautiful lemma due to Loomis [1946], is elegantly presented in Garsia [1970].

Three other proofs deserve mention: the original proof in Riesz [1927] or in Zygmund [1955]; Calderón's refinement in Calderón [1950a] or Zygmund [1968]; and the Green's theorem proof, due to P. Stein [1933], which is given in Duren [1970] and also in Zygmund [1968].

The best possible constants A_p in Theorem 2.3 were determined by Pichorides [1972] and independently by B. Cole (unpublished). Using Brownian motion, B. Davis [1974, 1976] found the sharp constants in Kolmogoroff's inequalities (N.1) and (N.2). Later Baernstein [1979] gave a classical proof of the Davis results. The idea is to reduce norm inequalities about conjugate functions to a problem about subharmonic functions on the entire plane. By starting with nonnegative functions, we could work with harmonic functions on a half plane instead. The cost is higher constants and some loss in generality. Gamelin's recent monograph [1979] gives a nice exposition of Cole's beautiful theory of conjugate functions in uniform algebras and derives the sharp constants for several theorems.

The Burkholder–Gundy–Silverstein theorem, Theorem 3.1, is now a fundamental result. On the one hand it explains why conjugate functions obey the same inequalities that maximal functions do. On the other hand, the case $p \leq 1$ of the theorem shows that the nontangential maximal function is more powerful than the Hardy–Littlewood maximal function. It is not the harmonicity but the smoothness of the Poisson kernel that is decisive here. For example, let $\varphi(t)$ be a positive, Dini continuous, compactly supported function on $\mathbb{R}$. Let $f \in L^1(\mathbb{R})$ be real valued, and, in analogy with the Poisson formula, define

$$U(x, y) = \frac{1}{y} \int \varphi\left(\frac{x-t}{y}\right) f(t)\, dt, \quad y > 0.$$

Then for $0 < p \leq 1$, $f = \mathrm{Re}\, F$, $F \in H^p$, if and only if

$$\sup_{y>0} |U(x, y)| \in L^p(\mathbb{R}).$$

This theorem, and another proof of Theorem 3.1 for $\mathbb{R}^n$, are in Fefferman and Stein [1972]. Incidentally, the exact necessary and sufficient conditions for a kernel φ to characterize H^p remains an unsolved problem (see Weiss [1979]).

Another route to Theorem 3.1 is through the atomic theory of H^p spaces. See Coifman [1974] for the case of $\mathbb{R}^1$, Latter [1978] for the generalization to $\mathbb{R}^n$, and Garnett and Latter [1978] for the case of the ball of $\mathbb{C}^n$. The theory of atoms is briefly discussed in Exercise 11 of Chapter VI.

The elementary proof of Theorem 3.1 given in the text is from Koosis [1978]. It makes this book 10 pages shorter.

Exercises and Further Results

1. (Peak sets for the disc algebra). Let $E \subset T$ be a compact set of measure zero.

(a) (Fatou [1906]) There exists $u \in L^1(T)$ such that $u : T \to [-\infty, 0]$ continuously, $u^{-1}(-\infty) = E$, and u is C^1 on $T \backslash E$. Then

$$g = u + i\tilde{u}$$

is continuous on $\overline{D} \backslash E$ and g has range in the left half plane. The function

$$f_0 = e^g$$

is in the disc algebra $A_o = H^\infty \cap C(T)$ and $f_0(z) = 0$ if and only if $z \in E$. Thus any closed set of measure zero is a *zero set* for A_o. Conversely, if $E \subset T$ is the zero set of some nonzero $f \in A_o$, then $|E| = 0.(\int \log |f| d\theta > -\infty.)$

(b) (F. and M. Riesz [1916]) Let g be the function from part (a) and let

$$f_1 = g/(g - 1).$$

Then $f_1 \in A_o$, $f_1 \equiv 1$ on E, and $|f_1(z)| < 1$, $z \in \overline{D} \backslash E$. This means E is a *peak set* for A_o. Conversely, any peak set for A_o has measure zero. (If E is a peak set and if $f \equiv 1$ on E, $|f| < 1$ on $\overline{D} \backslash E$, then $1 - f$ has zero set E.)

(c) Use part (b) to prove the F. and M. Riesz theorem on the disc: If μ is a finite complex Borel measure on T such that $\int e^{in\theta} d\mu(\theta) = 0, n = 1, 2, \dots ,$ then μ is absolutely continuous. (If not, there is compact E, $|E| = 0$ such that $\int_E e^{i\theta} d\mu \neq 0$. But

$$\lim_{n \to \infty} \int f_1^n e^{i\theta} d\mu = 0.)$$

This is the original proof of the theorem.

(d) (Rudin [1956], Carleson [1957]) If $E \subset T$ is a compact set of measure zero, then E is a *peak interpolation* set for A_o : Given $h \in C(E)$ there exists $g \in A_o$ such that the restriction to E is h and such that

$$\|g\|_\infty = \|h\| = \sup_E |h(z)|.$$

By Runge's theorem there are $g_n \in A_o$ such that $g_n \to h$ uniformly on E. Take n_j so that $|g_{n_{j+1}} - g_{n_j}| < 2^{-j}$ on some open neighborhood V_j of E, and take

k_j so that

$$|g_{n_{j+1}} - g_{n_j}\|f_1^{k_j}| < 2^{-j}$$

on $T \backslash V_j$, where f_1 is the function from part (b). Then

$$g = g_{n_1} + \sum_{j=1}^{\infty} f_1^{k_j}(g_{n_{j+1}} - g_{n_j})$$

is in A_o and $g = h$ on E. By being a little more careful, one can also obtain $\|g\|_\infty = \|h\|$.

(e) The result in (d) can also be derived using a duality. The restriction map $S : A_o \to C(E)$ has adjoint $S^* : M(E) \to M(T)/A_o^\perp$, where $M(E) = C(E)^*$ is the space of finite complex Borel measures on E and where $A_o^\perp \subset M(T)$ is the subspace of measures orthogonal to A_o. The F. and M. Riesz theorem from Chapter II shows that S^* is an isometry. If I is the ideal $\{f \in A_o : S(f) = 0\}$, then by a theorem in functional analysis the induced map $\tilde{S} : A_o/I \to C(E)$ is also an isometry and $\tilde{S}$ maps onto $C(E)$. Thus if $h \in C(E)$ and if $\varepsilon > 0$, there exists $g \in A_o$ such that $g = h$ on E and $\|g\|_\infty \le (1 + \varepsilon)\|h\|$. With m_k chosen correctly,

$$g_0 = \sum 2^{-k} f_1^{m_k} g$$

then satisfies $\|g_0\| = \|h\|$ and $S(g_0) = h$ (see Bishop [1962] and Glicksberg [1962]).

★(f) (Gamelin [1964]) Let $p(z)$ be any positive continuous function on $\overline{D}$. If $h \in C(E)$ and if $|h(z)| \le p(z), z \in E$, then there exists $g \in A_o$ such that $g = h$ on E and $|g(z)| \le p(z), z \in D$.

2. We shall later make use of the following application of 1(b) and 1(c) above. Let μ be a finite complex Borel measure on T. Assume $d\mu$ is singular to $d\theta$. Then there are analytic polynomials $p_n(z)$ such that

(i) $|p_n| \le 1, |z| \le 1$,
(ii) $p_n d\mu \to \|\mu\|$, and
(iii) $p_n \to 0$ almost everywhere $d\theta$.

Condition (iii) can also be replaced by $p_n \to -1$ almost everywhere $d\theta$. (Hint: There are E_n compact, $E_n \subset E_{n+1}, |E_n| = 0$ such that μ is supported on $\bigcup E_n$. Let $h_n \in C(E_n), \|h_n\| < 1, \int_{E_n} h_n d\mu \to \|\mu\|$; let $g_n \in A_o$ interpolate $h_n, \|g_n\| < 1$; and let p_n be a polynomial approximation of g_n.)

★★**3.** Let A^m denote the algebra of functions f such that f and its first m derivatives belong to A_o, and let $A^\infty = \cap A^m$. Let $A_\alpha, 0 < \alpha \le 1$ be the space of functions $f \in A_o$ such that

$$|f(z) - f(w)| \le K|z - w|^\alpha.$$

(a)   Let E be a closed subset of the unit circle having measure zero, and let $\{l_j\}$ be the lengths of the arcs complementary to E. Necessary and sufficient for E to be the zero set of a function in A^∞ or A_α is the condition $\Sigma_j l_j \log(1/l_j) < \infty$ (see Carleson [1952]).

(b)   On the other hand, there is $f \in A^m$ such that $f(z) = 1, z \in E$, and $|f(z)| < 1, z \in \overline{D}\backslash E$, if and only if E is a finite set (see Taylor and Williams [1970]; the papers of Alexander, Taylor, and Williams [1971] and Taylor and Williams [1971] have further information on related questions).

4.   Theorem 1.3 is sharp in the following sense. Let $\omega(\delta)$ be increasing and continuous on $[0, 2\pi]$, $\omega(\delta_1 + \delta_2) \leq \omega(\delta_1) + \omega(\delta_2)$, $\omega(0) = 0$. Suppose

$$\int_0^1 \frac{\omega(\delta)}{\delta} d\delta = \infty.$$

(a)   There exists $f \in C(T)$, f real, such that

$$\omega_f(\delta) \leq \omega(\delta), \quad \delta < \delta_0,$$

but such that $\tilde{f}$ is not continuous. (Let $f(t) = \omega(t), 0 < t < \delta_0, f(t) = 0, -1 < t < 0$.)

(b)   There is $g \in C(T)$, g real, such that

$$\omega_g(\delta) \geq \omega(\delta)$$

but such that $\tilde{g}$ is continuous. Here, in outline, is one approach. If K is the Cantor ternary set, then $\{x - y, x \in K, y \in k\} = [-1, 1]$. (Try drawing a picture of $K \times K$.) Let $h \in C(K)$ be real, $\omega_h(\delta) \geq \omega(\delta)$, and find $g \in H^\infty$ with continuous boundary values such that $g = h$ on $\{e^{i\theta} : \theta \in K\}$.

5.   If $f \in C(T)$ is Dini continuous, then $S_n f \to f$ uniformly. There is a sharper result: $S_n f \to f$ uniformly if $\omega(\delta) \log(1/\delta) \to 0 (\delta \to 0)$ (see Zygmund [1968]).

6.   If $1 < p < \infty$, and if $f \in L^p$, then

$$\|S_n f - f\|_p \to 0.$$

7.   Let $f \in L^1(T)$. If $\int |f| \log(2 + |f|) d\theta < \infty$, then $\tilde{f} \in L^1$. If $f \geq 0$ and $\tilde{f} \in L^1$, then

$$\int |f| \log(2 + |f|) d\theta < \infty.$$

The first assertion is due to Zygmund [1929], the second is due to M. Riesz; see Zygmund [1968].

8.   Prove the two Kolmogoroff estimates

$$|\{|\tilde{f}| > \lambda\}| \leq (A/\lambda)\|f\|_1 \quad \text{and} \quad \|\tilde{f}\|_p \leq A_p\|f\|_1, \quad 0 < p < 1.$$

The first is valid on both the line and the circle, the second on the circle.

9. Let E be a subset of $\mathbb{R}$ of finite measure and define

$$H\chi_E(x) = \lim_{\varepsilon \to 0} \frac{1}{\pi} \int_{|x-t|>\varepsilon} \frac{\chi_E(t)}{x-t}\, dt.$$

Then the distribution of $H\chi_E$ depends only on $|E|$. More precisely,

$$|\{|H\chi_E(x)| > \lambda\}| = \frac{2|E|}{\sinh \pi\lambda}$$

(see Stein and Weiss [1959]).

10. (More jump theorem). Let $f \in L^1(T)$ and define

$$F(z) = \frac{1}{2\pi i} \int_{|\zeta|=1} \frac{f(\zeta)d\zeta}{\zeta - z}, \quad |z| \neq 1.$$

From inside D, $F(z)$ has nontangential limit called $f_1(\zeta)$ at almost every $\zeta \in T$, and from $|z| > 1$, $F(z)$ has nontangential limit $f_2(\zeta)$ at almost every $\zeta \in T$. Moreover,

$$f_1(\zeta) - f_2(\zeta) = f(\zeta)$$

almost everywhere.

★11. 'Let $f \in L^p(\mathbb{R})$, $1 \leq p < \infty$. Write

$$H_\varepsilon f(x) = \frac{1}{\pi} \int_{|x-t|>\varepsilon} \frac{f(t)}{x-t}\, dt, \qquad \varepsilon > 0.$$

and

$$H^* f(x) = \sup_{\varepsilon>0} |H_\varepsilon f(x)|.$$

$H^* f$ is the *maximal Hilbert transform*.
  (a) Show

$$H^* f \leq c_1 M(Hf) + c_2 Mf,$$

where M is the maximal function, and $Hf = \lim_{\varepsilon \to 0} H_\varepsilon f$. Then conclude by means of (1.10) that $\|H^* f\|_2 \leq c\|f\|_2$.
  (b) Assuming that H^* is weak-type 1–1, show by interpolation that $\|H^* f\|_p \leq c_p \|f\|_p$, $1 < p < 2$.
  (c) With f fixed, there is a measurable function $\varepsilon(x)$, $0 < \varepsilon(x) < \infty$, such that

$$|H_{\varepsilon(x)} f(x)| \geq \tfrac{1}{2} H^* f(x).$$

(This process is called *linearization.*) Suppose $\varepsilon(x)$ is a simple function $\sum_{j=1}^{N} \varepsilon_j \chi_{E_j}$, $E_j \cap E_\ell = \varnothing$, then for $f \in L^p$, $g \in L^q$, $2 < p < \infty$,

$$\begin{aligned}
&\left| \int H_{\varepsilon(x)} f(x) g(x) d(x) \right| \\
&\quad = \left| \int g(x) \sum_j \chi_{E_j}(x) \int_{|t-x| \geq \varepsilon_j} \frac{f(t)}{x-t} dt dx \right| \\
&\quad \leq \left| \sum_j \int f(t) \int_{|t-x| \geq \varepsilon_j} \frac{\chi_{E_j}(x) g(x)}{t-x} dx dt \right| \\
&\quad \leq \|f\|_p \| \sum_j H^*(\chi_{E_j} g) \|_q .
\end{aligned}$$

(d) From (b) and (c) it is almost trivial that $Hf = \lim_{\varepsilon \to 0} H_\varepsilon f$ satisfies

$$\|Hf\|_p \leq C_p \|f\|_p .$$

(e) We turn to the weak-type estimate

$$|\{x : H^* f(x) > 3\lambda\}| \leq (C/\lambda) \|f\|_1 . \tag{E.1}$$

Let $\Omega = \{x : Mf(x) > \lambda\}$ and let $F = \mathbb{R} \backslash \Omega$. Write $\Omega = \bigcup_{j=1}^{\infty} I_j$ where the closed intervals I_j satisfy

$$\text{dist}\,(F, I_j) = |I_j| \qquad \text{(see Figure III.2).}$$

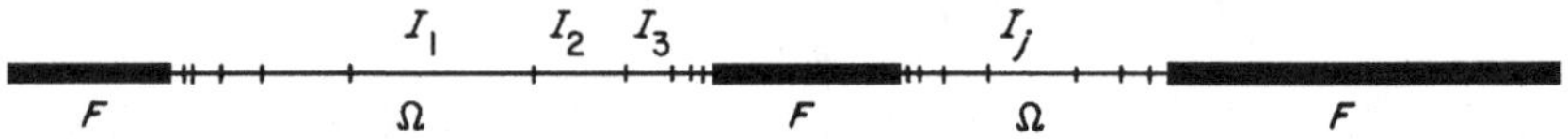

Figure III.2. The decomposition $\Omega = \bigcup I_j$.

Notice that

$$\frac{1}{|I_j|} \int_{I_j} |f(t)|\, dt \leq 2\lambda$$

because I_j is contained in an interval twice as large that touches F. Let

$$g(x) = f(x)\chi_F(x) + \sum_j \left(\frac{1}{|I_j|} \int_{I_j} f(t)\, dt \right) \chi_{I_j}(x)$$

and

$$b(x) = f(x) - g(x) = \sum_j \left(f(x) - \frac{1}{|I_j|} \int_{I_j} f(t)\, dt \right) \chi_{I_j}(x).$$

Real analysts call $g(x)$ the good function, $b(x)$ the bad function. The reason $g(x)$ is good is that $\|g\|_2$ is not too large,

$$\|g\|_2^2 = \int_F |f|^2 dx + \sum_j \left(\frac{1}{|I_j|} \int_{I_j} f(t)\,dt \right)^2 |I_j|$$

$$\leq \lambda \int_F |f|\,dx + 2\lambda \sum_j \int_{I_j} |f(t)|\,dt \leq 2\lambda \|f\|_1.$$

Therefore the L^2 estimate from part (a) yields

$$|\{x : H^* g(x) > \lambda\}| \leq \frac{C}{\lambda^2} \|g\|_2^2 \leq \frac{2C}{\lambda} \|f\|_1$$

and (E.1) will follow if we can show

$$|\{x : H^* b(x) > 2\lambda\}| \leq \frac{C}{\lambda} \|f\|_1.$$

By the maximal theorem, $|\Omega| \leq (C/\lambda)\|f\|_1$, so that we only have to prove

(E.2) $$|\{x \in F : H^* b(x) > 2\lambda\}| \leq \frac{C}{\lambda} \|f\|_1.$$

Write the bad function $b(x)$ as $\sum b_j(x)$, where

$$b_j(x) = \left(f(x) - \frac{1}{|I_j|} \int_{I_j} f(t)\,dt \right) \chi_{I_j}(x),$$

and note that $b_j(x)$ has the good cancellation property $\int b_j(x)\,dx = 0$. Fix $x \in F$ and let ε. Then

$$H_\varepsilon b(x) = \sum_j \int_{|x-t|>\varepsilon} \frac{b_j(t)}{x-t}\,dt$$

$$= \sum_{\text{dist}(x,I_j)>\varepsilon} \int_{I_j} \frac{b_j(t)}{x-t}\,dt + \sum_{\text{dist}(x,I_j)\leq\varepsilon} \int_{|x-t|>\varepsilon} \frac{b_j(t)}{x-t}\,dt.$$

$$= A_\varepsilon(x) + B_\varepsilon(x).$$

Let t_j be the center of I_j. When dist $(x, I_j) > \varepsilon$, we have

$$\int_{I_j} \frac{b_j(t)}{x-t}\,dt = \int_{I_j} b_j(t) \left(\frac{1}{x-t} - \frac{1}{x-t_j} \right) dt$$

$$= \int_{I_j} \frac{b_j(t)(t-t_j)}{(x-t)(x-t_j)}\,dt$$

since $\int b_j(t)\, dt = 0$. For $x \in F$ and $t \in I_j$, $|x - t|/|x - t_j|$ is bounded above and below, so that

$$\left| \int_{I_j} \frac{b_j(t)}{x - t}\, dt \right| \le c|I_j| \int_{I_j} \frac{|b_j(t)|}{|x - t|^2}\, dt.$$

Therefore

$$\sup_{\varepsilon > 0} |A_\varepsilon(x)| \le A(x) = \sum_j c|I_j| \int_{I_j} \frac{|b_j(t)|}{|x - t|^2}\, dt.$$

The trick here was to use the cancellation of $b_j(t)$ to replace the first-order singularity by a second-order singularity. Now we have

$$\int_F A(x)dx \le C \sum_j c|I_j| \int_{I_j} |b_j(t)| \int_{|I_j|}^\infty \frac{ds}{s^2}\, dt$$

$$\le C \sum \|b_j\|_1 \le 2C\|f\|_1$$

and

$$|\{x \in F : \sup |A_\varepsilon(x)| > \lambda\}| \le \frac{C}{\lambda} \|f\|_1.$$

When dist $(x, I_j) \le \varepsilon$, we have $|I_j| = \mathrm{dist}(F, I_j) \le \varepsilon$, so that

$$|B_\varepsilon(x)| \le \frac{c}{\varepsilon} \int_{\varepsilon < |x-1| < 2\varepsilon} |\sum b_j(t)|\, dt \le cMb(x).$$

Since $\|b\|_1 \le 2\|f\|_1$, this yields

$$|\{x \in F : \sup_{\varepsilon > 0} |B_\varepsilon(x)| > \lambda\}| \le \frac{C}{\lambda} \|f\|_1.$$

Now (E.2) follows from the weak-type estimates we have established for $\sup_\varepsilon |A_\varepsilon(x)|$ and $\sup_\varepsilon |B_\varepsilon(x)|$ (see Calderón and Zygmund [1952] and Stein [1970]).

IV

Some Extremal Problems

We begin with the basic duality relation,

$$(0.1) \qquad \inf_{g \in H^p} \| f - g \|_p = \sup \left\{ \left| \int F f \frac{d\theta}{2\pi} \right| : F \in H_0^q, \| F \|_q = 1 \right\}.$$

This relation is derived from the Hahn–Banach theorem in Section 1. Then, rather than continuing with a general theory, we use (0.1) to study three important and nontrivial problems.

1. Determining when a continuous function on the circle T has continuous best approximation in H^∞. This problem is discussed in Section 2.

2. Characterizing the positive measures μ on T for which

$$\int | \tilde{p} |^2 \, d\mu \le K \int | p |^2 \, d\mu$$

for all trigonometric polynomials $p(\theta)$. This topic will reappear in Chapter VI, where the main result of Section 3 of this chapter will yield information about the real parts of H^∞ functions.

3. Solving the interpolation problem

$$(0.2) \qquad\qquad f(z_j) = w_j, \quad j = 1, 2, \ldots,$$

$f \in H^\infty$, with the additional restriction that $| f(e^{i\theta}) |$ be constant or that $\| f \|_\infty$ be as small as possible. This problem is treated in some detail, because Theorem 4.1 below will have important and striking applications later and because the rather precise results on this topic require ideas somewhat deeper than the duality relation (0.1).

To compare the duality approach to a more classical method, we conclude the chapter with Nevanlinna's beautiful solution of (0.2).

1. Dual Extremal Problems

Let X be a Banach space with dual space X^* and let Y be a closed subspace of X. Then

$$Y^\perp = \{x^* \in X^* : \langle y, x^* \rangle = 0 \text{ for all } y \in Y\}$$

is a closed subspace of X^*. The Hahn–Banach theorem gives us the isometric isomorphisms

(1.1) $$Y^* \cong X^*/Y^\perp,$$
(1.2) $$(X/Y)^* \cong Y^\perp.$$

These isometries can be rewritten as two equalities.

Lemma 1.1. *If $x^* \in X^*$, then*

(1.1′) $$\sup\{|\langle x^*, y\rangle| : y \in Y, \|y\| \leq 1\} = \inf\{\|x^* - k\| : k \in Y^\perp\}.$$

If $x \in X$, then

(1.2′) $$\inf\{\|x - y\| : y \in Y\} = \sup\{|\langle x, k\rangle| : k \in Y^\perp, \|k\| \leq 1\}.$$

Proof. The left side of (1.1′) is the norm of the restriction of x^* to the subspace Y and the right-hand side of (1.1′) is the norm of the coset $x^* + Y^\perp$ in $X^*/Y^\perp$. By (1.1) these quantities are equal. In the same way (1.2) implies (1.2′). □

On the circle T we have the Banach spaces $C \subset L^\infty \subset L^p \subset L^1$, where $C = C(T)$ is the space of continuous functions on T and where the L^p spaces are with respect to $d\theta/2\pi$. We are interested in the subspaces $A_o \subset H^\infty \subset H^p \subset H^1$, where A_o is the disc algebra $A_o = C \cap H^\infty$. We then have the accompanying table, in which $q = p/(p-1)$. In the table H_0^q denotes $\{g \in H^q : g(0) = 0\}$, $M(T)$ is the space of complex Borel measures on T with total variation norm, and H_0^1 identified with the closed subspace of $M(T)$ consisting of those absolutely continuous measures $F d\theta/2\pi$ having density F in H_0^1. The two blanks in the L^∞ row will be filled in Chapter V, but the spaces filling the blanks will not be too useful for our purposes. To obtain the last column of the table observe that if $F \in L^q, q > 1$, and if $\int e^{in\theta} F(\theta)\, d\theta = 0$ for $n = 0, 1, 2, \ldots$, then $F \in H_0^q$. The characterization of $A_o^\perp$ as H_0^1 is the disc version of the F. and M. Riesz theorem (Theorem II.3.8).

X	Y	X^*	$Y^\perp$
C	A_o	$M(T)$	H_0^1
L^∞	H^∞	—	—
$L^p, p < \infty$	H^p	L^q	H_0^q
L^1	H_0^1	L^∞	H^∞

Theorem 1.2. *Let $1 \le p < \infty$, and let $f \in L^p, \notin H^p$. Then the distance from f to H^p is*

(1.3)
$$\operatorname{dist}(f, H^p) = \inf_{g \in H^p} \| f - g \|_p$$
$$= \sup \left\{ \left| \int f F \frac{d\theta}{2\pi} \right| : F \in H_0^q, \| F \|_q \le 1 \right\}.$$

There exists unique $g \in H^p$ such that $\operatorname{dist}(f, H^p) = \| f - g \|_p$ and there exists unique $F \in H_0^q, \| F \|_q = 1$ such that

(1.4)
$$\int f F \frac{d\theta}{2\pi} = \operatorname{dist}(f, H^p).$$

Proof. The identity (1.3) follows from (1.2'). Let $g_n \in H^p$ be such that $\| f - g_n \|_p \to \operatorname{dist}(f, H^p)$. The Poisson integrals of $f - g_n$ are bounded on any compact subset of the disc, so by normal families there is an analytic function g on D such that $g_n(z) \to g(z), z \in D$, if we replace $\{g_n\}$ by a subsequence. Taking means over circles of radius $r < 1$, we see that $g \in H^p$ and that $\| f - g \|_p \le \lim \| f - g_n \|_p$. Thus $\| f - g \|_p = \operatorname{dist}(f, H^p)$ and there exists at least one best approximation $g \in H^p$.

Let $F_n \in H_0^q, \| F_n \|_q \le 1$, be such that $\int f F_n d\theta/2\pi \to \operatorname{dist}(f, H^p)$. Since $1 < q \le \infty$, the Banach–Alaoglu theorem can be used to obtain a weak-star limit point F of $\{F_n\}$. Then $\| F \|_q \le 1$, $F \in H_0^q$, and $\int f F d\theta/2\pi = \operatorname{dist}(f, H^p)$. Since $\int g F \, d\theta = 0$, we have

(1.5) $\operatorname{dist}(f, H^p) = \displaystyle\int (f - g) F \, d\theta/2\pi \le \| f - g \|_p \| F \|_q \le \| f - g \|_p.$

Equality must hold throughout this chain of inequalities. This means $\| F \|_q = 1$, and there exists a dual extremal function F for which (1.4) holds.

Now let $g \in H^p$ be any best approximation of f and let $F \in H_0^q$ be any dual extremal function. Because equality holds in (1.5), the conditions for equality in Hölder's inequality give us

$$\frac{f - g}{\| f - g \|_p} = \bar{F} |F|^{q-2} \quad \text{and} \quad F = \overline{(f - g)} \frac{|f - g|^{p-2}}{\| f - g \|_p^{p-1}}$$

when $p > 1$. When $p = 1$ we get

$$(f - g) F = |f - g|$$

instead.

Let $p > 1$. Since $|F| > 0$ almost everywhere, the first equation shows that g is unique. Since $F \in H_0^q$ is determined by its values on any set of positive measure, the second equation shows that F is unique.

Similarly, when $p = 1$, the third equation shows that F is unique. The third equation then shows that $\operatorname{Im}(gF)$ is unique. Since $gF \in H_0^1$, this determines gF uniquely, and then g is determined almost everywhere because $|F| > 0$ almost everywhere. $\square$

Note that the best approximation $g \in H^p$ and the dual extremal function $F \in H_0^q, \|F\|_q = 1$, are characterized by the relation

(1.6) $$\int (f - g)F \frac{d\theta}{2\pi} = \|f - g\|_p \|F\|_q = \|f - g\|_p,$$

which is (1.5) with equality. Sometimes it is possible to compute $\text{dist}(f, H^p)$ by finding solutions F and g of (1.6) (see Exercise 3).

When $p = \infty$ some of the conclusions of the theorem can be rescued if we use the bottom row of the table instead of the second row, and use $(1.1')$ instead of $(1.2')$.

Theorem 1.3. *If $f \in L^\infty$, then the distance from f to H^∞ is*

$$\text{dist}(f, H^\infty) = \inf_{g \in H^\infty} \|f - g\|_\infty = \sup \left\{ \left| \int fF \frac{d\theta}{2\pi} \right| : F \in H_0^1, \|F\|_1 \le 1 \right\}.$$

There exists $g \in H^\infty$ such that $\|f - g\|_\infty = \text{dist}(f, H^\infty)$. If there exists $F \in H_0^1, \|F\|_1 \le 1$, such that

$$\text{dist}(f, H^\infty) = \int fF \frac{d\theta}{2\pi},$$

then the best approximation $g \in H^\infty$ is unique and

$$|f - g| = \text{dist}(f, H^\infty)$$

almost everywhere.

Proof. The dual expression for the distance follows from $(1.1')$. Just as in the proof of Theorem 1.2, a normal families argument shows there is a best approximating function $g \in H^\infty$. If a dual extremal function $F \in H_0^1$ exists, then

$$\text{dist}(f, H^\infty) = \int (f - g)F \frac{d\theta}{2\pi} = \|f - g\|_\infty \|F\|_1,$$

so that

(1.7) $$\frac{(f - g)}{\text{dist}(f, H^\infty)} = \frac{(f - g)}{\|f - g\|_\infty} = \frac{\bar{F}}{|F|}$$

almost everywhere, and there is a unique best approximation g. $\square$

If a dual extremal function F exists, (1.7) does not imply that F is unique. But it does, of course, imply that the argument of F is unique.

Example 1.4. Let $f(0) = e^{-2i\theta}$. Taking $F_0(\theta) = e^{2i\theta}$ we see that

$$\frac{1}{2\pi} \int fF_0 d\theta = 1.$$

Hence dist$(f, H^\infty) = 1$ and F_0 is a dual extremal function. However

$$F_\alpha(z) = \frac{z(z + \alpha)(1 + \bar\alpha z)}{1 + |\alpha|^2}, \quad |\alpha| < 1,$$

is another dual extremal function. In this problem the best approximating function is $g = 0$.

Example 1.5. Let

$$f(\theta) = \begin{cases} 1, & 0 < \theta < \pi/2, \\ 0, & \pi/2 < \theta \le 3\pi/2, \\ -1, & 3\pi/2 < \theta \le 2\pi. \end{cases}$$

If there were $g \in H^\infty$ such that $\|f - g\|_\infty < 1$, then $\operatorname{Re} g > \delta > 0$ on $(0, \pi/2)$ and $\operatorname{Re} g < -\delta < 0$ on $(-\pi/2, 0)$. Hence

$$\lim_{\varepsilon \to 0} \frac{1}{\pi} \int_{|\theta| > \varepsilon} \frac{\operatorname{Re} g(\theta)}{\theta} d\theta = +\infty.$$

Except for bounded error terms, this integral represents $-\lim_{r \to 1} \operatorname{Im} g(r)$. Thus there is no such bounded g, and so dist$(f, H^\infty) = 1$ and $g = 0$ is one best approximation. Now let g be the conformal mapping of D onto the half disc $D \cap \{\operatorname{Im} z > 0\}$. We can arrange that $g(1) = 0$, $g(i) = 1$, and $g(-i) = -1$. Then by checking the values of g along the three arcs on which f is constant, we see that $\|f - g\|_\infty = 1$. Hence there is not a unique best approximation to $f \in H^\infty$. By Theorem 1.3 there is no dual extremal function $F \in H_0^1$.

It is interesting to notice where the proof of Theorem 1.2 breaks down in Example 1.5. By the bottom row of the table there are $F_n \in H_0^1$, $\|F_n\|_1 \le 1$, such that $\int f F_n d\theta/2\pi \to 1$. As linear functionals on L^∞, the F_n have some weak-star limit point $\sigma \in (L^\infty)^*$, and σ is orthogonal to H^∞. In Chapter V we shall see that σ is a complex measure on a compact Hausdorff space, the maximal ideal space of L^∞. However, σ is not weakly continuous on L^∞; that is, σ cannot be represented as $\sigma(h) = \int h F d\theta$ with $F \in L^1$. Otherwise we could conclude that $|F| = 1$ almost everywhere. In fact, in a sense to be made precise in Chapter V, σ is singular to $d\theta$, which means, in classical language, that $F_n(z) \to 0$, $z \in D$. The absence of a dual extremal function $F \in H_0^1$ is often the central difficulty with an H^∞ extremal problem. We confront this difficulty again in Section 4.

If the function $f \in L^\infty$ is continuous, there is a dual extremal function F and the best approximation g is unique.

Lemma 1.6. *If $f \in C$, then*

$$\operatorname{dist}(f, H^\infty) = \operatorname{dist}(f, A_o) = \inf_{g \in A_o} \|f - g\|_\infty.$$

Proof. There exists $g \in H^\infty$ such that $\|f - g\|_\infty = \operatorname{dist}(f, H^\infty)$. Let $f_r = f * P_r$ be the Poisson integral of f and let $g_r = g * P_r$. Since $\|P_r\|_1 = 1$ we have

$$\|f_r - g_r\|_\infty = \|(f - g) * P_r\|_\infty \leq \|f - g\|_\infty.$$

But $g_r \in A_o$, and $\|f - f_r\|_\infty < \varepsilon$ if $1 - r$ is small. Thus

$$\text{dist}(f, A_o) \leq \lim_r \|f - g_r\|_\infty \leq \|f - g\|_\infty = \text{dist}(f, H^\infty).$$

The reverse inequality is clear since $H^\infty \supset A_o$.   □

We write $H^\infty + C$ for the set of functions $g + h$, $g \in H^\infty, h \in C$.

Theorem 1.7.   *If $f \in H^\infty + C$, then there exists $F \in H_0^1, \|F\|_1 = 1$, such that*

(1.8)
$$\frac{1}{2\pi} \int f F \, d\theta = \text{dist}(f, H^\infty),$$

and there exists unique $g \in H^\infty$ such that $\|f - g\|_\infty = \text{dist}(f, H^\infty)$.

Proof.   Write $f = g + h$, $g \in H^\infty, h \in C$. Then $\text{dist}(f, H^\infty) = \text{dist}(h, H^\infty)$ and we can assume that f is continuous. By Theorem 1.3 there are $F_n \in H_0^1, \|F_n\|_1 \leq 1$, such that

$$\frac{1}{2\pi} \int f F_n d\theta \to \text{dist}(f, H^\infty).$$

Taking a subsequence we can assume $F_n(z) \to F(z), z \in D$, where $F \in H_0^1, \|F\|_1 \leq 1$. This gives convergence of the Fourier coefficients, so that for all trigonometric polynomials $p(\theta)$,

$$\int F_n p \frac{d\theta}{2\pi} \to \int F p \frac{d\theta}{2\pi}.$$

Taking $\|f - p\|_\infty$ small we see that

$$\text{dist}(f, H^\infty) = \int F f \frac{d\theta}{2\pi}.$$

Hence (1.8) holds and Theorem 1.3 now implies that f has a unique best approximation in H^∞.   □

As an application we reprove some results from Chapter I, Section 2.

Corollary 1.8.   *Let $z_1, z_2, \ldots, z_n$ be distinct points in D, and let $w_1, w_2, \ldots, w_n$ be complex numbers. Among all $f \in H^\infty$ such that*

(1.9)
$$f(z_j) = w_j, \quad 1 \leq j \leq n,$$

there is a unique function f of minimal norm. This function has the form $cB(z)$ where $B(z)$ is a Blaschke product of degree at most $n - 1$.

Corollary 1.9. *Let $z_{n+1} \in D$ be distinct from $z_1, z_2, \ldots, z_n$. Assume* (1.9) *has a solution $f \in H^\infty$ with $\|f\|_\infty \leq 1$. Among such solutions let f_0 he one for which $|f(z_{n+1})|$ is largest. Then f_0 is uniquely determined by its value $f_0(z_{n+1})$ and f_0 is a Blaschke product of degree at most n.*

Corollary 1.9 is a simple consequence of Corollary 1.8. By normal families there exists an extremal function f_0, $\|f_0\|_\infty \leq 1$. Let $w_{n+1} = f_0(z_{n+1})$. If $f \in H^\infty$ interpolates (1.9) and if also

$$f(z_{n+1}) = w_{n+1},$$

then $\|f\|_\infty \geq 1$, because otherwise for some λ, $|\lambda|$ small,

$$g = f + \lambda \prod_{j=1}^{n} \frac{z - z_j}{1 - \bar{z}_j z}$$

is a function satisfying (1.9) such that $\|g\| \leq 1$ and $|g(z_{n+1})| > |f_0(z_{n+1})|$. Hence f_0 has minimal norm among the functions interpolating $w_1, \ldots, w_{n+1}$ at $z_1, \ldots, z_{n+1}$.

Proof of Corollary 1.8. Let B_0 be the Blaschke product with zeros $z_1, \ldots z_n$, and let f_0 be a polynomial that does the interpolation (1.9). The minimal norm of the functions in H^∞ satisfying (1.9) is

$$\inf_{g \in H^\infty} \|f_0 - B_0 g\|_\infty = \inf_{g \in H^\infty} \|\bar{B}_0 f_0 - g\|_\infty.$$

Since $\bar{B}_0 f_0 \in C$, there is a unique interpolating function $f \in H^\infty$ of minimal norm and there is $F \in H_0^1$, $\|F\|_1 = 1$, such that

$$\int f \bar{B}_0 F \frac{d\theta}{2\pi} = \|f\|_\infty.$$

Hence $|f| = |f \bar{B}_0| = \|f\|_\infty$ almost everywhere, and

(1.10) $$f F / B_0 \geq 0$$

almost everywhere.

Lemma 1.10. *If $G \in H^1$ is real almost everywhere on an arc $I \subset T$, then G extends analytically across I.*

Proof. On $|z| > 1$ define $G(z) = \overline{G(1/z)}$. This is an H^1 function on $|z| > 1$ with nontangential limits $G(\theta)$ at almost every point of I. Let $\zeta \in I$ and center at ζ a disc Δ so small that $\Delta \cap T \subset I$. Let $V = \Delta \cap D$, $W = \Delta \cap \{|z| > 1\}$. Since we are dealing with H^1 functions, we have for $w \in \Delta \backslash T$,

$$\frac{1}{2\pi i} \int_{\partial \Delta} \frac{G(z)}{z - w} dz = \frac{1}{2\pi i} \int_{\partial V} \frac{G(z)}{z - w} dz + \frac{1}{2\pi i} \int_{\partial W} \frac{G(z)}{z - w} dz = G(w).$$

This shows that G can be continued across I. □

To complete the proof of Corollary 1.8, notice that $G = fF/B_0$ is an H^1 function on the annulus $r < |z| < 1$ if $r > |z_j|$. Using (1.10) and using the lemma locally, we see that G is analytic across T, and that G is in fact a rational function. Since B_0G is analytic across T, and since $f/\|f\|_\infty$ is an inner function, Theorem II.6.3 shows that f is analytic across T. Consequently $f/\|f\|_\infty$ is a Blaschke product of finite degree, and F is a rational function. Now B_0 has n zeros in D and F has a zero at $z = 0$. By (1.10) and the argument principle, it follows that f has at most $n - 1$ zeros. $\square$

In Corollary 1.8 the extremal function $f(z) = cB(z)$ can have fewer than $n - 1$ zeros. For example, suppose g is the function of minimum norm interpolating

$$g(z_j) = w_j, \quad 1 \le j \le n - 1,$$

and take $w_n = g(z_n)$. Then $f = g$ has at most $n - 2$ zeros. Similarly, the extremal function in Corollary 1.9 can have fewer than n zeros. However, if the interpolation (1.9) has two distinct solutions f_1, f_2 with $\|f_1\|_\infty \le 1$, $\|f_2\|_\infty \le 1$, then the extremal function f_0 in Corollary 1.9 is a Blaschke product of degree n. For the proof, notice that (1.9) then has a solution f with $\|f\|_\infty < 1$, by the uniqueness asserted in Corollary 1.8. Then by Rouché's theorem, f_0 and $f_0 - f$ have the same number of zeros in $|z| < 1$. Since $f_0(z_j) = f(z_j)$, $1 \le j \le n$, it follows that f_0 has at least n zeros in $|z| < 1$.

Corollary 2.4 Chapter I, contains more information than we have obtained here, but the duality methods of this section apply to a wider range of problems.

2. The Carleson–Jacobs Theorem

Let $f(\theta) \in C$. From Theorem 1.7 we know there is a unique function $g \in H^\infty$ such that

$$\|f - g\|_\infty = \text{dist}(f, H^\infty).$$

We want to know when the best approximation $g(\theta)$ is continuous on T. Although necessary and sufficient conditions on $f(\theta)$ are not known, there is a sharp result parallel to Theorem 1.3 of Chapter III about the continuity of conjugate functions. Recall that the modulus of continuity of $f(\theta)$ is

$$\omega(\delta) = \omega_f(\delta) = \sup\{|f(\theta) - f(\varphi)| : |\theta - \varphi| < \delta\}.$$

Theorem 2.1. *If $f(\theta)$ is Dini continuous, that is, if, for some $\varepsilon > 0$*

$$\int_0^t \frac{\omega(t)}{t} dt < \infty,$$

then the best approximation $g \in H^\infty$ to f is continuous on T.

Theorem 2.2. *Let $\omega(t)$ be a continuous nondecreasing function such that $\omega(0) = 0$ and $\omega(t_1 + t_2) \le \omega(t_1) + \omega(t_2)$. If $\int_0 \omega(t)/t \, dt = \infty$, then there is*

$f(\theta) \in C$ such that $\omega_f(\delta) \le \omega(\delta)$ but such that the best approximation to f in H^∞ is not continuous.

We prove Theorem 2.1 first. In doing so we can assume

$$\|f - g\|_\infty = \text{dist}(f, H^\infty) = 1.$$

By Theorem 1.7, there is $F \in H_0^1$, $\|F\|_1 = 1$, such that

(2.1) $(f - g)F = |F|$ a.e

All information used in the proof will follow from (2.1) and the continuity of f. We need two lemmas.

Lemma 2.3.   Let $G = u + iv \in H^1$ and let I be an arc on T such that almost everywhere on I

$$u > 0, \quad |v| \le \alpha u,$$

where $\alpha > 0$. Let J be a relatively compact subarc of I and let V be the domain $\{re^{i\theta} : r_0 < r < 1, 0 \in J\}$. Then $G \in H^p(V)$ if

$$\arctan \alpha < \pi/2p.$$

The statement $G \in H^p(V)$ means that $|G|^p$ has a harmonic majorant in V.

Proof.   Recall from Corollary 2.5 in Chapter III that an H^1 function whose boundary values lie in the sector $S = \{x > 0, |y| < \alpha x\}$ is in H^p if $\arctan \alpha < \pi/2p$. Enlarging J, we may assume G has a finite radial limit at the endpoints of J. Then $M = \sup\{|G(z)| : z \in \partial V, |z| < 1\}$ is finite, and

$$g = M(1 + 1/\alpha) + G$$

has, at almost every point of ∂V, boundary value in the cone S (see Figure IV.1).

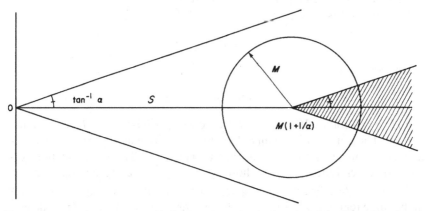

Figure IV.1. $S = \{|y| < \alpha x, x > 0\}$. On ∂V, g has values in the union of the disc and the shaded cone.

Because $g \in H^1(V)$, the Poisson integral formula (applied after conformally mapping V onto the unit disc) shows that $g(V) \subset S$. By Corollary III.2.5 and the conformal invariance of the definition of $H^p(V)$, we see that $g \in H^p(V)$ and hence that $G \in H^p(V)$. $\square$

Lemma 2.4. *Let $f(\theta) \in C$, let $g(\theta) \in H^\infty$ and let $F(\theta) \in H_0^1$ be functions such that (2.1) holds. Then*

    (a) *$F \in H^p$ for all $p < \infty$, and*
    (b) *if $\tau \in [0, 2\pi]$ and if*

$$f_\tau = f - f(\tau), \quad g_\tau = g - f(\tau),$$

then there is $\delta > 0$ and $r_0 > 0$ such that $|g_r(z)| \geq \frac{1}{2}$ on

$$W_\tau = \{re^{i\theta} : |\theta - \tau| < \delta/2, r_0 < r < 1\}.$$

Proof. To prove (a) let $p < \infty$ and let $\varepsilon > 0$ satisfy $\arctan(\varepsilon/(1 - \varepsilon)) < \pi/2p$. Choose δ so that $|f(\theta) - f(\tau)| < \varepsilon$ when $|\theta - \tau| < \delta$. Then on $I_\tau = \{\theta : |\theta - \tau| < \delta\}$, we have by (2.1)

$$-g_\tau F = (f - g)F - f_\tau F = |F| - f_\tau F.$$

Consequently, $\mathrm{Re}(-g_\tau F) > 0$ and $|\mathrm{Im}\, g_\tau F| < \varepsilon/(1 - \varepsilon)\, \mathrm{Re}(-g_\tau F)$ almost everywhere on I_τ. By Lemma 2.3 we have $g_\tau F \in H^p(W_\tau)$. If we replace W_τ by the intersection of two discs whose boundaries cross inside I_τ, a simple conformal mapping can be used to show

$$\int_{|\theta - \tau| < \delta/2} |g_\tau F|^p d\theta < \infty.$$

Since $|g_\tau| \geq \|f - g\|_\infty - \omega(\delta) \geq 1 - \varepsilon$ on I_τ, this means $\int_{|\theta - \tau| < \delta/2} |F|^p d\theta < \infty$, so that $F \in L^p$. Hence $F \in H^p$ by Corollary 11.4.3.

To prove (b) first take ε so small that on I_τ

$$-g_\tau F = \exp(u + iv),$$

where $\|v\|_\infty < \pi/4$. then

$$h = \exp(-iv\chi_{I_\tau} + \widetilde{v\chi_{I_\tau}})$$

is in H^2 by Corollary III.2.6. The H^1 function $g_\tau Fh$ is real on I_τ. By Lemma 1.10 and Theorem II.6.3, the inner factor of $g_\tau Fh$ is analytic across I_τ. Consequently, the inner factor of g_τ is analytic across I_τ. The representation formula for outer functions then shows $|g_\tau(z)| \geq \frac{3}{4}$ in a region $\{r_0(\tau) < r < 1, |\theta - \tau| < \delta/2\}$. Since $g_\sigma(z) = g_\tau(z) - (f(\sigma) - f(\tau))$, we have $|g_\sigma(z)| > \frac{1}{2}$ on the same region if $|\sigma - \tau| < \delta$. This means we may choose $r_0(\tau)$ independent of τ, and thus (b) is proved. $\square$

Proof of Theorem 2.1. By Lemma 2.4(b), $g_\tau(z)$ has a single-valued logarithm on W_τ, and the logarithm is defined by

$$(2.2) \qquad \log g_\tau(z) = \frac{1}{2\pi} \int_{|\theta - \tau| < \delta} \log |g_\tau(\theta)| \frac{e^{i\theta} + z}{e^{i\theta} - z} \, d\theta + R_\tau(z),$$

where $R_\tau(z)$ is the same integral over $|\theta - \tau| \geq \delta$ plus the logarithm of the inner factor of $g_\tau(z)$. Since $|g_\tau(z)| \geq \frac{1}{2}$ on W_τ, the inner factor is bounded below on W_τ and it is analytic across $\{e^{i\theta} : |\theta - \tau| < \delta/2\}$. Hence there is $r_1 > 0$ such that $R_\tau(z)$ is bounded and analytic on $\Delta_\tau = \{|z - e^{i\tau}| < r_1\}$. The radius r_1 and the bound sup $\{|R_\tau(z)| : z \in \Delta_\tau\}$ are independent of τ.

Because $|f_\tau - g_\tau| = 1$, we have

$$1 - |g_\tau|^2 = |f_\tau|^2 - 2 \operatorname{Re} f_\tau \bar{g}_\tau \leq A\omega(|\theta - \tau|),$$

so that

$$|\log |g_\tau(\theta)|| \leq c\omega(|\theta - \tau|).$$

Let $\Gamma(\tau)$ be the truncated cone

$$\Gamma(\tau) = \left\{ z : \frac{|z - e^{i\tau}|}{1 - |z|} \leq 2, r_2 < |z| < 1 \right\},$$

where $r_2 > 0$ is such that $\overline{\Gamma(\tau)} \subset \Delta_\tau$. For $z \in \Gamma_\tau$,

$$\left| \frac{e^{i\theta} + z}{e^{i\theta} - z} \right| \leq \frac{C}{|\theta - \tau|},$$

so that (2.2) yields

$$|\log g_\tau(z) - R_\tau(z)| \leq C \int_0^\delta \frac{\omega(t)}{t} dt, \quad z \in \Gamma(\tau).$$

By Schwarz's lemma and the uniform bound on $R_\tau(z)$, $|R_\tau(z) - R_\tau(w)| \leq c|z - w|, z, w \in \Gamma(\tau)$. Therefore

$$|g_\tau(z) - g_\tau(w)| \leq C|z - w| + \eta(\delta),$$

$z, w \in \Gamma_\tau$, where $\eta(\delta) \to 0 (\delta \to 0)$.

Now let τ and σ be so close together that there is a point $z \in \Gamma(\tau) \cap \Gamma(\sigma)$. Taking nontangential limits, we then obtain (when $1 - |z|$ is small)

$$\begin{aligned}
|g(e^{i\tau}) - g(e^{i\sigma})| &\leq |g(e^{i\tau}) - g(z)| + |g(z) - g(e^{i\sigma})| \\
&\leq |g_\tau(e^{i\tau}) - g_\tau(z)| + |g_\sigma(z) - g_\sigma(e^{i\sigma})| \\
&\leq 2\eta(\delta) + |e^{i\tau} - z| + |e^{i\sigma} - z| \\
&\leq 2\eta(\delta) + C|\sigma - \tau|.
\end{aligned}$$

Fixing $\delta > 0$ with $2\eta(\delta) < \varepsilon$, we conclude that

$$\overline{\lim_{\sigma \to \tau}} |g(e^{i\sigma}) - g(e^{i\tau})| < \varepsilon,$$

which means that g is continuous. $\quad\square$

Now let $\omega(t)$ satisfy the hypotheses of Theorem 2.2.

Lemma 2.5. *Let $\delta > 0$. Let*

$$f(e^{it}) = \begin{cases} \omega(t), & 0 \le t \le \delta, \\ 0, & -\delta \le t < 0. \end{cases}$$

Extend $f(e^{it})$ to be smooth on $\delta < |t| < \pi$ and continuous on $[-\pi, \pi]$ with $f(-\pi) = f(\pi)$. Let $g \in H^\infty$ be the best approximation of f. If g is continuous, then

$$g(1) = \pm i \|f - g\|_\infty.$$

Proof. Since f is real, $f \notin H^\infty$ and $\|f - g\|_\infty > 0$. We may suppose $\|f - g\|_\infty = 1$, so that $|f - g| = 1$ on T. We must prove $\operatorname{Re} g(1) = 0$. Since $|f - g| = 1$, $\log|g(t)| = 0$ on $-\delta < t < 0$. Suppose $\operatorname{Re} g(1) > 0$. Then for $0 < t < \delta$,

$$\log|g| = -\log\left|\frac{g-f}{g}\right| \le -\log\left(1 - \frac{f \operatorname{Re} g}{|g|^2}\right) \le -C\omega(t).$$

Similarly, if $\operatorname{Re} g(1) < 0$, then on $0 < t < \delta$, $\log|g| \ge C\omega(t)$. Moreover, g has a continuous logarithm on $\{|z - 1| < \delta, |z| \le 1\}$, again because $|f - g| = 1$. However, if $\operatorname{Re} g(1) \ne 0$, then by Lemma III.1.2,

$$\lim_{r \to 1} |\arg g(r)| = \left|c + \lim_{\varepsilon \to 0} \frac{1}{\pi} \int_{|t| > \varepsilon} \frac{\log|g(t)|}{t} dt\right|$$

$$\le c + C \lim_{\varepsilon \to 0} \frac{1}{\pi} \int_{|t| > \varepsilon} \frac{\omega(t)}{t} dt = +\infty.$$

This contradiction shows $g(1) = \pm i$. $\quad\square$

To prove Theorem 2.2 let $\delta_n = 2^{-n}$, and let

$$\omega_n(t) = \omega\left(\frac{\delta_n}{2} - \left|\frac{\delta_n + \delta_{n-1}}{2} - t\right|\right)$$

on $[\delta_n, \delta_{n-1}]$, $\omega_n(t) = 0$ off $[\delta_n, \delta_{n-1}]$. Let

$$f(e^{it}) = \sum_{k=1}^\infty \frac{1}{2}\omega_{4k+1}(t) + i\sum_{k=1}^\infty \frac{1}{2}\omega_{4k+3}(t).$$

Then f^2 is real, so that $f \notin H^\infty$. It is not hard to verify that $\omega_f(\delta) \leq \omega(\delta)$. If g is the best approximation of f, then Lemma 2.5 says that

$$g(e^{i\delta_{4k+1}}) = \pm i \| f - g \|_\infty, \quad g(e^{i\delta_{4k+3}}) = \pm \| f - g \|_\infty.$$

Thus $g(e^{it})$ is not continuous at $t = 0$. $\quad\square$

3. The Helson–Szegö Theorem

Let $p(\theta)$ be a trigonometric polynomial on the circle T and let $\tilde{p}(\theta)$ be its conjugate function. Then $\tilde{p}$ is another trigonometric polynomial, normalized to have mean zero. In this section we give the Helson–Szegö characterization of those positive measures μ on T for which

$$\int |\tilde{p}(\theta)|^2 \, d\mu(\theta) \leq K \int |p(\theta)|^2 \, d\mu(\theta)$$

for every trigonometric polynomial $p(\theta)$. This of course means that conjugation extends to a bounded operator on $L^2(\mu)$. In Chapter VI a completely different characterization of such measures μ will be given, and these two results will be merged to provide a description of the uniform closure in L^∞ of the space of real parts of H^∞ functions.

First we need the famous and beautiful theorem of Szegö. Let F be the set of polynomials in z vanishing at $z = 0$. Restricted to the circle, F coincides with the set of trigonometric polynomials of the form $\sum_{n>0} a_n e^{in\theta}$.

Theorem 3.1 (Szegö). *Let $d\mu$ be a finite positive measure on the circle. Write*

$$d\mu = w \, d\theta/2\pi + d\mu_s,$$

where $d\mu_s$ is singular to $d\theta$. Then

(3.1) $$\inf_{f \in F} \int |1 - f|^2 d\mu = \exp \frac{1}{2\pi} \int \log w \, d\theta.$$

Notice that if the infimum in (3.1) is zero, then F is dense in $L^2(\mu)$, because by induction $e^{-i\theta}, e^{-2i\theta}, \ldots$ also then lie in the closure of F .

Proof. We first dispense with the singular part μ_s. By Exercise 2, Chapter III, there are polynomials p_n in F such that $\| p_n \|_\infty \leq 1$, $p_n \to 1$ a.e. $d\mu_s$ and $p_n \to 0$ a.e. $d\theta$. For any $f_0 \in F$ we have $f_0 + p_n(1 - f_0) \in F$, so that

$$\inf_F \int |1 - f|^2 d\mu \leq \lim_n \int |1 - (f_0 + p_n(1 - f_0))|^2 d\mu$$

$$= \lim_n \int |1 - f_0|^2 |1 - p_n|^2 d\mu$$

$$= \int |1 - f_0|^2 w \frac{d\theta}{2\pi}.$$

Since $d\mu \geq w\, d\theta/2\pi$, the reverse inequality is trivial, so that

$$\inf_F \int |1 - f|^2 d\mu = \inf_F \int |1 - f|^2 w \frac{d\theta}{2\pi},$$

and we can assume μ is absolutely continuous.

Write $\varphi = \log w$. Assume for the moment that $\int \varphi\, d\theta > -\infty$. Let $\psi = \tilde{\varphi}$ be the conjugate function and let

$$G = \exp \frac{1}{2}(\varphi + i\psi).$$

Then G is an outer function in H^2, $|G|^2 = w$ almost everywhere, and

$$G^2(0) = \exp \int \varphi \frac{d\theta}{2\pi}$$

is the right side of (3.1). For $f \in F$ have

$$\frac{1}{2\pi} \int |1 - f|^2 w\, d\theta = \frac{1}{2\pi} \int |(1 - f)^2 G^2|\, d\theta$$

$$\geq |(1 - f(0))^2 G^2(0)| = \exp \frac{1}{2\pi} \int \varphi\, d\theta.$$

because $(1 - f)^2 G^2$ is in H^1.

Now G is an outer function, and so by Beurling's theorem there are polynomials $p_n(z)$ such that $p_n G$ converges to the constant function $G(0)$ in H^2. Then $p_n(0)G(0) \to G(0) \neq 0$, so that $p_n(0) \to 1$ and we can assume $p_n(0) = 1$. Thus we can take $p_n = 1 - f_n$, $f_n \in F$. But then

$$\lim_n \frac{1}{2\pi} \int |1 - f_n|^2 w\, d\theta = \lim_n \frac{1}{2\pi} \int |p_n G|^2\, d\theta$$

$$= \frac{1}{2\pi} \int |G(0)|^2\, d\theta = \exp \frac{1}{2\pi} \int \varphi\, d\theta.$$

Therefore (3.1) is proved when $\log w \in L^1$.

Now assume $\int \log w\, d\theta = -\infty$. For $\varepsilon > 0$, $\log(w + \varepsilon)$ is integrable, and so by the preceding

$$\inf_{f \in F} \frac{1}{2\pi} \int |1 - f|^2 w\, d\theta \leq \inf_{f \in F} \frac{1}{2\pi} \int |1 - f|^2 (w + \varepsilon)\, d\theta$$

$$= \exp \frac{1}{2\pi} \int \log(w + \varepsilon)\, d\theta.$$

The last expression tends to zero with ε and we have

$$\inf_{f \in F} \frac{1}{2\pi} \int |1 - f|^2 w\, d\theta = 0$$

if $\int \log w\, d\theta = -\infty$. $\square$

Turning to the work of Helson and Szegö, we again let

$$d\mu = w \, d\theta/2\pi + d\mu_s$$

be a finite positive measure on T. Let G be the space of conjugate analytic trigonometric polynomials

$$g = b_0 + b_1 e^{-i\theta} + b_2 e^{-2i\theta} + \dots ,$$

and let

$$\rho = \sup \left| \int f \bar{g} \, d\mu \right| ,$$

where f and g range over F and G respectively but where f and g are constrained by

$$\int |f|^2 \, d\mu \le 1, \qquad \int |g|^2 \, d\mu \le 1.$$

It is clear that $0 \le \rho \le 1$. The spaces F and G are orthogonal in $L^2(\mu)$ if and only if $\rho = 0$. If there is a nonzero vector in both the $L^2(\mu)$ closures $\overline{F}$ and $\overline{G}$, then $\rho = 1$. If $\rho < 1$ the closed subspaces $\overline{F}$ and $\overline{G}$ of $L^2(\mu)$ are said to be at *positive angle*, and the angle between the subspaces is $\cos^{-1} \rho$. When F and G are at positive angle, $\overline{F} + \overline{G}$ is closed in $L^2(\mu)$ and

$$\|f\|^2 + \|g\|^2 \le (1 - \rho)^{-1} \|f + g\|^2,$$

$f \in F$, $g \in G$, so that $\overline{F} + \overline{G}$ is the Banach space direct sum of $\overline{F}$ and $\overline{G}$. Examples exist of closed subspaces F and G such that $F \cap G = \{0\}$ but such that $\rho = 1$ and $F + G$ is not closed. (See Exercise 9.)

Theorem 3.2 (Helson–Szegö). *The subspaces $\overline{F}$ and $\overline{G}$ are at positive angle if and only if*

(3.2) $\mu_s = 0,$

and

(3.3) $\log w = u + \tilde{v}.$

where $u \in L^\infty$, $v \in L^\infty$ and $\|v\|_\infty < \pi/2$.

Proof.   First let us show that (3.2) is necessary. Suppose $d\mu_s > 0$. By Exercise 2, Chapter III, there are $p_n \in F$ such that $|p_n| \le 1$, $p_n \to 1$ almost everywhere $d\mu_s$, and $p_n \to 0$ almost everywhere $d\theta$. Then $g_n = \bar{p}_n \in G$, and

$$\int p_n \bar{g}_n d\mu \to \int d\mu_s,$$

while

$$\int |p_n|^2 \, d\mu = \int |g_n|^2 d\mu \to \int d\mu_s.$$

Scaling p_n and g_n to have unit norm in $L^2(\mu)$, we see that $\rho = 1$.

We can now assume μ is absolutely continuous, $\mu = w \, d\theta/2\pi$. We can also assume that

$$\int \log w \frac{d\theta}{2\pi} > -\infty,$$

because otherwise F is dense in $L^2(\mu)$ by Szegö's theorem, so that $G \subset \overline{F}$ and $\rho = 1$, while on the other hand, (3.3) obviously fails.

Let $\varphi = \log w$, let $\psi = \tilde{\varphi}$ and let $G = \exp \frac{1}{2}(\varphi + i\psi)$ be as in the proof of Theorem 3.1. Let $H = G^2$. Then $G \in H^2$, $H \in H^1$, G and H are outer, and

$$|G|^2 = |H| = w.$$

Then we have

$$\rho = \sup \left| \frac{1}{2\pi} \int (fG)(\bar{g}G)e^{-i\psi} \, d\theta \right|,$$

where the supremum is taken over all $f \in F$ and $g \in G$ such that

$$\frac{1}{2\pi} \int |fG|^2 d\theta = \frac{1}{2\pi} \int |\bar{g}G|^2 \, d\theta = 1.$$

By Beurling's theorem the set $\{fG : f \in F\}$ is dense in H_0^2 and the set $\{\bar{g}G : g \in G\}$ is dense in H^2. Since every $F \in H_0^1$ can be factored as $F = F_1 F_2$, where $F_1 \in H_0^2$, $F_2 \in H^2$, and

$$\|F_1\|_2 = \|F_2\|_2 = \|F\|_1,$$

we have

$$\rho = \sup \left\{ \left| \int F e^{-i\psi} \frac{d\theta}{2\pi} \right| : F \in H_0^1, \|F\|_1 \leq 1 \right\}.$$

By duality this means

$$\rho = \inf_{g \in H^\infty} \|e^{-i\psi} - g\|_\infty.$$

Lemma 3.3. *If ψ is a real measurable function, then*

$$\inf_{g \in H^\infty} \|e^{-i\psi} - g\| < 1$$

if and only if there are $\varepsilon > 0$ and $h \in H^\infty$ such that

(3.4) $|h| \geq \varepsilon$ a.e.

(3.5) $|\psi + \arg h| \leq \pi/2 - \varepsilon$ (modulo 2π).

Proof. Notice that if $g \in H^\infty$ and $\|e^{-i\psi} - g\|_\infty < 1$ then g satisfies (3.4) and (3.5). On the other hand, if (3.4) and (3.5) hold for h, then for a small $\lambda > 0$, $\|e^{-i\psi} - \lambda h\|_\infty < 1$. (The proof is illustrated in Figure IV.2.) $\square$

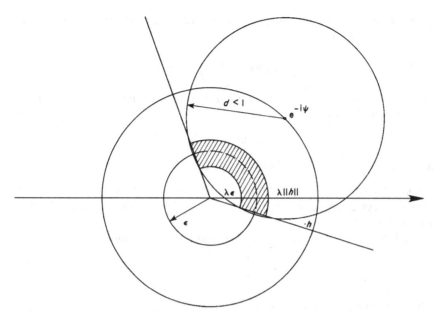

Figure IV.2. The proof of Lemma 3.3. The values of h lie in the cone $|\psi + \arg z| < \pi/2 - \varepsilon$, and in $|z| > \varepsilon$. The shaded region includes all values λh and is contained in $|z - e^{i\psi}| < d < 1$.

Now assume $\rho < 1$. Then there is $h \in H^\infty$ with (3.4) and (3.5). Since arg $H = \psi$ we have

$$| \arg(hH) | < \frac{\pi}{2} - \varepsilon,$$

and the H^1 function hH has a well-defined logarithm. Letting $v = - \arg(hH)$, we have $\|v\|_\infty < \pi/2 - \varepsilon$ and

$$\log |hH| = \tilde{v}.$$

By (3.4) $u = - \log |h|$ is bounded. Hence

$$\log w = \log |H| = \log |hH| - \log |h| = u + \tilde{v},$$

and (3.3) holds.

Conversely, suppose that μ is absolutely continuous and that $\log w$ has the form (3.3). We can then assume $u = 0$, because the property $\rho = 1$ is not changed by multiplying w by a positive bounded function bounded away from zero. (See Exercise 10.) Then in the above discussion

$$\psi = \tilde{\varphi} = -v.$$

Since $\|v\|_\infty < \pi/2$, the constant function $g = \cos(\|v\|_\infty)$ has

$$\|e^{-i\psi} - g\|_\infty < 1,$$

so that $\rho < \|e^{-i\psi} - g\| < 1$. $\square$

Theorem 3.4. *Let μ be a positive finite measure on the circle. Then there is a constant K such that*

$$(3.6) \qquad \int |\tilde{p}|^2 d\mu \le K^2 \int |p^2| d\mu$$

for all trigonometric polynomials if and only if μ is absolutely continuous, $d\mu = w \, d\theta/2\pi$, and

$$\log w = u + \tilde{v},$$

where $u \in L^\infty, v \in L^\infty$, and $\|v\|_\infty < \pi/2$.

Proof. We show the conjugation operator is bounded in $L^2(\mu)$ if and only if the subspaces $\overline{F}$ and $\overline{G}$ are at positive angle in $L^2(\mu)$. With Theorem 3.2 that will prove the result.

Let T be the operator defined on trigonometric polynomials by

$$T(p) = \frac{1}{2}\{(p - a_0) + i\,\tilde{p}\},$$

where $a_0 = a_0(p) = \int p \, d\theta/2\pi$. Thus

$$T(\sum a_n e^{in\theta}) = \sum_{n>0} a_n e^{in\theta}.$$

Since $(p - a_0(p)) = (-\tilde{p})^{\sim}$, we see that T is bounded with respect to the $L^2(\mu)$ norm whenever (3.6) holds. Conversely if

$$(3.7) \qquad \int |T(p)|^2 d\mu \le C^2 \int |p|^2 \, d\mu,$$

then (3.6) is true for real trigonometric polynomials, and therefore for all trigonometric polynomials. Thus (3.7) is equivalent to (3.6).

Every trigonometric polynomial $p(\theta)$ has the form $p = f - g$, where $f \in F, g \in G$, and

$$T(p) = T(f - g) = f.$$

Hence if T is bounded and if $\int |f|^2 d\mu = \int |g|^2 d\mu = 1$, then by (3.7) we have

$$C^{-2} \le \int |f - g|^2 d\mu = 2 - 2 \operatorname{Re} \int f\tilde{g} \, d\mu.$$

Letting $\int f\tilde{g} \, d\mu$ approach ρ, we calculate that $\rho < 1 - 1/2C^2$.

On the other hand, if $\rho < 1$ and if $f \in F, g \in G$, then

$$\int |f - g|^2 d\mu \ge \int |f|^2 d\mu + \int |g|^2 d\mu - 2\rho \left(\int |f|^2 \, d\mu\right)^{1/2} \left(\int |g|^2 \, d\mu\right)^{1/2}$$

$$\ge (1 - \rho)\left(\int |f|^2 \, d\mu + \int |g|^2 \, d\mu\right)$$

and

$$\int |f|^2 \, d\mu = \int |T(f - g)|^2 \, d\mu \le (1 - \rho)^{-1} \int |f - g|^2 \, d\mu.$$

Hence T is bounded if $\rho < 1$. $\quad\square$

4. Interpolating Functions of Constant Modulus

Let $\{z_j\}$ be a sequence of distinct points in D, and let $\{w_j\}$ be a sequence of complex numbers. Assume there is $f_0 \in H^\infty$ solving the interpolation problem

(4.1) $$f(z_j) = w_j, \qquad j = 1, 2, \dots .$$

If there are finitely many points z_j then by Pick's theorem or by Corollary 1.8 the interpolation (4.1) can be solved with $f = cB$ where B is a finite Blaschke product and where c is the minimum norm of all interpolating functions. In this section we consider the case in which there are infinitely many z_j. We assume

$$\sum (1 - |z_j|) < \infty,$$

a condition necessary and sufficient for (4.1) to have more than one solution in H^∞.

Theorem 4.1. (Nevanlinna). *If there are two distinct functions of unit norm in H^∞ that do the interpolation (4.1), then there is an inner function that also satisfies (4.1).*

If f_0 fulfills (4.1) and if $\|f_0\| < 1$, then the two function hypothesis is trivially satisfied. Indeed, let $B(z)$ be the Blaschke product with zeros $\{z_j\}$. Then for some $s > 0$ and for some $t < 0$, $f_0 + sB$ and $f_0 + tB$ are two distinct interpolating functions of norm 1. If there is exactly one function of unit norm satisfying (4.1), then this reasoning (and the theorem) show that interpolation is always possible with a function of constant modulus $1 + \varepsilon$, for any $\varepsilon > 0$. Before proving the theorem, we mention an example indicating that there may not be an inner function solving (4.1) when it is merely assumed that there exists one interpolating function of norm 1.

Example 4.2. Let $\lim z_j = 1$. Let I be an open are on T containing $z = 1$. Let $f \in (H^\infty)^{-1}$ have $\|f\| = 1$, $|f| = 1$ on I but $|f| \not\equiv 1$. Let $w_j = f(z_j)$. Then every other interpolating function for this problem has the form $f - Bg$, $g \in H^\infty$, where B is the Blaschke product with zeros z_j. We claim $g = 0$ if $\|f - Bg\| \le 1$. This means there is only one interpolating function of unit norm, namely f, and that function is not an inner function.

So assume $g \in H^\infty$ and $\|f - Gg\| \le 1$. Then

$$|1 - Bg/f| \le 1$$

almost everywhere on I, so that

$$\text{Re } Bg/f \geq 0$$

almost everywhere on I. By Exercise 14, Chapter II, the inner factor of Bg/f is analytic across I. If $g \not\equiv 0$, this inner factor is a multiple of B, which is not analytic across I. So $g \equiv 0$.

Theorem 4.1 is a consequence of the following theorem due to Adamyan, Arov, and Krein [1968].

Theorem 4.3. *Let $h_0 \in L^\infty$. If the coset $h_0 + H^\infty$ of L^∞/H^∞ contains two functions of unit norm, then it contains a function $h \in L^\infty$ such that $|h| = 1$ almost everywhere.*

To derive Nevanlinna's theorem from Theorem 4.3, let $h_0 = \bar{B} f_0$, where B is the Blaschke product with zeros $\{z_j\}$. If $h \in \bar{B} f_0 + H^\infty$ and if $|h| = 1$, then Bh is an inner function such that $(Bh)(z_j) = f_0(z_j)$, $j = 1, 2, \ldots$.

The proof of Theorem 4.3 will be divided into two cases, although the first case can be subsumed under the more difficult second case. The strategy for the proof has already been suggested by Corollary 1.9. To obtain a unimodular function we maximize a linear functional over

$$K = \{h \in h_0 + H^\infty : \|h\| = 1\}.$$

Write

$$\|h_0 + H^\infty\| = \inf\{\|h_0 - g\| : g \in H^\infty\}$$

for the norm of our coset in L^∞/H^∞.

Case 1. $\|h_0 + H^\infty\| < 1$. Consider the extremal problem

$$(4.2) \qquad a = \sup\left\{\left|\frac{1}{2\pi}\int h\,d\theta\right| : h \in K\right\}.$$

By a normal family argument there is an extremal function $h \in K$ such that $|(1/2\pi)\int h\,d\theta| = a$. We claim $|h| = 1$ almost everywhere.

Notice that

$$(4.3) \qquad \text{dist}(h, H_0^\infty) = \inf\{\|h - g\| : g \in H^\infty, g(0) = 0\} = 1,$$

since otherwise $h - g + \varepsilon \int h\,d\theta/2\pi$, $g \in H_0^\infty$, $\varepsilon > 0$, would be a function in K with a larger mean. On the other hand, by hypothesis we have

$$(4.4) \qquad \text{dist}(h, H^\infty) = \|h_0 + H^\infty\| < 1.$$

Since $(H^1)^\perp = H_0^\infty$, (4.3) gives

$$(4.3') \qquad \sup\left\{\left|\int hF\frac{d\theta}{2\pi}\right| : F \in H^1, \|F\|_1 \leq 1\right\} = 1,$$

while (4.4) gives

$$(4.4') \qquad \sup\left\{\left|\int hF\frac{d\theta}{2\pi}\right| : F \in H_0^1, \|F\|_1 \leq 1\right\} < 1.$$

By (4.3') there are $F_n \in H^1$, $\|F_n\|_1 \leq 1$ such that

$$(4.5) \qquad \int hF_n\frac{d\theta}{2\pi} \to 1.$$

This means

$$(4.6) \qquad \varlimsup_n |F_n(0)| > 0,$$

because otherwise we would have a subsequence for which

$$\int h\frac{F_n - F_n(0)}{\|F_n - F_n(0)\|_1}\frac{d\theta}{2\pi} \to 1,$$

and this contradicts (4.4').

Now suppose there were a measurable set E of positive measure such that $|h| < \lambda < 1$ on E. As $\|F_n\|_1 \leq 1$, (4.5) then yields

$$\int_E |F_n|\frac{d\theta}{2\pi} \to 0.$$

Since the logarithm is concave, Jensen's inequality now gives

$$\int_E \log|F_n|\frac{d\theta}{2\pi} \to -\infty.$$

and as

$$\log|F_n(0)| \leq \int_E \log|F_n|\frac{d\theta}{2\pi} + \int_{T/E} |F_n|\frac{d\theta}{2\pi},$$

we obtain $\log|F_n(0)| \to -\infty$, in contradiction to (4.6).

Before turning to Case 2 we digress somewhat in order to obtain further information when $\|h_0 + H^\infty\| < 1$.

Theorem 4.4. *If $\|h_0 + H^\infty\| < 1$, there is $h \in h_0 + H^\infty$ and there is $F \in H^1$, $F \neq 0$ such that*

$$(4.7) \qquad h = \bar{F}/|F|$$

almost everywhere.

Proof. Let h and $\{F_n\}$ be as in the above discussion. We claim the sequence $\{F_n\}$ has a subsequence converging weakly in L^1. If F is any weak limit point, then by (4.5) $\int hF \, d\theta/2\pi = 1$. Since $\|h\| \leq 1$, this implies (4.7).

If the sequence $\{F_n\}$ has no weakly convergent subsequence, then there are measurable sets $E_k \subset T$ such that

$$|E_k| \to 0$$

but such that

(4.8) $$\left| \int_{E_k} F_k \frac{d\theta}{2\pi} \right| \geq \beta > 0,$$

where $\{F_k\}$ denotes a subsequence of the F_n (see Dunford and Schwartz [1958, p. 292]).

Lemma 4.5. *If $\{E_k\}$ is a sequence of measurable subsets of T such that $|E_k| \to 0$, then there is a sequence $\{g_k\}$ of functions in H^∞ such that*

    (i) $\sup_{E_k} |g_k| \to 0,$
    (ii) $g_k(0) \to 1,$ *and*
    (iii) $|g_k| + |1 - g_k| \leq 1 + \varepsilon_k,$

where $\lim_k \varepsilon_k = 0$.

Let us assume the lemma for the moment and finish proving Theorem 4.4 Let

$$G_k = \frac{g_k F_k}{1 + \varepsilon_k}, \qquad H_k = \frac{(1 - g_k) F_k}{1 + \varepsilon_k}.$$

Then $G_k, H_k \in H^1$. Since $\varepsilon_k \to 0$, (4.5) gives us

(4.9) $$\|G_k\|_1 \int \frac{h G_k}{\|G_k\|_1} \frac{d\theta}{2\pi} + \|H_k\|_1 \int \frac{h H_k}{\|H_k\|_1} \frac{d\theta}{2\pi} \to 1.$$

Now for k sufficiently large, $\|H_k\|_1 \geq \beta/2$ by (4.8) and condition (i). By condition (iii) $\|G_k\|_1 + \|H_k\|_1 \leq 1$, and (4.9) then yields

$$\int \frac{h H_k}{\|H_k\|_1} \frac{d\theta}{2\pi} \to 1.$$

However, condition (ii) implies that $\lim H_k(0) = 0$, and we again have a contradiction to (4.4′). □

Proof of Lemma 4.5. Choose $A_k \to \infty$ so slowly that $A_k |E_k| \to 0$. Let f_k be the Poisson integral of

$$A_k \chi_{E_k} + i A_k \tilde\chi_{E_k}.$$

Then f_k is an analytic function in D and f_k takes values only in the right half plane. Also $f_k(0) = A_k |E_k|$ and Re $f_k = A_k$ almost everywhere on E_k. Let $h_k = (1 + f_k)^{-1}$. Then h_k maps D into the disc

(4.10) $$|w - \tfrac{1}{2}| < \tfrac{1}{2},$$

and $h_k(0) \to 1$ while $\sup_{E_k} |h_k| \to 0$. The disc defined by (4.10) is compressed into the ellipse defined by (iii) under the mapping $w \to w^\delta$, if $\delta > 0$ is small. We can choose $\delta_k \to 0$ so slowly that $g_k = h_k^{\delta_k}$ then satisfies conditions (i)–(iii). $\square$

We return to the proof of Theorem 4.3.

Case 2. $\|h_0 + H^\infty\| = 1$. By the hypothesis, there are two distinct functions h_1 and h_2 in the coset $h_0 + H^\infty$ such that $\|h_1\| = \|h_2\| = 1$. Since $h_1 \neq h_2$ there is a point $z \in D$ such that

$$\int h_1 P_z \frac{d\theta}{2\pi} \neq \int h_2 P_z \frac{d\theta}{2\pi}.$$

Using a Möbius transformation, we can suppose $z = 0$, and rotating h_0 we can assume

$$(4.11) \qquad \mathrm{Re} \int h_1 \frac{d\theta}{2\pi} \neq \mathrm{Re} \int h_2 \frac{d\theta}{2\pi}.$$

As in Case 1, the function $h \in h_0 + H^\infty$ with $|h| = 1$ almost everywhere will be found by maximizing

$$\mathrm{Re} \int h \frac{d\theta}{2\pi}$$

over $K = \{h \in h_0 + H^\infty, \|h\| \leq 1\}$. However, this time we must use the proof of the Hahn–Banach theorem instead of that theorem itself.

Since L^∞ / H^∞ is the dual space of H_0^1, the elements $h \in h_0 + H^\infty$ with $\|h\| = 1$ correspond naturally to the norm-preserving extensions to L^1 of the linear functional

$$H_0^1 \ni F \to \int h_0 F \frac{d\theta}{2\pi}.$$

Now (4.11) says that this functional has two norm-preserving extensions whose real parts disagree at the constant function $1 \in L^1 / H_0^1$. The proof of the Hahn–Banach theorem therefore gives us

$$\sup_{F \in H_0^1} \left\{ -\|1 + F\| - \mathrm{Re} \int h_0 F \frac{d\theta}{2\pi} \right\} = m,$$

$$\inf_{F \in H_0^1} \left\{ \|1 + F\| - \mathrm{Re} \int h_0 F \frac{d\theta}{2\pi} \right\} = M,$$

and

$$m < M.$$

We may take $h \in h_0 + H^\infty$, $\|h\| = 1$, so that $\mathrm{Re} \int h \, d\theta / 2\pi = M$. It will turn out that $|h| = 1$ almost everywhere. The last two displayed identities can now be rewritten as

$$(4.12) \qquad \inf_{F \in H_0^1} \left\{ \|1 + F\| + \operatorname{Re} \int h(1 + F) \frac{d\theta}{2\pi} \right\} = M - m > 0$$

and

$$(4.13) \qquad \inf_{F \in H_0^1} \left\{ \|1 + F\| - \operatorname{Re} \int h(1 + F) \frac{d\theta}{2\pi} \right\} = 0.$$

The left sides of (4.12) and (4.13) differ only in a change of sign in one place. By (4.13) there are $F_n \in H_0^1$ such that

$$(4.14) \qquad \frac{1}{2\pi} \int |1 + F_n| \left\{ 1 - \operatorname{Re} \left(h \frac{(1 + F_n)}{|1 + F_n|} \right) \right\} \frac{d\theta}{2\pi} \to 0.$$

Suppose there is a measurable set $E \subset T$ with $|E| > 0$ such that $|h| < \lambda < 1$ on E. Then by (4.14)

$$(4.15) \qquad \frac{1}{2\pi} \int_E |1 + F_n| d\theta \to 0.$$

Lemma 4.6. *If $E \subset T$ is a set of positive measure, then there is $g \in H^\infty$ such that $g(0) = 1$ and such that g is real and negative on $T \backslash E$.*

Accepting the lemma temporarily, write $1 + G_n = g(1 + F_n)$. Then $G_n \in H_0^1$, and by (4.15)

$$\frac{1}{2\pi} \int_E |1 + G_n| \left\{ 1 + \operatorname{Re} \left(h \frac{(1 + G_n)}{|1 + G_n|} \right) \right\} \frac{d\theta}{2\pi} \to 0.$$

On the other hand,

$$\operatorname{Re} \left(h \frac{(1 + G_n)}{|1 + G_n|} \right) = -\operatorname{Re} \left(h \frac{(1 + F_n)}{|1 + F_n|} \right)$$

on $T \backslash E$, so by (4.14)

$$\frac{1}{2\pi} \int_{T \backslash E} |1 + G_n| \left\{ 1 + \operatorname{Re} \left(h \frac{(1 + G_n)}{|1 + G_n|} \right) \right\} d\theta \to 0.$$

Hence

$$\int \{ |1 + G_n| + \operatorname{Re}(h(1 + G_n)) \} \frac{d\theta}{2\pi} \to 0,$$

and this contradicts (4.12)   □

Proof of Lemma 4.6. Let G be the outer function such that $|G| = e$ on $E, |G| = 1$ on $T \backslash E$. Then $G(0) = \exp |E| > 1$ and G has values in the annulus $\{ 1 < |w| < e \}$. The function

$$\varphi(w) = w + (1/w) - 2$$

maps this annulus into the domain bounded by the slit $[-4, 0]$ and an ellipse, and the circle $|w| = 1$ is mapped onto the slit $[-4, 0]$. Also $\varphi(G(0)) > 0$. Therefore

$$g = \frac{\varphi \circ G}{\varphi \circ G(0)}$$

has $g(0) = 1$ and g is real and negative almost everywhere on $T \backslash E$. □

5. Parametrization of K

We continue our discussion of a coset $h_0 + H^\infty \in L^\infty / H^\infty$, under the assumption that $K = \{h \in h_0 + H^\infty : \|h\| \le 1\}$ contains at least two functions. Our objective is a beautiful formula describing all functions in K, due to Adamyan, Arov, and Krein [1968]. Before beginning we need two results of deLeeuw and Rudin [1958] concerning the geometry of the unit ball of H^1.

A point x in a convex set K is an *extreme point* of K if x cannot be written as a proper convex combination

$$x = tx_1 + (1 - t)x_2, \quad 0 < t < 1,$$

with $x_1, x_2 \in K$ and $x_1 \ne x_2$.

Theorem 5.1. *A function F is an extreme point of the closed unit ball of H^1 if and only if F is an outer function and $\|F\|_1 = 1$. If $F \in H^1$, $\|F\|_1 = 1$ and if F is not an outer function, then*

$$(5.1) \qquad\qquad F = \frac{F_1 + F_2}{2},$$

where F_1 and F_2 are outer functions, and $\|F_1\|_1 = \|F_2\|_1 = 1$.

Proof. Suppose $F \in H^1$, $\|F\|_1 \le 1$. If $\|F\|_1 < 1$, then with $t = \|F\|_1$,

$$F = t(F/\|F\|_1) + (1 - t)0$$

and F is no extreme point of ball(H^1). For the rest of this proof we take $\|F\|_1 = 1$.

Assume F is not outer and write $F = uG$, with u inner and G outer. Choose λ, $|\lambda| = 1$, so that

$$(5.2) \qquad\qquad \frac{1}{2\pi} \int |F| \text{Re}(\lambda u) d\theta = 0,$$

and put $u_0 = \lambda u$. Since u_0 is inner,

$$2 \text{ Re } u_0 = u_0 + \overline{u_0} = u_0 + \frac{1}{u_0} = \frac{1 + u_0^2}{u_0}.$$

Consequently

$$(5.3) \quad J(e^{i\theta}) = F(e^{i\theta})\operatorname{Re}(u_0(e^{i\theta})) = \{\tfrac{1}{2}\bar{\lambda}G(e^{i\theta})u_0(e^{i\theta})\}\{2\operatorname{Re} u_0(e^{i\theta})\}$$
$$= \tfrac{1}{2}\bar{\lambda}G(e^{i\theta})(1 + u_0^2(e^{i\theta})),$$

and $J \in H^1$. Almost everywhere we have

$$|F \pm J| = |F|(1 \pm \operatorname{Re} u_0),$$

so that (5.2) gives

$$\|F + J\|_1 = \|F - J\|_1 = \|F\|_1 = 1.$$

Since F is not outer, u_0 is not constant, and $J \neq 0$. Then

$$F = \frac{F + J}{2} + \frac{F - J}{2}$$

and F is not an extreme point. Furthermore,

$$J \pm F = \tfrac{1}{2}\bar{\lambda}G(1 \pm 2u_0 + u_0^2) = \tfrac{1}{2}\bar{\lambda}G(1 \pm u_0)^2$$

is an outer function, because $1 \pm u_0$ is outer, and hence (5.1) holds.

Now assume F is an outer function, $\|F\|_1 = 1$. If $F = tF_1 + (1 - t)F_2$, $F_1, F_2 \in \text{ball}(H^1), 0 < t < 1$, then

$$1 = \|F\|_1 = t \int \frac{\bar{F}}{|F|} F_1 \frac{d\theta}{2\pi} + (1 - t) \int \frac{\bar{F}}{|F|} F_2 \frac{d\theta}{2\pi}$$
$$\leq t\|F_1\|_1 + (1 - t)\|F_2\|_1 \leq 1,$$

and equality must hold throughout these inequalities. Hence $\|F_1\|_1 = \|F_2\|_1 = 1$, and

$$\int \frac{\bar{F}}{|F|} F_j \frac{d\theta}{2\pi} = \|F_j\|_1, \quad j = 1, 2.$$

Since $|F| > 0$ almost everywhere, that means $F_j = k_j F$, where $k_j > 0$ and $tk_1 + (1 - t)k_2 = 1$. But then by the subharmonicity of $\log|F_j|$,

$$|F_j(0)| \leq |F(0)| \exp\left(\int \log k_j \frac{d\theta}{2\pi}\right),$$

because F is outer. Since $|F(0)| \leq t|F_1(0)| + (1 - t)|F_2(0)|$, Jensen's inequality now yields

$$1 \leq t \exp\left(\int \log k_1 \frac{d\theta}{2\pi}\right) + (1 - t)\exp\left(\int \log k_2 \frac{d\theta}{2\pi}\right)$$
$$\leq t \int k_1 \frac{d\theta}{2\pi} + (1 - t)\int k_2 \frac{d\theta}{2\pi} = 1.$$

Because the exponential is strictly convex we conclude that k_j is constant and $k_j = \|F_j\|_1/\|F\|_1 = 1$. Hence $F_1 = F_2 = F$ and F is an extreme point. $\square$

It is unfortunately the case that an H^1 function F of unit norm is not determined by its argument (which is defined modulo 2π almost everywhere, because $|F| > 0$ almost everywhere). For example, if F is not outer, then the two outer functions in (5.1) have the same argument as F (and as each other). That follows from the construction of F_1 and F_2 or from the observation that by (5.1),

$$\int \frac{\bar{F}}{|F|} F_j \frac{d\theta}{2\pi} = 1 = \|F_j\|_1, \quad j = 1, 2,$$

which means $(\bar{F}/|F|)F_j = |F_j|$ almost everywhere. (See Example 1.4 above for another counterexample.) When $h \in L^\infty$ and $|h| = 1$ almost everywhere, we define

$$S_h = \{F \in H^1 : \|F\|_1 = 1, F/|F| = h \text{ a.e}\}.$$

Geometrically, S_h is the intersection of ball(H^1) and the hyperplane

(5.4)
$$\left\{ F : \int \bar{h} F \frac{d\theta}{2\pi} = 1 \right\},$$

and so S_h is a convex set. Of course, sometimes S_h is empty, but we are interested in the case $S_h \neq \varnothing$. When S_h contains exactly one function F, the hyperplane (5.4) touches ball(H^1) only at F, which means F is an *exposed point* of ball(H^1). There is no good characterization of the exposed points of ball(H^1), or equivalently of those $F \in$ ball(H^1) such that $S_{F/|F|} = \{F\}$. However, if S_h contains two functions, then S_h is very large.

Theorem 5.2. *Let $h \in L^\infty$, $|h| = 1$ almost everywhere, and assume S_h contains at least two distinct functions. Let $z_0 \in D$. Then $\{F(z_0) : F \in S_h\}$ contains a disc centered at the origin.*

Proof. When $|z_0| \leq 1, |z_1| \leq 1$,

$$\frac{(z - z_1)(1 - \bar{z}_1 z)}{(z - z_0)(1 - \bar{z}_0 z)} = \frac{(z - z_1)z(\bar{z} - \bar{z}_1)}{(z - z_0)z(\bar{z} - \bar{z}_0)}$$

is real and nonnegative on T. If there exists $F \in S_h$ having a zero of order k at z_0, then

$$F_{z_1}(z) = \left(\frac{(z - z_1)(1 - \bar{z}_1 z)}{(z - z_0)(1 - \bar{z}_0 z)} \right)^k F(z)$$

is in S_h and the values $F_{z_1}(z_0)$ fill a disc about 0.

It remains to show there is $F \in S_h$ with $F(z_0) = 0$. If the convex set S_h contains two functions, then by Theorem 5.1 it contains a function $F = uG$

with nonconstant inner factor $u(z)$. Let $J(z)$ be the function defined in (5.3). Then for $0 < t < 1$,

$$F + tJ = F(1 + t \operatorname{Re} u_0)$$

satisfies $\|F + tJ\|_1 = 1$, because of (5.2), and $F + tJ \in S_h$. By (5.3) we also have

$$F + tJ = \tfrac{1}{2}\bar{\lambda} t G \left(1 + \frac{2u_0}{t} + u_0^2\right).$$

When $0 < t < 1$, the equation

$$\zeta^2 + \frac{2\zeta}{t} + 1 = 0$$

has a root $\zeta(t)$, $|\zeta(t)| < 1$, and these roots fill the segment $(-1, 0)$. If $F + tJ$ has no zero in D for each $t \in [0, 1)$, then the range of u_0 is disjoint from the segment $(-1, 0]$. But then by Corollary 4.8, Chapter II, u_0 is both an inner function and an outer function, so that u_0 is constant. Thus there exists t such that $F + tJ$ has a zero at some point $z_1 \in D$. Then

$$\frac{(z - z_0)(1 - \bar{z}_0 z)}{(z - z_1)(1 - \bar{z}_1 z)}(F + tJ)$$

is a function in S_h having a zero at z_0. $\square$

Returning to the topic of the previous section, we fix a coset $h_0 + H^\infty$ of L^∞ / H^∞ and we assume

$$K = \{h \in h_0 + H^\infty : \|h\| \leq 1\}$$

contains more than one function. By Theorem 4.3, K then contains a function, which we call h_0, such that $|h_0| = 1$ almost everywhere. We will need to recall that after a change of coordinates h_0 is an extremal function:

(5.5) $$\operatorname{Re} \int h_0 \frac{d\theta}{2\pi} = \sup_{h \in K} \operatorname{Re} \int h \frac{d\theta}{2\pi}.$$

(See the proof of Theorem 4.3.) The Adamyan, Arov, and Krein parametrization of K which we have been seeking is (5.7) below.

Theorem 5.3. *There exists a unique outer function $F \in H^1$, $\|F\|_1 = 1$, such that*

(5.6) $$h_0 = F/|F|.$$

Define $\chi \in H^\infty$ by

$$\frac{1 + \chi(z)}{1 - \chi(z)} = \frac{1}{2\pi} \int \frac{e^{i\theta} + z}{e^{i\theta} - z} |F(e^{i\theta})| d\theta.$$

Then

$$(5.7) \qquad K = \left\{ h_0 - \frac{F(1 - \chi)(1 - w)}{1 - \chi w} : w(z) \in H^\infty, \|w\|_\infty \leq 1 \right\}.$$

As a consequence of the parametrization (5.7) we see that, for $z \in D$,

$$\left\{ \int h P_z \frac{d\theta}{2\pi} : h \in K \right\}$$

is a nondegenerate closed disc. Another corollary of (5.7) is the description by Nevanlinna [1929] of all solutions $f \in H^\infty$, $\|f\| \leq 1$, of the interpolation problem

$$f(z_j) = w_j, \quad j = 1, 2, \ldots$$

(see Section 6 below).

Condition (5.6) may seem to contradict Theorem 4.4, because when $\|h_0 + H^\infty\| < 1$, we have claimed that the extremal function h_0 has the two forms

$$h_0 = \bar{F}_1/|F_1|, \quad h_0 = F_2/|F_2|$$

with $F_1, F_2 \in H^1$, $\|F_j\|_1 = 1$. However, if $F_1 \in H^1$ and $g \in H^\infty$, and if

$$\left\| \frac{\bar{F}_1}{|F_1|} - g \right\|_\infty = \alpha < 1,$$

then $|\arg(g F_1)| \leq \sin^{-1}(1 - \alpha) < \pi/2$, and Corollary III.2.5 shows that $(g F_1)^{-1} \in H^1$. Hence $F_1^{-1} = g(g F_1)^{-1} \in H^1$ and (5.6) holds with $F_2 = F_1^{-1}/\|F_1^{-1}\|_1$.

The proof of Theorem 5.3 requires three lemmas; the first uses an idea from Koosis [1973].

Lemma 5.4. *There exists an outer function $F \in H^1$, $\|F\|_1 = 1$, such that*

$$h_0 = F/|F|$$

almost everywhere.

Proof. We know $|h_0| = 1$ almost everywhere and we know there is $g \in H^\infty$, $g \not\equiv 0$, such that $\|h_0 - g\|_\infty \leq 1$. Then $|1 - \bar{h}_0 g| \leq 1$ almost everywhere. Let $\alpha = \arg \bar{h}_0 g$. Then $|\alpha| \leq \pi/2$ and

$$|g| = |\bar{h}_0 g| \leq 2 \cos \alpha$$

as shown in Figure IV.3. Let $\varphi = e^{\tilde{\alpha} - i\alpha}$. Then $\varphi \in H^p$, $p < 1$, by Theorem III.2.4. However, we also have $g\varphi \in H^1$, because

$$(5.8) \qquad |\varphi(e^{i\theta}) g(e^{i\theta})| \leq 2|\varphi(e^{i\theta}) \cos \alpha(e^{i\theta})| = 2 \, \mathrm{Re} \, \varphi(e^{i\theta}),$$

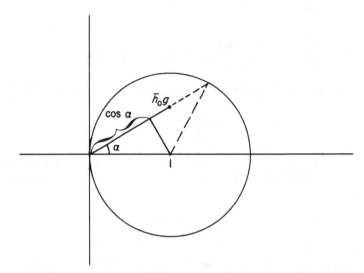

Figure IV.3. Why $|g| \le 2\cos\alpha$.

and since $\mathrm{Re}\,\varphi \ge 0$,

$$\frac{1}{2\pi}\int \mathrm{Re}\,\varphi\,d\theta \le \lim_{r\to 1}\int \mathrm{Re}\,\varphi\,(re^{i\theta})\frac{d\theta}{2\pi} \le \mathrm{Re}\,\varphi(0).$$

Hence $\varphi g \in H^p \cap L^1 = H^1$. Then $F_0 = \varphi g/\|\varphi g\|_1$ is in ball(H^1) and $\bar{h}_0 F_0 \ge 0$, which means that

$$h_0 = F_0/|F_0|.$$

Finally, by the remarks following Theorem 5.1 there is an outer function $F \in$ ball(H^1) such that (5.6) holds.   □

Lemma 5.5. *Suppose $F(z)$ is any H^1 function, $F \not\equiv 0$. Define $\chi \in H^\infty$ by*

(5.9)
$$\frac{1+\chi(z)}{1-\chi(z)} = \frac{1}{2\pi}\int \frac{e^{i\theta}+z}{e^{i\theta}-z}|F(e^{i\theta})|d\theta.$$

If $w(z) \in H^\infty$, $\|w\|_\infty \le 1$, then

(5.10)
$$g(z) = \frac{F(z)(1-\chi(z))(1-w(z))}{1-\chi(z)w(z)}$$

is in H^∞ and

(5.11)
$$\left\|\frac{F}{|F|} - g\right\|_\infty \le 1.$$

Proof. Note that

$$\text{Re} \frac{1 + \chi(z)}{1 - \chi(z)} = \int P_z(\theta) |F(\theta)| \frac{d\theta}{2\pi} > 0.$$

Let $|w(z)| < 1$ and set

(5.12) $$\varphi(z) = \frac{1 + \chi(z)}{1 - \chi(z)} + \frac{1 + w(z)}{1 - w(z)} = \frac{2(1 - \chi(z)w(z))}{(1 - \chi(z))(1 - w(z))}.$$

Then φ is holomorphic and since

$$\text{Re} \frac{1 + w(z)}{1 - w(z)} \geq 0,$$

we have

(5.13) $$\text{Re}\, \varphi(z) \geq \int P_z(\theta) |F(\theta)| \frac{d\theta}{2\pi} \geq |F(z)| \geq 0.$$

A simple calculation from (5.10) and (5.12) gives

$$g(z) = 2F(z)/\varphi(z),$$

so that $g \in H^\infty$. When $w \not\equiv 1$, $g = 0$, and when $w \not\equiv 1$, φ has nontangential limits almost everywhere satisfying

$$|\varphi(e^{i\theta})| \geq |F(e^{i\theta})|.$$

Notice that

$$\left| \frac{F(e^{i\theta})}{|F(e^{i\theta})|} - g(e^{i\theta}) \right| \leq 1$$

if and only if

(5.14) $$\left| \frac{1}{|F(e^{i\theta})|} - \frac{g(e^{i\theta})}{F(e^{i\theta})} \right| \leq \frac{1}{|F(e^{i\theta})|}.$$

The transformation $\zeta \to 2/\zeta$ maps the half plane $\text{Re}\, \zeta \geq |F(e^{i\theta})|$ onto the disc $\||F(e^{i\theta})|^{-1} - w| \leq |F(e^{i\theta})|^{-1}$. By (5.13), $\text{Re}\,(2F/g) = \text{Re}\, \varphi \geq |F|$ almost everywhere, so that we have (5.14) and consequently (5.11). $\square$

Lemma 5.6.   *The function $F \in H^1$, $\|F\|_1 = 1$, such that*

$$h_0 = F/|F|$$

is unique.

Proof.   Lemma 5.4 shows there exists at least one such F. What we must show is that S_{h_0} does not contain two functions. But if S_{h_0} contains two functions, then by Theorem 5.2, there is $F_1 \in S_{h_0}$ with $\text{Re}\, F_1(0) < 0$. Since $\|F_1\|_1 = 1$, the function χ associated with F_1 by (5.9) satisfies $\chi(0) = 0$. Taking $w \equiv -1$

in (5.10), we obtain $g \in H^\infty$ such that Re $g(0) < 0$, and by (5.11), $h_0 - g \in K$. Then

$$\text{Re} \int (h_0 - g)\frac{d\theta}{2\pi} > \text{Re} \int h_0 \frac{d\theta}{2\pi},$$

contradicting (5.5). □

Conclusion of the proof of Theorem 5.3. By Lemmas 5.4 and 5.6, there is a unique outer function $F \in H^1$, $\|F\|_1 = 1$, such that (5.6) holds. By Lemma 5.5 every function of the form

$$(5.15) \qquad\qquad h_0 - \frac{F(1 - \chi)(1 - w)}{1 - \chi w},$$

$w \in H^\infty$, $\|w\|_\infty \leq 1$, lies in K. Now let $g \in H^\infty$, $g \not\equiv 0$, be such that $\|h_0 - g\| \leq 1$. We must show $h_0 - g$ has the form (5.15) for $w \in \text{ball}(H^\infty)$. The idea is in the proof of Lemma 5.4. Setting $\alpha = \arg \bar{h}_0 g$, and $\varphi = e^{\bar{\alpha} - i\alpha}$, we have $\varphi g \in H^1$ and, by its uniqueness,

$$F = \frac{\varphi g}{\|\varphi g\|_1}.$$

Now Re $\varphi(z) > 0$ and by (5.8)

$$\frac{\text{Re } 2\varphi(e^{i\theta})}{\|\varphi g\|_1} \geq \frac{|\varphi(e^{i\theta})g(e^{i\theta})|}{\|\varphi g\|_1} = |F(e^{i\theta})|.$$

The positive harmonic function Re $2\varphi(z)/\|\varphi g\|_1$ is the Poisson integral of a positive measure with absolutely continuous part exceeding $|F(e^{i\theta})|$. Consequently

$$\frac{2\varphi(z)}{\|\varphi g\|_1} - \frac{1 + \chi(z)}{1 - \chi(z)} = k(z)$$

has Re $k(z) > 0$, so that

$$k(z) = \frac{1 + w(z)}{1 - w(z)},$$

$w \in H^\infty$, $\|w\|_\infty \leq 1$. A calculation then gives

$$g = \frac{\|\varphi g\|_1 F}{\varphi} = 2F\left(\frac{2\varphi}{|\varphi g|_1}\right)^{-1} = \frac{2F}{\left(\dfrac{1 + \chi}{1 - \chi}\dfrac{1 + w}{1 - w}\right)}$$

$$= F\frac{(1 - \chi)(1 - w)}{1 - \chi w},$$

and (5.15) holds. □

6. Nevanlinna's Proof

In the special case $h = \bar{B}f$, where $f \in B = \text{ball}(H^\infty)$ and B is a Blaschke product with distinct zeros $\{z_j\}$, formula (5.7) describes all $f \in B$ such that

$$(6.1) \qquad f(z_j) = w_j, \quad j = 1, 2, \ldots,$$

(provided it is assumed that (6.1) has two distinct solutions in B). Already in [1929], Nevalinna had a similar formula for the solutions of (6.1), and his original proof of Theorem 4.1 in the same paper came as an application of his formula. Having used so much theory in the two preceding sections, we should include Nevanlinna's elementary approach to these results.

The idea is to carefully iterate the invariant form of Schwarz's lemma. Suppose $B = \text{ball}(H^\infty)$ contains two solutions f_1 and f_2 of (6.1). There is a point $z_0 \in D$ such that $f_1(z_0) \neq f_2(z_0)$. After a Möbius transformation, we can take $z_0 = 0$. Write

$$E_n = \{f \in B : f(z_j) = w_j, 1 \leq j \leq n\}.$$

We are seeking a parameterization of

$$E_\infty = \bigcap_n E_n,$$

and we want to use that parametrization to show E_∞ contains inner functions.

Fix c_1, $|c_1| < 1$, to be determined later. If $f \in E_1$, then

$$(6.2) \qquad \frac{f - w_1}{1 - \bar{w}_1 f} = \frac{f_1 + c_1}{1 + \bar{c}_1 f_1} \frac{z - z_1}{1 - \bar{z}_1 z}$$

for some $f_1 \in B$. Conversely, whenever $f_1 \in B$, (6.2) defines a function f in E_1. We rewrite (6.2):

$$(6.3) \qquad f(z) = \frac{A_1(z) + B_1(z)f_1(z)}{C_1(z) + D_1(z)f_1(z)}$$

in which, by a calculation,

$$A_1(z) = w_1(1 - \bar{z}_1 z) + c_1(z - z_1),$$
$$B_1(z) = \bar{c}_1 w_1(1 - \bar{z}_1 z) + (z - z_1),$$
$$C_1(z) = (1 - \bar{z}_1 z) + c_1 \bar{w}_1(z - z_1),$$
$$D_1(z) = \bar{c}_1(1 - \bar{z}_1 z) + \bar{w}_1(z - z_1).$$

Then (6.3) is the parametrization we seek for the simple interpolation problem $f(z_1) = w_1$, $f \in B$.

Now suppose further that $f \in E_n$ for $n \geq 2$. Then (6.2) determines $f_1(z_j)$ for $2 \leq j \leq n$. Solving (6.2), we obtain

$$(6.4) \qquad f_1(z_j) = w_j^{(1)}, \quad 2 \leq j \leq n.$$

Clearly $|w_j^{(1)}| \leq 1$, because $f_1 \in B$. Moreover, the case $|w_j^{(1)}| = 1$ for some $j, 2 \leq j \leq n$, is not possible, because then $f_1(z) \equiv w_j^{(1)}$, $|z| < 1$, and by (6.2) E_n contains exactly one function. Thus we have

$$|w_j^{(1)}| < 1, \quad j = 2, 3, \dots .$$

Something has been gained, however, because (6.2) allows us to disregard the value $f_1(z_1)$, and E_n is now described by the $n - 1$ equations (6.4), instead of by the original n equations.

Repeating the above reasoning, we fix c_2, $|c_2| < 1$, and write

(6.5)
$$\frac{f_1 - w_2^{(1)}}{1 - \overline{w}_2^{(1)} f_1} = \frac{f_2 + c_2}{1 + \overline{c}_2 f_2} \frac{z - z_2}{1 - \overline{z}_2 z}.$$

Now $f(z) \in E_2$ if and only if $f_1(z_2) = w_2^{(1)}$, and that happens if and only if (6.5) holds for some $f_2 \in B$. Furthermore, when $n > 2$, $f \in E_n$ if and only if

$$f_2(z_j) = w_j^{(2)}, \quad 3 \leq j \leq n,$$

where $w_j^{(2)}$ is determined by (6.5). We also have $|w_j^{(2)}| < 1$, for the same reason that we had $|w_j^{(1)}| < 1$.

Continue by induction, always assuming E_n contains more than one function. For $k \leq n$, $f \in E_n$ if and only if there are

$$f_0, \quad f_1, \quad \dots, \quad f_k$$

in B such that $f_0 = f$, such that

$$f_k(z_j) = w_j^{(k)}, \quad k + 1 \leq j \leq n,$$

where $|w_j^{(k)}| < 1$, and $w_j^{(0)} = w_j$, and such that

(6.6)
$$\frac{f_{k-1} - w_k^{(k-1)}}{1 - \overline{w}_k^{(k-1)} f_{k-1}} = \frac{f_k + c_k}{1 + \overline{c}_k f_k} \frac{z - z_k}{1 - \overline{z}_k z},$$

where $|c_k| < 1$, c_k to be determined. (The explicit values of the $w_j^{(k)}$ are not important here.)

We now iterate (6.6) and obtain a one-to-one mapping of B onto E_n. Rewrite (6.6)

(6.7)
$$f_{k-1}(z) = \frac{\alpha_k(z) + \beta_k(z) f_k(z)}{\gamma_k(z) + \delta_k(z) f_k(z)},$$

in which

$$
\begin{aligned}
\alpha_k(z) &= w_k^{(k-1)}(1 - \bar{z}_k z) + c_k(z - z_k), \\
\beta_k(z) &= \bar{c}_k w_k^{(k-1)}(1 - \bar{z}_k z) + (z - z_k), \\
\gamma_k(z) &= (1 - \bar{z}_k z) + c_k \overline{w}_k^{(k-1)}(z - z_k), \\
\delta_k(z) &= \bar{c}_k(1 - \bar{z}_k z) + \overline{w}_k^{(k-1)}(z - z_k).
\end{aligned}
$$

(6.8)

By induction, (6.3) and (6.7) yield

(6.9)
$$
f(z) = \frac{A_n(z) + B_n(z) f_n(z)}{C_n(z) + D_n(z) f_n(z)},
$$

in which

$$
\begin{aligned}
A_n &= \gamma_n A_{n-1} + \alpha_n B_{n-1}, \\
B_n &= \delta_n A_{n-1} + \beta_n B_{n-1}, \\
C_n &= \gamma_n C_{n-1} + \alpha_n D_{n-1}, \\
D_n &= \delta_n C_{n-1} + \beta_n D_{n-1}
\end{aligned}
$$

(6.10)

are polynomials of degree at most n in z. Then we have proved

Lemma 6.1. *Suppose the polynomials $A_n(z)$, $B_n(z)$, $C_n(z)$, $D_n(z)$ are defined by (6.8) and (6.10). Then $f(z) \in E_n$ if and only if $f(z)$ satisfies (6.9) for some $f_n(z) \in B$.*

The polynomials $A_n, B_n, C_n,$ and D_n depend on the parameters $c_1, c_2, \ldots, c_n$. Now fix $c_k = \bar{z}_k w_k^{(k-1)}$. This makes

$$
\delta_n(0) = D_n(0) = 0
$$

in (6.8) and (6.10), which will facilitate the convergence argument below.

When $|z| < 1$, (6.9) shows that $\{f(z) : f \in E_n\}$ is a closed disc $\Delta_n(z)$ contained in $\bar{D}$ and defined by

(6.11)
$$
\left\{ \frac{A_n(z) + B_n(z)\zeta}{C_n(z) + D_n(z)\zeta} : |\zeta| \leq 1 \right\}.
$$

When $|z| = 1$, this formula for $\Delta_n(z)$ makes sense, although some $f \in E_n$ are not defined at z. Take $\Delta_n(z) = \{f(z) : f \in A_o \cap E_n\}$, A_o the disc algebra. The disc $\Delta_n(z)$ degenerates into a point if and only if the determinant

$$
A_n(z) D_n(z) - B_n(z) C_n(z) = 0.
$$

By (6.10) and induction

(6.12)

$$
A_n D_n - B_n C_n = (\gamma_n \beta_n - \alpha_n \delta_n)(A_{n-1} D_{n-1} - B_{n-1} C_{n-1})
$$
$$
= \prod_{k=1}^{n} (1 - |z_k|^2 |w_k^{(k-1)}|^2)(1 - |w_k^{(k-1)}|^2)(z - z_k)(1 - \bar{z}_k z),
$$

where $w_k^{(0)} = w_k$. Thus $\Delta_n(z)$ reduces to a point if and only if $z = z_j$, $1 \leq j \leq n$. An elementary computation with (6.11) shows that $\Delta_n(z)$ has radius

$$\rho_n(z) = \frac{|A_n D_n - B_n C_n|}{|C_n|^2 - |D_n|^2}.$$

Lemma 6.2. *If* $E_n \neq \varnothing$, *then when* $|z| = 1$, $\Delta_n(z) = \bar{D}$, $\rho_n(z) = 1$, *and*

$$\begin{aligned}
|B_n(z)| &= |C_n(z)|, \\
|A_n(z)| &= |D_n(z)|, \\
A_n(z)/C_n(z) &= \overline{(D_n(z)/B_n(z))} = \lambda_n(z),
\end{aligned}$$

(6.13)

with $|\lambda_n(z)| < 1$.

Proof. The conditions (6.13) follow from the assertion that $\Delta_n(z) = \bar{D}$, by the well-known characterization of the coefficients of the linear fractional transformations

$$\zeta \rightarrow \frac{A + B\zeta}{C + D\zeta}$$

that map $\bar{D}$ onto itself. That $\Delta_n(z) = \bar{D}$, $|z| = 1$, is proved by induction. Fix $z = e^{i\theta}$ and $\zeta \in \bar{D}$. For $n = 1$ there exists a constant $f_1 \in B$ such that

$$\frac{f_1(e^{i\theta}) + c_1}{1 + \bar{c}_1 f_1(e^{i\theta})} = \left(\frac{\zeta - w_1}{1 - \overline{w}_1 \zeta}\right) \bigg/ \left(\frac{e^{i\theta} - z_1}{1 - \bar{z}_1 e^{i\theta}}\right)$$

(6.14)

and (6.2) then produces $f \in E_1$ such that $f(e^{i\theta}) = \zeta$. For $n > 1$ the set

$$\{f_1 \in B : f_1(z_j) = w_j^{(1)}, \ j = 2, \ldots, n\}$$

is nonempty, because $E_n \neq \varnothing$, and by induction this set contains a function f_1 satisfying (6.14). As before, (6.2) then gives us $f \in E_n$ such that $f(e^{i\theta}) = \zeta$. □

Lemma 6.3. $C_n(z)$ *has no zeros in* $|z| \leq 1$.

Proof. If $C_n(z) = 0$, then by taking $\zeta = 0$ in (6.11), we see that $A_n(z) = 0$. Then

$$A_n(z)D_n(z) - B_n(z)C_n(z) = 0$$

and by (6.12), $z = z_j$ for some $j = 1, 2, \ldots, n$. Because by (6.12) the determinant $A_n D_n - B_n C_n$ has only simple zeros, we also have

$$A_n'(z_j)D_n(z_j) - B_n(z_j)C_n'(z_j) \neq 0,$$

when $C_n(z_j) = A_n(z_j) = 0$. Still taking $\zeta = 0$ in (6.11), we then obtain

$$w_j = \lim_{z \to z_j} A_n(z)/C_n(z) = \lim_{z \to z_j} A_n'(z)/C_n'(Z).$$

If $C'_n(z_j) = 0$, this means that $A'_n(z_j) = 0$, while if $C'_n(z_j) \neq 0$, it means that $A'_n(z_j)/C'_n(z_j) = w_j$. Thus we have

$$A'_n(z_j) = w_j C'_n(z_j).$$

On the other hand, taking $\zeta = 1$ in (6.11) gives

$$B_n(z_j) = w_j D_n(z_j).$$

Therefore

$$A'_n(z_j)D_n(z_j) - B_n(z_j)C'_n(z_j) = 0,$$

a contradiction. □

It will be convenient to renormalize (6.9). Let

$$\psi_n(z) = \left(\prod_{k=1}^{n} \frac{|z_k|}{-\bar{z}_k}(1 - |z_k|^2|w_k^{(k-1)}|^2)(1 - |w_k^{(k-1)}|^2)(1 - \bar{z}_k z)^2 \right)^{1/2}.$$

Since $\psi_n(z)$ has no zeros in $\bar{D}$,

$$P_n = \frac{A_n}{\psi_n}, \quad Q_n = \frac{B_n}{\psi_n}, \quad R_n = \frac{C_n}{\psi_n}, \quad \text{and} \quad S_n = \frac{D_n}{\psi_n}$$

are rational functions analytic in $|z| \leq 1$. From (6.9) we obtain

(6.15) $$f(z) = \frac{P_n(z) + Q_n(z)f_n(z)}{R_n(z) + S_n(z)f_n(z)}, \quad f_n \in B,$$

as our new parametrization of E_n. The advantage of this normalization is that by (6.12) the determinant is now

$$P_n(z)S_n(z) - Q_n(z)R_n(z) = \Pi_n(z) = \prod_{k=1}^{n} \frac{(-\bar{z}_k)}{|z_k|} \frac{z - z_k}{1 - \bar{z}_k z},$$

the Blaschke product having zeros $\{z_1, z_2, \ldots, z_n\}$ and normalized by $\Pi_n(0) > 0$. We also have

$$\rho_n(z) = \frac{|\Pi_n(z)|}{|R_n(z)|^2 - |S_n(z)|^2},$$

so that by Lemma 6.2,

(6.16) $$|R_n(z)|^2 - |S_n(z)|^2 = 1, \quad |z| = 1.$$

Now by Lemma 6.3, R_n has no zeros in $|z| \leq 1$, and (6.16) gives

$$\frac{1}{|R_n(z)|} \leq 1, \quad |z| \leq 1.$$

Then (6.13) and the maximum principle yield

$$|P_n(z)| \le |R_n(z)|,$$
(6.17)
$$|Q_n(z)| \le |R_n(z)|,$$
$$|S_n(z)| \le |R_n(z)|$$

on $|z| \le 1$.

Theorem 6.4. *Assume* $E_\infty = \bigcap E_n$ *contains two functions with different values at* $z = 0$. *Then there are* $n_j \to \infty$ *such that the limits*

$$P(z) = \lim_j P_{n_j}(z),$$

$$Q(z) = \lim_j Q_{n_j}(z),$$

$$R(z) = \lim_j R_{n_j}(z),$$

$$S(z) = \lim_j S_{n_j}(z)$$

all exist. The limits do not all vanish identically; in fact,

$$P(z)S(z) - Q(z)R(z) = \Pi(z),$$

the Blaschke product with zeros $\{z_n\}$. *If* $f(z) \in B$, *then* $f \in E_\infty$ *if and only if*

(6.18)
$$f(z) = \frac{P(z) + Q(z)f_\infty(z)}{R(z) + S(z)f_\infty(z)}, \quad |z| < 1,$$

for some $f_\infty \in B$.

Proof. The hypothesis that $\bigcap E_n$ contains two functions with different values at $z = 0$ implies that

$$\lim_n \rho_n(0) > 0.$$

(This limit exists because $\rho_{n+1} \le \rho_n$ since $E_{n+1} \subset E_n$.) Since $D_n(0) = 0$, by the choice of the constants c_n in (6.6), we have

$$|R_n(0)|^2 = |\Pi_n(0)|/\rho_n(0)$$

and hence

(6.19)
$$\lim_{n \to \infty} |R_n(0)| < \infty.$$

Choose n_j so that the bounded sequence $\{1/R_{n_j}(z)\}$ converges on D. Because $1/R_n(z)$ has no zeros in $|z| < 1$, the limit function either vanishes identically on D or it has no zeros on D. By (6.19) the limit does not vanish at $z = 0$, and so the limit is zero free on D. Consequently $R_{n_j}(z)$ converges uniformly on compact subsets of $|z| < 1$ to an analytic function $R(z)$, which has no zeros on $|z| < 1$. By (6.17) we can take a finer subsequence $\{n_j\}$ so that

$$\lim_j P_{n_j}(z) = P(z),$$
$$\lim_j Q_{n_j}(z) = Q(z),$$

and

$$\lim_j S_{n_j}(z) = S(z)$$

also all exist. Then

$$P(z)S(z) - Q(z)R(z) = \lim_j \Pi_{n_j}(z) = \Pi(z),$$

and so the limit functions are not all identically zero.

If $f \in \bigcap E_n$, then by Lemma 6.1 there is $f_n \in B$ such that

$$f = \frac{P_n + Q_n f_n}{R_n + S_n f_n}.$$

Refine the subsequence $\{n_j\}$ so that $f_{n_j}(z) \to f_\infty(z)$, $|z| < 1$, with $f_\infty \in B$. Then (6.18) follows. Conversely, if $f_\infty \in B$, then

$$f^{(n)} = \frac{P_n + Q_n f_\infty}{R_n + S_n f_\infty}.$$

is in E_n, and $f^{(n_j)}$ has limit $f \in E_\infty$. $\quad\square$

It turns out that P, Q, R, and S do not depend on the subsequence $\{n_j\}$. They are uniquely determined by the original interpolation problem (6.1) and by the choice of the constants c_n in (6.6). We shall not make use of this fact, however.

Before turning to Nevanlinna's proof that E_∞ contains inner functions, we mention a simple consequence of (6.18). Because $E_{n+1} \subset E_n$, the discs $\Delta_n(z)$, $|z| < 1$, decrease to a limit disc $\Delta_\infty(z) = \{f(z) : f \in E_\infty\}$. By (6.18),

$$\Delta_\infty(z) = \left\{ \frac{P(z) + Q(z)\zeta}{R(z) + S(z)\zeta} : |\zeta| < 1 \right\}.$$

Since the determinant of the coefficients for this mapping is $P(z)S(z) - Q(z)R(z) = \Pi(z)$, $\Delta_\infty(z)$ is nontrivial whenever $z \notin \{z_j\}$. The radius of $\Delta_\infty(z)$ is

$$\rho_\infty(z) = \lim_n \rho_n(z) = \frac{\Pi(z)}{|R(z)^2| - |S(z)|^2}.$$

Returning to $\Delta_n(z_0)$, $|z_0| < 1$, and to its parametrization (6.11), we see that $f \in E_n$ solves the extremal problem $f(z_0) \in \partial\Delta_n(z_0)$ if and only if

$$f(z) = \frac{A_n(z) + B_n(z)e^{i\varphi}}{C_n(z) + D_n(z)e^{i\varphi}} = \frac{P_n(z) + Q_n(z)e^{i\varphi}}{R_n(z) + S_n(z)e^{i\varphi}}$$

for some unimodular constant $e^{i\varphi}$. Thus $f(z)$ is a rational function of degree at most n. On two other occasions (Corollary 1.9 of this chapter and Corollary 2.4, Chapter I), we have seen that the extremal function $f(z)$ is a finite Blaschke product. This fact also follows from the reasoning in this section, because by Lemma 6.2, $|f(z)| = 1$ for $|z| = 1$. Further analysis of the polynomials $A_n(z)$, $B_n(z)$, $C_n(z)$, and $D_n(z)$ shows that the Blaschke product $f(z)$ has degree n. (See Exercise 20.)

The preceding discussion suggests that we might find an inner function in E_∞ by setting $f_\infty = e^{i\varphi}$ in (6.18). That is how Nevanlinna first proved Theorem 4.1. The proof above of Theorem 4.4 was based on the same idea: The unimodular function $h_0 \in K$ was obtained by maximizing the linear functional $\text{Re} \int h \, d\theta/2\pi, h \in K$.

We return to Theorem 4.1, which we restate as follows.

Theorem 6.5. *If E_∞ contains two distinct functions, and if $e^{i\varphi}$ is any unimodular constant, then*

(6.20)
$$f(z) = \frac{P(z) + Q(z)e^{i\varphi}}{R(z) + S(z)e^{i\varphi}}$$

is an inner function in E_∞.

Proof. By Theorem 6.4, $f \in E_\infty$, and in particular $\|f\|_\infty \leq 1$. We suppose there exists $E \subset T, |E| \equiv \int \chi_E(\theta)d\theta/2\pi > 0$, such that

(6.21)
$$|f(e^{i\theta})| \leq \alpha < 1, \quad \theta \in E,$$

and we argue toward a contradiction.

Reindex so that $R_n(z) \to R(z)$, $P_n(z) \to P(z)$, $Q_n(z) \to Q(z)$, and $S_n(z) \to S(z)$. Fix $M > |R(0)|$. Since by Lemma 6.3 $R_n(z)$ has no zeros on $\bar{D}$, we have

$$\log M > \log |R_n(0)| = \frac{1}{2\pi} \int \log |R_n(e^{i\theta})| d\theta$$

for large n, say $n > n_0$. By (6.16), $\log |R_n(e^{i\theta})| \geq 0$, and hence

$$|\{\theta : \log |R_n(e^{i\theta})| > 2 \log M/|E|\}| \leq |E|/2,$$

$n > n_0$. Therefore

$$E_n = \{\theta \in E : |R_n(e^{i\theta})| \leq M^{2/|E|}\}$$

satisfies

$$|E_n| \geq |E|/2, \quad n > n_0.$$

By (6.16) and (6.13), we also have

$$\left|\frac{P_n(e^{i\theta})}{R_n(e^{i\theta})}\right|^2 = 1 - \frac{1}{|R_n(e^{i\theta})|^2},$$

which gives

(6.22) $$\left| \frac{P_n(e^{i\theta})}{R_n(e^{i\theta})} \right| \le \{1 - M^{-4/|E|}\}^{1/2} = \beta < 1,$$

for $\theta \in E_n, n > n_0$.

Because $f \in E_n$ there exists $f_n \in B$ such that

$$f(z) = \frac{P_n(z) + Q_n(z)f_n(z)}{R_n(z) + S_n(z)f_n(z)}.$$

Since the rational functions P_n, Q_n, R_n, and S_n are continuous on $\bar{D}$ and since $|P_n S_n - Q_n R_n| = 1$ on T, $f_n(z)$ has radial limit $f_n(e^{i\theta})$ whenever $f(z)$ has radial limit $f(e^{i\theta})$. Thus at $e^{i\theta} \in E$, $f_k(e^{i\theta})$ exists and

$$f(e^{i\theta}) = \frac{P_n(e^{i\theta}) + Q_n(e^{i\theta})f_n(e^{i\theta})}{R_n(e^{i\theta}) + S_n(e^{i\theta})f_n(e^{i\theta})}.$$

By Lemma 6.2, the mapping

$$\zeta \to \frac{P_n(e^{i\theta}) + Q_n(e^{i\theta})\zeta}{R_n(e^{i\theta}) + S_n(e^{i\theta})\zeta},$$

which is an automorphism of D, preserves the pseudohyperbolic distance. Hence

$$|f_n(e^{i\theta})| = \rho(f_n(e^{i\theta}), 0) = \rho(f(e^{i\theta}), P_n(e^{i\theta})/Q_n(e^{i\theta})).$$

For $n > n_0$ and for $e^{i\theta} \in E_n$, (6.21) and (6.22) then yield

$$\begin{aligned}|f_n(e^{i\theta})| &\le \frac{|f(e^{i\theta})| + |P_n(e^{i\theta})|/|R_n(e^{i\theta})|}{1 + |f(e^{i\theta})||P_n(e^{i\theta})|/|R_n(e^{i\theta})|} \\ &\le \frac{\alpha + \beta}{1 + \alpha\beta} = \gamma < 1.\end{aligned}$$

Consequently

$$|f_n(0)| \le \frac{1}{2\pi} \int |f_n(e^{i\theta})|d\theta \le (\gamma|E_n| + (1 - |E_n|))$$

$$\le \left\{ \gamma \frac{|E|}{2} + \left(1 - \frac{|E|}{2}\right) \right\} = \eta < 1.$$

Take a subsequence $\{f_{n_j}\}$ so that $f_{n_j}(0) \to \zeta, |\zeta| \le \eta < 1$. Then

$$f(0) = \lim_j \frac{P_{n_j}(0) + Q_{n_j}(0)f_{n_j}(0)}{R_{n_j}(0) + S_{n_j}(0)f_{n_j}(0)} = \frac{P(0) + Q(0)\zeta}{R(0) + S(0)\zeta}.$$

Since $P(0)S(0) - Q(0)R(0) \ne 0$, we conclude from (6.20) that $\zeta = e^{i\varphi}$, a contradiction. $\square$

An interesting apparently unsolved problem is to determine whether or when E_∞ contains a Blaschke product. See page vii.

Notes

The first systematic treatment of dual extremal problems is the paper of Macintyre and Rogosinski [1950]. The functional analytic methods were introduced by Havinson [1949,1951] and by Rogosinski and Shapiro [1953]. Duren's book [1970] contains a slightly different treatment and some references to the older literature. See also Goluzin [1952] and Landau [1916].

Section 2 is from the paper of Carleson and Jacobs [1972]. Kahane [1974] discusses a number of related questions. An interesting open problem is to find intrinsic necessary and sufficient conditions for a function in L^∞ to have unique best approximation in H^∞. Exercise 17 gives a partial answer.

The primary references for Section 3 are Szegö [1920] and Helson and Szegö [1960]. Exercises 8 and 14 outline results similar to the Helson–Szegö theorem.

Most of the material in Sections 4 and 5 originates with Adamyan, Arov, and Krein [1968], but the proofs in the text are considerably different from their spectral theory approach. In [1971] Adamyan, Arov, and Krein extend their results to the matrix valued case. Theorem 5.1 and Theorem 5.2 are from deLeeuw and Rudin [1958].

Section 6 is from Nevanlinna's paper [1929], which includes a number of other classical results, all derived ultimately from Schwarz's lemma. It is a fundamental paper long overlooked. Schur's [1917] treatment of the coefficient problem is very similar. It is outlined in Exercise 21.

Exercises and Further Results

1. If $f \in L^p(\mathbb{R})$, $1 \le p \le \infty$, then

$$\inf_{g \in H^p} \|f - g\|_p = \sup_{\substack{G \in H^q \\ \|G\|_q = 1}} \left| \int fG \, dx \right|.$$

2. Let $f(z)$ be meromorphic on $|z| < 1$, and suppose that on some annulus $R < |z| < 1$, $f(z)$ is analytic and of class H^1 (i.e. $|f(z)|$ possesses a harmonic majorant). Then $f(e^{i\theta})$ exists almost everywhere. Prove $f(e^{i\theta}) \ge 0$ almost everywhere if and only if $f(z)$ is a rational function of the form

$$c \prod_{j=1}^{n} (z - \alpha_j)(1 - \bar{\alpha}_j z) \bigg/ \prod_{j=1}^{n} (z - \beta_j)(1 - \beta_j z),$$

where $|\alpha_j| \le 1$, $|\beta_j| \le 1$, $c > 0$.

3. Let $c_0, c_1, \ldots, c_N$ be given complex numbers and consider the maximum problem

$$M = \sup_{\substack{f \in H^\infty \\ \|f\| \le 1}} \left| \sum_{j=0}^{N} c_j a_j \right|,$$

where $f(z) = \sum a_j z^j$.

(a) The dual extremal problem is

$$M = \inf_{g \in H_0^1} \|k - g\|_1,$$

where $k(z) = \sum_{j=0}^{n} c_j z^{-j}$. It is equivalent to the minimum problem

$$M = \inf\{\|h\|_1 : h \in H^1 : h = c_N + c_{N-1}z + \cdots + c_0 z^N + \cdots\}.$$

(b) The original extremal problem has unique extremal function f_0 and the dual problem has unique minimizing function g_0 Moreover

$$f_0(z)(k(z) - g_0(z)) = |k(z) - g_0(z)|, \quad |z| = 1,$$

so that

$$f_0(z)(k(z) - g_0(z)) = cz^q \prod_{j=1}^{n} (z - \alpha_j)(1 - \bar{\alpha}_j z) \Big/ z^N$$

with $n + q = N, 0 < |\alpha_j| \le 1, c > 0$. Reindexing $\alpha_1, \ldots, \alpha_n$, there is $s, 0 \le s \le n$ such that $|\alpha_j| < 1, j \le s$, and

$$f_0(z) = \gamma z^q \prod_{j=1}^{s} \left(\frac{z - \alpha_j}{1 - \bar{\alpha}_j z} \right)$$

and

$$k(z) - g_0(z) = c\bar{\gamma} \prod_{j=1}^{n} (1 - \alpha_j z)^2 \prod_{s+1}^{n} \left(\frac{z - \alpha_j}{1 - \bar{\alpha}_j z} \right) \Big/ z^N,$$

where $|\gamma| = 1$.

(c) For $|z|$ small write

$$(c_N + c_{N-1}z + \cdots + c_0 z^N)^{1/2} = \sum_{j=0}^{\infty} \lambda_j z^j$$

and set

$$P_N(z) = \sum_{j=0}^{N} \lambda_j z^j.$$

Assume $P_N(z)$ has no zeros in $|z| < 1$. Then there exist $\alpha_1, \ldots, \alpha_n$, $n \leq N$, such that $0 < |\alpha_j| \leq 1$ and such that

$$P_N(z) = \lambda_0 \prod_{j=1}^{n}(1 - \bar{\alpha}_j z).$$

Then

$$z^{-N} P_N^2 = c_N z^{-N} \prod_{j=1}^{n}(1 - \bar{\alpha}_j z)^2$$

has the form $k - g$, $g \in H^1$. Setting

$$f_0(z) = \frac{\bar{c}_N}{|c_N|} z^{N-n} \prod_{j=1}^{n}\left(\frac{z - \alpha_j}{1 - \bar{\alpha}_j z}\right),$$

we obtain $f_0 P_N^2 / z^N \geq 0$ on $|z| = 1$. Consequently f_0 is the extremal function, $k(z) - g_0(z) = P_N^2(z)/z^N$, and

$$M = \|P_N^2\|_1 = \sum_{j=0}^{N}|\lambda_j|^2.$$

(d) Landau [1913, 1916] determined the extremum M in the special case $c_j = 1$ by a different method. In this case $P_N = \Sigma_0^N \lambda_j z^j$, where

$$\lambda_j = (-1)^j \binom{-\frac{1}{2}}{j} = \frac{(2_j)!}{4^j (j!)^2},$$

and by Wallis's formula

$$M \sim \log N/\pi \quad (N \to \infty).$$

Thus M has the same order of magnitude as the Lebesgue constants $\|D_N\|_1$, where $D_N = \Sigma_{-N}^N e^{ij\theta}$ is Dirichlet's kernel for partial sums.

The dual problem in part (a) was first treated by F. Riesz [1920] using a variational argument.

4. A function $f \in C(T)$ is called *badly approximable* if its nearest function g in H^∞ satisfies $g \equiv 0$. Show f is badly approximable if and only if $|f(e^{i\theta})|$ is a positive constant and $f(e^{i\theta})$ has negative winding number (see Poreda [1972] or Gamelin, Garnett, Rubel, and Shields [1976]).

5. Let $g \in H^\infty$ be the best approximation to $f \in C(T)$. It conjectured that g is continuous if the conjugate function $\hat{f}$ is continuous. (See page vii.) The converse of this is false. Use Exercise 4 above to find a badly approximable function whose conjugate function is not continuous (D. Sarason, unpublished).

6. Let $f \in C(T)$, $f \notin H^\infty$, let $g \in H^\infty$ be the best approximation to f, and let $F \in H_0^1$ be a dual extremal function, so that

$$(f - g)F = \|f - g\|_\infty |F|$$

almost everywhere.

(a) F can be chosen so that $F(z)/z$ is outer.

(b) If $F_1 \in H_0^1$ is another dual extremal function, then F/F_1 is a rational function.

(c) F is unique if and only if $z/F \in H^1$. In that case $z/F \in H^p$ for all $p < \infty$.

(d) More generally, if $f \in L^\infty \setminus H^\infty$, if F exists, and if $z/F \in H^1$, then F is unique.

(See deLeeuw and Rudin [1958] and Carleson and Jacobs [1972].)

7. If $f \in H^\infty$, then

$$\mathrm{dist}(f, A_o) \le 2\,\mathrm{dist}(f, C(T)),$$

where $A_o = H^\infty \cap C(T)$ is the disc algebra. The constant 2 is sharp. (See Davie, Gamelin, and Garnett [1973].)

★8. Let $d\mu = w\, d\theta/2\pi + d\mu_s$ be a positive measure on the circle. Then

$$\inf_{f \in \mathrm{F}} \int |1 - \mathrm{Re} f|^2 d\mu = \left(\int w^{-1} \frac{d\theta}{2\pi} \right)^{-1},$$

where F is the set of trigonometric polynomials $\Sigma_{n>0}\, a_n e^{in\theta}$ (see Grenander and Rosenblatt [1957]). The result is due to Kolmogoroff [1941].

9. Find two closed subspaces F and G of a Hilbert space such that

$$\mathrm{F} \cap \mathrm{G} = \{0\}$$

but such that

$$1 = \sup\{|\langle f, g \rangle| : f \in \mathrm{F}, g \in \mathrm{G}, \|f\| = \|g\| = 1\}.$$

10. If F and G are subspaces of $L^2(\mu)$, then $\rho = 1$ if and only if

$$0 = \inf \left\{ \int |f - g|^2 d\mu : f \in \mathrm{F}, g \in \mathrm{G}; \|f\| \ge 1, \|g\| \ge 1 \right\}.$$

Hence the property $\rho = 1$ is unchanged if $d\mu$ is replaced by $w\, d\mu$, w bounded above and below.

11. If a weight function w satisfies the Helson–Szegö condition (3.3), then for some $\varepsilon > 0$, $w \in L^{1+\varepsilon}$ and $w^{-1} \in L^{1+\varepsilon}$.

12. Let w be a weight function. Assume $\psi = \widetilde{(\log w)}$ is continuous except for jumps at a finite number of points. Then w satisfies (3.3) if and only if each jump is less than π (Helson and Szegö [1960]).

13. Let α be real. There is a constant K_α such that

$$\int_{-\pi}^{\pi} |\tilde{p}(\theta)|^2 |\theta|^\alpha d\theta \le K_\alpha \int_{-\pi}^{\pi} |p(\theta)|^2 |\theta|^\alpha d\theta$$

for all trigonometric polynomials if and only if $-1 < \alpha < 1$ (see Hardy and Littlewood [1936]).

★★**14.** Let F_n be the set of all trigonometric polynomials $\sum_{k \ge n} a_k e^{ik\theta}$ and let $G_n = \overline{F}_n$ be the set of trigonometric polynomials of the form $\sum_{k \ge n} b_k e^{-ik\theta}$. When μ is a positive finite measure on T write

$$\rho_n = \sup \left| \int f \bar{g} \, d\mu \right|,$$

where $f \in F_n, g \in G_n, \int |f|^2 d\mu \le 1$, and $\int |g|^2 d\mu \le 1$. Wright

$$d\mu = w \, d\theta/2\pi + d\mu_s$$

with $d\mu_s$, singular to $d\theta$.
  (a) If $d\mu_s \ne 0$ or if $\log w \notin L^1$, then $\rho_n = 1$ for all n.
  (b) Now assume $d\mu = w \, d\theta/2\pi$ with

$$\varphi = \log w \in L^1.$$

Let W be the set of weights w such that

$$\lim_{n \to \infty} \rho_n = 0,$$

where $d\mu = w \, d\theta/2\pi$. Then $w \in W$ if and only if $e^{-i\tilde{\varphi}} \in H^\infty + C$.
  (c) $H^\infty + C$ is a closed subalgebra of L^∞.
  (d) Let W_0 be the set of positive weights w such that, for every $\varepsilon > 0$,

$$\varphi = \log w = r + \tilde{s} + t$$

with $\|r\|_\infty < \varepsilon$, $\|s\|_\infty < \varepsilon$, and $t \in C$. Then $W_0 \subset W$ and $w \in W$ if and only if $w = |p|^2 w_0$, where $w_0 \in W_0$ and p is a trigonometric polynomial. (The set $\log W_0$ is the space VMO to be studied in Chapter VI.) (See Helson and Sarason [1967].)

15. Let u be a unimodular function in L^∞ and set

$$S_u = \{G \in H^1 : \|G\|_1 = 1, G/|G| = u\}.$$

  (a) There exist functions u for which $S_u = \varnothing$.
  (b) If $F \in S_u$, and if $1/F \in H^1$, then $S_u = \{F\}$.
  (c) If $F \in S_u = \{F\}$, then for all $\alpha, |\alpha| = 1, (z - \alpha)^{-2} F(z) \notin H^1$. (See deLeeuw and Rudin [1958].)

16. Let $h \in L^\infty, |h| = 1$ almost everywhere.
  (a) If $\text{dist}(h, H^\infty) < 1$, then $h = F/|F|$ for some $F \in H^1$. Moreover, the outer factor of F is invertible in H^1. (See Lemma 5.4.)

(b) If dist(h, H^∞) < 1 but dist (h, H_0^∞) $= 1$, then $h = \bar{F}/|F|$ for some $F \in H^1$. (See Case 1 of the proof of Theorem 4.4.) It follows that dist($\bar{h}$, H^∞) < 1.

17. (a) Let $h \in L^\infty$. If the coset $h + H^\infty$ is an extreme point of ball (L^∞/H^∞) then there is $g \in H^\infty$ such that $|h + g| = 1$ almost everywhere.

(b) Now suppose $|h| = 1$ almost everywhere. Then $h + H^\infty$ is an extreme point of ball(L^∞/H^∞) if and only if $\|h + g\| > 1$ for all $g \in H^\infty$, $g \neq 0$. Thus the extreme points of ball(L^∞/h^∞) are the cosets containing exactly one function of unit modulus (see Koosis [1971]).

(c) If $|h| = 1$ almost everywhere, then $\|h + g\|_\infty > 1$ for all $g \in H^\infty$, $g \neq 0$, if and only if h cannot be written as $F/|F|$, $F \in H^1$. (See Lemmas 5.4 and 5.5.) This is a weak generalization of Exercise 5. If $|h| = 1$, then h is badly approximable by $H^\infty(\|h - g\| > 1$ if $g \in H^\infty$, $g \neq 0)$ if and only if h is not the argument of an H^1 function. Example 4.2 shows there exist badly approximable functions not having constant modulus.

(d) L^∞/H^∞ is a dual space. Hence ball(L^∞/H^∞) is the weak-star closed convex hull of the set of cosets $\{h + H^\infty : |h| = 1, h$ badly approximable$\}$.

18. Let $F \in H^1$, $\|F\|_1 = 1$. Then every $g \in H^\infty$ such that

$$\|F/|F| - g\|_\infty \leq 1$$

is of the form (5.10) if and only if $\mathsf{S}_{F/|F|} = \{F\}$.

19. Let $f \in H^\infty$, $\|f\|_\infty = 1$. Then f is an extreme point of ball(H^∞) if and only if

$$\int \log(1 - |f(e^{i\theta})|)d\theta = -\infty$$

(see deLeeuw and Rudin [1958]).

20. In the notation of Section 6, and with $c_k = \bar{z}_k w_k^{(k-1)}$, $A_n(z)$ has degree at most $n - 1$ and $B_n(z)$ has degree n. When $e^{i\varphi}$ is a unimodular constant, the polynomials $A_n(z) + B_n(z)e^{i\varphi}$ and $C_n(z) + D_n(z)e^{i\varphi}$ have no common zero. Thus

$$f(z) = \frac{A_n(z) + B_n(z)e^{i\varphi}}{C_n(z) + D_n(z)e^{i\varphi}}$$

is a Blaschke product of degree n.

★21. Fix complex numbers $c_0, c_1, \ldots$ and set

$$E_n = \{f \in \mathrm{B} : f(z) = c_0 + c_1 z + \cdots + c_n z^n + \cdots\}.$$

Write $\gamma_0 = c_0$ and assume $|\gamma_0| \leq 1$.

(a) If $|\gamma_0| = 1$ then E_0 contains exactly one function, $f = \gamma_0$. Suppose $|\gamma_0| < 1$ and write $f \in E_0$

(E.1)
$$f = \frac{zf_1 + \gamma_0}{1 + \bar{\gamma}_0 z f_1},$$

with $f_1 \in B$. So E_0 is in one-to-one correspondence with B. Then $f \in E_1$ if and only if $f_1(0) = \gamma_1 = c_1/(1 - |c_0|^2)$. In particular $E_1 \neq \varnothing$ if and only if $|\gamma_1| \leq 1$, and E_1 consists of exactly one function if and only if $|\gamma_1| = 1$.

(b) Continue by induction. We obtain $\gamma_0, \gamma_1, \dots, \gamma_n, \dots$. If $|\gamma_0| < 1, |\gamma_1| < 1, \dots, |\gamma_n| < 1$, then E_n is in one-to-one correspondence with B through the formulas $f_0 = f$,

$$f_{k-1} = \frac{zf_k + \gamma_{k-1}}{1 + \bar{\gamma}_{k-1} z f_k}, \quad 1 \leq k \leq n,$$

$f_k \in \beta$. If $|\gamma_k| = 1$ but $|\gamma_j| < 1$, $j < k$, then E_k consists of exactly one function, a Blaschke product of degree k.

(c) Suppose $E_n \neq \varnothing$. Then the $(n + 1)$th coefficients c_{n+1} of the functions in E_n fill a closed disc with radius

(E.2)
$$\omega_n = (1 - |\gamma_0|) \cdots (1 - |\gamma_n|).$$

(Use (E.1) and induction.)

(d) Suppose $\bigcap E_n \neq \varnothing$. Then of course $\bigcap E_n = \{f\}$, $f = \sum_0^\infty c_n z^n$. Prove

$$\lim_{n \to \infty} \omega_n = \exp \frac{1}{2\pi} \int \log(1 - |f|^2) \, d\theta.$$

(Hint: By (E.1) and (E.2) we have, on $|z| = 1$,

$$1 - |f|^2 = \frac{(1 - |\gamma_0|^2)(1 - |f_1|^2)}{|1 + \bar{\gamma}_0 f_1|^2} = \frac{\omega_n(1 - |f_{n+1}|^2)}{\prod_{j=0}^n |1 + \bar{\gamma}_j z f_{j+1}|^2}.$$

Because the denominator has no zeros on D, this gives

$$\exp \frac{1}{2\pi} \int \log(1 - |f|^2) d\theta = \omega_n \exp \frac{1}{2\pi} \int \log(1 - |f_{n+1}|^2) d\theta \leq \omega_n.$$

For the reverse inequality use Szegös theorem. For $\varepsilon > 0$ there is $P(z) = 1 + b_1 z + \cdots + b_N z^N$ such that

$$\int |P|^2 (1 - |f|^2) \frac{d\theta}{2\pi} \leq \varepsilon + \exp \frac{1}{2\pi} \int \log(1 - |f|^2) d\theta.$$

Let $f^* \in E_n$ be obtained by setting $f_{n+1} = 0$. Then by Szegös theorem

$$\omega_n = \exp \int \log(1 - |f^*|^2) \frac{d\theta}{2\pi} \leq \int |P|^2 (1 - |f^*|^2) \frac{d\theta}{2\pi}.$$

By the Parseval relation,

$$\lim_{n \to \infty} \int |P|^2 (1 - |f^*|^2) \frac{d\theta}{2\pi} \leq \int |P|^2 (1 - |f|^2) \frac{d\theta}{2\pi}.$$

(e)   Let A_n be the matrix

$$\begin{bmatrix} c_0 & c_1 & \cdots & & c_n \\ 0 & c_0 & & & c_{n-1} \\ \vdots & & & & \vdots \\ 0 & & \cdots & 0 & c_0 \end{bmatrix}$$

and let I_n be the $(n + 1) \times (n + 1)$ identity matrix. Then $E_n \neq \varnothing$ if and only if $I_n - A_n^* A_n$ is nonnegative definite. If $I_n - A_n^* A_n$ is positive definite, then E_n is infinite. More precisely,

$$\det(I_n - A_n^* A_n) = \omega_0 \, \omega_1 \cdots \omega_n.$$

(See Schur [1917], except for part (d), which is due to Boyd [1979].)

V

Some Uniform Algebra

This chapter develops the background from uniform algebra theory which will be needed for our analysis of H^∞ below. Our treatment is quite brief. For a complete picture of the general theory the reader is referred to the books of Browder [1969], Gamelin [1969], and Stout [1971].

However, two topics special to H^∞ will be covered in detail. In Section 2 we prove Marshall's theorem that the Blaschke products generate H^∞. In Section 5, three theorems on the predual of H^∞ are proved by representing linear functionals as measures on the Šilov boundary of H^∞.

1. Maximal Ideal Spaces

A *Banach algebra* is a complex algebra A which is also a Banach space under a norm satisfying

$$(1.1) \qquad \|fg\| \le \|f\| \|g\|, \quad f, g \in A.$$

We always assume that A is commutative ($fg = gf$, $f, g \in A$) and that there is a unit $1 \in A (1 \cdot f = f, f \in A)$. The correspondence $\lambda \to \lambda \cdot 1$ identifies the complex field $\mathbb{C}$ as a subalgebra of A. We say $f \in A$ is *invertible* if there is $g \in A$ such that $gf = 1$. The unique inverse is denoted by $g = f^{-1}$. Write

$$A^{-1} = \{f \in A : f^{-1} \text{ exists}\}$$

for the set of invertible elements of A. A *complex homomorphism*, or *multiplicative linear functional*, is a nonzero homomorphism $m : A \to \mathbb{C}$ from A into the complex numbers. Trivially $m(1) = 1$.

Theorem 1.1. *Every complex homomorphism of A is a continuous linear functional with norm at most one,*

$$\|m\| = \sup_{\|f\| \le 1} |m(f)| \le 1.$$

Proof. Because m is linear, we only have to prove that $\|m\| \leq 1$. If m is unbounded, or if $\|m\| > 1$, then there is $f \in A$ such that $\|f\| < 1$ but such that $m(f) = 1$. By (1.1) the series

$$\sum_{n=0}^{\infty} f^n$$

is norm convergent. Its sum satisfies

(1.2) $$(1 - f) \sum_{n=0}^{\infty} f^n = 1,$$

so that $1 - f \in A^{-1}$. But then

$$1 = m(1) = m((1 - f)^{-1})(m(1) - m(f)) = 0,$$

a contradiction. $\quad\square$

Theorem 1.2. *Suppose M is a maximal (proper) ideal in A. Then M is the kernel of a complex homomorphism $m : A \to \mathbb{C}$.*

Proof. There are two steps. First we show M is closed. Now, if the closure $\bar{M}$ of M is proper, that is, if $\bar{M} \neq A$, then $\bar{M}$ is also an ideal in A. Therefore M is closed if $\bar{M} \neq A$, because M is maximal. However, if $g \in M$, then $g \notin A^{-1}$ and (1.2), applied to $f = 1 - g$, shows that $\|1 - g\| \geq 1$. Hence $1 \notin \bar{M}$ and M is closed.

The second step is to show the quotient algebra $B = A/M$ satisfies

$$B = \mathbb{C} \cdot 1,$$

where $1 = 1 + M$ now denotes the unit in B. The quotient mapping will then define the complex homomorphism with kernel M. Since M is maximal, $B = A/M$ is a field, and since M is closed, B is complete in the quotient norm

$$\|f + M\| = \inf_{g \in M} \|f + g\|,$$

which also satisfies (1.1).

Suppose there exists $f \in B \backslash \mathbb{C} \cdot 1$. Then $f - \lambda \in B^{-1}$ for all $\lambda \in \mathbb{C}$, because B is a field. On the disc $|\lambda - \lambda_0| < 1/\|(f - \lambda_0)^{-1}\|$, the series

(1.3) $$\sum_{n=0}^{\infty} (\lambda - \lambda_0)^n ((f - \lambda_0)^{-1})^{n+1}$$

converges in norm to $(f - \lambda)^{-1}$, because of the identity

$$\frac{1}{f - \lambda} = \frac{1}{(f - \lambda_0)} \frac{1}{[1 - (\lambda - \lambda_0)/(f - \lambda_0)]}.$$

Clearly $f^{-1} \neq 0$, and by the Hahn–Banach theorem there is a bounded linear functional L on B such that $\|L\| = 1$ and

$$(1.4) \qquad\qquad\qquad L(f^{-1}) \neq 0.$$

Since (1.3) is norm convergent and since $\|L\| = 1$, we have

$$F(\lambda) = L((f - \lambda)^{-1}) = \sum_{n=0}^{\infty} L(((f - \lambda_0)^{-1})^{n+1})(\lambda - \lambda_0)^n,$$

when $|\lambda - \lambda_0| < 1/\|(f - \lambda_0)^{-1}\|$. Because λ_0 is arbitrary, this means that $F(\lambda)$ is an entire analytic function. Now for $|\lambda|$ large, (1.2) yields

$$\|(f - \lambda)^{-1}\| = \frac{1}{|\lambda|} \left\| \left(1 - \frac{f}{\lambda}\right)^{-1} \right\| \leq \frac{1}{|\lambda|} \sum_{n=0}^{\infty} \frac{\|f\|^n}{|\lambda|^n}.$$

Consequently,

$$|F(\lambda)| = |L((f - \lambda)^{-1})| \leq C/|\lambda|, \qquad |\lambda| \text{ large,}$$

and by Liouville's theorem $F \equiv 0$. Hence

$$L(f^{-1}) = F(0) = 0,$$

contradicting (1.4).   □

The set $\mathfrak{M}_A$ of complex homomorphisms of A is called the *spectrum* or *maximal ideal space* of A. By Theorem 1.1, $\mathfrak{M}_A$ is contained in the unit ball of the dual Banach space A^*. Give $\mathfrak{M}_A$ the weak-star topology of A^*, in which a basic neighborhood V of $m_0 \in \mathfrak{M}_A$ is determined by $\varepsilon > 0$ and by $f_1, f_2, \ldots, f_n \in A$:

$$V = \{m \in \mathfrak{M}_A : |m(f_j) - m_0(f_j)| < \varepsilon, 1 \leq j \leq n\}.$$

This topology on $\mathfrak{M}_A$ is called the *Gelfand topology*. With the Gelfand topology $\mathfrak{M}_A$ is a weak-star closed subset of ball(A^*), because

$$\mathfrak{M}_A = \{m \in \text{ball}(A^*) : m(fg) = m(f)m(g), f, g \in A\}.$$

By the Banach–Alaoglu theorem, which says ball(A^*) is weak-star compact, $\mathfrak{M}_A$ is a compact Hausdorff space. Writing

$$\hat{f}(m) = m(f), \quad f \in A, \quad m \in \mathfrak{M}_A,$$

we have a homomorphism $f \to \hat{f}$ from A into $C(\mathfrak{M}_A)$, the algebra of continuous complex functions on $\mathfrak{M}_A$. This homomorphism is called the *Gelfand transform*. By Theorem 1.1, the Gelfand transfrom is norm decreasing:

$$\|\hat{f}\| = \sup_{m \in \mathfrak{M}_A} |\hat{f}(m)| \leq \|f\|.$$

By Theorem 1.2, $f \in A^{-1}$ if and only if $\hat{f}(m)$ is nowhere zero. Indeed, if $f \notin A^{-1}$, then by Zorn's lemma the ideal $\{fg : g \in A\}$ is contained in a (proper) maximal ideal.

The Banach algebra A is called a *uniform algebra* if the Gelfand transform is an isometry, that is, if

$$\|\hat{f}\| = \|f\|, \qquad f \in A.$$

Theorem 1.3. *The Gelfand transform is an isometry if and only if*

(1.5) $$\|f^2\| = \|f\|^2$$

for all $f \in A$.

Proof. Since $\|\hat{f}\|$ is a supremum, $\|\hat{f}^2\| = \|\hat{f}\|^2$ and (1.5) holds for any uniform algebra.

Now assume (1.5). By Theorem 1.1, we have $\|\hat{f}\| \leq \|f\|$. To complete the proof we take $f \in A$ with $\|\hat{f}\| = 1$, we fix $\varepsilon > 0$, and we show $\|f\| \leq 1 + \varepsilon$. By Theorem 1.2, $f - \lambda \in A^{-1}$ when $|\lambda| > 1 = \|\hat{f}\|$. By (1.3) the A-valued function $(f - \lambda)^{-1}$ is analytic on $|\lambda| > 1$. This means that whenever $L \in A^*$, the scalar function $F(\lambda) = L((f - \lambda)^{-1})$ is analytic on $|\lambda| > 1$. By compactness,

$$\sup_{|\lambda|=1+\varepsilon} \|(f - \lambda)^{-1}\| = K$$

is finite. By (1.2),

$$F(\lambda) = L\left(\frac{-1}{\lambda}\left(1 - \frac{f}{\lambda}\right)^{-1}\right) = -\sum_{n=0}^{\infty} \frac{L(f^n)}{\lambda^{n+1}}, \qquad |\lambda| > \|f\|.$$

This series must also represent $F(\lambda)$ on $|\lambda| > 1$. Taking $\|L\| = 1$ we obtain from Cauchy's theorem,

$$L(f^n) = \left|-\frac{1}{2\pi i}\int_{|\lambda|=1+\varepsilon} F(\lambda)\lambda^n d\lambda\right| \leq (1 + \varepsilon)^{n+1} K.$$

Consequently, by the Hahn–Banach theorem,

$$\|f^n\| = \sup_{\|L\|=1} |L(f^n)| \leq (1 + \varepsilon)^{n+1} K.$$

Setting $n = 2^k$ and using (1.5), we conclude that

$$\|f\| \leq \lim_{k \to \infty} (K(1 + \varepsilon)^{2^k+1})^{1/2^k} = 1 + \varepsilon,$$

as desired. □

When A is a uniform algebra, the range $\hat{A}$ of the Gelfand transform is a uniformly closed subalgebra of $C(\mathfrak{M}_A)$, and $\hat{A}$ is isometrically isomorphic to

A. In that case we identify f with $\hat{f}$ and write

$$f(m) = m(f) = \hat{f}(m), \quad f \in A, \quad m \in \mathfrak{M}.$$

Thus we view A as a uniformly closed algebra of continuous functions on $\mathfrak{M}_A$. Note that A separates the points of $\mathfrak{M}_A$ and that A contains the constant functions on $\mathfrak{M}_A$.

Example 1. Suppose A is any algebra of continuous complex functions on a compact Hausdorff space Y. If A has the uniform norm, $\|f\| = \sup_{y \in Y} |f(y)|$ and if A is complete, then A is a uniform algebra. If A contains the constant functions and separates the points of Y, then Y is homeomorphic to a closed subset of $\mathfrak{M}_A$, and we say that A is a *uniform algebra on Y*. This is the generic example, because any uniform algebra A is clearly a uniform algebra on its spectrum $Y = \mathfrak{M}_A$. If $A = C(Y)$, then $\mathfrak{M}_A = Y$. (See Exercise 5.)

Example 2. Let l^∞ denote the space of bounded complex sequences. With the norm $\|x\| = \sup_n |x_n|$ and with the pointwise multiplication $(xy)_n = x_n y_n$, l^∞ is a uniform algebra, by Theorem 1.3. The maximal ideal space of l^∞ has the special name $\beta\mathbb{N}$, the *Stone–Čech compactification* of the positive integers $\mathbb{N}$.

The Gelfand transform of l^∞ is $C(\beta\mathbb{N})$. To see this, note that if $x \in l^\infty$ is real, that is, if $x_n \in \mathbb{R}$ for all n, then $\hat{x}(m)$ is real on $\mathfrak{M}_{l^\infty} = \beta\mathbb{N}$, because then $(x - \lambda)^{-1} \in l^\infty$ whenever Im $\lambda \neq 0$. It now follows from the Stone–Weierstrass theorem that $l^\infty = C(\beta\mathbb{N})$.

Since the functional $m_n(x) = x_n$ is multiplicative on l^∞, $\mathbb{N}$ can be identified with a subset of $\beta\mathbb{N}$, and the Gelfand topology is defined in such a way that $\mathbb{N}$ is homeomorphic to its image in $\beta\mathbb{N}$. Moreover, $\mathbb{N}$ is dense in $\beta\mathbb{N}$, because every function in $C(\beta\mathbb{N}) = l^\infty$ is completely determined by its behavior on $\mathbb{N}$.

The Stone–Čech compactification $\beta\mathbb{N}$ can also be characterized functorially.

Theorem 1.4. *Let Y be a compact Hausdorff space and let $\tau : \mathbb{N} \to Y$ be a continuous mapping. Then the mapping τ has a unique continuous extension $\tilde{\tau} : \beta\mathbb{N} \to Y$.*

If $\tau(\mathbb{N})$ is dense in Y and if the images of disjoint subsets of $\mathbb{N}$ have disjoint closures in Y, then the extension $\tilde{\tau}$ is a homeomorphism of $\beta\mathbb{N}$ onto Y.

Proof. The mapping

$$T : C(Y) \to l^\infty,$$

defined by $Tf(n) = f \circ \tau(n)$, is a homomorphism from $C(Y)$ into l^∞. Because T is continuous, the adjoint mapping $T^* : (l^\infty)^* \to (C(Y))^*$ is weak-star to weak-star continuous. If $m \in (l^\infty)^*$ is a multiplicative linear functional, that is, if $m \in \beta\mathbb{N}$, then since $T(fg) = T(f)T(g)$, $T^*(m)$ is also a multiplicative linear functional on $C(Y)$, and by Exercise 5, $T^*(m) \in Y = \mathfrak{M}_{C(Y)}$. Restricting T^* to $\beta\mathbb{N}$, we have a mapping $\tilde{\tau}(m) = T^*(m)$ from $\beta\mathbb{N}$ into Y, and by the definition of T, $\tilde{\tau}(n) = \tau(n)$, $n \in \mathbb{N}$. Since $\beta\mathbb{N}$ has the weak-star topology of

$(l^\infty)^*$ and $Y = \mathfrak{M}_{C(Y)}$ has the weak-star topology of $(C(Y))^*$, the mapping $\tilde{\tau}$ is continuous. Because $\mathbb{N}$ is dense in $\beta\mathbb{N}$, $\tilde{\tau}$ is the unique continuous extension of τ to $\beta\mathbb{N}$.

Now suppose that $\overline{\tau(S_1)} \cap \overline{\tau(S_2)} = \varnothing$ wherever $S_1 \subset \mathbb{N}$, $S_2 \subset \mathbb{N}$ and $S_1 \cap S_2 = \varnothing$. Then there is $f \in C(Y)$ such that $f = 1$ on $\overline{\tau(S_1)}$ and $f = 0$ on $\overline{\tau(S_2)}$. We can assume that $S_2 = \mathbb{N} \backslash S_1$, so that $Tf = \chi_{S_1}$. Because linear combinations of characteristic functions are dense in l^∞, this means the range of T is dense in l^∞. If we further assume that $\tau(\mathbb{N})$ is dense in Y, then we have $\|Tf\| = \|f\|$, $f \in C(Y)$, and the mapping T is an isometry. Consequently the range of T is norm closed in l^∞. Hence the range of T is both dense and closed and T maps $C(Y)$ onto l^∞. Now because $\|Tf\| = \|f\|$, the homomorphism T is one-to-one, and so T is an algebra isomorphism from $C(Y)$ onto l^∞. The adjoint T^* then defines a homeomorphism from $\beta\mathbb{N}$ onto Y. $\square$

Theorem 1.4 determines the space $\beta\mathbb{N}$ up to a homeomorphism, because if Z is another compact Hausdorff space and if Z contains a dense sequence $\{z_n\}$ homeomorphic to $\mathbb{N}$ such that Theorem 1.4 holds with Z in place of $\beta\mathbb{N}$, then the correspondence

$$n \leftrightarrow z_n$$

extends to a homeomorphism between $\beta\mathbb{N}$ and Z.

The space $\beta\mathbb{N}$ is extremely huge. It can be mapped onto any separable compact Hausdorff space. No point of $\beta\mathbb{N} \backslash \mathbb{N}$ can be exhibited concretely.

Example 3. The space L^∞ of essentially bounded, measurable functions on the unit circle is a uniform algebra when it is given the pointwise multiplication and the essential supremum norm

$$\|f\| = \inf\{\alpha : |f| \le \alpha \text{ almost everywhere}\}.$$

We fix the notation X for the maximal ideal space of L^∞, because this space will be reappearing from time to time. Under the Gelfand transform, L^∞ is isomorphic to $C(X)$, the algebra of continuous complex functions on X. This has the same proof as the corresponding result on l^∞. If $f \in L^\infty$ is real, then $(f - \lambda)^{-1} \in L^\infty$ whenever Im $\lambda \ne 0$, so that $\hat{f}$ is real on X. The Stone-Weierstrass theorem then shows that $L^\infty = C(X)$.

Like $\beta\mathbb{N} \backslash \mathbb{N}$, the space X is large and intractable. We cannot construct a single point of X. Nevertheless, the space X is quite useful in the theory of bounded analytic functions. Some of the intricacies of X are outlined in Exercise 8. See Hoffman's book [1962a] for further details.

Let A be a uniform algebra on $\mathfrak{M}_A$. A closed subset K of $\mathfrak{M}_A$ is called a *boundary for A* if

$$\|f\| = \sup_{m \in K} |f(m)|$$

for all $f \in A$.

Theorem 1.5.   *There is a smallest closed boundary K_0, which is contained in every boundary K.*

This smallest boundary is called the *Šilov boundary* of A. Note that A is a uniform algebra on its Šilov boundary.

Proof.   Let K_0 be the intersection of all boundaries. We must show K_0 is a boundary for A.

Lemma 1.6.   *Let $f_1, f_2, \ldots, f_n \in A$ and set*

$$U = \{m : |f_j(m)| < 1, j = 1, \ldots, n\}.$$

Then either $U \cap K \neq \varnothing$ for every boundary K, or else $K \setminus U$ is a boundary for every boundary K.

Accepting Lemma 1.6 for a moment, we prove Theorem 1.5. Suppose $f \in A$ and $|f| < 1$ on K_0. Set $J = \{m : |f(m)| \geq 1\}$. If we show $J = \varnothing$ for every such f, then we will have proved that K_0 is a boundary. Since $J \cap K_0 = \varnothing$, each $m \in J$ has (by the definition of K_0) a neighborhood U of the form in the lemma such that $U \cap K_m = \varnothing$ for some boundary K_m. Cover J by finitely many such neighborhoods U_i, $1 \leq i \leq N$. Then by the lemma, $K \setminus U_i$ is a boundary whenever K is a boundary. Consequently, by induction

$$K_1 = \mathfrak{M}_A \setminus \bigcup_{i=1}^{N} U_i = ((\mathfrak{M}_A \setminus U_1) \setminus U_2) \setminus \ldots \setminus U_N)$$

is a boundary. Since $|f| < 1$ on K_1, this means $\|f\| < 1$ and $J = \varnothing$, so that K_0 is a boundary for A. $\square$

Proof of Lemma 1.6.   We suppose that K is a boundary but that $K \setminus U$ is not a boundary, and we show that U intersects every boundary for A. By hypothesis there is $f \in A$ such that $\|f\| = 1$ but such that $\sup_{K \setminus U} |f(m)| < 1$. Replacing f by a power f^n, we can assume that $\sup_{K \setminus U} |f(m)| < \varepsilon$, where $\varepsilon \|f_j\| < 1$, $j = 1, \ldots, n$, and where $f_1, \ldots, f_n$ are the functions defining U. Then $|ff_j| < 1$ on U by the definition of U, while $|ff_j| < 1$ on $K \setminus U$ by the choice of ε. Since K is a boundary, that means $\|ff_j\| < 1$, $j = 1, \ldots, n$. Hence

$$\{m : |f(m)| = 1\} \subset \bigcap_{j=1}^{n} \{m : |f_j(m)| < 1\} = U.$$

Since $\{m : |f(m)| = 1 = \|f\|\}$ meets every boundary, this implies that U also meets every boundary. $\square$

Of course, if $A = C(Y)$, then A has Šilov boundary Y. A point $x \in \mathfrak{M}_A$ is a *peak point* for A if there is $f \in A$ such that

$$f(x) = 1,$$
$$|f(y)| < 1, \quad y \in \mathfrak{M}_A, \; y \neq x.$$

Clearly, every peak point for A is in the Šilov boundary of A.

Example 4.   The *disc algebra* is the algebra of functions continuous on the closed disc $\overline{D}$ and analytic on the open disc D. We reserve the notation A_o for this algebra. With the supremum norm $\|f\| = \sup_{z \in D} |f(z)|$, A_o is a uniform algebra. Because analytic polynomials separate the points of $\overline{D}$, A_o is a uniform algebra on $\overline{D}$. By the maximum principle, the unit circle T is a boundary for A_o. If $\lambda \in T$, then $f(z) = (1 + \bar{\lambda}z)/2$ satisfies $f(\lambda) = 1, |f(z)| < 1, z \neq \lambda$. Thus λ is a peak point for A_o and T is the Šilov boundary of A_o. The maximal ideal space of A_o is $\overline{D}$. (See Exercise 9.)

Example 5. H^∞ is a uniform algebra with pointwise multiplication and with the supremum norm

$$\|f\| = \sup_{z \in D} |f(z)|.$$

We shall always write $\mathfrak{M}$ for the maximal ideal space of H^∞.

For each point $\zeta \in D$ there exists $m_\zeta \in \mathfrak{M}$ such that $m_\zeta(z) = \zeta$, where z denotes the coordinate function, because $(z - \zeta) \notin (H^\infty)^{-1}$. Now whenever $f \in H^\infty$, we have $(f - f(\zeta))/(z - \zeta) \in H^\infty$, and

$$f = f(\zeta) + (z - \zeta)\left(\frac{f - f(\zeta)}{z - \zeta}\right).$$

But then

$$m_\zeta(f) = f(\zeta) + m_\zeta(z - \zeta)m_\zeta\left(\frac{f - f(\zeta)}{z - \zeta}\right) = f(\zeta),$$

so that the point $m_\zeta \in \mathfrak{M}$ is uniquely determined by the condition $m_\zeta(z) = \zeta$. Hence

$$\zeta \to m_\zeta$$

defines an embedding of D into $\mathfrak{M}$. By the definition of the topology of $\mathfrak{M}$, this embedding is a homeomorphism. We now identify ζ with m_ζ and regard D as a subset of $\mathfrak{M}$. Then D is an open subset of $\mathfrak{M}$ because

$$D = \{m \in \mathfrak{M} : |\hat{z}(m)| < 1\}.$$

From now on we identify the uniform algebra H^∞ with its Gelfand transform and we think of H^∞ as a subalgebra of $C(\mathfrak{M})$. There is no ambiguity in doing this because by the discussion above

$$\hat{f}(m_\zeta) = f(\zeta), \qquad \zeta \in D.$$

Now suppose $|\zeta| = 1$. Then $(z - \zeta) \notin (H^\infty)^{-1}$ and there exist points $m \in \mathfrak{M}$ such that $\hat{z}(m) = \zeta$. As we shall see in a moment, however, the *fiber* $\mathfrak{M}_\zeta = \{m : \hat{z}(m) = \zeta\}$ is very large when $\zeta \in \partial D$.

The Gelfand transform of the coordinate function z defines a map

$$\hat{z} : \mathfrak{M} \to \overline{D}.$$

Having identified the open disc D with $\hat{z}^{-1}(D)$, we can write

$$\mathfrak{M} = D \cup \bigcup_{|\zeta|=1} \mathfrak{M}_\zeta.$$

Thus we can imagine $\mathfrak{M}$ as the open disc D with the large compact space $\mathfrak{M}_\zeta = \hat{z}^{-1}(\zeta)$ lying above $\zeta \in \partial D$. The fibers $\mathfrak{M}_\zeta$ over points $\zeta \in \partial D$ are homeomorphic to one another because the rotation $\tau(z) = \zeta z, |\zeta| = 1$, induces an automorphism $f \to f \circ \tau$ of H^∞, and the adjoint of this automorphism maps $\mathfrak{M}_1$ onto $\mathfrak{M}_\zeta$.

To see just how large $\mathfrak{M}_\zeta$ is, take $\zeta = 1$ and consider the singular function $S(z) = \exp((z + 1)/(z - 1))$. In Chapter II we showed that the cluster set of $S(z)$ at $\zeta = 1$ is the closed unit disc. That is, whenever $|w| \leq 1$ there exists a sequence $\{z_n\}$ in D such that $z_n \to 1$ and $S(z_n) \to w$. By the compactness of $\mathfrak{M}$ the sequence $\{z_n\}$ has a cluster point $m \in \mathfrak{M}_1$ and $S(m) = w$. Hence S maps $\mathfrak{M}_1$ onto the closed unit disc.

Moreover, there exists a sequence $\{z_n\}$ in D such that $\lim z_n = 1$ and such that every interpolation problem

$$f(z_n) = \alpha_n, \qquad n = 1, 2, \ldots,$$

$\{\alpha_n\} \in l^\infty$, has solution $f \in H^\infty$. Such sequences, which are called *interpolating sequences*, will be discussed in Chapter VII, and a simple example of an interpolating sequence is given in Exercise 11. Here we only want to make this observation: If $\{z_n\}$ is an interpolating sequence, then by Theorem 1.4 the map $n \to z_n$ extends to define a homeomorphism from $\beta\mathbb{N}$ onto the closure of $\{z_n\}$ in $\mathfrak{M}$. Since $\lim_n z_n = 1$ (as a sequence in the plane), we now see that $\mathfrak{M}_1$ contains a homeomorphic copy of $\beta\mathbb{N}\backslash\mathbb{N}$.

By Fatou's theorem, H^∞ is a closed subalgebra of L^∞. Now H^∞ separates the points of $X = \mathfrak{M}_{L^\infty}$, because every real L^∞ function has the form $u = \log|f|$ with $f \in (H^\infty)^{-1}$ (see Theorem 4.5, Chapter II), and because $u(m) = \log|f(m)|$ when u and f are viewed as elements of $C(X)$. Hence by compactness, the continuous map $X \to \mathfrak{M}$, which is defined by restricting each multiplicative linear functional on L^∞ to H^∞, is a homeomorphism. Accordingly, we think of X as a closed subset of $\mathfrak{M}$. Since the injection $H^\infty \subset L^\infty$ is isometric, X is a boundary for H^∞. Moreover, if K is a proper closed subset of X, then since $C(X) = L^\infty = \log|(H^\infty)^{-1}|$, there is $f \in H^\infty$ such that

$$\sup_K \log|f| < \sup_X \log|f|,$$

and so K is not a boundary for H^∞. We have proved the following theorem:

Theorem 1.7. *The Šilov boundary of H^∞ is $X = \mathfrak{M}_{L^\infty}$.*

Every inner function has unit modulus on X, because it has unit modulus when viewed as an element of L^∞. So the singular function $S(z) = \exp((z + 1)/(z - 1))$ satisfies $|S| = 1$ on X. Now $S \notin (H^\infty)^{-1}$, but S has no zeros on D.

Therefore

$$\mathfrak{M} \neq D \cup X.$$

Carleson's corona theorem states that D is dense in $\mathfrak{M}$. In other words, the corona $\mathfrak{M} \setminus \bar{D}$ is empty. This famous result will be proved in Chapter VIII. For the present we only translate its statement into classical language.

Theorem 1.8. *The open disc D is dense in $\mathfrak{M}$ if and only if the following condition holds: if $f_1, \ldots, f_n \in H^\infty$ and if*

(1.6) $$\max_{1 \leq j \leq n} |f_j(z)| \geq \delta > 0$$

for all $z \in D$, then there exist $g_1, \ldots, g_n \in H^\infty$ such that

(1.7) $$f_1 g_1 + \cdots + f_n g_n = 1.$$

Proof. Suppose D is dense in $\mathfrak{M}$. Then by continuity we have

$$\max_{1 \leq j \leq n} |f_j(m)| \geq \delta$$

for all $m \in \mathfrak{M}$, so that $\{f_1, \ldots, f_n\}$ is contained in no proper ideal of H^∞. Hence the ideal J generated by $\{f_1, \ldots, f_n\}$ contains the constant 1. But

$$J = \{f_1 g_1 + \cdots + f_n g_n : g_j \in H^\infty\},$$

and so (1.7) holds.

Conversely, suppose D is not dense in $\mathfrak{M}$. Then some point $m_0 \in \mathfrak{M}$ has a neighborhood disjoint from D. This neighborhood has the form

$$V = \bigcap_{j=1}^{n} \{m : |f_j(m)| < \delta\},$$

where $\delta > 0$, and where $f_1, \ldots, f_n \in H^\infty$, $f_j(m_0) = 0$. The functions $f_1, \ldots, f_n$ satisfy (1.6) because $V \cap D = \varnothing$, but they do not satisfy (1.7) with $g_1, \ldots, g_n \in H^\infty$, because they all lie in the ideal $\{f : f(m_0) = 0\}$. $\square$

For the disc algebra A_o the "corona theorem," that $\mathfrak{M}_A = D$, is a very easy consequence of the Gelfand theory (see Exercise 9). Therefore, whenever (1.6) holds for $f_1, \ldots, f_n \in A_o$, there exist $g_1, \ldots, g_n \in A_o$ such that (1.7) holds. Now suppose we knew the "corona theorem with bounds" for A_o. In other words, suppose that whenever $f_1, \ldots, f_n \in A_o$ satisfied (1.6), we could find $g_1, \ldots, g_n \in A_o$ that solved (1.7) and in addition satisfied

(1.8) $$\|g_j\| \leq C\left(n, \delta, \max_j \|f_j\|\right).$$

Then the corona theorem for H^∞ would follow by a simple normal families argument: Given $f_1, \ldots, f_n \in H^\infty$ having (1.6), and given $r < 1$, we could take $g_1^{(r)}, \ldots, g_n^{(r)}$ in A_o such that

$$\sum f_j(rz) g_j^{(r)}(z) = 1, \quad z \in D,$$

and such that $\|g_j^{(r)}\|_\infty \le C(n, \delta, \max_j \|f_j\|)$. For some sequence $r_k \to 1$,

$$g_j(z) = \lim_{k \to \infty} g_j^{(r_k)}(z)$$

would then provide H^∞ solutions of (1.7). In other words, we would get the corona theorem for H^∞ if we had a proof of the easy "corona theorem" for A_o that was constructive enough to include the bounds (1.8).

2. Inner Functions

Recall that a function $u \in H^\infty$ is an inner function if $|u(e^{i\theta})| = 1$ almost everywhere. Every Blaschke product is an inner function, and by Frostman's theorem, Theorem II.6.4, every inner function is the uniform limit of a sequence of Blaschke products. In this section we prove that the inner functions, and therefore the Blaschke products, generate H^∞ as a uniform algebra.

Theorem 2.1 (Douglas–Rudin). *Suppose U is a unimodular function in L^∞, $|U(e^{i\theta})| = 1$ almost everywhere. For any $\varepsilon > 0$ there exist inner functions u_1, u_2 in H^∞ such that*

$$\|U - u_1/u_2\| < \varepsilon.$$

Before proving Theorem 2.1 let us consider the corresponding result for continuous functions on the circle T. It can be proved in a few lines. Suppose $U \in C(T)$, $|U| \equiv 1$. Write $U = z^n V^2$, where n is an integer (the winding number of U) and $V \in C(T)$, $|V| \equiv 1$. By Weierstrass's approximation theorem there is a rational function $h(z)$, analytic on T, such that

$$|V(z) - h(z)| < \varepsilon, \qquad z \in T.$$

Since $|V| \equiv 1$, we also have

$$|V(z) - (1/\overline{h(1/\bar{z})})| < \varepsilon/(1 - \varepsilon), \qquad z \in T.$$

The rational function

$$g(z) = z^n h(z)/\overline{h(1/\bar{z})}$$

is then a quotient of finite Blaschke products, because $|g| \equiv 1$ on T, and we have

$$|U(z) - g(z)| < \varepsilon + \varepsilon/(1 - \varepsilon), \qquad z \in T.$$

Proof of Theorem 2.1. Let E be a measurable subset of T with $|E| > 0$. We may assume

$$U = \alpha \chi_E + \beta \chi_{T \setminus E},$$

where $|\alpha| = |\beta| = 1, \alpha \neq \beta$, because the finite products of functions of this form are norm dense in the set of unimodular L^∞ functions.

Consider the circular arcs

$$A = \{e^{i\theta} : |e^{i\theta} - \alpha| \leq \varepsilon/2\}, \qquad B = \{e^{i\theta} : |e^{i\theta} - \beta| \leq \varepsilon/2\}.$$

We may suppose $|\alpha - \beta| > 2\varepsilon$, so that $A \cap B = \varnothing$. Let Ω be the complement of $A \cup B$ in the Riemann sphere. There exists $r, 0 < r < 1$, such that the annulus

$$V = \{r < |w| < 1/r\}$$

can be mapped conformally onto Ω. Moreover, the conformal mapping $\varphi : V \to \Omega$ extends continuously to $\overline{V}$ and

$$\varphi(|w| = r) \subset A, \qquad \varphi(|w| = 1/r) \subset B$$

(we can replace φ by $\varphi(1/w)$ if necessary). See Ahlfors [1966, p. 247]. The function φ is analytic on V except for a simple pole at some point p.

Now let $h \in (H^\infty)^{-1}$ be an outer function such that $|h| = r\chi_E + (1/r)\chi_{T \setminus E}$ almost everywhere. Then $h(D) \subset V$ and

$$\|U - \varphi \circ h\| \leq \varepsilon.$$

We shall show that $\varphi \circ h$ is a quotient of inner functions. Note that $\varphi \circ h$ is meromorphic on D with poles only on the set $h^{-1}(p)$. The function $\zeta(w) = w + 1/w$ maps V onto an ellipse W. Let $G(\zeta)$ be a conformal mapping from W to the unit disc, with $G(\zeta(p)) = 0$. Then $\psi(w) = G(\zeta(w))$ is analytic on V and continuous on $\overline{V}$. Moreover, $|\psi(w)| = 1$ on ∂V and $\psi(p) = 0$. Consequently $u_1 = (\varphi\psi) \circ h$ and $u_2 = \psi \circ h$ are inner functions for which we have

$$\|U - u_1/u_2\|_\infty \leq \varepsilon. \quad \square$$

Another proof of Theorem 2.1 is given in Exercise 13. Recall that the Šilov boundary of H^∞ is $X = \mathfrak{M}_{L^\infty}$.

Theorem 2.2 (D. J. Newman). *If $m \in \mathfrak{M}$, the maximal ideal space of H^∞, then the following conditions are equivalent:*

(a) $m \in X$, *the Šilov boundary*,
(b) $|u(m)| = 1$ *for every inner function $u(z)$, and*
(c) $|B(m)| > 0$ *for every Blaschke product $B(z)$.*

Proof. Since $X = \mathfrak{M}_{L^\infty}$, and since every inner function is a unimodular L^∞ function, (a) implies (b).

Now suppose (b) holds. We show m can be extended to a multiplicative linear functional on L^∞. By Theorem 2.1, the algebra

$$\mathcal{I} = \left\{ \sum_{j=1}^{n} \lambda_j u_j \bar{v} : \lambda_j \in \mathbb{C}, \, v, u_j \text{ inner functions} \right\}$$

is norm dense in L^∞. On $\mathcal{I}$ define

$$\tilde{m}\left(\sum_{j=1}^{n} \lambda_j u_j \bar{v} \right) = \sum_{j=1}^{n} \lambda_j m(u_j) \overline{m(v)}.$$

Then because of (b), $\tilde{m}$ is well defined, linear, and multiplicative on $\mathcal{I}$ and

$$\left| \tilde{m}\left(\sum_{j=1}^{n} \lambda_j u_j \bar{v} \right) \right| = \left| m\left(\sum_{j=1}^{n} \lambda_j u_j \right) \right| \leq \left\| \sum_{j=1}^{n} \lambda_j u_j \right\| = \left\| \sum_{j=1}^{n} \lambda_j u_j \bar{v} \right\|.$$

Hence $\tilde{m}$ is bounded and $\tilde{m}$ has a unique continuous extension to L^∞. So there is $x \in X$ such that $\tilde{m}(g) = g(x)$, $g \in \mathcal{I}$. We must show that $m(f) = f(x)$, $f \in H^\infty$. Choose $g = \sum \lambda_j u_j \bar{v} \in \mathcal{I}$ such that $\|g - f\|_\infty < \varepsilon$. Then

$$|\tilde{m}(g) - f(x)| = |g(x) - f(x)| < \varepsilon.$$

Since $vg \in H^\infty$, $vf \in H^\infty$, and $|m(v)| = 1$, we have

$$|m(f) - \tilde{m}(g)| = |m(v)m(f) - m(v)\tilde{m}(g)| = |m(vf) - m(vg)| < \varepsilon,$$

by the definition of $\tilde{m}$. Therefore $|m(f) - f(x)| < 2\varepsilon$ and so $m \in X$.

Trivially, (b) implies (c). Now suppose (b) does not hold. Then $|u(m)| < 1$ for some inner function $u(z)$, and by Frostman's theorem, $|B(m)| < 1$ for some Blaschke product $B(z)$. We may assume that m is not the evaluation functional for some point in D, so that $|m(B_0)| = 1$ for every finite Blaschke product B_0. Thus $B(z)$ is an infinite product

$$B(z) = z^k \prod_{j=1}^{\infty} \frac{-\bar{z}_j}{|z_j|} \frac{z - z_j}{1 - \bar{z}_j z}.$$

Choose $n_j \to \infty$ such that $\sum n_j(1 - |z_j|) < \infty$, and form the Blaschke product

$$B_1(z) = \prod_{j=1}^{\infty} \left(\frac{-\bar{z}_j}{|z_j|} \frac{z - z_j}{1 - \bar{z}_j z} \right)^{n_j}.$$

For N large we have

$$B_1 = B_0^{(N)} B_1^{(N)} B_2^{(N)},$$

where $B_0^{(N)}$ is some finite Blaschke product, where

$$B_1^{(N)} = \prod_{j=N}^{\infty} \left(\frac{-\overline{z_j}}{|z_j|} \frac{z - z_j}{1 - \overline{z}_j z} \right)^{n_N},$$

and where $B_2^{(N)}$ is a third Blaschke product. Since $m \notin D$, $|B_0^{(N)}(m)| = 1$, and $|B_1^{(N)}(m)| = |B(m)|^{n_N}$. Clearly $|B_2^{(N)}(m)| \leq 1$, so that

$$|B_1(m)| \leq |B(m)|^{n_N}, \qquad N = 1, 2, \ldots .$$

Therefore $|B_1(m)| = 0$ and (c) implies (b). $\quad\square$

Let S be a subset of a uniform algebra A. We say that S *generates* A if the linear combinations of products of functions in S are norm dense in A. If S is closed under multiplication, then S generates A if and only if A is the closed linear span of S.

Theorem 2.3 (A. Bernard). *Let A be a uniform algebra on a compact Hausdorff space Y and let*

$$U = \{u \in A : |u| = 1 \text{ on } Y\}$$

be the set of unimodular functions in A. If U generates A, then the unit ball of A is the norm closed convex hull of U.

Proof. Let $f \in A$, $\|f\| < 1$. We can suppose $f = \sum_{j=1}^{n} \lambda_j u_j$, $u_j \in U$, $\lambda_j \in \mathbb{C}$, because functions of this form are dense in ball(A). Write $u = \prod_{j=1}^{n} u_j \in U$. Then $\bar{f}u \in A$. Now

$$f = \frac{1}{2\pi} \int_0^{2\pi} \frac{f + e^{it} u}{1 + e^{it} \bar{f} u} dt$$

at every point of Y. For each fixed e^{it} the integrand is a function in U. Since $\|f\| < 1$, there is a sequence of Riemann sums which converges to the integral uniformly in $y \in Y$. These sums are convex combinations of elements of U and they converge in norm to f. $\quad\square$

Corollary 2.4. *The unit ball of the disc algebra A_0 is the closed convex hull of the set of finite Blaschke products.*

Proof. The inner functions in A_0 are the finite Blaschke products. They generate A_0 because the polynomials are dense in A_0. $\quad\square$

Theorem 2.5 (Marshall). *H^∞ is generated by the Blaschke products.*

Corollary 2.6. *The unit ball of H^∞ is the norm closed convex hull of the set of Blaschke products.*

The corollary follows immediately from Theorem 2.5, from Theorem 2.3, and from Frostman's theorem. Notice that the Carathéodory result for pointwise convergence (Chapter I, Theorem 2.1) is much easier than Corollary 2.6.

The proof of Theorem 2.5 is a clever combination of three ingredients: the Douglas–Rudin theorem, Bernard's trick from the proof of Theorem 2.3, and the Nevanlinna theorem from Chapter IV. The proof brings out a connection between interpolation problems and approximation problems which is recurrent throughout this theory.

Proof of Theorem 2.5. By Frostman's theorem it is enough to prove that the inner functions generate H^∞. Let J be the closed subalgebra of H^∞ generated by the inner functions and set

$$\mathfrak{N} = \{f \in H^\infty : \bar{f}u \in H^\infty \text{ for some inner function } u\}.$$

We must show $J = H^\infty$. We need two preliminary observations:

(i) *If* $f = \sum \lambda_j u_j$ *is a linear combination of inner functions, then* $f \in \mathfrak{N}$;
(ii) $\mathfrak{N} \subset J$.

To prove (i) let $u = \prod u_j$ and note that

$$\bar{f}u = \sum_{j=1}^{n} \bar{\lambda}_j \prod_{k \neq j} u_k \in H^\infty.$$

The proof of (ii) uses Bernard's idea from Theorem 2.3. Suppose $f \in \mathfrak{N}$, $\|f\| < 1$. If u is an inner function such that $\bar{f}u \in H^\infty$, then for all real t, $(f + e^{it}u)/(1 + e^{it}\bar{f}u)$ is an inner function. Then the integral

$$f = \frac{1}{2\pi} \int_0^{2\pi} \frac{f + e^{it}u}{1 + e^{it}\bar{f}u} dt$$

expresses f as a uniform limit of convex combinations of inner functions.

Now let $f \in H^\infty$. By the Douglas–Rudin theorem (Theorem 2.1) and by Frostman's theorem, there exist inner functions $u_1, \ldots, u_n$, and complex numbers $\lambda_1, \ldots, \lambda_n$, and there exists a Blaschke product $B(z)$ such that

(2.1) $$\left\| f - \sum \lambda_j u_j \bar{B} \right\| < \varepsilon.$$

We next use Theorem 4.1, Chapter IV, to unwind the antiholomorphic factor $\bar{B}$. Let $g = \sum \lambda_j u_j$. Then by (2.1)

$$\|Bf - g\| < \varepsilon,$$

so that $|g(z_n)| < \varepsilon$ at the zeros $\{z_n\}$ of $B(z)$. By Chapter IV, Theorem 4.1, there exists an inner function $v(z)$ such that

$$\varepsilon v(z_n) = g(z_n), \quad n = 1, 2, \ldots .$$

Hence

$$g - \varepsilon v = Bh$$

for some $h \in H^\infty$. Recalling that $g = \sum \lambda_j u_j$, we see that $Bh \in \mathfrak{N}$ because of (i). Thus there is an inner function $u(z)$ such that $\bar{B}\bar{h}u \in H^\infty$. Consequently $B\bar{B}\bar{h}u = \bar{h}u \in H^\infty$, so that $h \in \mathfrak{N}$, and from (ii) we conclude that $h \in J$. But then

$$\|f - h\| = \|Bf - Bh\| \le \|Bf - g\| + \|g - Bh\|$$
$$= \|Bf - g\| + \varepsilon\|v\| \le 2\varepsilon,$$

and therefore $f \in J$. $\square$

3. Analytic Discs in Fibers

Let $\{z_n\}$ be a Blaschke sequence in D, $\sum(1 - |z_n|) < \infty$. We assume $\lim z_n = 1$. If $B(z)$ is the Blaschke product with zeros $\{z_n\}$, then

$$B'(z_n) = \frac{-\bar{z}_n}{|z_n|} \frac{1}{1 - |z_n|^2} \prod_{k;k \ne n} \frac{-\bar{z}_n}{|z_k|} \frac{z_n - z_k}{1 - \bar{z}_k z_n}$$

and

$$(1 - |z_n|^2)|B'(z_n)| = \prod_{k;k \ne n} \left| \frac{z_n - z_k}{1 - \bar{z}_k z_n} \right|.$$

We replace $\{z_n\}$ by a subsequence such that

$$\inf_n \prod_{k;k \ne n} \left| \frac{z_n - z_k}{1 - \bar{z}_k z_n} \right| = \delta > 0,$$

which can easily be accomplished by a diagonalization process. Thus we assume

(3.1) $$\inf_n (1 - |z_n|^2)|B'(z_n)| = \delta > 0.$$

This condition will be very important to us later because it characterizes interpolating sequences. Here we are only going to use (3.1) to find non-constant analytic maps from the open unit disc into the fiber $\mathfrak{M}_1$ of $\mathfrak{M}$ over $z = 1$. Let

$$L_n(\zeta) = \frac{\zeta + z_n}{1 + \bar{z}_n \zeta}, \quad |\zeta| < 1.$$

Then $f \circ L_n \in H^\infty$ whenever $f \in H^\infty$ and $f \circ L_n(0) = f(z_n)$. We regard L_n as a map from $\mathrm{D} = \{|\zeta| < 1\}$ into $\mathfrak{M}$. (The notation $\mathrm{D} = \{|\zeta| < 1\}$ is used here to distinguish the domain of L_n from its range $D \subset \mathfrak{M}$.) Now the space $\mathfrak{M}^\mathrm{D}$ of mappings (continuous or not) from D into $\mathfrak{M}$ is a compact Hausdorff space in the product topology. From the sequence $\{L_n\} \subset \mathfrak{M}^D$ we take a convergent subnet (L_{n_j}) with limit $L \in \mathfrak{M}^D$. (Nets are needed because $\mathfrak{M}^D$ is not a metric space.) We get a map

$$L : D \to \mathfrak{M}$$

such that for all $f \in H^\infty$

$$(3.2) \qquad f \circ L(\zeta) = \lim_j f \circ L_{n_j}(\zeta) = \lim_j f\left(\frac{\zeta + z_{n_j}}{1 + \bar{z}_{n_j}\zeta}\right).$$

Theorem 3.1. *The mapping $L : D \to \mathfrak{M}$ has the following three properties:*

(a) $L(D) \subset \mathfrak{M}_1$, *the fiber of $\mathfrak{M}$ over $z = 1$;*

(b) L *is an analytic mapping, that is, $f \circ L(\zeta)$ is analytic on D whenever $f \in H^\infty$;*

(c)   *the mapping L is not constant.*

Proof.   Take $f(z) = z$ in (3.2). Then

$$\hat{z}(L(\zeta)) = \lim_j \left(\frac{\zeta + z_{n_j}}{1 + \bar{z}_{n_j}\zeta}\right) = 1,$$

since $\lim z_n = 1$, and (a) holds. Property (b) follows from (3.2) and from the fact that the limit of any bounded net of functions analytic on D is an analytic function on D. It also follows from (3.2) that $(f \circ L)'(\zeta) = \lim_j (f \circ L_{n_j})'(\zeta)$, whenever $f \in H^\infty$. Hence

$$(B \circ L)'(0) = \lim_j (B \circ L_{n_j})'(0) = \lim_j B'(z_{n_j})L'_{n_j}(0)$$

$$= \lim_j (1 - |z_{n_j}|^2)B'(z_{n_j}),$$

so that by (3.1) we have

$$(3.3) \qquad\qquad |(B \circ L)'(0)| \geq \delta > 0.$$

Hence $B \circ L$ is not constant and (c) holds. $\square$

We will see in Chapter X that L is actually $1 - 1$ from D into $\mathfrak{M}_1$ and that $\mathfrak{M}_1$ contains uncountably many pairwise disjoint such analytic discs $L(D)$.

By (3.3) and by Schwarz's lemma (see Exercise 1, Chapter I), there exists $\eta = \eta(\delta) > 0$ such that $L(\zeta)$ is one-to-one on $|\zeta| < \eta$. Thus we have para-matrized the set $L(|\zeta| < \eta) \subset \mathfrak{M}_1$ as a disc on which all functions in H^∞ are analytic. The set $L(|\zeta| < \eta)$ is called an *analytic disc*.

The following remarks may help to make the analytic disc $L(|\zeta| < \eta)$ a little less mysterious. Consider the discs

$$\Delta_n = \left\{z : \rho(z, z_n) = \left|\frac{z - z_n}{1 - \bar{z}_n z}\right| < \eta\right\} = L_n(|\zeta| < \eta).$$

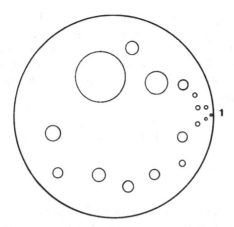

Figure V.1. The hyperbolically congruent discs Δ_n.

With respect to the Euclidean metric, the discs Δ_n converge to the point $z = 1$. But with respect to the hyperbolic metric, the discs Δ_n are all congruent to $\Delta = \{|\zeta| < \eta\}$. See Figure V.1. For $f \in H^\infty$, the functions

$$f_n(\zeta) = f(L_n(\zeta))$$

form a normal family, and the behavior of f on Δ_n is the same as the behavior of f_n on Δ. When $f = B$, $f_n'(0) = B'(z_n)(1 - |z_n|^2)$ and by (3.1) no limit of $\{f_n\}$ is constant on $|\zeta| < \eta$. Because the maximal ideal space $\mathfrak{M} = \mathfrak{M}_{H^\infty}$ is compact, there is a net of indices (n_j) such that $f_{n_j} = f(L_{n_j}(\zeta))$ converges for every $f \in H^\infty$. That is the same as saying that $L_{n_j} \to L$ in the space $\mathfrak{M}^D$. Consequently, we can think of the net of discs (Δ_{n_j}) as converging in $\mathfrak{M}$ to a limit set $L(|\zeta| < \eta)$. Because of (3.1) and (3.3) the limit set is a disc on which $B(z)$ is one-to-one, and on which all functions in H^∞ are analytic.

4. Representing Measures and Orthogonal Measures

Theorem 4.1.   *Let A be a uniform algebra on a compact Hausdorff space Y and let $m \in \mathfrak{M}_A$. Then*

$$\|m\| = \sup_{\|f\| \le 1} |m(f)| = 1.$$

There exists a positive Borel measure μ on Y such that $\mu(Y) = 1$ and such that

$$(4.1) \qquad m(f) = \int_Y f \, d\mu, \qquad f \in A.$$

A positive measure μ on Y for which (4.1) holds is called a *representing measure* for m. Any representing measure is a probability measure, that is, $\mu(Y) = 1$, because $\int 1 \, d\mu = m(1) = 1$.

Proof. We have $\|m\| \leq 1$ by Theorem 1.1. But since A is a uniform algebra we also have $\|1\| = 1$, so that $\|m\| \geq |m(1)| = 1$. Hence $\|m\| = 1$. By the Hahn–Banach theorem, the linear functional $m(f)$ has a norm-preserving extension to $C(Y)$, and by the Riesz representation theorem, this extension is given by integration against a finite complex Borel measure on Y. Thus (4.1) holds for some measure μ, and

$$1 = \|\mu\| = \sup \left\{ \left| \int f\, d\mu \right| : f \in C(Y), \|f\| \leq 1 \right\}.$$

But then $\|\mu\| = 1 = \int d\mu$, so that μ is positive and $\mu(Y) = 1$. $\square$

When using Theorem 4.1 one usually takes Y to be the Šilov boundary of A. For example, let A be the disc algebra A_0 and let $Y = T$, the unit circle. The maximal ideal space is the closed disc $\bar{D}$ (Exercise 9). When $|z| = 1$, the representing measure is the point mass δ_z at z. When $|z| < 1$, the representing measure is given by the Poisson kernel

$$d\mu = \frac{1}{2\pi} \frac{1 - |z|^2}{|e^{i\theta} - z|^2} d\theta.$$

Each point in $\bar{D}$ has a unique representing measure on T because

$$\text{Re}\, f(z) = \int_T \text{Re}\, f\, d\mu,$$

since μ is real, and because the real trigonometric polynomials, that is the real parts of analytic polynomials, are dense in the space $C_{\mathbb{R}}(T)$ of real continuous functions on T.

More generally, a uniform algebra A is called a *Dirichlet algebra on Y* if $\text{Re}\, A = \{\text{Re}\, f : f \in A\}$ is uniformly dense in $C_{\mathbb{R}}(Y)$. Thus every complex homomorphism of a Dirichlet algebra has a unique representing measure on Y. We say A is a *logmodular algebra on Y* if

$$\log |A^{-1}| = \{\log |f| : f \in A^{-1}\}$$

is dense in $C_{\mathbb{R}}(Y)$. Every Dirichlet algebra is a logmodular algebra, since $\log |e^f| = \text{Re}\, f$. However, H^∞ is not a Dirichlet algebra (see Exercise 15), but H^∞ is a logmodular algebra on its Šilov boundary X. Indeed, as we have observed earlier, if $u \in L^\infty_{\mathbb{R}}$, and if $u(z)$ is its Poisson integral, then $f(z) = \exp(u(z) + i\tilde{u}(z)) \in (H^\infty)^{-1}$ and $\log |f| = u$ almost everywhere.

Note that if A is a logmodular algebra on Y, then Y is the Šilov boundary of A. The proof is the same as the proof of Theorem 1.7.

Theorem 4.2. *Suppose A is a logmodular algebra on Y. Then each $m \in \mathfrak{M}_A$ has unique representing measure on Y.*

Proof.   Suppose μ_1 and μ_2 are representing measures on Y for $m \in \mathfrak{M}_A$. Let $f \in A^{-1}$. Then since μ_1 and μ_2 are probability measures, we have

$$|m(f)| \leq \int |f| \, d\mu_1, \qquad |m(f^{-1})| \leq \int |f|^{-1} \, d\mu_2,$$

so that

$$1 \leq \int |f| \, d\mu_1 \int |f|^{-1} \, d\mu_2.$$

By the density of $\log |A^{-1}|$ in $C_{\mathbb{R}}(Y)$, this yields

$$1 \leq \int e^u \, d\mu_1 \int e^{-u} \, d\mu_2, \qquad u \in C_{\mathbb{R}}(Y).$$

Now fix u and define

$$h(t) = \int e^{tu} \, d\mu_1 \int e^{-tu} \, d\mu_2, \qquad t \in \mathbb{R}.$$

Then $h(0) = 1$ and $h(t) \geq 1$. Differentiating $h(t)$ at $t = 0$, we obtain

$$\int u \, d\mu_1 - \int u \, d\mu_2 = 0,$$

so that $\mu_1 = \mu_2$. $\square$

Since H^∞ is a logmodular algebra, each $m \in \mathfrak{M}$ has a unique representing measure μ_m on X. When $m \in X$ the unique measure must be the point mass δ_m. What is the measure on X representing $f \to f(0)$? We have $f(0) = (1/2\pi) \int f(e^{i\theta}) \, d\theta$, $f \in H^\infty$, but strictly speaking, $d\theta/2\pi$ is not a measure on the big compact space X. However, the linear functional

$$L^\infty \ni g \to \int_T g \, d\theta/2\pi,$$

does, by the Riesz representation theorem, determine a Borel probability measure μ_0 on X, and μ_0 is the representing measure we are seeking. If $E \subset T$ is a measurable set of positive measure, then χ_E is an idempotent (0 and 1 valued) function in L^∞. Thus $\hat{\chi}_E$ assumes only the values 0 and 1, and there is an open–closed subset $\tilde{E}$ of X such that $\hat{\chi}_E = \chi_{\tilde{E}}$. Then

(4.2) $$\mu_0(\tilde{E}) = \int_X \chi_{\tilde{E}} \, d\mu_0 = \int_T \chi_E \, d\theta/2\pi.$$

Because the simple functions are norm dense in L^∞, the open–closed sets are a base for the topology of X and (4.2) uniquely determines the Borel measure μ_0 on X.

Theorem 4.3.   *Let $0 < p \leq \infty$. Then the correspondence*

$$\chi_E \to \chi_{\tilde{E}}$$

extends to a unique positive isometric linear operator S from $L^p(T, d\theta/2\pi)$ onto $L^p(X, \mu_0)$.

Proof. For $f = \sum \alpha_j \chi_{E_j}$ a simple function, we define $Sf = \sum \alpha_j \chi_{\tilde{E}_j}$. This is the only possible definition of a linear extension of the mapping $\chi_E \to \chi_{\tilde{E}}$. By (4.2), $\|Sf\|_p = \|f\|_p$, $0 < p \leq \infty$, and by density, S extends to a positive isometric linear operator from $L^p(T, d\theta/2\pi)$ into $L^p(X, \mu_0)$.

We must show that S maps onto $L^p(X, \mu_0)$ It suffices to treat the case $p = \infty$, because $\|Sf\|_p = \|f\|_p$ and because $L^\infty(\mu_0)$ is dense in $L^p(\mu_0)$, $p < \infty$. Let $g \in L^\infty(\mu_0)$. Take $\{g_n\} \subset C(X)$ such that $\|g_n\|_\infty \leq \|g\|_\infty$ and $g_n \to g$ almost everywhere with respect to μ_0 Because $C(X)$ is the Gelfand transform of L^∞, and because simple functions are dense in L^∞, we can suppose that $g_n = S(f_n)$, where $f_n \in L^\infty$ is a simple function and $\|f_n\|_\infty = \|g_n\|_\infty$. Then because S is an isometry, $\|f_n - f_k\|_2 = \|g_n - g_k\|_2$, and $\{f_n\}$ converges in $L^2(d\theta/2\pi)$ to a limit f. Then $S(f) = g$. Moreover, $f \in L^\infty$ because $\|f_n\|_\infty \leq \|g\|_\infty$ and because some subsequence of $\|f_n\|$ converges to f almost everywhere. $\square$

When $p = \infty$, Theorem 4.3 shows that $L^\infty(X, \mu_0) = C(X)$, reflecting the fact that X is very big and disconnected.

By Theorem 4.3 the representing measure on X for the point $z \in D$ is $S(P_z)d\mu_0$, where P_z is the Poisson kernel for z. By Theorem 4.3 and the Radon–Nikodym theorem, the space of measures on X absolutely continuous to μ_0 can be identified with $L^1(d\theta/2\pi)$, and thus there is no real danger in regarding $P_z d\theta/2\pi$ as a measure on X.

When A is a uniform algebra on a compact Hausdorff space Y, we write $A^\perp$ for the space of finite complex Borel measures ν on Y which are orthogonal to A,

$$\int_Y f \, d\nu = 0, \quad \text{all} \quad f \in A.$$

Similarly, if $m \in \mathfrak{M}_A$ and if $A_m = \{f \in A : f(m) = 0\}$ is the corresponding maximal ideal, then $A_m^\perp = \{\nu : \int f \, d\nu = 0, \text{ all } f \in A_m\}$. In particular, if $H_0^\infty = \{f \in H^\infty : f(0) = 0\}$, then by Theorem 4.3 and the F. and M. Riesz theorem, $\{\nu \in (H_0^\infty)^\perp : \nu \ll \mu_0\}$ can be identified with the space H^1.

Theorem 4.4. *Let A be a uniform algebra on a compact Hausdorff space Y and let $m \in \mathfrak{M}_A$. Assume m has a unique representing measure μ on Y. If $\nu \in A_m^\perp$ and if $\nu = \nu_a + \nu_s$ is the Lebesgue decomposition of ν with respect to μ, then $\nu_a \in A_m^\perp$ and $\nu_s \in A^\perp$.*

This is a generalization of the F. and M. Riesz theorem. However, even in the case $A = H^\infty$ we cannot conclude that $\nu_s = 0$ as we can with the classical Riesz theorem. We can only obtain the weaker statement $\int d\nu_s = 0$. For the proof we need the following lemma:

Lemma 4.5. *Assume $m \in \mathfrak{M}_A$ has unique representing measure μ on Y. Let $E \subset Y$ be an F_σ set such that $\mu(E) = 0$. Then there are $f_n \in A$ such that*

(i) $\|f_n\| \le 1$,
(ii) $f_n(y) \to 0$, $y \in E$, *and*
(iii) $f_n \to 1$, *almost everywhere* $d\mu$.

Proof.   We need a preliminary observation. If $u \in C_{\mathbb{R}}(Y)$ then

(4.3) $\sup\{\text{Re } m(f) : f \in A, \ \text{Re } f < u\} = \inf\{\text{Re } m(f) : f \in A, \text{Re} f > u\}$

$$= \int u \, d\mu.$$

To establish (4.3), consider the positive linear functional on Re A defined by

$$\text{Re} f \to \text{Re } m(f).$$

Each representing measure for m is a positive extension of this functional to $C_{\mathbb{R}}(Y)$. By the proof of the Hahn–Banach theorem, this functional has a positive extension whose value at u is any number between the supremum and the infimum in (4.3). So when the representing measure is unique, we must have equality in (4.3).

Write $E = \bigcup E_n$, with E_n compact and $E_n \subset E_{n+1}$. Take $u_n \in C_{\mathbb{R}}(Y)$ such that

$$u_n(y) < 0, \quad u_n(y) > n, \quad y \in E_n,$$
$$\int u_n d\mu < 1/n^2.$$

These are possible because $\mu(E_n) = 0$. By (4.3) there are $g_n \in A$ such that

$$\text{Re } g_n > u_n, \quad \text{Re } g_n(m) < 1/n^2, \quad \text{Im } g_n(m) = 0.$$

Let $f_n = e^{-g_n}$, then (i) holds for f_n because Re $g_n > u_n > 0$, and (ii) holds because $|f_n(y)| \le e^{-u_n(y)} < e^{-n}$, $y \in E_n$ to prove (iii), note that

$$\int |1 - f_n|^2 d\mu = 1 + \int |f_n|^2 d\mu - 2 \text{ Re } f_n(m)$$

$$\le 2 - 2 \text{ Re } e^{-g_n(m)}$$

$$\le 2(1 - e^{-1/n^2}) \le 2/n^2,$$

because $f_n(m) = e^{-g_m(m)}$. Therefore $f_n \to 1$ almost everywhere $d\mu$.   □

Proof of Theorem 4.4.   Since v_s is singular to μ there is an F_σ set $E \subset Y$ such that $\mu(E) = 0$ but such that $|v_s|(Y \backslash E) = 0$. Let f_n be as in the lemma and let $dv_n = f_n \, dv$. Because A_m is an ideal we have $v_n \in A_m^\perp$. By the lemma and by dominated convergence v_n converges weak-star to $v_a = v - v_s$. Hence $v_a \in A_m^\perp$ and $v_s = v - v_a \in A_m^\perp$. Finally, by assertion (iii) of the lemma, $f_n(m) \to 1$, and by assertion (ii),

$$\int dv_s = \lim \int (f_n(m) - f_n) dv_s = 0.$$

Therefore $v_s \in A^\perp$.   □

As an application of Theorem 4.4, we reprove a result from Section 4, Chapter IV: Let $h_0 \in L^\infty \setminus H^\infty$. If $\|h_0 + H^\infty\| < 1$, there exists $F \in H^1$, $F \neq 0$, such that

$$(4.4) \qquad \overline{F}/|F| \in h_0 + H^\infty.$$

Proof. As in the proof of Theorem IV.4.4, we choose $h \in h_0 + H^\infty$, $\|h\| = 1$, such that

$$\left| \frac{1}{2\pi} \int h \, d\theta \right| = \sup \left\{ \left| \frac{1}{2\pi} \int g \, d\theta \right| : g \in h_0 + H^\infty, \|g\| \leq 1 \right\},$$

and we observe that dist$(h, H_0^\infty) = 1$ while dist $(h, H^\infty) < 1$. By duality there are $F_n \in H^1$, $\|F_n\|_1 \leq 1$, such that

$$(4.5) \qquad \int h F_n \frac{d\theta}{2\pi} \to 1$$

but such that

$$(4.6) \qquad \inf_n |F_n(0)| > 0.$$

Let $dv_n = S(F_n) d\mu_0$. Then $v_n \in (H_0^\infty)^\perp$ and $\|v_n\| \leq 1$. Let v be a weak-star cluster point of $\{v_n\}$. Then $v \in (H_0^\infty)^\perp$, and by Theorem 4.4, $dv = S(F) d\mu_0 + dv_s$ where $F \in H^1$ and $v_s \in (H^\infty)^\perp$. By (4.6), $F \not\equiv 0$, because $\int dv \neq 0$, but $\int dv_s = 0$. We also have $\|v\| \leq 1$ and $\int h \, dv = 1$, by (4.5), so that $\|v\| = 1$ and $\int h \, dv = \int |dv| = 1$. Because v_s singular to μ_0, that means $\int h \, dv_a = \int |dv_a|$. But then

$$\int h F \frac{d\theta}{2\pi} = \int |F| \frac{d\theta}{2\pi},$$

so that $h = \overline{F}/|F|$ and (4.4) holds. $\square$

A word of caution. Because $(H^\infty)^\perp$ contains measures singular to all representing measures (see Exercise 17), there is a dangerous curve in the above reasoning. Only because of (4.6) could we conclude that $v_a \neq 0$. In general, it can happen that every weak-star cluster point of a sequence $S(F_n) \, d\mu_0$, $F_n \in H^1$, is singular to μ_0. In that case $F_n(z) \to 0$, $|z| < 1$ (see Exercise 18 and Example 1.5, Chapter IV). However, there are some positive results along these lines (see Gamelin [1973, 1974] and Bernard, Garnett, and Marshall [1977]).

5. The Space L^1/H_0^1

H^∞ is the dual space of L^1/H_0^1, since $H^\infty = \{f \in L^\infty : \int f F \, d\theta = 0, F \in H_0^1\}$. The elements of L^1/H_0^1 are the most useful linear functionals on H^∞ because they have the concrete representation as integrable functions on T. In this section we prove three theorems about the predual L^1/H_0^1.

A closed subset $P \subset X$ is a *peak set* for H^∞ if there is $f \in H^\infty$ such that $\hat{f} \equiv 1$ on P and $|\hat{f}| < 1$ on $X \backslash P$. Our starting point is an elegant lemma of Amar and Lederer [1971].

Lemma 5.1. *Let v be a finite complex Borel measure on X. Assume v is singular with respect to μ_0. For any $\varepsilon > 0$ there is a peak set $P \subset X$ such that $\mu_0(P) = 0$ but such that $|v|(X \backslash P) < \varepsilon$.*

Proof. By hypothesis, the total variation $|v|$ is singular to μ_0 and there is a compact set $K \subset X$ such that

$$\mu_0(K) = 0 \quad \text{and} \quad |v|(X \backslash K) < \varepsilon.$$

Because the open–closed sets form a base for the topology of X, we can find open–closed subsets V_n of X such that $K \subset V_{n+1} \subset V_n$ and such that

(5.1) $$\sum n\mu_0(V_n) < \infty.$$

Let $u_n = n\chi_{V_n}$. Since V_n is open–closed there is a measurable set $E_n \subset T$ such that $V_n = \{\hat{\chi}_{E_n} = 1\}$, and we can think of u_n as an L^∞ function, or better yet, as the bounded harmonic function $u_n(z) = n \int_{E_n} P_z(\theta)d\theta/2\pi$. Let

$$g_n(z) = u_n(z) + i\tilde{u}_n(z), \quad G(z) = \sum_{n=1}^{\infty} g_n(z).$$

Then by (5.1), $G(z)$ is a finite analytic function on D and Re $G(z) > 0$. Then

$$f(z) = \frac{G(z)}{1 + G(z)}$$

is in H^∞, $\|f\|_\infty \leq 1$, and

$$|1 - f(z)| = \left| \frac{1}{1 + G(z)} \right| \leq \frac{1}{1 + u_n(z)}.$$

Consequently $|1 - f(e^{i\theta})| \leq 1/(1 + n)$ almost everywhere on E_n, so that

$$|1 - f(x)| \leq \frac{1}{1+n}, \quad x \in V_n.$$

The set $P = \{x : f(x) = 1\}$ is a peak set, because $g = (1 + f)/2$ satisfies $g(x) = 1, x \in P$, and $|g(x)| < 1, x \notin P$. Moreover we have $K \subset P$ because $\bigcap V_n \subset P$, and hence

$$|v|(X \backslash P) \leq |v|(X \backslash K) < \varepsilon. \quad \square$$

Theorem 5.2. (Mooney). *Let $\{\varphi_n\}$ be a sequence of integrable functions such that*

$$\lim_{n \to \infty} \int f\varphi_n \frac{d\theta}{2\pi} = L(f)$$

exists for all $f \in H^\infty$. Then there exists $\varphi \in L^1$ such that

$$(5.2) \qquad\qquad L(f) = \int f\varphi \frac{d\theta}{2\pi}, \quad f \in H^\infty.$$

A linear functional on H^∞ is called *weakly continuous* if it has the representation (5.2). If $F \in H_0^1$, then

$$\int f(\varphi + F) \frac{d\theta}{2\pi} = \int f\varphi \frac{d\theta}{2\pi}, \quad f \in H^\infty,$$

and so the set of weakly continuous linear functionals corresponds naturally to L^1/H_0^1. Thus the theorem asserts that every weak Cauchy sequence in L^1/H_0^1 has a weak limit in L^1/H_0^1. In other words, L^1/H_0^1 is weakly sequentially complete. The weak completeness of L^1 has been established much earlier (see Dunford and Schwartz [1958, p. 298]).

Proof. Write

$$L_n(f) = \int f\varphi_n \frac{d\theta}{2\pi}, \quad f \in H^\infty.$$

By the uniform boundedness principle, $\sup_n \|L_n\| < \infty$, and the limit L is a bounded linear functional on H^∞. By the Hahn–Banach theorem there is a finite complex measure σ on X such that

$$L(f) = \int f \, d\sigma, \quad f \in H^\infty.$$

Write $d\sigma = \Phi \, d\mu_0 + d\sigma_s$, with $\Phi \in L^1(\mu_0)$ and σ_s singular to μ_0. Then by Theorem 4.3 there is $\varphi \in L^1$ such that

$$\int f\Phi \, d\mu_0 = \int f\varphi \frac{d\theta}{2\pi}, \quad f \in L^\infty.$$

We are going to show that $\sigma_s \in (H^\infty)^\perp$. This will mean that

$$L(f) = \int f\Phi \, d\mu_0 = \int f\varphi \frac{d\theta}{2\pi}, \quad f \in H^\infty,$$

which is the assertion of the theorem. Replacing φ_n by $\varphi_n - \varphi$, we assume that

$$L(f) = \int_X f \, d\sigma_s, \quad f \in H^\infty.$$

Suppose $\sigma_s \notin (H^\infty)^\perp$. Then there exists $g \in H^\infty$ such that $\int g \, d\sigma_s \neq 0$. Applying Lemma 5.1 with $dv = g \, d\sigma_s$, we get a peak set $P \subset X$ such that

$$\mu_0(P) = 0$$

but such that

$$v(P) = \int_P g \, d\sigma_s \neq 0.$$

Let $f \in H^\infty$ be a function peaking on P. Then

(5.3) $$\lim_{k \to \infty} \int f^k dv = v(P) \neq 0,$$

and

(5.4) $$\lim_{k \to \infty} \int f^k g \varphi_n \frac{d\theta}{2\pi} = 0, \quad n = 1, 2, \ldots$$

(because $\mu_0(P) = 0$ and $f^k(x) \to 0$, $x \in X \backslash P$). On the other hand, we have

(5.5) $$\lim_{n \to \infty} \int f^k g \varphi_n \frac{d\theta}{2\pi} = L(f^k g) = \int f^k dv.$$

Inductively, we shall choose $n_j \to \infty$ and $m_j \to \infty$ in such a way that

(5.6) $$h(z) = \sum_{j=1}^{\infty} (-1)^j f^{m_j}(z) \in H^\infty$$

but such that $\{L_{n_j}(hg)\}$ does not converge. That will give a contradiction, thereby proving the theorem.

Fix $\varepsilon > 0$. Take $n_1 = 1$ and choose m_1 so that

$$\left| \int f^{m_1} g \varphi_{n_1} \frac{d\theta}{2\pi} \right| < \varepsilon \quad \text{and} \quad \left| \int f^{m_1} dv - v(P) \right| < \frac{\varepsilon}{2}.$$

These are possible by (5.4) and (5.3). Assuming $n_1, \ldots, n_{k-1}$ and $m_1, \ldots, m_{k-1}$ have been chosen, and writing

$$L_{n_k}\left(\sum_{j=1}^{k} (-1)^j f^{m_j} g \right) = \sum_{j=1}^{k-1} \int (-1)^j f^{m_j} g \varphi_{n_k} \frac{d\theta}{2\pi} + (-1)^k \int f^{m_k} g \varphi_{n_k} \frac{d\theta}{2\pi}$$

$$= A_k + B_k,$$

we first pick $n_k > n_{k-1}$ so that

(5.7) $$\left| A_k - \sum_{j=1}^{k-1} (-1)^j \int_X f^{m_j} dv \right| < \varepsilon.$$

This is possible because of (5.5). Having fixed n_k, we then use (5.3) and (5.4) to choose m_k so that

$$|B_k| < \varepsilon$$

and so that

(5.8)
$$\left| \int f^{m_k} dv - v(P) \right| < \varepsilon.$$

Notice that (5.7) and (5.8) imply

(5.9)
$$|A_k - A_{k+1}| \geq v(P) - 3\varepsilon.$$

Since $|f(z)| < 1, z \in D$, and since $\mu_0(P) = 0$, we can make m_k increase so fast that (5.6) holds and so that by (5.4)

$$C_k = \int \sum_{j=k+1}^{\infty} (-1)^j f^{m_j} g \varphi_{n_k} \frac{d\theta}{2\pi}$$

satisfies $|C_k| < \varepsilon, k = 1, 2, \ldots$. But then

$$L_{n_k}(hg) = A_k + B_k + C_k,$$

and by (5.9)

$$|L_{n_k}(hg) - L_{n_{k+1}}(hg)| \geq (v(P) - 3\varepsilon) - (|B_k| + |B_{k+1}| + |C_k| + |C_{k+1}|)$$
$$\geq v(P) - 7\varepsilon.$$

Taking $\varepsilon < v(P)/7$ now gives a contradiction. $\square$

Theorem 5.3.   *Let L be a bounded linear functional on H^∞. Then the following conditions are equivalent:*

(a)   *L is weakly continuous, there is $\varphi \in L^1$ such that*

$$L(f) = \int f\varphi \frac{d\theta}{2\pi}, \quad f \in H^\infty,$$

(b)   *L is continuous under bounded pointwise convergence: If $g_n \in H^\infty$, $\|g_n\| \leq M$ and if $g_n(z) \to g(z), z \in D$, then $L(g_n) \to L(g)$,*

(c)   *If $h_k \in H^\infty$ and if*

(5.10)
$$\sum |h_k(e^{i\theta})| \leq M < \infty$$

almost everywhere, then

(5.11)
$$L\left(\sum h_k\right) = \sum L(h_k).$$

Note that by (5.10), $\sum h_k$ is a well-defined function in H^∞.

Proof.   It is clear that (a) implies (b). If $\|g_n\| \leq M$ and if $g_n(z) \to g(z)$, then the only weak-star accumulation point of the sequence $\{g_n\}$ in $L^\infty = (L_1)^*$

is g, because Poisson kernels are in L^1. By the Banach–Alaoglu theorem, this means $g_n \to g$ weak-star, so that $\int g_n \varphi \, d\theta \to \int g\varphi \, d\theta$.

Trivially, (b) implies (c), because if (5.10) holds, the partial sums $g_n = \sum_{k=1}^n h_k(z)$ converge pointwise and boundedly to $\sum_{k=1}^\infty h_k$.

Now suppose L satisfies (c). Represent L by a measure σ on the space X and write $d\sigma = \Phi \, d\mu_0 + d\sigma_s$, with $d\sigma_s$ singular to $d\mu_0$. Subtracting $\Phi d\mu_0$, we can assume $\sigma = \sigma_s$. Our task is to then use (c) to prove that $\sigma \in (H^\infty)^\perp$.

Suppose there is $g \in H^\infty$ such that $\int g \, d\sigma \neq 0$. Set $dv = g \, d\sigma$ and choose a peak set $P \subset X$ such that $\mu_0(P) = 0$ but

$$(5.12) \qquad |v|(X \backslash P) < \left| \int g \, d\sigma \right|$$

by using Lemma 5.1. Let $f \in H^\infty$ be a function peaking on P. Replacing f by $1 - (1 - f)^{1/2}$, also a peaking function, we can assume that f has range in a cone with vertex at 1, so that almost everywhere

$$\frac{|f(e^{i\theta}) - 1|}{1 - |f(e^{i\theta})|} \leq A < \infty.$$

Define $h_0 = g, h_j = f^j g - f^{j-1} g, j \geq 1$. Then $\sum h_j = 0$, while

$$\sum_{j=0}^\infty |h_j| \leq |g| + |g| \sum_{j=1}^\infty |f|^{j-1} |f - 1|$$

$$\leq |g| \left(\frac{|f - 1|}{1 - |f|} \right) \leq A\|g\|_\infty < \infty,$$

almost everywhere. But since $f = 1$ on P, dominated convergence yields

$$\sum_{j=1}^\infty L(h_j) = \int_{X \backslash P} \sum_{j=1}^\infty f^{j-1}(1 - f)g \, d\sigma = \int_{X \backslash P} g \, d\sigma$$

so that by (5.12)

$$\left| \sum_{j=1}^\infty L(h_j) \right| \leq |v|(X \backslash P) < \left| \int g \, d\sigma \right|.$$

Hence

$$\left| \sum_{j=0}^\infty L(h_j) \right| \geq \left| \int g \, d\sigma \right| - |v|(X \backslash P) > 0,$$

which contradicts (5.11). $\quad\square$

Theorem 5.4. *Let E be a complex Banach space with dual space E^*. If E^* is isometrically isomorphic to H^∞, then E is isometrically isomorphic to L^1/H_0^1.*

In other words, H^∞ has a unique isometric predual. The Banach space l^1 does not have a unique isometric predual (see Bessaga and Pelczynski [1960]).

Proof. By hypothesis, there is a linear mapping T from H^∞ into E^* which is isometric, $\|T(f)\|_{E^*} = \|f\|_\infty$. The adjoint $T^* : E^{**} \to (H^\infty)^*$ maps $E \subset E^{**}$ isometrically onto the closed subspace $T^*(E)$ of $(H^\infty)^*$. We shall show $T^*(E)$ coincides with L^1/H_0^1 by verifying condition (c) of Theorem 5.3 for $L = T^*(\varphi), \varphi \in E$.

Suppose $\{h_k\} \subset H^\infty$ satisfies (5.10) and let $h = \sum_{k=1}^\infty h_k$. By (5.10) and by the logmodular property there is $h_0 \in H^\infty$ such that

$$(5.13) \qquad |h_0| + \sum_{k=1}^\infty |h_k| = M + 1.$$

Let φ^* be any weak-star cluster point of bounded sequence $\{\sum_1^n T(h_k)\}$ in E^*. We shall be done provided we can show $\varphi^* = T(h)$, because we should then have

$$L(h) = \lim_{n \to \infty} \sum_1^n L(h_k)$$

whenever $L = T^*(\varphi), \varphi \in E$.

For any choice of constants $\{\varepsilon_k\}, |\varepsilon_k| = 1, k = 1, 2, \ldots$, we have

$$\left\| \sum_0^N \varepsilon_k T(h_k) + \varphi^* - \sum_1^N T(h_k) \right\|_E \leq \varlimsup_{n \to \infty} \left\| \sum_0^N \varepsilon_k T(h_k) + \sum_{N+1}^n T(h_k) \right\|_E$$

$$= \varlimsup_{n \to \infty} \left\| \sum_0^N \varepsilon_k h_k + \sum_{N+1}^n h_k \right\|$$

$$\leq M + 1,$$

because T is an isometry. Hence

$$\left\| \sum_0^N \varepsilon_k h_k + T^{-1}(\varphi^*) - \sum_1^N h_k \right\| \leq M + 1,$$

and if for each $e^{i\theta}$ we choose ε_k adroitly, we obtain

$$\sum_0^N |h_k(e^{i\theta})| + \left| T^{-1}(\varphi^*)(e^{i\theta}) - \sum_1^N h_k(e^{i\theta}) \right| \leq M + 1$$

almost everywhere. From (5.13) we conclude that

$$T^{-1}(\varphi^*) = \sum_1^\infty h_k = h$$

almost everywhere, so that $\varphi^* = T(h)$. $\square$

We have based the results in this section on Lemma 5.1 to illustrate the power and beauty of the abstract techniques. These results can also be obtained by classical methods which are perhaps easier and more informative. See Exercises 19 and 20, for example.

Notes

The books of Browder [1969], Gamelin [1969], and Stout [1971] provide much more complete discussions of general uniform algebra theory, and Hoffman [1962a] gives more details on the spectrum of L^∞ and on the fibers $\mathfrak{M}_\zeta$.

See Kelley [1955] for a different approach to Stone–Čech compactifications. The proof of Theorem 1.5 is from Hörmander [1966]. Theorem 2.1 is from Douglas and Rudin [1969]; the corresponding result for $C(T)$ was noted in Helson and Sarason [1967]. Newman proved Theorem 2.2 in [1959b], where he rediscovered a slightly weaker form of Frostman's Theorem. Corollary 2.4 is due to Fisher [1968]. See also Phelps [1965], Rudin [1969], and Fisher [1971]. Theorem 2.5 was first published by Marshall in [1976a]. The paper of Bernard, Garnett, and Marshall [1977] contains some generalizations and further references, as well as Bernard's proof of Theorem 2.3. In Marshall's thesis [1976c] some of the ideas of Section 2 are carried further. Analytic discs were first embedded in the fibers of $\mathfrak{M}$ in the paper of Schark [1961].

In Section 4 we have only touched the surface of the general theory of abstract Hardy spaces. See Hoffman [1962b], Gamelin [1969], Lumer [1969], Stout [1971], and the references therein, for the complete picture. The idea of using the proof of the Hahn–Banach theorem, as in Lemma 4.5 and as in Chapter IV, Theorem 4.3, comes from Lumer [1965].

See Amar and Lederer [1971] for Lemma 5.1 Mooney [1973] gave the first proof of Theorem 5.2. The proof in the text is from Amar [1973]; the sliding hump idea was introduced into this context by Kahane [1967]. Havin [1973] published a more elementary proof outlined in Exercise 20. The implication (c) $\Rightarrow$ (a) in Theorem 5.3 is from Barbey [1975], and Theorem 5.4 is due to Ando [1977]. See Havin [1974], Chaumat [1978], and Pelczynski [1977] for more information on L^1/H_0^1.

The Banach space structure of H^∞ is not yet well understood. In particular, it is not known if H^∞ has the *Banach approximation property*. That is, does there exist a net $\{T_\alpha\}$ of bounded linear operators on H^∞ such that T_α has finite dimensional range and such that $\|T_\alpha f - f\|_\infty \to 0$ uniformly on compact subsets of H^∞? Cesàro means give the analogous result for weak-star convergence and for norm convergence on the disc algebra. See Pelczynski [1977] for details and for related open questions. Theorem 2.5 might be helpful for this problem.

Exercises and Further Results

1. The set A^{-1} of invertible elements of a Banach algebra with unit is an open set on which $f \to f^{-1}$ is continuous. In fact, $f \to f^{-1}$ is analytic, that is, it has a norm convergent power series expansion in a ball about each $f \in A^{-1}$.

2. Let A be the algebra of absolutely convergent Fourier series

$$f(e^{i\theta}) = \sum_{-\infty}^{\infty} a_n e^{in\theta}$$

such that

$$\|f\| = \sum_{-\infty}^{\infty} |a_n| < \infty.$$

With the pointwise multiplication of functions,

$$(fg)(e^{i\theta}) = f(e^{i\theta})g(e^{i\theta}),$$

A is a commutative Banach algebra with unit.

(a) Prove that $\mathfrak{M}_A$ can be identified with the unit circle in the natural way.

(b) If $f(e^{i\theta})$ has absolutely convergent Fourier series and if $f(e^{i\theta})$ is nowhere zero, then $1/f$ has absolutely convergent Fourier series. This result is due to Wiener. Gelfand's proof, using part (a) and Theorem 1.2, drew considerable attention to Banach algebra theory.

(c) The range of the Gelfand map is not closed in $C(T)$. In particular, A is not a uniform algebra under any equivalent norm.

(d) Obtain similar results for the algebra of absolutely convergent Taylor series

$$f = \sum_{n=0}^{\infty} \lambda_n z^n, \qquad \|f\| = \sum_{n=0}^{\infty} |\lambda_n|.$$

3. Let $m_1, m_2, \ldots, m_n$ be distinct points in the spectrum of a Banach algebra A and let $\alpha_1, \alpha_2, \ldots, \alpha_n$ be distinct complex numbers. Then there is $f \in A$ such that $\hat{f}(m_j) = \alpha_j, j = 1, \ldots, n$.

4. If A is a Banach algebra and $f \in A$, then

$$\|\hat{f}\| = \lim_{n \to \infty} \|f^n\|^{1/n}.$$

In particular, the limit exists. This is called the *spectral radius formula*.

5. The maximal ideal space of $C(Y)$ is Y. More generally, let A be a uniform algebra and say $f \in A$ is real if $f - \lambda \in A^{-1}$ whenever Im $\lambda \neq 0$. If the real elements of A separate the points of $\mathfrak{M}_A$, then $A = C(\mathfrak{M}_A)$.

6. A sequence $\{p_n\}$ in the spectrum of a uniform algebra A is called an interpolating sequence if

$$f(p_n) = \alpha_n, \qquad n = 1, 2, \ldots,$$

has solution $f \in A$ whenever $\{\alpha_n\} \in l^\infty$. If $\{p_n\}$ is an interpolating sequence, then the mapping $n \to p_n$ extends to define a homeomorphism from $\beta \mathbb{N}$ into the closure of $\{p_n\}$ in $\mathfrak{M}_A$. (See the proof of Theorem 1.4.)

7. Let A be a uniform algebra with maximal ideal space $\mathfrak{M}_A$. Then A is separable if and only if $\mathfrak{M}_A$ is metrizable.

8. Let X be the maximal ideal space of L^∞.

(a) X is not metrizable. (Use Exercise 7 above.)

(b) X is *extremely disconnected*, that is, if U is an open subset of X, its closure $\overline{U}$ is also open. This holds because every bounded subset of $L^\infty_{\mathbb{R}}$ has a least upper bound.

(c) For $|\zeta| = 1$, let $X_\zeta = \{m \in X : m(z) = \zeta\}$ be the fiber of X over ζ. Let $f \in L^\infty$. Then $w \in \hat{f}(X_\zeta)$ if and only if

$$|\{\theta : |e^{i\theta} - \zeta| < \varepsilon, |f(e^{i\theta}) - w| < \varepsilon\}| > 0$$

for all $\varepsilon > 0$.

★(d) If $\hat{f}$ has a zero on X_α, then $\hat{f} = 0$ on a relatively open subset of X_α (Hoffman [1962a]).

9. (a) Let $m \in \mathfrak{M}_{A_o}$, A_o the disc algebra, and let $\zeta = \hat{z}(m) = m(z)$. Then $|\zeta| \leq 1$. If $f(\zeta) = 0$, then f can be uniformly approximated by function of the form $(z - \zeta)g(z)$, $g \in A_o$, and hence $m(f) = 0$. Consequently $\mathfrak{M}_{A_o} = \overline{D}$.

(b) A_o consists of those $f \in C(T)$ such that

$$\frac{1}{2\pi} \int_{-\pi}^{\pi} e^{int} f(e^{it}) dt = 0, \qquad n = 1, 2, \ldots.$$

(c) (Wermer's maximality theorem) Let B be a closed subalgebra of $C(T)$ containing A_o. Then either $B = A_o$ or $B - C(T)$. If $z \in B^{-1}$, then $\bar{z} \in B$ and $B = C(T)$. If $z \notin B^{-1}$ then there is $m \in \mathfrak{M}_B$ such that $m(z) = 0$. Then

$$m(f) = \frac{1}{2\pi} \int_{-\pi}^{\pi} f(e^{i\theta}) d\theta$$

because the restriction of m to A_o has a unique representing measure on T. This means $B = A_o$, by part (b). (See Wermer [1953] and Hoffman and Singer [1957].)

(d) Cohen [1961] gave an elementary proof of Wermer's theorem. If $B \neq A_o$, there is $f \in B$ such that

$$\frac{1}{2\pi} \int e^{i\theta} f(e^{i\theta}) d\theta = 1.$$

Write

$$zf = 1 + zp + \overline{z}\overline{q} + h$$

with p, q polynomials in A_o and with $\|h\| \leq \frac{1}{2}$. Then $zq - \overline{z}\overline{q}$ is pure imaginary, so that for $\delta > 0$,

$$\|1 + \delta(zq - \overline{z}\overline{q})\| \leq (1 + \delta^2 M^2)^{1/2},$$

where $M = \|zq - \overline{z}\overline{q}\|$. But

$$\delta\overline{z}\overline{q} = \delta(zf - 1 - zp) - \delta h = \delta zg - \delta - \delta h,$$

where $g = f - p \in B$. Consequently,

$$\|(1 + \delta) - z(\delta g - \delta q)\| = \|1 + \delta(zq - \overline{z}\overline{q}) - \delta h\| \leq (1 + \delta^2 M^2)^{1/2} + \delta/2.$$

If $\delta > 0$ is small, this means $z(g - \delta q) \in B^{-1}$, so that $z \in B^{-1}$ and $B = C(T)$.

★**10.** A closed subset K of the spectrum of a uniform algebra A is called a *peak set* if there exists $f \in A$ such that $f(x) = 1$, $x \in K$, but $|f(x)| < 1$, $x \in \mathfrak{M}_A \backslash K$.

(a) A countable intersection of peak sets is again a peak set.

(b) Suppose E is an intersection of (perhaps uncountably many) peak sets. If $g \in A$, then there exists $G \in A$ such that $G = g$ on E and $\|G\| = \sup\{|g(x)| : x \in E\}$.

(c) Consequently, if $|\lambda| = 1$,

$$\left\{ x \in E : g(x) = \lambda \sup_{y \in E} |g(y)| \right\}$$

is either another intersection of peak sets or the empty set.

(d) By Zorn's lemma, $\mathfrak{M}_A$ contains sets which are minimal intersections of peak sets. By part (c), such a set consists of a single point, called a *strong boundary point*.

(e) By part (c) the closure of the set of strong boundary points is the Šilov boundary of A.

(See Bishop [1959].)

11. In the upper half plane let $z_n = i + 10n$, $n = 1, 2, \ldots$. Then $\{z_n\}$ is an interpolating sequence for H^∞. Suppose $|\alpha_n| \leq 1$. Let

$$f_1(z) = -4 \sum_{n=1}^{\infty} \frac{\alpha_n}{(z - \overline{z}_n)^2}.$$

Then $\|f_1\| \leq A$ and $|f_1(z_n) - \alpha_n| \leq \lambda < 1$, $n = 1, 2, \ldots$. Repeat with

$$f_2 = -4 \sum_{n=1}^{\infty} \frac{(\alpha_n - f_1(z_n))}{(z - \overline{z}_n)^2}$$

and continue. We obtain $f = \sum_{k=1}^{\infty} f_k$, $\|f_k\| \le A\lambda^{k-1}$, such that $f(z_n) = \alpha_n$, $n = 1, 2, \ldots$.

★**12.** Let $\mathfrak{M}_\zeta$ be the fiber of $\mathfrak{M}_{H^\infty}$ at $\zeta \in \partial D$, and let $\mathrm{Cl}(f, \zeta)$ be the cluster set of $f \in H^\infty$ at ζ.

(a) Then $\mathrm{Cl}(f, \zeta) \subset \hat{f}(\mathfrak{M}_\zeta)$.

(b) If f is continuous on $D \cup \{\zeta\}$, that is, if $\mathrm{Cl}(f, \zeta) = \{\lambda\}$, then $\hat{f}(\mathfrak{M}_\zeta) = \lambda$. (Approximate $f(z) - \lambda$ by $(z - \zeta)g(z)$, $g \in H^\infty$.)

(c) $\hat{f}(\mathfrak{M}_\zeta) \subset \mathrm{Cl}(f, \zeta)$. Suppose $0 \notin \mathrm{Cl}(f, \zeta)$. Then there is a disc $\tilde{\Delta} = \{|z - \zeta| < 2\delta\}$, $\delta > 0$, such that $|f(z)| > a > 0$ on $D \cap \tilde{\Delta}$. Let $\varphi \in C^\infty$ satisfy

$$
\begin{aligned}
\varphi &= 1 && \text{on } \Delta = \{|z - \zeta| < \delta\}, \\
\varphi &= 0 && \text{off } \tilde{\Delta}, \\
0 &\le \varphi \le 1,
\end{aligned}
$$

$$
\left| \frac{\partial \varphi}{\partial \bar{z}} \right| = \frac{1}{2} \left| \frac{\partial \varphi}{\partial x} + i \frac{\partial \varphi}{\partial y} \right| \le \frac{4}{\delta}.
$$

By Green's theorem,

$$
G(w) = \frac{\varphi(w)}{f(w)} + \frac{1}{\pi} \iint\limits_{D \cap \tilde{\Delta}} \frac{1}{f(z)} \frac{\partial \varphi}{\partial \bar{z}} \frac{dx\, dy}{z - w}
$$

is in H^∞. Inspection of the integral shows that

$$
g(w) = G(w) - 1/f(w)
$$

has an analytic extension to Δ. Thus g is continuous at ζ. Then by (b)

$$
f(w)(G(w) - g(\zeta)) \in H^\infty
$$

has no zeros on $\mathfrak{M}_\zeta$, so that $0 \notin \hat{f}(\mathfrak{M}_\zeta)$. This proof works on arbitrary plane domains. (See Gamelin [1970] and Garnett [1971a]; an easier proof for the disc is in Chapter 10 of Hoffman [1962a].)

13. Here is another proof of Theorem 2.1. It was discovered by J. P. Rosay, and independently by D. E. Marshall. Write $[f, g]$ for the closed subalgebra of L^∞ generated by f and g.

(a) Let $E \subset T$, $|E| > 0$, and let $h = \exp(\chi_E + i\tilde{\chi}_E)$. Then there exist inner functions $u_1, u_2 \in H^\infty$ such that

$$
[u_1, u_2] = [h, 1/h].
$$

For the proof, notice that h maps D onto the annulus $\{1 < |w| < e\}$, so that

$$
h_1 = e^{1/2}/h + h/e^{1/2}
$$

maps D onto an ellipse W_1. Let G_1 be a conformal map from W_1 onto D. Then $u_1 = G_1 \circ h_1$ is an inner function. On $\overline{W}_1$, $G_1(w)$ is a uniform limit of polynomials, and hence $u_1 \in [h, 1/h]$. On $\bar{D}$, G_1^{-1} is a uniform limit of polynomials, and so $h_1 \in [u_1]$. Now

$$h_2 = e^{1/2}/h - h/e^{1/2}$$

maps D onto another ellipse W_2. If G_2 is a conformal map from W_2 to D, then $u_2 = G_2 \circ h_2$ is an inner function and $u_2 \in [h, 1/h]$, $h_2 \in [u_2]$. Consequently $[u_1, u_2] = [h, 1/h]$.

   (b) It follows from (a) that L^∞ is the closed algebra generated by the quotients of inner functions.

   (c) Let U be a unimodular function in L^∞. Write $U = V^2$. By part (b) there are inner functions v and $u_1, \ldots, u_n$ and there are complex numbers $\lambda_1, \ldots, \lambda_n$ such that

$$\left\| V - \bar{v} \sum \lambda_j u_j \right\|_\infty < \varepsilon.$$

Let $u = \prod u_j$, $g = \sum \lambda_j u_j \in H^\infty$. Then $\bar{g}u \in H^\infty$ and $g = v_1 G$, $\bar{g}u = v_2 G$, with G outer and v_1, v_2 inner. Then

$$\| V - \bar{v}v_1 G \| < \varepsilon \quad \text{and} \quad \| 1/\bar{V} - u\bar{v}/v_2 G \|_\infty < \varepsilon/(1 - \varepsilon)$$

so that

$$\| U - uv_1/v^2 v_2 \| < 2\varepsilon/(1 - \varepsilon).$$

14.   Let

$$\mathfrak{R} = \{ f \in H^\infty : \bar{f}u \in H^\infty \text{ for some inner function } u \}.$$

Then $f \in \mathfrak{R}$ if and only if $f \in H^\infty$ and $f(e^{i\theta})$ is the nontangential limit of a meromorphic function from the Nevanlinna class (quotients of bounded analytic functions) on $\{|z| > 1\}$. Indeed, if $\bar{f}u = g \in H^\infty$, then

$$f(e^{i\theta}) = \lim_{r \to 1} \overline{g(1/\bar{z})}/\overline{u(1/\bar{z})}, \quad z = re^{i\theta}, \quad r > 1,$$

almost everywhere. For the converse note that $\{h \in H^2 : \bar{f}h \in H^2\}$ is a non-void invariant subspace of H^2 and use Beurling's theorem. The function e^z is not in $\mathfrak{R}$. (See Shapiro [1968] and Douglas, Shapiro, and Shields [1970].)

15. H^∞ is not a Dirichlet algebra on X. If $g(e^{i\theta}) = \theta$, $-\pi \le \theta < \pi$, then

$$\inf_{f \in H^\infty} \| g - \operatorname{Re} f \| = \pi$$

because if $\| u - g \| < \pi$, then the Hilbert transform of u cannot be bounded at $\theta = \pi$. See Example 1.5, Chapter IV.

16. Because each $\mathfrak{M} \in \mathfrak{M}_{H^\infty}$ has a unique representing measure μ_m on X, $\mathfrak{M}$ is homeomorphic to the weak-star compact set of probability measures on X satisfying

$$\int fg \, d\mu = \int f \, d\mu \int g \, d\mu, \qquad f, g \in H^\infty.$$

Consequently each $u \in L^\infty$ has a continuous extension to $\mathfrak{M}$ defined by

$$u(m) = \int u \, d\mu_m.$$

On D this extension reduces to the Poisson integral representation.

★17. (a) Let m be a multiplicative linear functional on a uniform algebra A. Then m has a representing measure μ_m, supported on the Šilov boundary, for which the subharmonicity inequality

(E.1) $$\log |f(m)| \leq \int_Y \log |f| d\mu_m, \qquad f \in A,$$

holds. (Proof: Let Q be the set of $u \in C_\mathbb{R}(Y)$ for which there exists $\alpha > 0$ and $f \in A$, $f(m) = 1$, such that

$$u > \alpha \log |f|.$$

Then Q is a convex cone in $C_\mathbb{R}(Y)$ containing all strictly positive functions. But $0 \notin Q$ and the separation theorem for convex sets (Dunford and Schwartz [1958 p. 417]) yields a probability measure μ on Y such that $\int u \, d\mu \geq 0$ for all $u \in Q$. It follows that μ is a representing measure μ_m for m and that (E.1) holds (Bishop [1963]).)

(b) Let A be a logmodular algebra on its Šilov boundary Y and let μ_m be the unique representing measure on Y for $m \in \mathfrak{M}_A$. Then (E.1) holds for μ_m. Suppose $G \in L^1(\mu_m)$ is real and orthogonal to A,

$$\int_Y fG \, d\mu_m = 0, \qquad f \in A.$$

Then

$$\int \log |1 + G| d\mu_m \geq 0.$$

Indeed, if $f \in A^{-1}$, then by (E.1)

$$\int \log |f| d\mu_m = \log |f(m)| = \log \left| \int f(1 + G) d\mu_m \right|$$

$$\leq \log \int |f||1 + G| d\mu_m,$$

so that

$$\int u \, d\mu_m \leq \log \int e^u |1 + G| \, d\mu_m, \qquad u \in C_{\mathbb{R}}(Y).$$

Approximating

$$u = -\log |1 + G| + \int \log |1 + G| \, d\mu_m$$

by functions from $C_{\mathbb{R}}(Y)$ then yields the desired inequality.

(c) If μ is a probability measure, if $G \in L^1(\mu)$ is real, and if

$$\int \log |1 + tG| \, d\mu = 0, \qquad t \in \mathbb{R},$$

then $G = 0$ almost everywhere. On the upper half plane,

$$U(z) = \int \log |1 + zG| \, d\mu$$

is harmonic and nonnegative. Show that $\lim_{y \to \infty} U(iy)/y = 0$ and conclude that $U(z) = 0$. Hence $\int \log(1 + G^2) d\mu = 0$ and $G = 0$ almost everywhere.

(Part (c) is due to R. Arens. See Hoffman [1962b] or Stout [1971] for more details regarding parts (b) and (c).)

(d) From (b) and (c) we see that if μ_m is a representing measure for a logmodular algebra A and if $\nu \in A^\perp$ is real and absolutely continuous to μ_m, then $\nu = 0$. Since H^∞ is not a Dirichlet algebra, there exists a real measure ν on X orthogonal to X. This orthogonal measure is singular to every representing measure. It can also be chosen to be an extreme point of the compact convex set ball $(H^\infty)^\perp$ (see Glicksberg [1967]).

18. If $\nu \in (H^\infty)^\perp$, then ν is singular to μ_0 if and only if

$$\int \bar{z}^k \, d\nu = 0, \qquad k = 1, 2, \ldots.$$

Let $\{F_n\}$ be a bounded sequence in H_0^1. Then every weak-star cluster point of the sequence $F_n \, d\theta$ in $(L^\infty)^*$ is singular to μ_0 if and only if $F_n(z) \to 0$ for all $z \in D$. By Example 1.5, Chapter IV, such sequences do exist.

19. In Theorem 5.3 the implication (b) $\Rightarrow$ (a) has an easier proof. Let L be a linear functional on H^∞, and suppose $L(g_n) \to L(g)$ whenever $g_n \in H^\infty$, $\|g_n\| \leq M$, and $g_n(z) \to g(z), z \in D$. Let μ be a measure on the unit circle such that

$$L(f) = \int f \, d\mu, \qquad f \in A_o = H^\infty \cap C(T).$$

Let $E \subset T$ by any closed set such that $|E| = 0$. Then E is a peak set for A_o. If $f \in A_0$ peaks on E then $f^n(z) \to 0$, $z \in D$, so that

$$0 = \lim_{n \to \infty} \int f^n \, d\mu_0 = \mu(E).$$

Hence μ is absolutely continuous, $\mu = \varphi \, d\theta/2\pi$, $\varphi \in L^1$. Now let $g \in H^\infty$ and take $g_n(z) = g(r_n z)$, where r_n increases to 1. Then $L(g) = \lim_n L(g_n) = \lim_n \int g_n \varphi \, d\theta/2\pi = \int g\varphi \, d\theta/2\pi$.

20. Havin proved Theorem 5.2 using the easy part of Theorem 5.3 (or Exercise 19 above) and a variant of Lemma 4.5, Chapter IV. Give ball(H^∞) the L^1 metric, $d(f, g) = \int |f - g| d\theta/2\pi$. In this metric ball($H^\infty$) is complete and the functionals

$$L_n(f) = \frac{1}{2\pi} \int f\varphi_n \, d\theta$$

are continuous. By the Baire category theorem, there exists $b \in$ ball(H^∞) at which $L(f) = \lim_n L_n(f)$ is continuous with respect to this metric.

Suppose $f_k \in$ ball(H^∞) satisfy $f_k \to 0$ almost everywhere. By Theorem 5.3, part (b), it is enough to show $L(f_k) \to 0$. Fix $\varepsilon > 0$ and set

$$E_k = \{e^{i\theta} : |f_k(e^{i\theta})| > \varepsilon\}.$$

Then $|E_k| \to 0$. Let the functions $g_k(z)$ and the constants $\varepsilon_k \to 0$ be as in Lemma 4.5, Chapter IV. Then $\|1 - g_k\|_1 \to 0$, so that

$$d\left(\frac{g_k b + (1 - g_k)f_k}{1 + \varepsilon_k}, b\right) \to 0 \quad \text{and} \quad d\left(\frac{g_k b}{1 + \varepsilon_k}, b\right) \to 0.$$

The limit functional L is linear and bounded. Consequently,

$$L(f_k) = L(g_k f_k) + L(g_k b + (1 - g_k)f_k) - L(g_k b)$$

and

$$\overline{\lim_{k \to \infty}} |L(f_k)| = \overline{\lim_{k \to \infty}} L(g_k f_k) \leq \|L\| \overline{\lim_{k \to \infty}} \|g_k f_k\|_\infty.$$

But since $\sup_{E_k} |g_k| \to 0$ and $\varepsilon_k \to 0$, we have

$$\overline{\lim_{k \to \infty}} \|g_k f_k\|_\infty \leq \varepsilon,$$

so that $\lim_{k \to \infty} |L(f_k)| \leq \varepsilon \|L\|$ (see Havin [1973]).

★**21.** (a) A point x in the unit ball of a Banach space B is an *exposed point* of ball(B) if there is $x^* \in B^*$ such that $\|x^*\| = x^*(x) = 1$ but such that $|x^*(y)| < 1$, for all $y \in$ ball(B), $y \neq x$. A function $f \in H^\infty$ is an exposed point of ball(H^∞) if and only if $\| f \| = 1$ and

$$|\{e^{i\theta} : |f(e^{i\theta})| = 1\}| > 0.$$

Thus not every extreme point is an exposed point. (See Fisher [1969b] and Amar and Lederer [1971]. The case of the disc algebra is discussed in Phelps [1965].)

(b) By part (a) and by the theorem of Bishop and Phelps [1961], $\{f \in H^\infty : |f| = \|f\|$ on a set of positive measure$\}$ is norm dense in H^∞ (Fisher [1969b]).

VI

Bounded Mean Oscillation

The space BMO of functions of bounded mean oscillation is the real dual space of the real Banach space H^1. As a complex Banach space, H^1 has dual L^∞/H_0^∞, and so BMO has a close connection to H^∞.

Moreover, some of the ideas developed in this chapter will be important to us later. Among these ideas we single out three.

(i)   The "stopping time argument" introduced with the Calderón–Zygmund lemma in Section 2 to prove the John–Nirenberg theorem. The same stopping time procedure will play an incisive role in the corona construction and its applications.

(ii)   The conformal invariance of BMO. This is the real reason underlying the frequent occurence of BMO in the function theory. Closely related to this invariance is the invariant property of Carleson measures described in Section 3.

(iii)   The Littlewood–Paley integral formula, which permits us to replace certain line integrals by area integrals that are easier to estimate. This method is used in Section 4 to prove the duality theorem; it will be used again in Chapters VIII and IX.

The chapter ends with a discussion of weighted norm inequalities for conjugate functions. This result is then used to obtain a sharp estimate on the distance from $f \in L_{\mathbb{R}}^\infty$ to Re H^∞. This last section of the chapter is a little more difficult than what we have done before, and its techniques, while very important analysis, will not be used below. Thus less experienced readers might prefer to read further into the book before digesting Section 6.

1. Preliminaries

A measurable function $\varphi(t)$ on $\mathbb{R}$ is *locally integrable*, $\varphi \in L_{\text{loc}}^1$ if $|\varphi|$ is integrable over any compact set. If $\varphi \in L_{\text{loc}}^1$ and if I is a bounded interval,

215

write

$$\varphi_I = \frac{1}{|I|} \int_I \varphi \, dt$$

for the average of φ over I. If $\varphi \in L^1_{\text{loc}}$, and if

(1.1) $$\sup_I \frac{1}{|I|} \int_I |\varphi - \varphi_I| \, dt = \|\varphi\|_* < \infty,$$

where the supremum is over all bounded intervals, then we say φ is of *bounded mean oscillation*, $\varphi \in \text{BMO}$. The bound $\|\varphi\|_*$ in (1.1) is the BMO norm of φ. Because constant functions have BMO norm zero, we identify $\varphi \in \text{BMO}$ with $\varphi + \alpha$, α constant, and we view BMO as a subset of $L^1_{\text{loc}}/\{\text{constants}\}$. It is then immediate from the definition that $\| \ \|_*$ is a norm on BMO.

It is not important that we subtract exactly φ_I in (1.1). Suppose that for each bounded interval I there is a constant α_I such that

(1.2) $$\frac{1}{|I|} \int_I |\varphi - \alpha_I| \, dt \leq M.$$

Then trivially $|\varphi_I - \alpha_I| \leq M$, so that $\|\varphi\|_* \leq 2M$.

It is clear that $L^\infty \subset \text{BMO}$ (or more precisely, that $L^\infty/\mathbb{C} \subset \text{BMO}$), and that for $\varphi \in L^\infty$,

$$\|\varphi\|_* \leq \sup_I \left(\frac{1}{|I|} \int_1 |\varphi - \varphi_I|^2 \, dx \right)^{1/2} \leq \sup_I \left(\frac{1}{|I|} \int_I |\varphi|^2 \, dx \right)^{1/2} \leq \|\varphi\|_\infty.$$

Since $\|\varphi - \alpha\|_* = \|\varphi\|_*$, α constant, this means

$$\|\varphi\|_* \leq \inf_\alpha \|\varphi - \alpha\|_\infty.$$

The function $\log|t|$ is an unbounded function in BMO. If $-b < a < b$, then

$$\frac{1}{b-a} \int_a^b |\log|t| - \log b| \, dt \leq C,$$

and because $\log|t|$ is an even function we conclude that $\log|t| \in \text{BMO}$. In a sense to be made more precise in Section 2, $\log|t|$ is typical of the unbounded functions in BMO. Notice that $\log|t|\chi_{\{t>0\}}(t)$ is not in BMO, because condition (1.1) fails for small intervals centered at 0.

If $\varphi \in \text{BMO}$ and if I and J are intervals such that $I \subset J$, $|J| \leq 2|I|$, then

(1.3) $$|\varphi_I - \varphi_J| \leq \frac{1}{|I|} \int_I |\varphi - \varphi_J| \, dt$$

$$\leq \frac{2}{|J|} \int_J |\varphi - \varphi_J| \, dt \leq 2\|\varphi\|_*.$$

Lemma 1.1. *Let $\varphi \in$ BMO and let I and J be bounded intervals.*

(a) *If $I \subset J$ and $|J| > 2|I|$, then*

$$|\varphi_I - \varphi_J| \le c \log(|J|/|I|)\|\varphi\|_*.$$

(b) *If $|I| = |J|$, then*

$$|\varphi_I - \varphi_J| \le c \log(2 + \text{dist}(I, J)/|I|)\|\varphi\|_*.$$

Proof. For part (a) let

$$I = I_1 \subset I_2 \subset \cdots \subset I_n = J,$$

where $|I_{k+1}| \le 2|I_k|$ and where $n \le c \log(|J|/|I|)$. Then (1.3) gives

$$|\varphi_I - \varphi_J| \le 2n\|\varphi\|_* = c \, \log(|J|/|I|)\|\varphi\|_*.$$

Part (b) follows from part (a) by letting K be the smallest interval containing I and J and by comparing each of φ_I and φ_J separately to φ_K. $\quad\square$

Theorem 1.2. *Let $\varphi \in L^1_{\text{loc}}$. Then $\varphi \in$ BMO if and only if*

(1.4)
$$\int \frac{|\varphi(t)|}{1 + t^2} dt < \infty$$

and

(1.5)
$$\sup_{\text{Im}z>0} \int |\varphi(t) - \varphi(z)| P_z(t) dt = A < \infty,$$

where $\varphi(z) = \int P_z(t)\varphi(t) \, dt$ is the Poisson integral of φ. There are constants c_1 and c_2 such that

$$c_1\|\varphi\|_* \le A \le c_2\|\varphi\|_*,$$

where A is the supremum in (1.5).

Condition (1.4) implies that $\int |\varphi(t)| P_z(t) dt < \infty$, so that $\varphi(z)$ exists at each point of H . Condition (1.5) is very similar to the definition (1.1) of BMO. The only difference is that the Poisson kernel is used in (1.5) while the box kernel $(1/2y)\chi_{\{|t-x|<y\}}(t)$ appears in (1.1).

Proof. Suppose φ satisfies (1.4) and (1.5). Let I be a bounded interval, and let $z = x + iy$, where x is the center of I and $y = \frac{1}{2}|I|$. Then

$$\frac{\chi_I(t)}{|I|} \le \pi P_y(x - t),$$

and by (1.5),

$$\frac{1}{|I|} \int_I |\varphi(t) - \varphi(z)| dt \le \pi A.$$

By (1.2) this means $\|\varphi\|_* \le 2\pi A$.

Now assume $\varphi \in$ BMO and let $z = x + iy \in H$. Let I_0 be the interval $\{|t - x| < y\}$ and let I_k be the interval $\{|t - x| < 2^k y\}, k = 1, 2, \ldots$. Then $|I_k| = 2^{k+1} y$ and

$$P_z(t) \le C/y, \quad t \in I_0,$$
$$P_z(t) \le C/2^{2k} y, \ t \in I_k \backslash I_{k-1}.$$

Also, Lemma 1.1 gives

$$|\varphi_{I_k} - \varphi_{I_0}| \le ck \|\varphi\|_*.$$

Consequently,

$$\int |\varphi(t) - \varphi_{I_0}| P_z(t) \, dt \le \frac{c}{y} \int_{I_0} |\varphi - \varphi_{I_0}| \, dt + \sum_{k=1}^{\infty} \frac{c}{2^{2k} y} \int_{I_k \backslash I_{k-1}} |\varphi - \varphi_{I_k}| \, dt$$
$$+ \sum_{k=1}^{\infty} \frac{c}{2^{2k} y} \int_{I_k \backslash I_{k-1}} |\varphi_{I_k} - \varphi_{I_0}| \, dt$$
$$\le 2c\|\varphi\|_* + \sum_{k=1}^{\infty} \frac{2c}{2^k} \|\varphi\|_* + \sum_{k=1}^{\infty} \frac{ck}{2^k} \|\varphi\|_*,$$

and hence

$$\int |\varphi - \varphi_{I_0}| P_z(t) \, dt \le C\|\varphi\|_*.$$

This clearly implies (1.4). Moreover, this shows $|\varphi(z) - \varphi_{I_0}| \le C\|\varphi\|_*$ and hence

$$\int |\varphi - \varphi(z)| P_z(t) \, dt \le 2C\|\varphi\|_*. \quad \square$$

There is also a BMO space of functions on the circle T. We shall immediately see that this space is the image of BMO($\mathbb{R}$) under a conformal transformation. Let $\psi \in L^1(T)$. We say $\psi \in$ BMO(T) if

$$\sup_I \frac{1}{|I|} \int_I |\psi - \psi_I| d\theta/2\pi = \|\psi\|_* < \infty,$$

where I denotes any arc on T, $|I| = \int_I d\theta/2\pi$ is the length of I, and

$$\psi_I = \frac{1}{|I|} \int_I \psi \, d\theta/2\pi$$

is the average of ψ over I. The proof of Theorem 1.2 shows that

(1.6) $$\sup_{z \in D} \int |\psi(\theta) - \psi(z)| P_z(\theta) \, d\theta/2\pi = B,$$

where $\psi(z) = \int \psi P_z d\theta/2\pi$, defines an equivalent norm on BMO(T). That is,

$$(1.7) \qquad c_1 \|\psi\|_* \leq B \leq c_2 \|\psi\|_*$$

for constants c_1 and c_2 (not necessarily the same as the constants in Theorem 1.2). Map D to the upper half plane by

$$z(w) = i\frac{1-w}{1+w}, \qquad w \in D.$$

For $w = e^{i\theta}$, $\theta \neq \pi$, write $t(\theta) = z(w)$. Then as in Chapter I, Section 3,

$$P_{w_0}(\theta)d\theta/2\pi = P_{z_0}(t)\,dt$$

when $z_0 = z(w_0)$, $|w_0| < 1$. Consequently, if $\varphi \in L^1_{loc}(\mathbb{R})$ and if $\psi(\theta) = \varphi(t(\theta))$, then comparing (1.5) and (1.6) we see that $\varphi \in$ BMO($\mathbb{R}$) if and only if $\psi \in$ BMO(T). We also see that

$$(1.8) \qquad c_1 \|\varphi\|_* \leq \|\psi\|_* \leq c_2 \|\varphi\|_*$$

for some constants c_1 and c_2.

Corollary 1.3. *Under the conformal mapping*

$$z = i\frac{1-w}{1+w}, \qquad |w| < 1,$$

BMO ($\mathbb{R}$) *and* BMO(T) *are transformed into each other. The norms of $\varphi \in$* BMO($\mathbb{R}$) *and its image $\psi \in$* BMO(T) *are related by* (1.8).

Condition (1.6) also says that BMO(T) has an equivalent norm invariant under Möbius transformations.

Corollary 1.4. *Let $\psi \in L^1(T)$ and let $\tau(z)$ be a Möbius transformation. Then $\psi \in$* BMO(T) *if and only if $\psi \circ \tau \in$* BMO(T). *There is a constant C independent of τ such that*

$$\|\psi \circ \tau\|_* \leq C\|\psi\|_*.$$

Moreover, there are constants c_1 and c_2 such that

$$(1.9) \qquad c_1 \|\psi\|_* \leq \sup_\tau \int |\psi \circ \tau - \psi \circ \tau(0)| d\theta/2\pi \leq c_2 \|\psi\|_*,$$

where

$$\psi \circ \tau(0) = \int \psi \circ \tau \, d\theta/2\pi.$$

Proof. Let

$$\|\psi\|'_* = \sup_{z \in D} \int |\psi - \psi(z)| P_z(\theta) \, d\theta/2\pi.$$

By (1.6) and (1.7), $\|\psi\|'_*$ is an equivalent norm on BMO(T). Now the transformation rule (3.1) from Chapter I shows that

$$\|\psi \circ \tau\|'_* = \|\psi\|'_*$$

for every Möbius transformation τ. That means $\|\psi \circ \tau\|_* \le C\|\psi\|_*$. The same transformation rule also shows that

$$\|\psi\|'_* = \sup_\tau \int |\psi \circ \tau - \psi \circ \tau(0)| d\theta/2\pi,$$

and hence (1.9) holds. $\quad\square$

It is useful to interpret BMO in the following way. By (1.9) $\psi \in$ BMO if and only if the distances

$$\inf_\alpha \|\psi \circ \tau - \alpha\|_{L^1(T)}$$

have a bound not depending on the Möbius transformation τ. In this way $\| \ \|'_*$ can be viewed as a conformal invariant version of the norm of the quotient space $L^1(T)/\mathbb{C}$. The conformal invariance of this norm suggests that BMO is more closely related to L^∞ than it is to any other L^p space, $p < \infty$. These observations also hold for BMO($\mathbb{R}$). Indeed, (1.5) shows that the norm on BMO is not seriously increased by the mapping $\varphi \to \varphi \circ \tau$, when $\tau(t) = (t-x)/y$ is a conformal self-map of H that fixes the point of ∞. The invariance of BMO ($\mathbb{R}$) under the full group of conformal self-maps of H follows from Corollaries 1.3 and 1.4. It can also be proved directly by examining $\varphi(-1/t)$, $\varphi \in$ BMO($\mathbb{R}$).

Theorem 1.5. *If $\varphi \in L^\infty$, then the conjugate function $\tilde\varphi$ is in BMO, and*

$$\|\tilde\varphi\|_* \le C\|\varphi\|_\infty,$$

for some universal constant C.

Proof. Because of Corollary 1.3 it makes no difference whether we prove the theorem on the line or on the circle. On the circle the proof is quite transparent if we use Corollary 1.4. Let τ be any Möbius transformation. The normalization $\tilde\psi(0) = 0$ in the definition of the conjugate function means that

$$(\psi \circ \tau)\tilde{} = \tilde\psi \circ \tau - \tilde\psi(\tau(0)).$$

Parseval's theorem and Hölder's inequality then give

$$\int |\tilde\psi \circ \tau(\theta) - \tilde\psi(\tau(0))| \frac{d\theta}{2\pi} \le \left\{ \int |\tilde\psi \circ \tau(\theta) - \tilde\psi(\tau(0))|^2 \frac{d\theta}{2\pi} \right\}^{1/2}$$

$$\le \left\{ \int |\tilde\psi \circ \tau|^2 \frac{d\theta}{2\pi} \right\}^{1/2} \le \|\psi \circ \tau\|_\infty = \|\psi\|_\infty.$$

By (1.9) we obtain $\|\tilde\psi\|_* \le C\|\psi\|_\infty$. $\quad\square$

The real-variables proof of Theorem 1.5 is also instructive. We give the argument on the line.

Let $\varphi \in L^\infty(\mathbb{R})$ and let I be a fixed bounded interval on $\mathbb{R}$. Let $J = \tilde{I}$ be the interval concentric with I having length $|J| = 3|I|$. Write $\varphi = \varphi_1 + \varphi_2$, where

$$\varphi_1 = \varphi \chi_J, \quad \varphi_2 = \varphi - \varphi \chi_J.$$

Writing

$$H\varphi_1(x) = \lim_{\varepsilon \to 0} \frac{1}{\pi} \int_{|x-t|>\varepsilon} \frac{\varphi_1(t)}{x - t} \, dt,$$

$$H\varphi_2(x) = \frac{1}{\pi} \int_{\mathbb{R}\setminus J} \varphi_2(t) \left(\frac{1}{x - t} - \frac{1}{x_0 - t} \right) dt,$$

when $x \in I$ and x_0 is the center of I, we have

$$\tilde{\varphi} = H\varphi_1 + H\varphi_2 + c, \quad x \in I,$$

where c is some unimportant constant depending on I. By (1.10) of Chapter III and Hölder's inequality,

$$\frac{1}{|I|} \int_I |H\varphi_1| dx \leq \left\{ \frac{1}{|I|} \int_I |H\varphi_1|^2 \, dx \right\}^{1/2} \leq \frac{1}{|I|^{1/2}} \|\varphi_1\|_2$$

$$\leq \left(\frac{|J|}{|I|} \right)^{1/2} \|\varphi_1\|_\infty \leq 3^{1/2} \|\varphi\|_\infty.$$

When $x \in I$ and $t \notin J$,

$$\left| \frac{1}{x - t} - \frac{1}{x_0 - t} \right| \leq \frac{C|I|}{|x_0 - t|^2},$$

so that for $x \in I$,

$$|H\varphi_2(x)| \leq \frac{C|I|}{\pi} \int_{\mathbb{R}\setminus J} \frac{|\varphi(t)|}{|t - x_0|^2} \, dt \leq C \|\varphi\|_\infty.$$

Consequently,

$$\frac{1}{|I|} \int_I |H\varphi_2(x)| dx \leq C \|\varphi\|_\infty,$$

and by (1.2) we see that $\|\tilde{\varphi}\|_* \leq (3^{1/2} + C) \|\varphi\|_\infty$.

Let σ be a finite signed measure on the upper half plane. The *balayage* or *sweep* of σ is the function

$$S\sigma(t) = \int P_z(t) \, d\sigma(z).$$

Fubini's theorem shows that $S\sigma(t)$ exists almost everywhere and that

$$\int |S\sigma(t)|\, dt \le \int d|\sigma| = \|\sigma\|.$$

The operator S is the adjoint of the operator $f \to f(z) = \int f(t)P_z(t)\, dt$, since if $f \in L^\infty(R)$, Fubini's theorem gives

$$\int f(z)d\sigma(z) = \int f(t)S\sigma(t)\, dt.$$

Now let σ be a finite signed measure whose total variation $|\sigma|$ is a Carleson measure:

$$(1.10) \qquad\qquad |\sigma|(Q) \le N(\sigma)h,$$

where $Q = \{x_0 < x < x_0 + h, 0 < y < h\}$. Theorem II.3.9, on Carleson measures, then shows that

$$\left| \int f(t)S\sigma(t)\, dt \right| \le N(\sigma)\|f\|_{H^1}$$

whenever $f \in H^1$, but this as yet does not tell us much about the function $S\sigma(t)$.

Theorem 1.6. *If σ is a finite measure on the upper half plane and if $|\sigma|$ is a Carleson measure, then $S\sigma \in$ BMO and*

$$\|S\sigma\|_* \le CN(\sigma),$$

where $N(\sigma)$ is the constant in (1.10) and where C is some universal constant.

Proof. The argument is very similar to the real-variables proof of Theorem 1.5. Writing $\sigma = \sigma^+ - \sigma^-$, where $|\sigma| = \sigma^+ + \sigma^-, \sigma^+ \ge 0, \sigma^- \ge 0$, we can suppose $\sigma \ge 0$. Let I_0 be a fixed interval and let I_n be the concentric interval with length $|I_n| = 2^n|I_0|, n = 1, 2, \dots$. Let Q_n be the square with base I_n:

$$Q_n = \{z : x \in I_n, 0 < y < |I_n|\}.$$

Then

$$\int_{I_0} \int_{Q_1} P_z(t)d\sigma(z)\, dt \le \sigma(Q_1) \le 2N(\sigma)|I_0|.$$

For $z \in Q_n \backslash Q_{n-1}, n \ge 2$, and $t \in I_0, t_0 \in I_0$, we have

$$|P_z(t) - P_z(t_0)| \le \frac{c}{2^{2n}|I_0|}.$$

Hence

$$\int_{I_0} \left| \int_{Q_n \backslash Q_{n-1}} (P_z(t) - P_z(t_0))d\sigma(z) \right| dt \le \frac{c\sigma(Q_n)}{2^{2n}} \le \frac{cN(\sigma)|I_0|}{2^n}.$$

Summing, we see that by (1.2),

$$\frac{1}{|I_0|} \int_{I_0} |S\sigma - (S\sigma)_{I_0}| \, dt \leq CN(\sigma)$$

and $\|S\sigma\|_* \leq CN(\sigma)$. $\square$

In Theorem 1.6 we have assumed that $|\sigma|$ is finite only so that we can be sure that $S\sigma$ exists. The theorem is also true if $|\sigma|$ is infinite but if

$$|\sigma|(\{y > y_0\}) = 0$$

for some y_0. However, some hypothesis is required if the balayage defining $S\sigma$ is to converge. If $\sigma = \sum 2^n \delta_{2^n I}$, then $S\sigma \equiv \infty$.

The main theorem about BMO states that BMO is the dual space of the real Banach space H^1. This theorem implies that every $\varphi \in$ BMO has the form

$$\varphi = f + Hg, \quad f, g \in L^\infty,$$

and thus gives the converse of Theorem 1.5. This theorem will be proved in Section 4 after we discuss the John–Nirenberg theorem and introduce some important quadratic integrals. Theorem 1.6 also has a converse, which is equivalent to the converse of Theorem 1.5 (see Exercise 7).

2. The John–Nirenberg Theorem

Theorem 2.1. *Let $\varphi \in$ BMO($\mathbb{R}$) and let I be an interval. Then for any $\lambda > 0$,*

$$(2.1) \qquad \frac{|\{t \in I : |\varphi(t) - \varphi_I| > \lambda\}|}{|I|} \leq C \exp\left(\frac{-c\lambda}{\|\varphi\|_*}\right).$$

The constants C and c do not depend on φ or λ.

Condition (2.1) says that the distribution of $|\varphi - \varphi_I|$, relative to the normalized Lebesgue measure on I, is not worse than the distribution of $\log 1/t$ relative to $[0, 1]$. The converse of the John–Nirenberg theorem is trivial, and we see that in terms of distribution functions $\log |t|$ is typical of the unbounded BMO functions.

The only reason a BMO function satisfies the very strong condition (2.1) is that the BMO condition says something about the behavior of a function on all subintervals of I. The key to the proof is a basic lemma due to Calderón and Zygmund.

Lemma 2.2. *Let I be a bounded interval, let $u \in L^1(I)$ and let*

$$\alpha > \frac{1}{|I|} \int_I |u| \, dt.$$

Then there is a finite or infinite sequence $\{I_j\}$ of pairwise disjoint open subintervals of I such that

(2.2) $|u| \leq \alpha$ *almost everywhere on* $I \setminus \bigcup I_j$,

(2.3) $\alpha \leq \dfrac{1}{|I_j|} \int_{I_j} |u| \, dt < 2\alpha$,

(2.4) $\sum |I_j| \leq \dfrac{1}{\alpha} \int_I |u| \, dt$.

Proof.   We may assume $I = (0, 1)$. Partition I into two intervals $\omega_0 = (0, \frac{1}{2})$ and $\omega_1 = (\frac{1}{2}, 1)$. For each interval ω there are two cases.

$$\text{Case(i):}\quad \frac{1}{|\omega|} \int_\omega |u| \, dt < \alpha;$$

$$\text{Case(ii):}\quad \frac{1}{|\omega|} \int_\omega |u| \, dt \geq \alpha.$$

Case (i) applies to the initial interval I by hypothesis. In Case (i) we partition ω into two disjoint (open) subintervals of length $|\omega|/2$. For each of these two subintervals we apply Case (i) or Case (ii).

When we get a Case (i) interval we repeat the partition process. However, whenever we reach a Case (ii) interval ω we stop and put ω in the sequence $\{I_j\}$, and we do not partition ω. Since no interval in $\{I_j\}$ is partitioned, the selected intervals I_j are pairwise disjoint.

If $x \in I \setminus \bigcup I_j$, then every dyadic interval containing x is a Case (i) interval. By the theorem of Lebesgue on differentiating the integral, this means $|u(x)| \leq \alpha$ for almost every $x \in I \setminus \bigcup I_j$, and (2.2) holds.

Each selected interval I_j is contained in a unique dyadic interval I_j^* such that $|I_j^*| = 2|I_j|$. The larger interval I_j^* was not selected and hence was a Case (i) interval. Therefore

$$\alpha > \frac{1}{|I_j^*|} \int_{I_j^*} |u| \, dt \geq \frac{1}{2|I_j|} \int_{I_j} |u| \, dt$$

and (2.3) holds.

Because the I_j are pairwise disjoint Case (ii) intervals, we have

$$\sum |I_j| \leq \sum \frac{1}{\alpha} \int_{I_j} |u| \, dt \leq \frac{1}{\alpha} \int_I |u| \, dt$$

and (2.4) holds. $\square$

The intervals I_j were selected according to the following rule: *Among the dyadic intervals $\omega = (k2^{-n}, (k+1)2^{-n})$ contained in $(0, 1)$, select the maximal* *ω for which*

$$\frac{1}{|\omega|} \int_\omega |u| \, dt \geq \alpha.$$

This method of selecting intervals is the simplest example of what is called a *stopping time argument*. This is something probabilists often use more generally with martingales. We shall frequently use similar simple stopping time arguments in the remaining chapters.

Proof of Theorem. The homogeneity in (2.1) is such that we can assume $\|\varphi\|_* = 1$. Fix an interval I and apply Lemma 2.2 to $u = |\varphi - \varphi_I|$ with $\alpha = \frac{3}{2}$. We obtain intervals $I_j^{(1)}$ such that $|\varphi - \varphi_I| \leq \frac{3}{2}$ almost everywhere on $I \setminus \bigcup I_j^{(1)}$, such that

(2.5) $$|\varphi_{I_j^{(1)}} - \varphi_I| < 3$$

by (2.3), and such that

(2.6) $$\sum |I_j^{(1)}| \leq \tfrac{2}{3}|I|$$

by (2.4).

On each $I_j^{(1)}$ we again apply Lemma 2.2 to $|\varphi - \varphi_{I_j^{(1)}}|$ with $\alpha = \frac{3}{2}$. We obtain intervals $I_j^{(2)}$ such that each $I_j^{(2)}$ is contained in some $I_j^{(1)}$. By (2.5) and (2.2), we have $|\varphi - \varphi_I| < \frac{3}{2} + 3 < 6$ almost everywhere on $I \setminus \bigcup I_j^{(2)}$. Also, by (2.3) and (2.5) we have

$$|\varphi_{I_j^{(2)}} - \varphi_I| < 6$$

and by (2.4) and (2.6),

$$\sum_j |I_j^{(2)}| \leq \tfrac{2}{3} \sum_j |I_j^{(1)}| \leq \left(\tfrac{2}{3}\right)^2 |I|.$$

Continue this process indefinitely. At stage n we get intervals $I_j^{(n)}$ such that $|\varphi - \varphi_I| < 3n$ almost everywhere on $I \setminus \bigcup I_j^{(n)}$, and such that

$$\sum_j |I_j^{(n)}| \leq \left(\tfrac{2}{3}\right)^n |I|.$$

If $3n < \lambda \leq 3n + 3, n > 1$, then

$$|\{t \in I : |\varphi(t) - \varphi_I| > \lambda\}| \leq \sum |I_j^{(n)}| \leq \left(\tfrac{2}{3}\right)^n |I| \leq e^{-c\lambda}|I|$$

for $c = \frac{1}{6} \log \frac{3}{2}$. Thus (2.1) holds if $\lambda \geq 3$.

If $0 < \lambda < 3$, then trivially

$$|\{t \in I : |\varphi(t) - \varphi_I| > \lambda\}| \leq |I| < e^{3c}e^{-c\lambda}|I|,$$

and taking $C = e^{3c}$ we obtain (2.1) for all λ. $\square$

The John–Nirenberg theorem has a number of interesting corollaries, the first being a magical reverse of Hölder's inequality.

Corollary 2.3. *Let $\varphi \in L^1_{loc}(\mathbb{R})$. If*

$$\sup_I \frac{1}{|I|} \int_I |\varphi - \varphi_I| \, dt = \|\varphi\|_* < \infty,$$

then for any finite $p > 1$,

$$\sup_I \left(\frac{1}{|I|} \int_I |\varphi - \varphi_I|^p \, dt \right)^{1/p} \leq A_p \|\varphi\|_*,$$

where the constant A_p depends only on p.

Proof. By hypothesis, $\varphi \in$ BMO and we have (2.1) at our disposal. Fix I and write

$$m(\lambda) = \frac{|\{t \in I : |\varphi(t) - \varphi_I| > \lambda\}|}{|I|}$$

for the distribution function of $|\varphi - \varphi_I|$. Then

$$\frac{1}{|I|} \int_I |\varphi - \varphi_I|^p \, dt = p \int_0^\infty \lambda^{p-1} m(\lambda) \, d\lambda$$

and (2.1) yields

$$\frac{1}{|I|} \int_I |\varphi - \varphi_I|^p \, dt \leq Cp \int_0^\infty \lambda^{p-1} \exp \frac{-c\lambda}{\|\varphi\|_*} d\lambda = Cp\Gamma(p) \frac{\|\varphi\|_*^p}{c^p}. \quad \square$$

The constant in Corollary 2.3 has the form

$$A_p \sim pA \quad (p \to \infty).$$

for some constant A. Indeed, with Stirling's formula, the proof of the corollary shows that $A_p \leq pA$ when p is large, and the remarks following Theorem 2.3 of Chapter III show that this estimate on A_p is sharp as $p \to \infty$.

Corollary 2.4. *Let $\varphi \in L^1_{loc}(\mathbb{R})$. Then $\varphi \in$ BMO if and only if*

$$\int \frac{|\varphi(t)|^2}{1 + t^2} \, dt < \infty$$

and

$$(2.7) \qquad \sup_{\text{Im} z > 0} \int |\varphi - \varphi(z)|^2 P_z(t) \, dt = B_2 < \infty,$$

where $\varphi(z)$ is the Poisson integral of φ. There are constants c_1 and c_2 such that

$$c_1 \|\varphi\|_* \leq B_2^{1/2} \leq c_2 \|\varphi\|_*,$$

where B_2 is the supremum in (2.7).

Before giving the proof we note the trivial but useful identity

$$(2.8) \qquad \inf_\alpha \int |\varphi - \alpha|^2 P_z(t)\, dt = \int |\varphi - \varphi(z)|^2 P_z(t)\, dt$$

$$= \int |\varphi|^2 P_z(t)\, dt - |\varphi(z)|^2,$$

which holds because in the Hilbert space $L^2(P_z(t)\, dt)$ the orthogonal projection of φ onto the constants is $\varphi(z)$.

Proof. If the Poisson integral of $|\varphi|^2$ converges, and if (2.7) holds, then Hölder's inequality and Theorem 1.2 show that $\varphi \in \mathrm{BMO}$ with

$$\|\varphi\|_* \le c_2 B_2^{1/2}.$$

The proof of the converse resembles the proof of Theorem 1.2. Suppose $\varphi \in \mathrm{BMO}$ and fix $z = x + iy$, $y > 0$. Let I_k be the interval $\{|t - x| < 2^k y\}$, $k = 0, 1, 2, \ldots$. Then $|I_0| = 2y$ and

$$P_z(t) \le \frac{c}{y}, \qquad t \in I_0,$$

$$P_z(t) \le \frac{c}{2^{2k} y}, \qquad t \in I_k \backslash I_{k-1} \quad k = 1, 2, \ldots .$$

Also, by Lemma 1.1,

$$|\varphi - \varphi_{I_0}|^2 \le 2|\varphi - \varphi_{I_k}|^2 + 2|\varphi_{I_K} - \varphi_{I_0}|^2 \le 2|\varphi - \varphi_{I_k}|^2 + 2c^2 k^2 \|\varphi\|_*^2.$$

Hence

$$\int |\varphi - \varphi_{I_0}|^2 P_z(t)\, dt \le \frac{2c}{|I_0|} \int_{I_0} |\varphi - \varphi_{I_0}|^2\, dt$$

$$+ \sum_{k=1}^{\infty} \frac{2c}{2^k |I_k|} \int_{I_k \backslash I_{k-1}} |\varphi - \varphi_{I_k}|^2\, dt$$

$$+ \sum_{k=1}^{\infty} \frac{2c^2}{2^k |I_k|} k^2 \|\varphi\|_*^2 |I_k|$$

$$\le C_2^2 \|\varphi\|_*^2$$

and (2.7) holds. □

Corollary 2.5. *If $\varphi \in \mathrm{BMO}$, then the conjugate function $\tilde{\varphi} \in \mathrm{BMO}$, and*

$$c_1 \|\varphi\|_* \le \|\tilde{\varphi}\|_* \le c_2 \|\varphi\|_*$$

for some constants c_1 and c_2 not depending on φ.

Proof. By Corollary 2.4 it is enough to prove

$$\int |\tilde{\varphi} - \tilde{\varphi}(z)|^2 P_z(t)\, dt = \int |\varphi - \varphi(z)|^2 P_z(t)\, dt$$

for any fixed $z \in H$. This reduces to the identity

$$\frac{1}{2\pi} \int |\tilde{g}|^2 \, d\theta = \frac{1}{2\pi} \int |g|^2 \, d\theta,$$

$g \in L^2(T)$, $\int g \, d\theta = 0$, by means of the conformal mapping from H to D that sends z to the origin. □

3. Littlewood–Paley Integrals and Carleson Measures

We begin with a classical identity of Littlewood–Paley type. Let $g(e^{i\theta})$ be an integrable function on $T = \partial D$, and let $g(z)$, $z \in D$, denote the Poisson integral of g. The gradient $\nabla g(z)$ is the complex vector $(\partial g/\partial x, \partial g/\partial y)$ and its squared length is

$$|\nabla g(z)|^2 = \left| \frac{\partial g}{\partial x} \right|^2 + \left| \frac{\partial g}{\partial y} \right|^2.$$

In this notation we have

$$|\nabla g(z)|^2 = 2|g'(z)|^2$$

if $g(z)$ is analytic.

Let Ω be a plane domain with smooth boundary and let $u(z)$ and $v(z)$ be C^2 functions on $\overline{\Omega}$. Then Green's theorem states that

$$\iint_{\Omega} (v \, \Delta u - u \, \Delta v) \, dx \, dy = \int_{\partial \Omega} \left(v \frac{\partial u}{\partial n} - u \frac{\partial v}{\partial n} \right) ds,$$

where Δ is the Laplacian, $\partial/\partial n$ is differentiation in the outward normal direction, and ds is arc length on $\partial \Omega$.

Lemma 3.1. *If $g(e^{it}) \in L^1(T)$, and if $g(0) = (1/2\pi) \int g(\theta) \, d\theta$ is its mean value, then*

$$(3.1) \qquad \frac{1}{\pi} \iint_{D} |\nabla g(z)|^2 \log \frac{1}{|z|} dx \, dy = \frac{1}{2\pi} \int |g(e^{i\theta}) - g(0)|^2 d\theta.$$

Proof. We may assume that $g(0) = 0$. Notice that

$$2|\nabla g(z)|^2 = \Delta(|g(z)|^2).$$

For $r < 1$, Green's theorem, with $u(z) = |g(z)|^2$ and $v(z) = \log(r/|z|)$, now yields

$$\iint\limits_{|z|<r} \Delta(|g(z)|^2) \log \frac{r}{|z|} dx\, dy = \int_{|z|=r} |g(z)|^2 \frac{ds}{r}$$

$$- \lim_{\varepsilon \to 0} \int_{|z|=\varepsilon} \left(\frac{|g(z)|^2}{\varepsilon} - \log \frac{r}{\varepsilon} \frac{\partial}{\partial|z|} |g(z)|^2 \right) ds.$$

Since $g(0) = 0$ and since $\nabla g(z)$ is bounded in $|z| < \frac{1}{2}$, the above limit is zero, and we have

$$\iint\limits_{|z|<r} |\nabla g(z)|^2 \log \frac{r}{|z|} dx\, dy = \frac{1}{2} \int |g(re^{i\theta})|^2\, d\theta.$$

By monotone convergence the left side of this equality tends to

$$\iint\limits_{D} |\nabla g(z)|^2 \log \frac{1}{|z|} dx\, dy$$

as $r \to 1$, while the right side has limit $\frac{1}{2} \int |g(e^{i\theta})|^2 d\theta$. That proves the lemma. $\square$

A slightly different form of (3.1) is sometimes easier to use.

Lemma 3.2. *If $g(e^{i\theta}) \in L^1(T)$, then*

$$(3.2) \quad \frac{1}{\pi} \iint\limits_{D} |\nabla g(z)|^2(1 - |z|^2)dx\, dy \le \frac{1}{\pi} \int |g(e^{i\theta}) - g(0)|^2 d\theta$$

$$\le \frac{C}{\pi} \iint\limits_{D} |\nabla g(z)|^2(1 - |z|^2)dx\, dy,$$

where C is some absolute constant.

Proof.   The leftmost inequality in (3.2) follows from (3.1) and the simple fact that $1 - |z|^2 \le 2\log(1/|z|)$, $|z| < 1$. To prove the other inequality, suppose that the integral on the right is finite and normalize $g(z)$ so that

$$\frac{1}{\pi} \iint |\nabla g(z)|^2(1 - |z|^2)\, dx\, dy = 1.$$

For $|z| > \frac{1}{4}$, we have the reverse inequality $\log(1/|z|) \le c_1(1 - |z|^2)$, which yields

$$\frac{1}{\pi} \iint\limits_{1/4<|z|<1} |\nabla g(z)|^2 \log \frac{1}{|z|} dx\, dy \le \frac{c_1}{\pi} \iint\limits_{D} |\nabla g(z)|^2(1 - |z|^2)\, dx\, dy.$$

For $|z| \leq \frac{1}{4}$, the subharmonicity of $|\nabla g(z)|^2$ gives $(\zeta = \xi + i\eta)$

$$|\nabla g(z)|^2 \leq \frac{16}{\pi} \iint\limits_{|\zeta - z| < 1/4} |\nabla g(\zeta)|^2 d\xi \, d\eta$$

$$\leq \frac{32}{\pi} \int\limits_{|\zeta| < 1/2} |\nabla g(\zeta)|^2 (1 - |\zeta|^2) \, d\xi \, d\eta \leq 32.$$

Hence

$$\frac{1}{\pi} \iint\limits_{|z| < 1/4} |\nabla g(z)|^2 \log \frac{1}{|z|} dx \, dy \leq \frac{32}{\pi} \iint\limits_{|z| < 1/4} \log \frac{1}{|z|} dx \, dy = c_2.$$

Using (3.1), we conclude that

$$\frac{1}{2\pi} \int |g(e^{i\theta}) - g(0)|^2 \, d\theta \leq \frac{c_1 + c_2}{\pi} \iint |\nabla g(z)|^2 (1 - |z|^2) dx \, dy$$

and (3.2) is proved. □

It is also possible to use Fourier series to prove (3.2), and the Fourier series proof gives the sharp constant $C = 4$ in (3.2). For some problems (3.2) is preferable to (3.1) because of the logarithmic singularity in (3.1), but the equality (3.1) has the advantage that it can be polarized, whereas (3.2) cannot.

To study BMO we should use the conformally invariant forms of (3.1) and (3.2). Let $z_0 \in D$ and let $\varphi \in L^1(T)$. The identity

$$(3.3) \quad \frac{1}{2\pi} \int |\varphi - \varphi(z_0)|^2 P_{z_0}(\theta) d\theta = \frac{1}{\pi} \iint\limits_{D} |\nabla \varphi(z)|^2 \log \left| \frac{1 - \bar{z}_0 z}{z - z_0} \right| dx \, dy$$

has the same proof as (3.1). It can also be derived from (3.1) using the change of variable $z \to (z - z_0)/(1 - \bar{z}_0 z)$, because the differential form

$$|\nabla \varphi(z)|^2 \, dx \, dy$$

is a conformal invariant. Using the identity

$$1 - \left| \frac{z - z_0}{1 - \bar{z}_0 z} \right|^2 = \frac{(1 - |z|^2)(1 - |z_0|^2)}{|1 - \bar{z}_0 z|^2},$$

we similarly obtain

$$(3.4) \qquad \frac{1}{\pi} \iint_D |\nabla\varphi(z)|^2 \frac{(1 - |z|^2)(1 - |z_0|^2)}{|1 - \bar{z}_0 z|^2} \, dx \, dy$$

$$\leq \frac{1}{\pi} \int |\varphi - \varphi(z_0)|^2 P_{z_0}(\theta) d\theta$$

$$\leq \frac{C}{\pi} \iint_D |\nabla\varphi(z)|^2 \frac{(1 - |z|^2)(1 - |z_0|^2)}{|1 - \bar{z}_0 z|^2} \, dx \, dy,$$

which is the invariant version of (3.2). Now by Corollary 2.4, $\varphi \in \mathrm{BMO}(T)$ if and only if

$$\sup_{z_0 \in D} \frac{1}{2\pi} \int |\varphi - \varphi(z_0)|^2 P_{z_0}(\theta) d\theta < \infty,$$

and the supremum of these expressions is the square of an equivalent norm on BMO. Thus the supremum, over $z_0 \in D$, of the double integral in either (3.3) or (3.4) also determines a norm on BMO.

A positive measure λ on D is a *Carleson measure* if there is a constant $N(\lambda)$ such that

$$(3.5) \qquad \lambda(S) \leq N(\lambda)h$$

for every sector

$$S = \{re^{i\theta} : 1 - h \leq r < 1, |\theta - \theta_0| \leq h\}.$$

We include the case $h = 1$, so that $\lambda(D) \leq 4N(\lambda)$. From Chapters I and II we know that λ is a Carleson measure if and only if $\int |f(z)|^p \, d\lambda \leq C_p \|f\|_p^p$ for all $f \in L^p$, $1 < p < \infty$, or for all $f \in H^p$, $0 < p < \infty$. Here we want to notice the conformally invariant character of Carleson measures.

Lemma 3.3. *A positive measure λ on the disc is a Carleson measure if and only if*

$$(3.6) \qquad \sup_{z_0 \in D} \int \frac{1 - |z_0|^2}{|1 - \bar{z}_0 z|^2} \, d\lambda(z) = M < \infty.$$

The constant M in (3.6) satisfies

$$C_1 N(\lambda) \leq M \leq C_2 N(\lambda)$$

for absolute constants C_1 and C_2.

Proof. Suppose (3.6) holds, and let S be any sector, $S = \{1 - h \leq r < 1, |\theta - \theta_0| < h\}$. Since (3.6) with $z_0 = 0$ shows that $\lambda(D) \leq M$, we can suppose $h < \frac{1}{4}$.

Take $z_0 = (1 - \frac{1}{2}h)e^{i\theta_0}$. For $z \in S$, we have

$$\frac{1 - |z_0|^2}{|1 - \bar{z}_0 z|^2} \geq \frac{C}{1 - |z_0|^2},$$

and hence

$$\lambda(S) \leq C_1(1 - |z_0|) \int \frac{1 - |z_0|^2}{|1 - \bar{z}_0 z|^2} d\lambda \leq C_1 M(1 - |z_0|) = C_1 M h/2.$$

Conversely, suppose λ is a Carleson measure and let $z_0 \in D$. If $|z_0| < \frac{3}{4}$, we have the trivial estimate

$$\int \frac{1 - |z_0|^2}{|1 - \bar{z}_0 z|^2} d\lambda(z) \leq C\lambda(D) \leq C'N(\lambda).$$

If $|z_0| > \frac{3}{4}$, we let

$$E_n = \{z \in D : |z - (z_0/|z_0|)| < 2^n(1 - |z_0|)\}.$$

Then by (3.5), $\lambda(E_n) \leq CN(\lambda)2^n(1 - |z_0|)$, $n = 1, 2, \ldots$. We have

$$\frac{1 - |z_0|^2}{|1 - \bar{z}_0 z|^2} \leq \frac{C}{1 - |z_0|}, \qquad z \in E_1,$$

and for $n \geq 2$,

$$\frac{1 - |z_0|^2}{|1 - \bar{z}_0 z|^2} \leq \frac{C}{2^{2n}(1 - |z_0|)}, \qquad z \in E_n \backslash E_{n-1}.$$

Consequently,

$$\int \frac{1 - |z_0|^2}{|1 - \bar{z}_0 z|^2} d\lambda \leq \int_{E_1} \frac{1 - |z_0|^2}{|1 - \bar{z}_0 z|^2} d\lambda(z) + \sum_{n=2}^{\infty} \int_{E_n \backslash E_{n-1}} \frac{1 - |z_0|^2}{|1 - \bar{z}_0 z|^2} d\lambda(z)$$

$$\leq \sum_{n=1}^{\infty} \frac{2C\lambda(E_n)}{2^{2n}(1 - |z_0|)} \leq C'N(\lambda) \sum_n 2^{-n} = C'N(\lambda),$$

and (3.6) is proved. $\square$

In the upper half plane, Carleson measures are defined by requiring that

(3.7) $\lambda(\{x_0 \leq x \leq x_0 + h, 0 < y \leq h\}) \leq N(\lambda)h,$

which is (3.5) with squares instead of sectors. The analog of (3.6) is

(3.8) $\displaystyle\sup_{z_0 \in \mathbb{H}} \int \frac{y_0}{|z - \bar{z}_0|^2} d\lambda(z) = M < \infty$

and the proof that (3.7) and (3.8) are equivalent, with bounds relating the constants $N(\lambda)$ and M, is geometrically even simpler than the proof of Lemma 3.3. The next theorem, the main result of this section, is now virtually trivial.

Theorem 3.4.   *Let $\varphi \in L^1(T)$, and let*

$$d\lambda_\varphi = |\nabla\varphi(z)|^2 \log \frac{1}{|z|} dx\, dy,$$

where $\nabla\varphi$ is the gradient of the Poisson integral $\varphi(z)$. Then $\varphi \in \mathrm{BMO}(T)$ if and only if λ_φ is a Carleson measure. There are universal constants C_1 and C_2 such that

$$C_1\|\varphi\|_*^2 \leq N(\lambda_\varphi) \leq C_2\|\varphi\|_*^2.$$

Proof.   By Corollary 2.4 and by the inequalities (3.4), we know that $\varphi \in \mathrm{BMO}(T)$ if and only if

$$(3.9) \qquad \sup_{z_0 \in D} \iint_D |\nabla\varphi(z)|^2 \frac{(1 - |z|^2)(1 - |z_0|^2)}{|1 - \bar{z}_0 z|^2} dx\, dy = M_1 < \infty$$

and that $M_1 \approx \|\varphi\|_*^2$. By Lemma 3.3, (3.9) holds if and only if

$$d\mu_\varphi = |\nabla\varphi(z)|^2 (1 - |z|^2)\, dx\, dy$$

is a Carleson measure, and $N(\mu_\varphi) \approx \|\varphi\|_*^2$. What remains to be proved are that μ_φ is a Carleson measure if and only if λ_φ is a Carleson measure and that $N(\lambda_\varphi) \approx N(\mu_\varphi)$. Half of this task is trivial because the inequality $1 - |z|^2 < 2\log(1/|z|)$ shows that $\mu_\varphi \leq 2\lambda_\varphi$. For $|z| > \frac{1}{4}$ we have the reverse inequality

$$\log(1/|z|) \leq C(1 - |z|^2),$$

which shows that

$$\lambda_\varphi(S) \leq C\mu_\varphi(S)$$

for sectors $S = \{1 - h \leq r < 1, |\theta - \theta_0| < h\}$ provided $h \leq \frac{3}{4}$. This will give $N(\lambda_\varphi) \leq CN(\mu_\varphi)$ if we can prove

$$(3.10) \qquad \lambda_\varphi(\{|z| \leq \frac{1}{4}\}) \leq C\mu_\varphi(\{|z| \leq \frac{1}{2}\}) \leq CN(\mu_\varphi).$$

However, we already touched on (3.10) in the proof of Lemma 3.2. Because $|\nabla\varphi(z)|^2$ is subharmonic, we have

$$\lambda_\varphi(\{|z| \leq \frac{1}{4}\}) \leq C \sup_{|z|<1/4} |\nabla\varphi(z)|^2 \leq \frac{32C}{\pi} \int_{|\zeta|<1/2} |\nabla\varphi(\zeta)|^2 (1 - |\zeta|^2)\, d\xi\, d\eta$$

$$\leq C'\mu_\varphi(\{|\zeta| \leq \frac{1}{2}\}).$$

That gives (3.10) and therefore $N(\lambda_\varphi) \leq CN(\mu_\varphi)$. $\square$

The measure λ_φ with the logarithm will be used when we polarize the identity (3.1) in the next section. The most important ingredient of Theorem 3.4 is the

John–Nirenberg theorem, which makes it possible to characterize BMO using quadratic expressions.

4. Fefferman's Duality Theorem

Let us first digress briefly to find the dual space of $H^p(dt)$, $1 \le p < \infty$. This was done for the disc in Chapter IV, Section 1, and the argument for the half plane is formally the same.

Lemma 4.1. *If $1 \le p < \infty$, if $q = p/(p-1)$, and if $g \in L^q$, then $g \in (H^p)^\perp$, that is,*

$$(4.1) \qquad\qquad \int fg \, dt = 0$$

for all $f \in H^p$ if and only if $g \in H^q$.

Proof. If $g \in H^q$, then by Hölder's inequality $fg \in H^1$, and (4.1) follows from Lemma II.3.7.

On the other hand if (4.1) holds then the Poisson integral of g is analytic on H . Indeed, if $z \in$ H and if $z_0 \in$ H is fixed, then

$$h_z(t) = \frac{1}{\bar{z} - t} - \frac{1}{\bar{z}_0 - t}$$

is in H^p and

$$P_z - P_{z_0} = \frac{1}{2\pi i}(h_z - \bar{h}_z).$$

Hence, by (4.1),

$$g(z) - g(z_0) = \int g(t)(P_z(t) - P_{z_0}(t)) \, dt$$

$$= -\frac{1}{2\pi i} \int g(t) \left(\frac{1}{z-t} - \frac{1}{z_0-t} \right) dt,$$

and $g(z)$ is analytic. Theorem I.3.5 now implies that $g \in H^p$. $\square$

With the Hahn–Banach theorem, the lemma yields

$$(4.2) \qquad\qquad (H^p)^* = L^q/H^q, \quad 1 \le p < \infty.$$

In (4.2), the pairing between $f \in H^p$ and a coset $g + H^q \in L^q/H^q$ is given by

$$\int fg \, dt.$$

Our object in this section is to represent $(H^p)^*$ as a space of functions rather than as a quotient space. Two Banach spaces X and Y are said to be *isomorphic* if there is a linear mapping T from X onto Y such that

$$c_1\|x\| \le \|Tx\| \le c_2\|x\|, \quad x \in X.$$

The isomorphism T from X onto Y is called an *isometry* if $\|Tx\| = \|x\|$ for all $x \in X$. When two spaces are isomorphic, one thinks of them as being the same space with two different, but equivalent, norms.

What we are looking for is an isomorphism between $(H^p)^*$ and some space of functions. For $1 < p < \infty$, the M. Riesz theorem on conjugate functions gives the isomorphism we want.

Theorem 4.2. *For $1 < p < \infty$, $(H^p)^*$ is isomorphic to $\bar{H}^q$, the space of complex conjugates of functions in H^q. The isomorphism $T : \bar{H}^q \to (H^p)^*$ is defined by*

$$(4.3) \qquad (Tg)(f) = \int fg\, dt, \quad f \in H^p, \quad g \in \bar{H}^q.$$

Proof. Since $1 < q < \infty$, the Hilbert transform H is bounded on L^q. The operator

$$S(g) = g - iHg$$

is then also bounded on L^q. The kernel of S is H^q and the range of S is $\bar{H}^q$. By the open mapping theorem, the induced mapping

$$S : L^q/H^q \to \bar{H}^q$$

is an isomorphism. By (4.2), $T = S^{-1}$ is an isomorphism of $\bar{H}^q$ onto $(H^p)^* = L^q/H^q$. When $g \in \bar{H}^q$, $S(g + H^q) = g$, and hence (4.3) holds. $\square$

When $p = 1$, the above argument does not apply because the Riesz theorem fails for L^∞. We shall see that the function space isomorphic to $(H^1)^* = L^\infty/\bar{H}^\infty$ is not $\bar{H}^\infty$ but BMO.

It will be convenient to regard H^1 as a Banach space over the real numbers only. Any complex Banach space X can of course be viewed as a real Banach space. However, the complex linear functionals on X can be recovered from the real linear functionals on X in a very simple way. If L is a real linear functional on X, then

$$L_{\mathbb{C}}(x) = L(x) - iL(ix)$$

defines a complex linear functional on X, and

$$L(x) = \operatorname{Re} L_{\mathbb{C}}(x).$$

Because $\|L_{\mathbb{C}}\| = \sup\{\operatorname{Re} L_{\mathbb{C}}(x) : \|x\| = 1\}$, the functionals L and $L_{\mathbb{C}}$ have the same norm. Hence, the correspondence $L \to L_{\mathbb{C}}$ is a real linear isometry between the space of continuous real linear functionals on X and the real Banach space of complex linear functionals on X.

For example, when $1 < p < \infty$, H^p is isomorphic, as a real Banach space, to $L_{\mathbb{R}}^p$, the space of real L^p functions. Again Riesz's theorem provides the isomorphism defined by

$$L_{\mathbb{R}}^p \ni u \to u + iHu.$$

Any real linear functional on H^p is therefore given by

$$L(u + iHu) = \int uv\, dt$$

for a unique $v \in L_{\mathbb{R}}^q$. The corresponding complex linear functional is

$$L_{\mathbb{C}}(u + iHu) = \int uv\, dt + i \int (Hu)v\, dt.$$

The identities

$$\int (Hu)v\, dt = -\int uHv\, dt, \qquad \int (Hu)(Hv)\, dt = \int uv\, dt,$$

which follow from Lemma 4.1 or from (1.9) and (1.10) of Chapter III, show that

$$L_{\mathbb{C}}(u + iHu) = \int (u + iHu)g\, dt,$$

where $g = (v - iHv)/2 \in \bar{H}^q$. This brings us back to (4.3), and we have, in fact, merely rephrased the proof of Theorem 4.2. It is for the case $p = 1$ that real linear functionals simplify the duality problem.

Now suppose $p = 1$. As a real Banach space, H^1 is isomorphic to the space

$$H_{\mathbb{R}}^1 = \{u \in L_{\mathbb{R}}^1 : Hu \in L_{\mathbb{R}}^1\}$$

provided that $H_{\mathbb{R}}^1$ is given the graph norm

$$\|u\|_{H^1} = \|u\|_1 + \|Hu\|_1.$$

This norm is chosen so that

$$H^1 \ni f \to \operatorname{Re} f$$

is an isomorphism of H^1 onto $H_{\mathbb{R}}^1$.

Recall from Chapter II, Corollary 3.3, that the subset

$$\mathfrak{A} = \{u \in H_{\mathbb{R}}^1 : (1 + t^2)|u + iHu| \in L^\infty\}$$

is norm dense in $H_{\mathbb{R}}^1$.

Lemma 4.3. *Let L be a continuous real linear functional on $H_{\mathbb{R}}^1$. Then there are φ_1 and φ_2 in $L_{\mathbb{R}}^\infty$ such that*

$$\|\varphi_1\|_\infty \leq \|L\|, \qquad \|\varphi_2\|_\infty \leq \|L\|$$

and such that

$$(4.4) \qquad L(u) = \int (u\varphi_1 - (Hu)\varphi_2)\, dt, \quad u \in H_{\mathbb{R}}^1.$$

If $u \in \mathfrak{A}$, *then*

$$(4.5) \qquad L(u) = \int u(\varphi_1 + H\varphi_2)\, dt.$$

Moreover, there is a unique real function $\varphi \in$ BMO *such that*

$$\|\varphi\|_* \le C\|L\|$$

for some universal constant C and such that

$$(4.6) \qquad L(u) = \int u\varphi\, dt, \quad u \in \mathfrak{A}.$$

By Theorems 1.2 and 1.5, the integrals in (4.5) and (4.6) are absolutely convergent when $u \in \mathfrak{A}$.

Proof. The space $H_{\mathbb{R}}^1$ is a closed subspace of $L_{\mathbb{R}}^1 \oplus L_{\mathbb{R}}^1$ when this latter space is given the norm $\|(u, v)\| = \|u\|_1 + \|v\|_1$. Extend L to a bounded real linear functional Φ on $L_{\mathbb{R}}^1 \oplus L_{\mathbb{R}}^1$ such that $\|\Phi\| = \|L\|$. Now Φ has the representation

$$\Phi(u, v) = \int (u\varphi_1 - v\varphi_2)\, dt,$$

where $(\varphi_1, \varphi_2) \in L_{\mathbb{R}}^\infty \oplus L_{\mathbb{R}}^\infty$, and

$$\|\Phi\| = \max(\|\varphi_1\|_\infty, \|\varphi_2\|_\infty),$$

because of the choice of the norm on $L_{\mathbb{R}}^1 \oplus L_{\mathbb{R}}^1$. Thus, $\|\varphi_1\|_\infty \le \|L\|$, $\|\varphi_2\|_\infty \le \|L\|$, and (4.4) holds. Since

$$\int (Hu)\varphi_2\, dt = -\int u(H\varphi_2)\, dt$$

when $u \in \mathfrak{A}$ and $\varphi_2 \in L^\infty$, (4.5) follows from (4.4). Also the function

$$\varphi = \varphi_1 + H\varphi_2$$

is in BMO, and $\|\varphi\|_* \le C\|L\|$, by Theorem 1.5. Now (4.6) is obvious, since it is just a restatement of (4.5).

It remains to show that φ is uniquely determined by L. The pair $(\varphi_1, \varphi_2) \in L_{\mathbb{R}}^\infty \oplus L_{\mathbb{R}}^\infty$ corresponding to L is by no means unique. A pair (φ_1, φ_2) induces the zero functional on $H_{\mathbb{R}}^1$ if and only if

$$(4.7) \qquad \int u(\varphi_1 + H\varphi_2)\, dt = 0, \quad u \in \mathfrak{A}.$$

Since any difference $P_z(t) - P_{z_0}(t)$ of Poisson kernels is in $\mathfrak{A}$, this holds if and only if $\varphi_1 + H\varphi_2$ has constant Poisson integral. Thus (4.7) occurs if and only if $\varphi_1 + H\varphi_2$ is constant. Since constants have zero BMO norm, this means that L gives rise to a unique BMO function φ such that (4.6) holds. $\quad\square$

As an aside, we note that $\varphi_1 + H\varphi_2$ is constant if and only if $\varphi_1 + i\varphi_2 \in H^\infty$. Thus the proof of the lemma again shows that $(H^1)^* = L^\infty/H^\infty$.

Theorem 4.4 (C. Fefferman). *The dual space of $H_{\mathbb{R}}^1$ is BMO. More precisely, if $\varphi \in$ BMO is real valued and if L is defined by*

$$(4.8) \qquad L(u) = \int u\varphi \, dt, \quad u \in \mathfrak{A},$$

then $|L(u)| \leq C_1\|\varphi\|_*\|u\|_{H^1}$. Conversely, if $L \in (H_{\mathbb{R}}^1)^*$, then there is a unique real $\varphi \in$ BMO with $\|\varphi\|_* \leq C\|L\|$ such that (4.8) holds.

Although the integral in (4.8) makes sense only for $u \in \mathfrak{A}$, the density of $\mathfrak{A}$ in $H_{\mathbb{R}}^1$ and the inequality $|L(u)| \leq C_1\|\varphi\|_*\|u\|_{H^1}$ enable us now to regard each $\varphi \in$ BMO as a continuous linear functional on $H_{\mathbb{R}}^1$.

Proof. By Lemma 4.3, each $L \in (H_{\mathbb{R}}^1)^*$ yields a unique $\varphi \in$ BMO with $\|\varphi\|_* \leq C\|L\|$ such that (4.8) holds. The important thing that remains to be proved is that

$$(4.9) \qquad \left|\int u\varphi \, dt\right| \leq C_1\|\varphi\|_*\|u\|_{H^1}, \quad u \in \mathfrak{A},$$

whenever $\varphi \in$ BMO. The proof of the corresponding inequality on the unit circle is a little less technical, and instead of (4.9) we prove

$$(4.10) \qquad \left|\int u\varphi \, \frac{d\theta}{2\pi}\right| \leq C_1\|\varphi\|_*\|u + i\tilde{u}\|_1$$

for $u \in L_{\mathbb{R}}^2(T)$ and for $\varphi \in$ BMO(T) with $\int \varphi \, d\theta = 0$. Now, if $u \in \mathfrak{A}$ and if $\varphi \in$ BMO$(\mathbb{R})$, then, with $e^{i\theta} = (t-i)/(t+i)$, $d\theta = 2\,dt/(1+t^2)$,

$$\int u\varphi \, dt = \frac{1}{2}\int U(\theta)\,\Phi(\theta)\,dt$$

where $U(\theta) = (1+t^2)\,u(t) \in L_{\mathbb{R}}^2(\mathbb{T})$ and $\Phi(\theta) = \varphi(t) \in$ BMO $(\mathbb{T})$. Then

$$(U + i\tilde{U})(e^{i\theta}) = (u+i\tilde{u})(t)\left(\frac{-4e^{i\theta}}{(e^{i\theta}-i)^2}\right) \in \mathrm{H}^2$$

has mean value zero. Hence $\|U + i\tilde{U}\|_1 = 2\|u + i\tilde{u}\|_1$ and by Corollary 1.3, inequality (4.10) implies inequality (4.9).

Now, let $u \in L_{\mathbb{R}}^2(T)$ and let $\varphi \in$ BMO(T). Notice that, by Corollary 2.4 and Corollary 1.3, the integral (4.10) converges absolutely. Let $f = u + iHu$.

Then $f \in H^1$, and

$$\int u\varphi \frac{d\theta}{2\pi} = \operatorname{Re} \int f\varphi \frac{d\theta}{2\pi}.$$

The advantage of working with f instead of with u is that we may suppose that $f = g^2$, $g \in H^2$. Indeed, write $f = BF$, where B is a Blaschke product, $F \in H^1$ has no zeros in D, and $\|F\|_1 = \|f\|_1$. Write $f_1 = (B-1)F$ and $f_2 = (B+1)F$. Then $\|f_1\|_1 \le 2\|f\|_1$, $\|f_2\|_1 \le 2\|f\|_1$, and $f = (f_1 + f_2)/2$. Moreover f_1 and f_2 have no zeros in D, so that f_1 and f_2 are both of the form g^2, $g \in H^2$. In estimating $\int f\varphi\, d\theta$, we may replace f by f_1 or by f_2. Thus, we assume $f = g^2$, $g \in H^2$.

We can also assume $\varphi(0) = 0$. Polarization of the Littlewood–Paley identity, Lemma 3.1, then yields

$$\int f\varphi \frac{d\theta}{2\pi} = \int (f - f(0))\varphi \frac{d\theta}{2\pi} = \frac{1}{\pi} \iint_D (\nabla f) \cdot (\nabla \varphi) \log \frac{1}{|z|} dx\, dy.$$

Since $f(z)$ is analytic, we have

$$\nabla f \cdot \nabla \varphi = f_x \varphi_x + f_y \varphi_y = f'(z)(\varphi_x + i\varphi_y) = 2g(z)g'(z)(\varphi_x + i\varphi_y),$$

and so

$$|\nabla f \cdot \nabla \varphi| \le 2|g(z)||g'(z)||\nabla \varphi(z)|.$$

The Cauchy–Schwarz inequality now gives us

$$\left| \int f\varphi \frac{d\theta}{2\pi} \right| \le \left(\frac{2}{\pi} \iint_D |g'(z)|^2 \log \frac{1}{|z|} dx\, dy \right)^{1/2}$$

$$\times \left(\frac{2}{\pi} \iint_D |g(z)|^2 |\nabla \varphi(z)|^2 \log \frac{1}{|z|} dx\, dy \right)^{1/2}.$$

By Lemma 3.1, the first factor is

$$\left(\frac{1}{2\pi} \int |g(e^{i\theta}) - g(0)|^2 d\theta \right)^{1/2} = \|f\|_1^{1/2}.$$

By Theorem 3.4 and by the disc version of the theorem on Carleson measures, the second factor is bounded by $C\|\varphi\|_*\|f\|_1^{1/2}$. These give us (4.10), and the theorem is proved. $\square$

To prove (4.9) without moving to the circle, see Exercise 6.

The complex linear functional on H^1 determined by $\varphi \in$ BMO by use of (4.8) is

$$L_{\mathbb{C}}(u + iHu) = \int (u + iHu)\varphi \, dt$$

because $\mathrm{Re}(-i(u + iHu)) = Hu$. We now have a real linear isomorphism between the complex Banach space $(H^1)^*$ and the real space of real BMO functions. In order to regard BMO as a complex space isomorphic to $(H^1)^*$, observe that

$$iL_{\mathbb{C}}(u + iHu) = L_{\mathbb{C}}(-Hu + iu) = \int (-Hu + iu)\varphi \, dt.$$

By (4.5) and (4.6), the last integral equals

$$\int (u + iHu)H\varphi \, dt,$$

at least when $u \in \mathfrak{A}$. If we were to define multiplication by i on the space of real BMO functions $i\varphi = H\varphi$, then real BMO would become a complex Banach space isomorphic to $(H^1)^*$, with the complex linear isomorphism from L^∞/H^∞ to real BMO given by $L^\infty/H^\infty \ni f \to \mathrm{Re}\, f + H(\mathrm{Im}\, f)$. This unusual notion of complex scalar multiplication, which is brought about by identifying $H^1_{\mathbb{R}}$ with H^1, is one reason that real linear functionals on H^1 are easier to discuss.

Corollary 4.5. *If $\varphi \in L^1_{\mathrm{loc}}$, then $\varphi \in$ BMO if and only if*

(4.11) $$\varphi = \varphi_1 + H\varphi_2 + \alpha,$$

where α is a constant and where φ_1 and φ_2 are L^∞ functions. When $\varphi \in$ BMO, φ_1 and φ_2 can be chosen so that

(4.12) $$\|\varphi_1\|_\infty \leq C\|\varphi\|_*, \quad \|\varphi_2\|_\infty \leq C\|\varphi\|_*$$

for some constant C.

Proof. This corollary is equivalent to the theorem. Theorem 1.5 tells us that every function of the form (4.11) is in BMO that

(4.13) $$\|\varphi\|_* \leq C(\|\varphi_1\|_\infty + \|\varphi_2\|_\infty).$$

If $\varphi \in$ BMO, then by (4.9), the functional $L(u) = \int u\varphi \, dt$, $u \in \mathfrak{A}$, is bounded against $\|u\|_{H^1}$. By Lemma 4.3, there are $\varphi_1, \varphi_2 \in L^\infty$ such that

$$L(u) = \int u(\varphi_1 + H\varphi_2) \, dt.$$

As we observed after the proof of Lemma 4.3, this means that $\varphi - (\varphi_1 + H\varphi_2)$ has constant Poisson integral. Consequently (4.11) and (4.12) hold. $\square$

A constructive proof of (4.11), and hence of the duality theorem itself, will be given in Chapter VIII.

Because of (4.12) and (4.13), the expression

$$\inf\{\|\varphi_1\|_\infty + \|\varphi_2\|_\infty : \varphi = \varphi_1 + H\varphi_2 + \alpha\}$$

defines a norm on BMO equivalent to $\|\varphi\|_*$.

Corollary 4.6. *Let $f \in L^\infty$. Then the distance*

$$\mathrm{dist}(f, H^\infty) = \inf_{g \in H^\infty} \|f - g\|_\infty$$

satisfies

(4.14) $\qquad C_1\|f - iHf\|_* \le \mathrm{dist}(f, H^\infty) \le C_2\|f - iHf\|_*$

for some absolute constants C_1 and C_2.

Proof. By (4.2) and the Hahn–Banach theorem, we have

$$\mathrm{dist}(f, H^\infty) = \sup\left\{\left|\int fF\, dt\right| : F \in H^1, \|F\|_1 \le 1\right\}.$$

By density, we may suppose $F = u + iHu$, where $u \in \mathfrak{A}$. Then,

$$\int fF\, dt = \int fu\, dt + i\int fHu\, dt = \int (f - iHf)u\, dt.$$

Taking the real and imaginary parts of the last integral, we see that (4.14) follows directly from the theorem. $\quad\square$

Corollary 4.7. *Let $f \in L^\infty$ be real valued. Then the distances*

$$\mathrm{dist}(f, \mathrm{Re}\, H^\infty) = \inf_{g \in H^\infty} \|f - \mathrm{Re}\, g\|_\infty$$

and

$$\mathrm{dist}_*(Hf, L^\infty) = \inf_{\varphi \in L^\infty} \|Hf - \varphi\|_*$$

satisfy

$$C_1\, \mathrm{dist}(f, \mathrm{Re}\, H^\infty) \le \mathrm{dist}_*(Hf, L^\infty) \le C_2\, \mathrm{dist}(f, \mathrm{Re}\, H^\infty)$$

for some absolute constants C_1 and C_2.

Proof. If $g \in H^\infty$, then $\mathrm{Im}\, g \in L^\infty$ and $Hf - \mathrm{Im}\, g = H(f - \mathrm{Re}\, g)$, so that by Theorem 1.5,

$$\|Hf - \mathrm{Im}\, g\|_* \le C_2\|f - \mathrm{Re}\, g\|_*.$$

Thus, we have

$$\mathrm{dist}_*(Hf, L^\infty) \le C_2\, \mathrm{dist}(f, \mathrm{Re}\, H^\infty).$$

The other inequality asserted in the corollary lies deeper and uses the duality theorem. If $\varphi \in L^\infty$, then by Corollary 4.5,

$$Hf - \varphi = \varphi_1 + H\varphi_2 + \alpha,$$

where

$$\|\varphi_1\|_\infty + \|\varphi_2\|_\infty \leq C\|Hf - \varphi\|_*.$$

Then

$$u = \varphi + \varphi_1 + \alpha \in L^\infty$$

and

$$Hu = H\varphi + H\varphi_1 = \varphi_2 - f,$$

so that $g = -Hu + iu$ is in H^∞. Also

$$\|f - \operatorname{Re} g\|_\infty = \|f + Hu\|_\infty = \|\varphi_2\|_\infty \leq C\|Hf - \varphi\|_*.$$

Thus, we have

$$C_1 \operatorname{dist}(f, \operatorname{Re} H^\infty) \leq \operatorname{dist}_*(Hf, L^\infty),$$

and Corollary 4.7 is proved. $\square$

A method of estimating $\operatorname{dist}_*(\varphi, L^\infty)$ in terms of the exponential integrability of $|\varphi - \varphi_I|$ will be given in Section 6.

5. Vanishing Mean Oscillation

Let $\varphi \in L^1_{\text{loc}}(\mathbb{R})$. For $\delta > 0$, write

$$(5.1) \qquad M_\delta(\varphi) = \sup_{|I| < \delta} \frac{1}{|I|} \int_I |\varphi - \varphi_I|\, dt,$$

where I denotes an interval. Then $\varphi \in$ BMO if and only if $M_\delta(\varphi)$ is bounded and $\|\varphi\|_* = \lim_{\delta \to \infty} M_\delta(\varphi)$. We say that φ has *vanishing mean oscillation*, $\varphi \in$ VMO, if

   (i) $\varphi \in$ BMO and
  (ii) $M_0(\varphi) = \lim_{\delta \to 0} M_\delta(\varphi) = 0.$

It is easy to see that VMO is a closed subspace of BMO. The relation between BMO and the subspace VMO is quite similar to the relation between L^∞ and its subspace of bounded uniformly continuous functions. Write UC for the space of uniformly continuous functions on $\mathbb{R}$ and write BUC for $L^\infty \cap$ UC.

Theorem 5.1. *For a function $\varphi \in$ BMO, the following conditions are equivalent:*

(a) $\varphi \in$ VMO.

(b)   *If $\varphi_x(t) = \varphi(t - x)$ is the translation of φ by x units, then*

$$\lim_{x \to 0} \|\varphi_x - \varphi\|_* = 0.$$

(c)   *If $\varphi(t, y) = P_y * \varphi$ is the Poisson integral of φ, then*

$$\lim_{y \to 0} \|\varphi(t) - \varphi(t, y)\|_* = 0.$$

(d)   *φ is in the BMO closure of* UC $\cap$ BMO;

$$\inf_{g \in \text{UC} \cap \text{BMO}} \|\varphi - g\|_* = 0.$$

Proof.   We verify the circle of implications (a) $\Rightarrow$ (b) $\Rightarrow$ (c) $\Rightarrow$ (d) $\Rightarrow$ (a). Assume (a) holds. Let $\delta > 0$ and partition $\mathbb{R}$ into intervals

$$I_j = (j\delta/2, (j+1)\delta/2), \quad -\infty < j < \infty,$$

of length $\delta/2$. By (5.1) we have

(5.2) $$|\varphi_{I_j} - \varphi_{I_{j+1}}| \le 2M_\delta(\varphi).$$

Define

$$h(t) = \sum_{j=-\infty}^{\infty} \varphi_{I_j} \chi_{I_j}(t).$$

We first show that

(5.3) $$\|\varphi - h\|_* \le 5M_\delta(\varphi).$$

If $|I| \le \delta$, then by (5.2) we have

$$\frac{1}{|I|} \int_I |h - h_I|\, dt \le 4M_\delta(\varphi),$$

so that

$$\frac{1}{|I|} \int_I |(\varphi - h) - (\varphi - h)_I|\, dt \le 5M_\delta(\varphi).$$

If $|I| > \delta$ and if J is the union of those I_j such that $I_j \cap I \ne \varnothing$, then $|J| \le 2|I|$. Writing $J = I_1 \cup I_2 \cup \cdots \cup I_N$, we have

$$\frac{1}{|I|} \int_I |(\varphi - h) - (\varphi - h)_I|\, dt \le \frac{2}{|I|} \int_I |\varphi - h|\, dt \le \frac{4}{|J|} \int_J |\varphi - h|\, dt$$

$$\le \frac{4}{N} \sum_{j=1}^{N} \frac{1}{|I_j|} \int_{I_j} |\varphi - \varphi_{I_j}|\, dt \le 4M_\delta(\varphi),$$

and (5.3) is proved. Now if $|x| < \delta$, then by (5.2), $\|h - h_x\|_\infty \le 2M_\delta(\varphi)$, so that $\|h - h_x\|_* \le 2M_\delta(\varphi)$. Consequently, if $|x| < \delta$, then

$$\|\varphi - \varphi_x\|_* \le \|\varphi - h\|_* + \|h - h_x\|_* + \|\varphi_x - h_x\|_*$$
$$= 2\|\varphi - h\|_* + \|h - h_x\|_* \le 12M_\delta(\varphi).$$

This proves that (a) implies (b).

Now assume (b) holds. Then

$$\varphi(t, y) = \int \varphi(t - x) P_y(x)\, dx$$

is an average of the translates φ_x of φ. By (b), $\|\varphi - \varphi_x\|_*$ is small if $|x|$ is small, say, if $|x| < \delta$, while for any x, $\|\varphi - \varphi_x\|_* \le 2\|\varphi\|_*$. But when y is small, most of the weight of $P_y(x)$ is given to $|x| < \delta$, and this means that $\|\varphi(t) - \varphi(t, y)\|_*$ is small. To be precise, by Fubini's theorem we have

$$\|\varphi(t) - \varphi(t, y)\|_* \le \int_{|x|<\delta} \|\varphi - \varphi_x\|_* P_y(x)dx + 2\|\varphi\|_* \int_{|x|>\delta} P_y(x)\, dx,$$

so that

$$\varlimsup_{y \to 0} \|\varphi(t) - \varphi(t, y)\|_* \le \sup_{|x|<\delta} \|\varphi - \varphi_x\|_*.$$

Hence (b) implies (c).

To prove that (c) implies (d), we use the estimate

$$(5.4) \qquad\qquad y|\nabla\varphi(x, y)| \le c\|\varphi\|_*,$$

which is easy to prove. When $\varphi \in L^\infty$, (5.4) can be derived from Harnack's inequality and a change of scale, and the extension to $\varphi \in$ BMO then follows from Corollary 4.5. However, there is also an elementary proof of (5.4) using Theorem 1.2 instead of the duality theorem. The simple inequality

$$y|\nabla P_y(x - t)| \le c P_y(x - t),$$

in which the derivatives are taken with respect to x and y, combines with Theorem 1.2 to give

$$y|\nabla\varphi(x, y)| \le y \int |\varphi(t) - \varphi(x, y)||\nabla P_y(x - t)|\, dt$$
$$\le c \int |\varphi(t) - \varphi(x, y)| P_y(x - t)\, dt \le c\|\varphi\|_*.$$

Now (5.4) shows that $\varphi(x, y)$ is a uniformly continuous function of x. Because $\varphi(x, y) \in$ BMO, we see that (d) follows from (c).

It is trivial that (d) implies (a), because UC $\cap$ BMO $\subset$ VMO and because VMO is closed in BMO.　□

Theorem 5.2. *If φ is a locally integrable function on $\mathbb{R}$, then $\varphi \in$ VMO if and only if*

$$(5.5) \qquad \varphi = \varphi_1 + H\varphi_2 + \alpha,$$

where $\varphi_1, \varphi_2, \in$ BUC and where α is a constant. When $\varphi \in$ VMO, φ_1 and φ_2 can be chosen in BUC so that (5.5) holds and so that

$$(5.6) \qquad \|\varphi_1\|_\infty + \|\varphi_2\|_\infty \le C\|\varphi\|_*,$$

where C is a constant not depending on φ.

Proof. Suppose that φ has the form (5.5), with $\varphi_1, \varphi_2 \in$ BUC. Then $\varphi \in$ BMO, and for $|x|$ small, $\|\varphi_j - (\varphi_j)_x\|_\infty < \varepsilon$, $j = 1, 2$. Therefore

$$\|\varphi_1 - (\varphi_1)_x\|_* < \varepsilon \quad \text{and} \quad \|(H\varphi_2) - (H\varphi_2)_x\|_* < \varepsilon$$

when $|x|$ is small, so that by Theorem 5.1(b) $\varphi \in$ VMO.

Conversely, if $\varphi \in$ VMO, then by Corollary 4.5,

$$\varphi = u_1 + Hu_2 + \alpha,$$

where $\|u_1\|_\infty \le C\|\varphi\|_*$, $\|u_2\|_\infty \le C\|\varphi\|_*$, and α is constant. By Theorem 5.1, there is $y_0 > 0$ such that $\|\varphi(x) - \varphi(x, y_0)\|_* < \|\varphi\|_*/2$. Let $\varphi_1^{(1)}(x) = u_1(x, y_0)$, $\varphi_2^{(1)}(x) = u_2(x, y_0)$. Then $\varphi_j^{(1)} \in$ BUC,

$$\|\varphi_j^{(1)}\|_\infty \le \|u_j\|_\infty \le C\|\varphi\|_*, \quad j = 1, 2,$$

and $\varphi(x, y_0) = \varphi_1^{(1)}(x) + H\varphi_2^{(1)}(x) + \alpha$, so that

$$\|\varphi - (\varphi_1^{(1)} + H\varphi_2^{(1)} + \alpha)\|_* = \|\varphi(x) - \varphi(x, y_0)\|_* \le \|\varphi\|_*/2.$$

Hence

$$R_1 = \varphi - (\varphi_1^{(1)} + H\varphi_2^{(1)} + \alpha) = (u_1 - \varphi_1^{(1)}) + H(u_2 - \varphi_2^{(1)})$$

is in VMO and $\|R_1\|_* \le \|\varphi\|_*/2$. Repeating the above argument with R_1 and iterating, we obtain

$$\varphi = \sum_{k=1}^\infty \varphi_1^{(k)} + H\left(\sum_{k=1}^\infty \varphi_2^{(k)}\right) + \alpha$$

with $\varphi_1^{(k)} \in$ BUC, $\varphi_2^{(k)} \in$ BUC, and

$$\sum_k \|\varphi_1^{(k)}\|_\infty + \sum_k \|\varphi_2^{(k)}\|_\infty \le 4C\|\varphi\|_*.$$

That proves Theorem 5.2. $\square$

For the circle the proofs of Theorems 5.1 and 5.2 show that VMO is the closure of $C = C(T)$ in BMO and that VMO $= C + \tilde{C}$.

In Chapter IX we shall see that VMO is an important tool in the study of the algebra $H^\infty + C$.

6. Weighted Norm Inequalities for Maximal Functions and Conjugate Functions

Let $1 < p < \infty$ and let μ be a positive Borel measure on $\mathbb{R}$, finite on compact sets. We consider two problems.

Problem 1. *When is the Hardy–Littlewood maximal operator bounded on $L^p(\mu)$? That is, when is there a constant B_p such that*

$$\int |Mf|^p d\mu \le B_p \int |f|^p d\mu$$

for all measurable functions $f(x)$, where

$$Mf(x) = \sup_{x \in I} \frac{1}{|I|} \int_I |f| \, dt?$$

Problem 2. *When is the Hilbert transform a bounded operator on $L^p(\mu)$? That is, when is there a constant C_p such that*

$$\int |Hf|^p d\mu \le C_p \int |f|^p d\mu$$

for all functions $f \in L^2(dx)$, where

$$Hf(x) = \lim_{\varepsilon \to 0} \frac{1}{\pi} \int_{|x-t|>\varepsilon} \frac{f(t)}{x - t} \, dt?$$

When $d\mu = dx$ we know that both operators are bounded for all p, $1 < p < \infty$. In the case $p = 2$, the Helson–Szegö theorem, Theorem IV.3.4, provides a necessary and sufficient condition that the Hilbert transform be bounded on $L^2(\mu)$. The condition is that μ must be absolutely continuous, $d\mu = w(x)\,dx$, and the density $w(x)$ must have the form

$$\log w = u + Hv$$

with $u \in L^\infty$ and $\|v\|_\infty < \pi/2$.

The proof of this theorem was given in Chapter IV for the unit circle, but by now the reader should have no difficulty transferring the theorem to the line. For $p \neq 2$ the Helson–Szegö method has not produced very satisfying answers to Problem 2.

On the other hand, the real-variables approach to these problems taken by Muckenhoupt and others has been quite successful. Pleasantly, both problems have the same answer: The measure must be absolutely continuous

$$d\mu = w(x)\,dx, \quad w \in L^1_{\text{loc}},$$

and the weight $w(x)$ must satisfy the (A_p) condition,

$$(6.1) \qquad \sup_I \left(\frac{1}{|I|} \int_I w \, dx \right) \left(\frac{1}{|I|} \int_I \left(\frac{1}{w} \right)^{1/(p-1)} dx \right)^{p-1} < \infty.$$

Theorem 6.1. *Let μ be a positive measure finite on compact sets, and let $1 < p < \infty$. Then*

$$\int |Mf|^p d\mu \le B_p \int |f|^p d\mu$$

with B_p independent of f if and only if μ is absolutely continuous, $d\mu = w(x)dx$, and the density $w(x)$ satisfies (6.1).

Theorem 6.2. *Let μ be a positive measure finite on compact sets and let $1 < p < \infty$. Then*

$$\int |Hf|^p d\mu \le C_p \int |f|^p d\mu$$

with C_p independent of f if and only if $d\mu = w(x) \, dx$ and the density $w(x)$ satisfies (6.1).

Before going further, let us try to understand how the (A_p) condition (6.1) comes about.

Lemma 6.3. *If $1 < p < \infty$ and if μ is a positive measure on $\mathbb{R}$ such that*

$$\int |Mf|^p d\mu \le C \int |f|^p d\mu$$

for all f, then μ is absolutely continuous, $d\mu = w(x) \, dx$, and $w(x)$ satisfies the (A_p) condition (6.1).

Proof. Let E be a compact set with $|E| = 0$, let $\varepsilon > 0$, and let V be an open neighborhood of E with $\mu(V \setminus E) < \varepsilon$. Then $f = \chi_{V \setminus E}$ has $\int |f|^p d\mu < \varepsilon$. On the other hand, $Mf(x) = 1$ when $x \in E$ because $|E| = 0$. Hence by hypothesis,

$$\mu(E) \le \int |Mf|^p d\mu \le C\|f\|_p^p \le C\varepsilon,$$

so that $\mu(E) = 0$ and μ is absolutely continuous.

Write $d\mu = w(x)dx$, when $w(x) \in L_{\text{loc}}^1$ and $w(x) \ge 0$. Fix an interval I and let $f(x) = w(x)^\alpha \chi_I(x)$, for some real α. Then

$$\int |f|^p d\mu = \int_I w(x)^{1+p\alpha} \, dx,$$

and if $x \in I$,

$$Mf(x) \ge \frac{1}{|I|} \int_I f \, dt \ge \frac{1}{|I|} \int_I w^\alpha \, dt.$$

Approximating $f(x)$ from below by bounded functions, we obtain

$$\left(\int_I w \, dx \right) \left(\int_I w^\alpha \, dx \right)^p \leq |I|^p \int_I |Mf|^p d\mu \leq C|I|^p \int_I w^{1+p\alpha} \, dx,$$

and taking $\alpha = -1/(p-1) = 1 + p\alpha$, we conclude that

$$\left(\int_I w \, dx \right) \left(\int_I \left(\frac{1}{w} \right)^{1/(p-1)} dx \right)^{p-1} \leq C|I|^p,$$

which is the (A_p) condition. $\square$

Lemma 6.4. *If $1 < p < \infty$ and if μ is a positive measure on $\mathbb{R}$, finite on compact sets, such that for all f*

$$\int |Hf|^p d\mu \leq C \int |f|^p d\mu,$$

then $d\mu = w(x)dx$, where $w(x)$ satisfies the (A_p) condition (6.1).

Proof. Testing the hypothesis of the lemma with $f = \chi_{[0,1]}$, we see that

$$(6.2) \qquad\qquad \int \frac{d\mu(x)}{(1+|x|)^p} < \infty.$$

Let $g \in L^q(\mu), q = p/(p-1)$, be real. Then by the hypothesis and by duality, there is $h \in L^q(\mu)$ such that

$$\int (Hf)g \, d\mu = \int fh \, d\mu, \quad f \in L^p(\mu).$$

If $f \in H^2 \cap L^p(\mu)$, then $Hf = -if$ and the above identity becomes

$$(6.3) \qquad\qquad \int f(g - ih) \, d\mu = 0.$$

By (6.2) and Hölder's inequality,

$$d\nu(x) = \frac{(g - ih)(x) \, d\mu(x)}{x + i}$$

is a finite measure, and by (6.3) $\int (x - z)^{-1} d\nu(x) = 0$ on $\text{Im } z < 0$. The F and M. Riesz theorem now implies that ν is absolutely continuous. Hence $g(x)d\mu(x) = \text{Re}((x + i)d\nu(x))$ is absolutely continuous. Since $g \in L^q(\mu)$ is arbitrary, we conclude that μ is absolutely continuous.

Write $d\mu = w(x)dx$. Let I be an interval, and split I into two equal pieces, $I = I_1 \cup I_2, |I_1| = |I_2| = \frac{1}{2}|I|$. Let $f \geq 0$, f supported in I_1. Then for $x \in I_2$ we have

$$(6.4) \qquad |Hf(x)| = \frac{1}{\pi} \int_{I_1} \frac{f(t)}{|x-t|} \, dt \geq \frac{1}{2\pi} \int_{I_1} \frac{f(t)}{|I_1|} \, dt = \frac{1}{2\pi} f_{I_1}.$$

Taking $f = \chi_{I_1}$, we get

$$\int_{I_2} w\, dx \le C \int_{I_1} w\, dx.$$

By symmetry we also have

$$\int_{I_1} w\, dx \le C \int_{I_2} w\, dx.$$

Taking $f = w(x)^\alpha \chi_{I_1}$ in (6.4), we then get

$$\left(\int_{I_2} w\, dx\right)\left(\frac{1}{|I_1|}\int_{I_1} w^\alpha\, dx\right)^p \le C \int_{I_1} w^{1+\alpha p}\, dx.$$

Setting $\alpha = -1/(p-1)$ now gives

$$\left(\int_{I_1} w\, dx\right)\left(\frac{1}{|I_1|}\int_{I_1} \left(\frac{1}{w}\right)^{1/(p-1)} dx\right)^p \le C \int_{I_1} \left(\frac{1}{w}\right)^{1/(p-1)} dx,$$

which is the (A_p) condition. $\quad\square$

Thus (A_p) is a necessary condition in both Theorem 6.1 and Theorem 6.2. Now suppose w satisfies (A_p) and write

$$\varphi = \log w, \quad \psi = \log\left(\left(\frac{1}{w}\right)^{1/(p-1)}\right) = \frac{-\varphi}{p-1}.$$

Then φ and ψ are locally integrable, because w and $1/w$ are. For any interval I, we have

$$e^{\varphi_I}(e^{\psi_I})^{p-1} = 1$$

trivially, so that (A_p) can be rewritten

(6.5) $$\sup_I \left(\frac{1}{|I|}\int_I e^{\varphi - \varphi_I} dx\right)\left(\frac{1}{|I|}\int_I e^{\psi - \psi_I} dx\right)^{p-1} < \infty.$$

By Jensen's inequality,

$$\frac{1}{|I|}\int_I e^{\varphi - \varphi_I} dx \ge 1 \quad\text{and}\quad \frac{1}{|I|}\int_I e^{\psi - \psi_I} dx \ge 1.$$

Consequently (A_p) holds if and only if each factor in (6.5) is bounded separately, and we have the following lemma.

Lemma 6.5. Let $w \ge 0$, and let $\varphi = \log w$. Then w has (A_p), $1 < p < \infty$, if and only if

$$\sup_I \frac{1}{|I|}\int_I e^{\varphi - \varphi_I} dx < \infty$$

and

$$\sup_I \frac{1}{|I|} \int_I e^{-(\varphi - \varphi_I)/(p-1)} dx < \infty.$$

Thus if w has (A_p), then $\varphi = \log w \in$ BMO. Conversely if $\log w \in$ BMO, then by the John–Nirenberg theorem, w^δ has (A_p) for some $\delta > 0$.

Before turning to the more difficult proof that (A_p) is sufficient, we mention two corollaries. The case $p = 2$ of Theorem 6.2 can be merged with the Helson–Szegö theorem to give concrete expressions for the BMO distance of $\varphi \in$ BMO to L^∞ and for the distance from $f \in L_{\mathbb{R}}^\infty$ to Re H^∞. The reason is that the Helson–Szegö condition is, by Theorem 6.2, equivalent to

$$(A_2): \qquad \sup_I \left(\frac{1}{|I|} \int_I w \, dx \right) \left(\frac{1}{|I|} \int_I \frac{1}{w} dx \right) < \infty.$$

If $\varphi \in$ BMO, then $\varphi = f + Hg + \alpha$ with $f \in L^\infty$, $g \in L^\infty$, and α constant, and

$$\|\varphi\|' = \inf\{\|f\|_\infty + \|f\|_\infty : \varphi = f + Hg + \alpha\}$$

defines a norm on BMO equivalent to $\|\varphi\|_*$. With respect to $\| \ \|'$ the distance from φ to L^∞ is

$$\text{dist}(\varphi, L^\infty) = \inf_{f \in L^\infty} \|\varphi - f\|' = \inf\{\|g\|_\infty : \varphi - Hg \in L^\infty\}.$$

By the John–Nirenberg theorem there are $\varepsilon > 0$ and $\lambda(\varepsilon) > 0$ such that

$$(6.6) \qquad \sup_I \frac{|\{x \in I : |\varphi - \varphi_I| > \lambda\}|}{|I|} \leq e^{-\lambda/\varepsilon}, \quad \lambda > \lambda(\varepsilon).$$

Write

$$\varepsilon(\varphi) = \inf\{\varepsilon > 0 : (6.6) \text{ holds}\}.$$

Clearly $\varepsilon(\varphi) = 0$ if $\varphi \in L^\infty$. The John–Nirenberg theorem shows that

$$\varepsilon(\varphi) \leq C\|\varphi\|_* \leq C'\|\varphi\|'.$$

Corollary 6.6. *If $\varphi \in$ BMO is real valued, then*

$$\text{dist}(\varphi, L^\infty) = (\pi/2)\varepsilon(\varphi).$$

Proof. Condition (6.6) implies that

$$(6.7) \qquad M = \sup_I \frac{1}{|I|} \int_I \exp|A\varphi - A\varphi_I| dx < \infty$$

whenever $A < 1/\varepsilon(\varphi)$, and Chebychev's inequality shows that $A \leq 1/\varepsilon(\varphi)$ whenever (6.7) holds for A. Thus

$$1/\varepsilon(\varphi) = \sup\{A : \varphi \text{ has } (6.7)\}.$$

By (6.7) we have

$$(6.8) \qquad 1 \le \frac{1}{|I|} \int_I e^{\pm A(\varphi - \varphi_I)} dx \le M$$

for any interval I. Conversely, since $e^{|x|} \le e^x + e^{-x}$ (6.8) implies (6.7). Thus (6.8) is equivalent to (6.7). Hence by Lemma 6.5, (6.7) holds for $A > 0$ if and only if the weight $w = e^{A\varphi}$ satisfies the (A_2) condition. We conclude that

$$1/\varepsilon(\varphi) = \sup\{A : e^{A\varphi} \text{ has } (A_2)\}.$$

Because the (A_2) condition is equivalent to the Helson–Szegö condition, we now have

$$1/\varepsilon(\varphi) = \sup\{A : A\varphi = f + Hg, f \in L^\infty, \|g\|_\infty < \pi/2\},$$

which is the same thing as

$$(\pi/2)\varepsilon(\varphi) = \inf\{\|g\|_\infty : \varphi = f + Hg, f \in L^\infty\}. \qquad \square$$

Corollary 6.7. *If $f \in L^\infty$ is real valued, then*

$$\text{dist}(f, \text{Re } H^\infty) = \inf_{F \in H^\infty} \|f - \text{Re } F\|_\infty$$

satisfies

$$\text{dist}(f, \text{Re } H^\infty) = (\pi/2)\varepsilon(Hf).$$

Proof. This is immediate from Corollary 6.6 and the proof of Corollary 4.7. $\square$

The distances in Corollaries 6.6 and 6.7 can also be related to the growth of the local L^p oscillations of $\varphi(x)$ by means of the identity

$$\frac{\varepsilon(\varphi)}{e} = \overline{\lim_{p \to \infty}} \frac{1}{p} \left(\sup_I \frac{1}{|I|} \int_I |\varphi - \varphi_I|^p dx \right)^{1/p}.$$

Establishing this identity is a recreation left for the reader. See Exercise 17.

To show that (A_p) implies the boundedness of M and H on $L^p(w \, dx)$, we need four consequences of

$$(A_p): \qquad \sup_I \left(\frac{1}{|I|} \int_I w \, dx \right) \left(\frac{1}{|I|} \int_I \left(\frac{1}{w} \right)^{1/(p-1)} dx \right)^{p-1} < \infty.$$

The first two consequences are quite trivial.

Lemma 6.8. *If $1 < p < \infty$ and if $w(x)'$ satisfies (A_p), then*

(a) *$w(x)$ satisfies (A_r) for all $r > p$, and*
(b) *the weight $(1/w)^{1/(p-1)}$ satisfies (A_q), where $q = p/(p-1)$.*

Proof. For (a) note that $1/(r - 1) < 1/(p - 1)$, so that by Hölder's inequality

$$\frac{1}{|I|} \int_I \left(\frac{1}{w}\right)^{1/(r-1)} dx \leq \left(\frac{1}{|I|} \int_I \left(\frac{1}{w}\right)^{1/(p-1)} dx\right)^{(p-1)/(r-1)}.$$

For (b) note that $1/(p - 1) = q - 1$, so that if $v = (1/w)^1/(p - 1)$, then $(1/v)^{1/(q-1)} = w$. $\square$

The other two consequences of (A_p) lie deeper, but they can be derived from a delightful inequality due to Gehring.

Theorem 6.9. Let $p > 1$. If $v(x) \geq 0$, and if

$$(6.9) \qquad \left(\frac{1}{|I|} \int_I v^p dx\right)^{1/p} \leq K \frac{1}{|I|} \int_I v \, dx$$

for all subintervals of some interval I_0, then

$$(6.10) \qquad \left(\frac{1}{|I_0|} \int_{I_0} v^r dx\right)^{1/r} \leq C(p, K, r) \left(\frac{1}{|I_0|} \int_{I_0} v \, dx\right)$$

for $p \leq r < p + \eta$, where $\eta = \eta(p, K) > 0$.

The inequality reverse to (6.9), with constant 1, follows trivially from Hölder's inequality. Therefore the constant K in (6.9) must obviously satisfy $K \leq 1$. Theorem 6.9 is a close relative of the John–Nirenberg theorem and, as with that theorem, it is crucial that (6.9) hold for many subintervals of I_0.

Proof. We can suppose that $I_0 = [0, 1]$ and that $\int_{I_0} v^p \, dx = 1$. For $\lambda > 0$ write

$$E_\lambda = \{x \in I_0 : v(x) > \lambda\}.$$

What we are going to prove is the estimate

$$(6.11) \qquad \int_{E_\lambda} v^p \, dx \leq A\lambda^{p-1} \int_{E_\lambda} v \, dx, \qquad \lambda \leq 1,$$

with some constant $A = A(p, K)$. But first let us observe how (6.11) easily gives (6.10). For $r > p$ we have

$$\int_{E_1} v^r \, dx = \int_{E_1} v^p v^{r-p} dx = \int_{E_1} v^p \left(1 + (r - p) \int_1^v \lambda^{r-p-1} d\lambda\right) dx$$

$$= \int_{E_1} v^p dx + (r - p) \int_1^\infty \lambda^{r-p-1} \int_{E_\lambda} v^p dx \, d\lambda.$$

By (6.11) the last term in the above expression does not exceed

$$A(r - p) \int_1^\infty \lambda^{r-2} \int_{E_\lambda} v \, dx \, d\lambda \le A(r - p) \int_{E_1} v \int_0^v \lambda^{r-2} \, d\lambda$$
$$= A \frac{(r - p)}{(r - 1)} \int_{E_1} v^r \, dx.$$

Hence

$$\left(1 - \frac{A(r - p)}{(r - 1)}\right) \int_{E_1} v^r \, dx \le \int_{E_1} v^p \, dx.$$

Taking $A > 1$, we have $A(r - p)/(r - 1) < 1$ if $r < p + (p - 1)/(A - 1) = p + \eta$, $\eta + 0$. For such values of r we then have

$$\int_{I_0} v^r \, dx \le \int_{\{v<1\}} v^p \, dx + C_r \int_{E_1} v^p \, dx.$$

Because of (6.9) and the normalizations $|I_0| = 1$, $\int I_0 v^p \, dx = 1$, this proves (6.10) and the theorem.

To prove (6.11) we set $\beta = 2K\lambda > \lambda \ge 1$. Since trivially

(6.12) $$\int_{E_\lambda \setminus E_\beta} v^p \, dx \le \beta^{p-1} \int_{E_\lambda \setminus E_\beta} v \, dx \le (2K)^{p-1}\lambda^{p-1} \int_{E_\lambda} v \, dx,$$

proving (6.11) really amounts to making an estimate of $\int_{E_\beta} v^p \, dx$. By the Calderón–Zygmund Lemma 2.2, there are pairwise disjoint subintervals $\{I_j\}$ of I_0 such that

(6.13) $$\beta^p \le \frac{1}{|I_j|} \int_{I_j} v^p \, dx < 2\beta^p$$

and such that $v \le \beta$ almost everywhere on $I_0 \setminus \bigcup I_j$. So except for a set of measure zero, we have $E_\beta \subset \bigcup I_j$. By (6.13) we have

(6.14) $$\int_{E_\beta} v^p \, dx \le \sum_j \int_{I_j} v^p \, dx \le 2\beta^p \sum_j |I_j|.$$

By (6.9) and the other inequality in (6.13), we also have

$$\beta \le \left(\frac{1}{|I|_j} \int_{I_j} v^p \, dx\right)^{1/p} \le \frac{K}{|I|_j} \int_{I_j} v \, dx.$$

This means that

$$|I_j| \le \frac{K}{\beta} \int_{I_j} v \, dx \le \frac{K}{\beta} \int_{I_j \cap E_\lambda} v \, dx + \frac{K\lambda}{\beta}|I_j|,$$

so that by our choice of β,

$$|I_j| \le \frac{1}{\lambda} \int_{I_j \cap E_\lambda} v \, dx.$$

Substituting this inequality into (6.14) gives

$$\int_{E_\beta} v^p \, dx \le \frac{2\beta^p}{\lambda} \int_{E_\lambda} v \, dx \le 2^{p+1} K^p \lambda^{p-1} \int_{E_\lambda} v \, dx,$$

and with (6.12) this yields the desired inequality (6.11). $\quad\square$

Corollary 6.10. *If $1 < p < \infty$ and if the weight $w(x)$ satisfies the (A_p) condition, then*

(a) *there are $\delta > 0$ and $C > 0$ such that, for any interval I,*

(6.15)
$$\left(\frac{1}{|I|} \int_I w(x)^{1+\delta} \, dx \right)^{1/(1+\delta)} \le \frac{C}{|I|} \int_I w(x) \, dx,$$

(b) *there is $\varepsilon > 0$ such that $w(x)$ also satisfies the $(A_{p-\varepsilon})$ condition.*

Proof. To prove part (a) we can assume that $p > 2$, because of Lemma 6.8(a). The Cauchy–Schwarz inequality shows that

$$1 \le \left(\frac{1}{|I|} \int_I \left(\frac{1}{w} \right)^{1/(p-1)} dx \right)^{1/(p-1)} \left(\frac{1}{|I|} \int_I w^{1/(p-1)} \, dx \right).$$

With this inequality (A_p) yields

$$\frac{1}{|I|} \int_I w \, dx \le K \left(\frac{1}{|I|} \int_I w^{1/(p-1)} \, dx \right)^{p-1}.$$

Since $p - 1 > 1$, we can now use Theorem 6.9 on the function $v = w^{1/(p-1)}$ to obtain

$$\left(\frac{1}{|I|} \int_I w^{r/(p-1)} \, dx \right)^{1/r} \le C \left(\frac{1}{|I|} \int_I w \, dx \right)^{1/(p-1)}$$

for $p - 1 < r < p - 1 + \eta, \eta > 0$. Taking $1 + \delta = r/(p-1)$, we have (6.15).

To prove part (b), we apply (6.15) to $(1/w)^{1/(p-1)}$, which is a weight function satisfying condition $(A_q), q = p/(p-1)$. We get

$$\left(\frac{1}{|I|} \int_I \left(\frac{1}{w} \right)^{(1+\delta)/(p-1)} dx \right)^{(p-1)/(1+\delta)} \le \left(\frac{C}{|I|} \int_I \left(\frac{1}{w} \right)^{1/(p-1)} dx \right)^{p-1}.$$

Setting $\varepsilon = (\delta/(1+\delta))(p-1) > 0$, so that $(p - \varepsilon) - 1 = (p-1)/(1+\delta)$, and multiplying both sides of the above inequality by $(1/|I|) \int_I w \, dx$ now yields

$$\left(\frac{1}{|I|}\int_I w\,dx\right)\left(\frac{1}{|I|}\int_I \left(\frac{1}{w}\right)^{1/(p-\varepsilon-1)}\right)^{p-\varepsilon-1}$$
$$\leq C\left(\frac{1}{|I|}\int_I w\,dx\right)\left(\frac{1}{|I|}\int_I \left(\frac{1}{w}\right)^{1/(p-1)}dx\right)^{p-1},$$

and w has $(A_{p-\varepsilon})$. □

Another proof of (b), closer in spirit to the proof of the John–Nirenberg theorem, is outlined in Exercise 15.

Proof of Theorem 6.1. We suppose $w(x)$ has (A_p) and, writing $d\mu = w(x)\,dx$, we prove

$$\int |Mf|^p d\mu \leq B_p \int |f|^p d\mu.$$

The converse of this was proved in Lemma 6.3.

Applying Hölder's inequality to $fw^{1/p}$ and $w^{-1/p}$ and noting that $q/p = 1/(p-1)$, we have

$$\frac{1}{|I|}\int_I |f|\,dx \leq \left(\frac{1}{|I|}\int_I |f|^p w\,dx\right)^{1/p}\left(\frac{1}{|I|}\int_I \left(\frac{1}{w}\right)^{1/(p-1)}dx\right)^{(p-1)/p}.$$

The second factor can be estimated using (A_p) to yield

$$\frac{1}{|I|}\int_I |f|\,dx \leq K\left(\frac{1}{|I|}\int_I w\,dx\right)^{-1/p}\left(\frac{1}{|I|}\int_I |f|^p d\mu\right)^{1/p}$$
$$= K\left(\frac{1}{\mu(I)}\int_I |f|^p d\mu\right)^{1/p};$$

since $d\mu = w(x)\,dx$. Writing

$$M_\mu g(x) = \sup_{x\in I}\frac{1}{\mu(I)}\int_I |g|\,d\mu$$

and taking the supremum in the last inequality over all intervals I containing x, we get

(6.16)　　　　$Mf(x) \leq K(M_\mu(|f|^p))^{1/p}.$

The covering lemma (Lemma 1.4.4) applies to the measure μ, and the proof of the maximal theorem (Theorem 1.4.3) can then be adapted without difficulty to yield

(6.17)　　　　$\int (M_\mu(g))^r d\mu \leq C_r \int |g|^r d\mu, \quad 1 < r < \infty$

(see Exercise 13, Chapter I). Now since $w(x)$ also satisfies $(A_{p-\varepsilon})$, (6.16) can be replaced by

$$Mf(x) \leq K'M_\mu(|f|^{p-\varepsilon})^{1/(p-\varepsilon)}.$$

Using this inequality and (6.17) with $r = p/(p - \varepsilon)$, we obtain

$$\int |Mf|^p d\mu \leq K' \int (M_\mu(|f|^{p-\varepsilon}))^{p/(p-\varepsilon)} d\mu \leq C_r K' \int |f|^p d\mu,$$

which proves Theorem 6.1.   □

To prove the remaining half of Theorem 6.2 we need one more lemma and one additional theorem. A measure $d\mu = w(x)\,dx$ is said to satisfy *condition* (A_∞) If

$$(6.18) \qquad\qquad \mu(E)/\mu(I) \leq C(|E|/|I|)^\alpha$$

whenever E is a Borel subset of an interval I. The constants $C > 0$ and $\alpha > 0$ in (6.18) are supposed to be independent of E and I.

Lemma 6.11. *If $w(x)$ satisfies (A_p) for some $p < \infty$, then $d\mu = w(x)\,dx$ satisfies (A_∞).*

Proof. For $E \subset I$, Hölder's inequality and Corollary 6.10(a) give

$$\frac{\mu(E)}{|I|} = \frac{1}{|I|}\int_E w\,dx \leq \left(\frac{1}{|I|}\int_I w^{1+\delta}\,dx\right)^{1/(1+\delta)}\left(\frac{|E|}{|I|}\right)^{\delta/(1+\delta)}$$

$$\leq \left(\frac{C}{|I|}\int_I w\,dx\right)\left(\frac{|E|}{|I|}\right)^{\delta/(1+\delta)} = C\frac{\mu(I)}{|I|}\left(\frac{|E|}{|I|}\right)^{\delta/(1+\delta)},$$

which is the (A_∞) condition (6.18) with $\alpha = \delta/(1 + \delta)$.   □

Theorem 6.12. *If a measure $d\mu = w(x)\,dx$ satisfies the (A_∞) condition, and if $1 < p < \infty$, then*

$$\int |Hf|^p w\,dx \leq C_p \int |Mf|^p w\,dx.$$

Our objective, Theorem 6.2, follows directly from Lemma 6.11, Theorem 6.12, and Theorem 6.1. Our only unfinished business now is the proof of Theorem 6.12.

Proof. The maximal Hilbert transform

$$H^*f(x) = \sup_\varepsilon \left|\int_{|x-t|>\varepsilon} \frac{f(t)}{x - t}\,dt\right|$$

satisfies the weak-type inequality

$$(6.19) \qquad\qquad |\{x : H^*f(x) > \lambda\}| \leq \frac{C}{\lambda}\int |f(x)|\,dx.$$

by Exercise 11 of Chapter III.

Combining (6.19) with the (A_∞) condition, we shall show that when $0 < \gamma < 1$,

(6.20)

$$\mu(\{x : H^* f(x) > 2\lambda \quad \text{and} \quad Mf(x) \le \gamma\lambda\}) \le A\gamma^\alpha \mu\{H^* f(x) > \lambda\},$$

where the constant A does not depend on γ. Now (6.20) easily implies the theorem, because

$$\int |H^* f|^p \, d\mu = p2^p \int_0^\infty \lambda^{p-1} \mu(\{H^* f(x) > 2\lambda\}) \, d\lambda$$

$$\le Ap2^p\gamma^\alpha \int_0^\infty \lambda^{p-1} \mu(\{x : H^* f(x) > \lambda\}) \, d\lambda$$

$$+ p2^p \int_0^\infty \lambda^{p-1} \mu(\{x : Mf(x) > \gamma\lambda\}) \, d\lambda$$

$$= A2^p\gamma^\alpha \int |H^* f|^p \, d\mu + 2^p\gamma^{-p} \int |Mf|^p \, d\mu.$$

Choosing $\gamma > 0$ so small that $A2^p\gamma^\alpha < 1$, we obtain

$$\int |H^* f|^p \, d\mu \le C_p \int |Mf|^p \, d\mu,$$

and since trivially $|Hf| \le H^* f$, this proves the theorem.

To prove (6.20), write the open set $U_\lambda = \{x : H^* f(x) > \lambda\}$ as the union of disjoint open intervals $\{J_k\}$. Partition each J_k into closed intervals $\{I_j^k\}$ with disjoint interiors such that

$$|I_j^k| = \text{dist}(I_j^k, \mathbb{R} \setminus J_k).$$

The family of intervals shown in Figure VI.1,

$$\{I_j\} = \bigcup_{k=1}^\infty \{I_j^k\},$$

is called the *Whitney decomposition* of U_λ because it has the three properties

$$U_\lambda = \bigcup_j I_j,$$

$$I_i^0 \cap I_j^0 = \varnothing, \quad i \ne j,$$

$$\text{dist}(I_i, \mathbb{R} \setminus (U_\lambda)) = |I_j|.$$

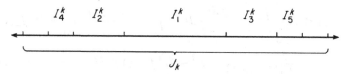

Figure VI.1.

The main step in the proof of (6.20) is to establish the inequality

(6.21) $|\{x \in I_j : H^* f(x) > 2\lambda \quad \text{and} \quad Mf(x) \le \gamma\lambda\}| \le B\gamma|I_j|.$

Indeed, (6.21) and the (A_∞) condition (6.18) then yield

$$\mu(\{x \in I_j : H^* f(x) > 2\lambda \quad \text{and} \quad Mf(x) \le \gamma\lambda\}) \le CB^\alpha \gamma^\alpha \mu(I_j),$$

and (6.20) follows by summation over $\{I_j\}$.

Thus the proof of Theorem 6.12 has been reduced by means of (A_∞) to proving (6.21), a condition not involving weight functions. No further relocations of the proof are necessary, and we conclude with the proof of (6.21). Because $\{I_j\}$ is the Whitney decomposition of U_λ, there is x_j such that

$$\text{dist}(x_j, I_j) = |I_j| \quad \text{and} \quad H^* f(x_j) \le \lambda.$$

We can suppose there is $\xi_j \in I_j$ with $Mf(\xi_j) \le \gamma\lambda$, because otherwise (6.21) is trivial for I_j. We can also suppose that γ is small, because (6.21) is obvious if $B\gamma > 1$. Let $\tilde{I}_j$ be that interval concentric with I_j having length $|\tilde{I}_j| = 3|I_j|$. Then $x_j \in \tilde{I}_j$. Also, let $I_j^* = \tilde{\tilde{I}}_j$ be that concentric interval with $|I_j^*| = 9|I_j|$. Write $f = f_1 + f_2$, where $f_1 = f\chi_{I_j^*}$, $f_2 = f\chi_{\mathbb{R}\setminus I_j^*}$. Since $\xi_j \in I_j^*$, we have

$$\frac{\|f_1\|_1}{|I_j^*|} = \frac{1}{|I_j^*|} \int_{I_j^*} |f| \, dx \le 2Mf(\xi_j) \le 2\gamma\lambda,$$

so that (6.19) gives

(6.22) $|\{x : H^* f_1(x) > \lambda/2\}| \le (2C/\lambda)\|f_1\|_1 \le 4C\gamma|I_j^*|.$

When $x \in I_j$ the integral $Hf_2(x)$ has no singularity x, and for any $\varepsilon > 0$ we have

$$\left| \frac{d}{dx} \int_{|t-x|>\varepsilon} \frac{f_2(t)}{x-t} \, dt \right| \le \int_{\mathbb{R}\setminus I_j^*} \frac{|f(t)|}{|x-t|^2} \, dt$$

$$\le C_1 \int_{|x-\xi_j|>4|I_j|} \frac{|f(s)|}{|s-\xi|^2} \, ds \le \frac{C_2}{|I_j|} Mf(\xi_j),$$

because the last integral can be viewed as a sum of averages over intervals centered of ξ_j. Consequently,

$$H^* f_2(x) \le \sup_\varepsilon \left| \int_{|x-t|>\varepsilon} \frac{f_2(t)}{x-t} \, dt - \int_{|x_j-t|>\varepsilon} \frac{f_2(t)}{x_j-t} \, dt \right| + H^* f_2(x_j)$$

$$\le \frac{C_2}{|I_j|} Mf(\xi_j)|x - x_j| + H^* f_2(x_j) \le 3C_2\gamma\lambda + H^* f_2(x_j).$$

On the other hand, since $\text{dist}(x_j, \mathbb{R}\backslash I_j^*) = 3|I_j|$, $H^* f_2(x_j)$ is not much larger than $H^* f(x_j)$. To be precise, we have

$$H^* f_2(x_j) = \sup_{\varepsilon > 3|I_j|} \left| \int_{|x_j - t| > \varepsilon} \frac{f_2(t)\,dt}{x_j - t} \right|$$

$$\leq \sup_{\varepsilon > 3|I_j|} \left| \int_{|x_j - t| > \varepsilon} \frac{f(t)}{x_j - t}\,dt \right| + \frac{1}{3|I_j|} \int_{3|I_j| < |x_j - t| < 6|I_j|} |f(t)|\,dt.$$

The first of these integrals is bounded by $H^* f(x_j) \leq \lambda$, while the second integral is bounded by $C_3 M f(x_j) \leq C_3 \gamma \lambda$. Hence we conclude that

$$(6.23) \qquad H^* f_2(x) \leq (3C_2 + C_3)\gamma\lambda + \lambda, \quad x \in I_j.$$

Since $H^* f \leq H^* f_1 + H^* f_2$, (6.22) and (6.23) yield the inequality

$$|\{x \in I_j : H^* f(x) > (3C_2 + C_3)\gamma\lambda + \tfrac{3}{2}\lambda\}| \leq |\{x : H^* f_1(x) > \tfrac{1}{2}\lambda\}|$$
$$\leq 12C\gamma|I_j|,$$

and this gives (6.21) when $(3C_2 + C_3)\gamma < \tfrac{1}{2}$. $\quad\square$

Notes

There is now a sizeable literature on BMO, with connections to univalent function theory, quasiconformal mappings, partial differential equations, and probability. To a large extent it is the duality theorem and the conformal invariance that make BMO important in so many areas. Some of this literature is cited in the bibliography.

The conformal invariance of BMO, which has been exploited by many authors, seems first to have appeared in Garsia's notes [1971]. Much of the material in Sections 1–4 comes from the fundamental paper of Fefferman and Stein [1972], where the duality theorem was first proved. Theorem 1.5 was first proved by Spanne [1966] and independently by Stein [1967].

Theorem 2.1 is from John and Nirenberg [1961]. Its simple proof has wide applications. Campanato [1963] and Meyers [1964] have given similar characterizations of Hölder classes.

The proof of the duality theorem would have been considerably simpler had it been the case that $|\nabla\varphi(x)|\,dx\,dy$ were a Carleson measure whenever $\varphi \in$ BMO (T). As explained in Exercise 9, this is not the case, even when φ is a Blaschke product. The Littlewood–Paley expression

$$|\nabla\varphi(z)|^2 \log(1/|z|)\,dx\,dy$$

is a useful device for overcoming this difficulty. A different device, due to Varopoulos [1977], is outlined in Exercises 12 and 13. The proof of the corona theorem provides yet another way around this difficulty (see Section 6 of Chapter VIII).

With the exception of the Fourier transform, the Hilbert transform is perhaps the most important operator in real or complex analysis. Much of the strength of the duality theorem lies in the fact that it characterizes BMO($\mathbb{R}$) as the set of functions of the form $\varphi = u + Hv, u, v \in L^\infty$. A more constructive proof of this decomposition will be given in Chapter VIII.

The results on VMO and their applications in Chapter IX below are from Sarason's paper [1975].

Theorem 6.1 is due to Muckenhoupt [1972] and Theorem 6.2 was first proved by Hunt, Muckenhoupt, and Wheeden [1973]. The absolute continuity of μ had been established earlier by Forelli [1963]. The proofs of Theorem 6.1 and Theorem 6.2 in the text follow Coifman and Fefferman [1974]. Gehring's inequality in Theorem 6.9 is in essence a result about maximal functions (see Gehring [1973]). Several other important inequalities hold when a weight satisfies (A_p) (see Gundy and Wheeden [1974], Muckenhoupt and Wheeden [1974]).

Nobody has ever found a direct proof of the equivalence of (A_2) with the Helson–Szegö condition. The papers of Garnett and Jones [1978], Jones [1980b], Uchiyama [1981], and Varopoulos [1980] shed some light on this problem and study the higher dimensional form of Corollary 6.6. A very interesting related problem has been posed by Baernstein: If $\varphi \in$ BMO satisfies

$$\sup_I (1/|I|)|\{x \in I : |\varphi(x) - \varphi_I| > \lambda\}| \le Ce^{-\lambda},$$

can $\varphi(x)$ be written $\varphi = u + Hv$ with $u \in L^\infty$ and $\|v\|_\infty \le \pi/2$? (See Nikolski, Havin, and Kruschev [1978, p. 230], but also page vii above.) The converse of Baernstein's conjecture is quite easy (see Exercise 18). Together, Theorem 6.2 and the Helson–Szegö theorem give another proof of the H^1–BMO duality on the line.

Inequality (6.21) explains why the Hilbert transform and the maximal function are often bounded on the same spaces. See Burkholder [1973] and Burkholder and Gundy [1972] for more about inequalities of this type, which are called "good λ inequalities." The argument in the proof of (6.20) is a powerful method that evolved from the real-variables proof that the Hilbert transform is weak type 1–1 (see Calderón and Zygmund [1952], Stein [1970]).

Most of the results in this chapter are really theorems about functions on Euclidean space $\mathbb{R}^n$ or even on the spaces of homogeneous type introduced by Coifman and Weiss [1971]. In fact, that is how the results are presented in many of the papers cited above. To keep matters as simple as possible and to keep in touch with the applications to follow, we have limited our discussion to $\mathbb{R}^1$ and T.

There are some beautiful connections among analytic BMO functions, univalent functions, and Bloch functions. Some of these results are outlined in

Exercises 22–25. Sarason's notes [1979] and Baernstein's lecture [1980] provide good surveys of this topic.

Exercises and Further Results

1.   If $\varphi \in$ BMO and if φ is real, then $\max(\varphi, 0)$ is in BMO.

2.   BMO is complete.

3.   (a)   If $\varphi \in$ BMO, $\chi_I \varphi \in$ BMO, and if

$$\|\chi_I \varphi\|_* \le C\|\varphi\|_*,$$

for every interval I, then φ is bounded and $\|\varphi\|_\infty < C'\|\varphi\|_*$.

   (b)   If h is measurable on T, then $h\varphi \in$ BMO (T) for all $\varphi \in$ BMO(T) if and only if $h \in L^\infty$ and

$$\sup_I \left\{ \frac{1}{|I|} \left(\log \frac{1}{|I|} \right) \int_I |h - h_I| \, dx \right\} < \infty$$

(Stegenga [1976]).

   (c)   Formulate and prove a similar result on the line.

   (d)   Suppose $\varphi \in$ BMO and suppose I is an interval such that $\varphi_I = 0$.

Let $\tilde{I}$ be the interval concentric with I having length $|\tilde{I}| = 3|I|$. Then there is $\psi \in$ BMO such that $\psi = \varphi$ on I, $\psi = 0$ on $\mathbb{R} \backslash \tilde{I}$, and $\|\psi\|_* \le C\|\varphi\|_*$. (Hint: Write $I = \bigcup_{n=0}^\infty J_n$ where dist $(J_n, \partial I) = |J_n|$, as in Figure VI.1. Suppose $|J_0| > |J_n|$, $n \ne 0$, so that J_0 is the middle third of I. For $n > 0$, let K_n be the reflection of J_n across the nearest endpoint of I and set $\psi(x) = \varphi_{J_n}, x \in K_n, \psi(x) = 0, x \notin I \cup \bigcup K_n$.)

4.   Let $f(x)$ be measurable on $\mathbb{R}$. Suppose there exist $\alpha < \frac{1}{2}$ and $\lambda > 0$ such that for each interval I there is some constant a_I such that

$$|\{x \in I : |f(x) - a_I| > \lambda\}| \le \alpha|I|.$$

Then $f \in$ BMO. The proof is like that of the John-Nirenberg theorem. The result is no longer true when $\frac{1}{2}$ is replaced by a larger number (Stromberg [1976]).

5.   On the circle, $(H^1_\mathbb{R})^* = $ BMO, with the pairing between $u \in H^1_\mathbb{R}$ and $\varphi \in$ BMO(T) given by

$$\frac{1}{2\pi} \int u\varphi \, d\theta,$$

provided constant BMO functions are not identified to zero, that is, provided BMO is normed by

$$\left| \frac{1}{2\pi} \int \varphi \, d\theta \right| + \|\varphi\|_*.$$

With the same norm BMOA $= H^2 \cap$ BMO is the dual of the classical space H^1 under the pairing

$$\int f\bar\varphi \frac{d\theta}{2\pi}, \qquad f \in H^1, \qquad \varphi \in \text{BMOA}.$$

6. For $\varphi \in L^1(\, dt/(1 + t^2))$ write

$$\nabla\varphi(z) = \left(\frac{\partial\varphi(z)}{\partial x}, \frac{\partial\varphi(z)}{\partial y} \right), \qquad z \in \mathrm{H} \, ,$$

where $\varphi(z)$ is the Poisson integral of $\varphi(t)$.
  (a) Show for $\varphi \in L^2$,

$$\int_{\mathbb{R}} |\varphi(t)|^2 \, dt = 2 \iint_{\mathrm{H}} y|\nabla\varphi(z)|^2 \, dx \, dy,$$

using either Green's theorem or the Fourier transform.
  (b) Prove $\varphi \in$ BMO if and only if $y|\nabla\varphi|^2 \, dx \, dy$ is a Carleson measure (not necessarily finite).
  (c) If $\varphi \in$ BMO and if f is a holomorphic function in H^1 having no zeros in H , then

$$\iint_{\mathrm{H}} y|\nabla f \cdot \nabla\varphi| \, dx \, dy \le C\|\varphi\|_* \|f\|_{H^1}.$$

Moreover, if $f \in \mathfrak{A}$, then

$$\left| \int f(\iota)\varphi(t) \, dt \right| = \left| \lim_{y\to 0} \int f(x + iy)\varphi(x + iy) \, dx \right|$$

$$\le C \iint_{\mathrm{H}} y|\nabla f \cdot \nabla\varphi| \, dx \, dy,$$

which proves the duality theorem on the line. Notice the last inequality does not follow directly from part (a) by polarization because it is not assumed that $\varphi \in L^2$. However, when φ and f are continuous, the identity $\Delta(\varphi(z)f(z)) = 2\nabla f(z) \cdot \nabla\varphi(z)$ and Green's theorem yield

$$\lim_{R\to\infty} \int_{-R}^{R} f(t)\varphi(t) \, dt = \lim_{R\to\infty} 2 \iint_{\mathrm{H} \cap \{|z|<R|\}} y\nabla f(z) \cdot \nabla\varphi(z) \, dx \, dy.$$

7. (a) Let T be the space of two tailed sequences

$$F = \{f_n : -\infty < n < \infty\}$$

of measurable functions on $\mathbb{R}$ such that

$$\int \sup_n |f_n(x)| \, dx = \|F\| < \infty.$$

Under this norm T is a Banach space.

(b) Let T_0 be the closure in T of the set of sequences F for which there exists $N = N(F)$ such that

$$f_n = f_N, \quad n \geq N; \quad f_n = 0, \quad n \leq -N.$$

When $F \in T_0$, $\lim_{n \to \infty} f_n = f$ exists in L^1. The dual space of T_0 consists of sequences $G = \{g_n : -\infty < n \leq \infty\}$ of L^∞ functions with norm

$$\|G\| = \left\| |g_\infty(x)| + \sum_{-\infty}^{\infty} |g_n(x)| \right\|_\infty$$

under the pairing

$$\langle F, G \rangle = \int f_\infty(x) g_\infty(x) \, dx + \sum_{-\infty}^{\infty} \int f_n(x) g_n(x) \, dx.$$

(c) Choose $\{y_n : -\infty < n < \infty\}$ such that

$$y_n \to 0 \ (n \to \infty), \ y_n \to \infty \ (n \to -\infty),$$

and

$$0 < y_n - y_{n+1} \leq \min(y_n^2, 1).$$

Let $S : H_\mathbb{R}^1 \to T$ be defined by

$$(S(u))_n(x) = u(x, y_n).$$

Then S maps $H_\mathbb{R}^1$ onto a closed subspace of T_0. (Use the vertical maximal function

$$u^+(x) = \sup_{y>0} |u(x, y)|.$$

(d) By parts (b) and (c), every bounded linear functional on $H_\mathbb{R}^1$ has the form

$$L(u) = \int u(x) g_\infty(x) \, dx + \sum_{n=-\infty}^{\infty} \int u(x, y_n) g_n(x) \, dx$$

with

$$\frac{1}{|I|} \int_I \left(|g_\infty(x)| + \sum_{n=-\infty}^{\infty} |g_n(x)| \right) dx \leq C \|L\|,$$

and conversely. With the duality theorem this gives the converse of Theorem 1.6. If $\varphi \in$ BMO, then

$$\varphi(t) = g_\infty(t) + \int_H P_y(x - t)d\sigma(x, y),$$

where $|\sigma|$ is a Carleson measure. Here

$$d\sigma = \sum_{-\infty}^{\infty} g_n(x)\, ds_n,$$

where ds_n is dx on the line $\{y = y_n\}$. Note that $|\sigma|$ is more than a Carleson measure since

$$|\sigma|(I \times (0, \infty)) \le C|I|.$$

This reflects the fact H^1 is determined by the vertical maximal function as well as by the nontangential maximal function. The proof sketched above is due to Fefferman (unpublished).

★★★(e) Carleson [1976] constructively obtained the decomposition

$$\varphi(t) = g_\infty(t) + \sum_{-\infty}^{\infty} \int P_{y_n}(x - t)g_n(t)\, dt,$$

where $|g_\infty| + \sum_{-\infty}^{\infty} |g_n| \in L^\infty$, for each $\varphi \in$ BMO.

(f) Assuming the result in (e), prove the maximal function characterization of H^1: If $u \in L^1 \cap L^2$, then

$$\|Hu\|_1 \le C\|u^+\|_1.$$

(Hint: Let $g \in L^2 \cap L^\infty$, $\|g\|_\infty = 1$, and let $\varphi = Hg \in$ BMO. Then

$$\left| \int (Hu)g\, dt \right| = \left| \int u\varphi\, dt \right| \le C\|u^+\|_1.$$

The left side has supremum $\|Hu\|_1$.)

(g) Derive the difficult half of the duality theorem from part (e).

8. Suppose $f = \sum_{n\ge0} a_n e^{in\theta}$ is in H^1. Use duality to prove

$$\sum_{k>0} |a_{2^k}|^2 \le C\|f\|_1^2 \quad \text{(Paley)}$$

and

$$\sum_{n\ge0} \frac{|a_n|}{n+1} \le C\|f\|_1 \quad \text{(Hardy)}.$$

9. There exist BMO functions such that $|\nabla u|\, dx\, dy$ is not a Carleson measure. There is even a Blaschke product $B(z)$ such that $\iint_D |B'(z)|\, dx\, dy = \infty$.

(a)   Let $f(e^{i\theta}) = \sum_{n\geq 1}(1/n)e^{i2^n\theta}$. Then $f \in$ BMO. Let A_n be the annulus $1 - 2^{-n} \leq |z| \leq 1 - 2^{-n-1}$. Then $\iint_{A_n} |f'(z)|\, dx\, dy \geq c/n$. Writing Re $f = u_1 + \tilde{u}_2$, show there is a bound harmonic function $u(z)$ such that $\iint_D |\nabla u|\, dx\, dy = \infty$. Then $F = \exp(u + i\tilde{u})$ is an H^∞ function with

$$\iint |F'(z)|\, dx\, dy = \infty.$$

(b)   If $g(e^{i\theta}) = \sum a_n e^{i2^n\theta}$, then $g \in$ BMO if and only if $\sum |a_n|^2 < \infty$, but $\iint |g'(z)|\, dx\, dy < \infty$ if and only if $\sum |a_n| < \infty$.

(c)   If $\sum |a_n|^2 < \infty$ there is $F \in H^\infty$ such that $\hat{F}(2^n) = a_n$ (see Fournier [1974]). If $\sum |a_n|^2 < \infty$, but $\sum |a_n| = \infty$ then $\iint |F'(z)|\, dx\, dy = \infty$.

(d)   There exists a Blaschke product $B(z)$ such that

$$\iint_D |B'(z)|\, dx\, dy = \infty.$$

The earliest example is due to Rudin [1955b].

10.   There exist $f_1(z)$ and $f_2(z)$ in H^∞ such that

$$\int_\Gamma (|f_1'(z)| + |f_2'(z)|)\, ds = \infty$$

for all smooth curves Γ in D terminating on ∂D. Consequently the mapping

$$z \to (z, f_1(z), f_2(z))$$

embeds the unit disc into $\mathbb{C}^3$. The embedded manifold is bounded and it is complete in the metric of Euclidean arc length.

To construct f_1 and f_2, first take

$$\varphi(re^{i\theta}) = \sum_{n=1}^{\infty} \frac{r^{10^n}}{n} \cos(10^n\theta).$$

Then

$$|\nabla\varphi(z)| \geq 10^n/100n \quad \text{on} \quad A_n = \{9 \cdot 10^{-n-1} \leq 1 - |z| \leq 11 \cdot 10^{-n-1}\},$$

so that

$$\int_\Gamma |\nabla\varphi(z)|\, ds = \infty$$

for every curve Γ in D which terminates on ∂D. On the other hand, $\varphi(z)$ is the Poisson integral of $\varphi(e^{i\theta}) \in$ BMO. Write $\varphi = u + i\tilde{v}$ and set $f_1 = e^{u+i\tilde{u}}$, $f_2 = e^{v+i\tilde{v}}$. This example is due to Jones [1979a]. See Yang [1977] for background on this problem.

11.   An atom is a function $a(x)$ supported on an interval I and satisfying

$$\int a(x)\,dx = 0, \quad |a(x)| \le 1/|I|.$$

If $\{a_j\}$ is a sequence of atoms and if $\sum |\lambda_j| < \infty$, then

$$f(x) = \sum \lambda_j a_j \in H_{\mathbb{R}}^1$$

and $\|f\|_{H^1} \le C \sum |\lambda_j|$. Conversely, by the duality theorem every function in $H_{\mathbb{R}}^1$ has the above form with $\sum |\lambda_j| \le C \|f\|_{H^1}$. See Coifman [1974] or Latter [1978] for direct proofs of the atomic decomposition, which in turn implies the H^1–BMO duality.

12.   (Dyadic BMO). A dyadic interval is an interval of the form $\omega = (j2^{-n}, (j+1)2^{-n})$ with n and j integers. For $\varphi \in L_{\text{loc}}^1$, define

$$\|\varphi\|_d = \sup_\omega \frac{1}{|\omega|} \int |\varphi - \varphi_\omega|\,dx,$$

the supremum taken only over dyadic intervals. The dyadic BMO space, BMO_d, consists of the functions φ with $\|\varphi\|_d$ finite.

(a)   BMO is a subset of BMO_d, but there are functions in BMO_d not in BMO.

(b)   If $\varphi \in \text{BMO}_d$, then $\varphi \in \text{BMO}$ if and only if

$$|\varphi_{\omega_1} - \varphi_{\omega_2}| \le A$$

whenever ω_1 and ω_2 are adjacent dyadic intervals of the same length. Then

$$C_1 \|\varphi\|_* \le A + \|\varphi\|_d \le C_2 \|\varphi\|_*.$$

(c)   Suppose $\varphi \in \text{BMO}_d$ has support $[0, 1]$. Then there exists a family G of dyadic intervals ω such that

(E.1) $$\sum_{\substack{\omega \subset I \\ \omega \in G}} |\omega| \le C \|\varphi\|_d |I|$$

for every dyadic interval I, and there exist weights α_ω, $\omega \in G$, such that $|\alpha_\omega| \le C\|\varphi\|_d$ and such that

$$\varphi = \psi + \sum_{\omega \in G} \alpha_\omega \chi_\omega$$

with $\psi \in L^\infty$, $\|\psi\|_\infty \le C\|\varphi\|_d$. Conversely, every function of this form is in BMO_d. To prove the decomposition, mimic the proof of the John–Nirenberg theorem. Suppose $\|\varphi\|_d = 1$. Take $G = \{I_j^{(n)}\}$ from that proof and take $\alpha_\omega = \varphi_{I_j^{(n)}} - \varphi_{I_k^{(n-1)}}$, where $I_k^{(n-1)} \supset I_j^{(n)}$.

(d) The dyadic maximal function is

$$f^{\mathrm{d}}(x) = \sup_{x \in \omega} \left| \frac{1}{|\omega|} \int_{\omega} f \, dt \right|,$$

where the supremum is taken over dyadic intervals. The space H_{d}^1 consists of all $f \in L^1([0, 1])$ for which

$$\|f^{\mathrm{d}}\|_1 = \int_0^1 |f^{\mathrm{d}}(x)| \, dx < \infty.$$

The norm in the dyadic H^1 space H_{d}^1 is $\|f^{\mathrm{d}}\|_1$. By part (c) the dual of H_{d}^1 is $\mathrm{BMO}_{\mathrm{d}}$.

The dyadic spaces H_{d}^1 and $\mathrm{BMO}_{\mathrm{d}}$ are special cases of the martingal H^1 and BMO spaces (see Garsia [1973]). Technically, $\mathrm{BMO}_{\mathrm{d}}$ is much easier to work with than BMO. For example, part (c) above is quite easy, but the direct proof of the analogous result for BMO, Exercise 7(e), is rather difficult.

13. (a) Suppose $\varphi \in \mathrm{BMO}$ has support contained in $[0, 1]$. Then there exists $F(x, y) \in C^\infty(\mathrm{H})$ such that $|\nabla F(x, y)| \, dx \, dy$ is a Carleson measure

(E.2) $$\iint_Q |\nabla F(x, y)| \, dx \, dy \leq A \|\varphi\|_* h,$$

$Q = [a, a + h] \times (0, h]$, such that

(E.3) $$\sup_{y>0} |F(x, y)| \in L^1$$

and such that

(E.4) $$\varphi(x) = \lim_{y \to 0} F(x, y) + \psi(x)$$

with $\|\psi\|_\infty \leq C \|\varphi\|_*$. By 12(c) we can suppose φ has the special form

$$\varphi(x) = \sum_{w \in G} \alpha_\omega \chi_\omega(x),$$

where the family G of dyadic intervals satisfies (E.1). Then (E.2) and (E.3) hold for

$$F_0(x, y) = \sum_{\omega \in G} \alpha_\omega \chi_\omega(x) \chi_{(0, |\omega|)}(y).$$

Although F_0 is not in C^∞, $|\nabla F_0(x, y)|$, taken as a distribution, is a Carleson measure. (To control $|\partial F_0 / \partial x|$, use 12(b) above.)

The smooth function $F(x, y)$ is a mollification of F_0. Let $h \in C^\infty(\mathbb{C})$ satisfy

$$h(z) \geq 0;$$
$$\int h \, dxdy = 1;$$
$$h(z) = 0, \quad |z| > \tfrac{1}{2}.$$

Set $h_n(z) = 2^{2n}h(2^n z)$, $F_n(z) = F_0(z)\chi_{\{2^{-n} < y < 2^{-n+1}\}}(t)$, and write

$$F(z) = \sum_{n=1}^{\infty}(h_n * F_n)(z).$$

Then $F \in C^\infty(H\ )$ satisfies (E.2)–(E.4) (Varopoulos [1977a]).
   (b)   Use the result in (a) to give another proof that $(H_{\mathbb{R}}^1)^* = $ BMO.

★**14.**   (a)   On the unit circle, the dual of VMO is $H_{\mathbb{R}}^1$.
   (b)   Let $f \in$ BMO. Then $f \in$ VMO if and only if

$$\int_a^{a+h} \int_0^h y|\nabla u(z)|^2 \, dx \, dy = o(h)$$

uniformly in $a \in \mathbb{R}$, where $u(z)$ is the Poisson integral of $f(t)$.

★**15.**   (a)   Suppose $\psi \in$ BMO, $\|\psi\|_* \leq B_0$. If

$$\sup_I \frac{1}{|I|} \int e^{\psi - \psi_I} \, dx = B_1 < \infty,$$

then there are $\delta = \delta(B_0, B_1) > 0$ and $B_2 = B_2(B_0, B_1)$ such that

$$\sup_I \frac{1}{|I|} \int e^{(1+\delta)(\psi - \psi_I)} \, dx \leq B_2.$$

The proof is a variation of the proof of the John–Nirenberg theorem. It is enough to show there is $\alpha > 1$ such that for $n = 1, 2, \ldots$

(E.5) $|\{x \in I : \psi(x) - \psi_I > n\alpha\}| \leq e^{-(1+2\delta)n\alpha}|I|$

for every interval I.
   Take $\lambda > 0$ so that

(E.6) $|\psi_I - \psi_j| \leq \tfrac{1}{2}\lambda$

if $I \subset J, |J| = 2|I|$, and so that

(E.7) $|\{x \in I : |\psi(x) - \psi_I| \geq \tfrac{1}{2}\lambda\}| \geq \tfrac{1}{2}|I|$

for all I. Fix an interval I_0 and assume $I_0 = [0, 1]$ and $\psi_{I_0} = 0$. For $n \geq 1$, let $\{I_{n,j}\}$ be the maximal dyadic intervals inside I_0 for which $\psi_{I_{n,j}} \geq n\lambda$. Then by (E.6)

$$n\lambda \leq \psi_{I_{n,j}} \leq (n + \tfrac{1}{2})\lambda,$$

and by (E.7) there exists $E_{n,j} \subset I'_{n,j}$ such that $|E_{n,j}| \geq \frac{1}{2}|I_{n,j}|$ and

$$\psi(x) \geq (n - \tfrac{1}{2})\lambda, \quad x \in E_{n,j}.$$

Hence for any chosen interval $I_{m,k}$ we have

$$\sum_{n \geq m} e^{-(m-n+1)\lambda} \sum_{I_{n,j} \subset I_{m,k}} \frac{|I_{n,j}|}{|I_{m,k}|} \leq 2 \sum_{n \geq m} e^{-(m-n+1)\lambda} |\bigcup E_{n,j}|/|I_{m,k}|$$

$$\leq \frac{C}{|I_{m,k}|} \int_0^\infty e^t |\{x \in I_{m,k} : \psi(x) - \psi_{I_{m,k}} > t\}| \, dt$$

$$= \frac{C}{|I_{m,k}|} \int_{I_{m,k}} e^{\psi - \psi_{I_{m,k}}} \, dx \leq B_1.$$

In particular, whenever $n \geq m$

(E.8)
$$\sum_{I_{n,j} \subset I_{m,k}} \frac{|I_{n,j}|}{|I_{m,k}|} \leq C B_1 e^{(m-n+1)\lambda}.$$

Let s be a positive integer, $s > 2$. For each $I_{m,k}$ we have n_0, $m < n_0 \leq m + s$, so that

$$e^{-(m-n_0+1)\lambda} \sum_{I_{n_0,j} \subset I_{m,k}} \frac{|I_{n_0,j}|}{|I_{m,k}|} \leq \frac{C B_1}{s},$$

because otherwise (E.8) would fail. Using (E.8) to compare $\sum_{I_{m+s,j}}$ to $\sum_{I_{n_0,j}}$, we therefore obtain

$$\sum_{I_{m+s,j} \subset I_{m,k}} \frac{|I_{m+s,j}|}{|I_{m,k}|} \leq \frac{C B_1^2}{s} e^{(-s+2)\lambda}.$$

Now choose s so that

$$C B_1^2 e^{2\lambda}/s \leq \tfrac{1}{2},$$

set $\alpha = s\lambda$, and $\delta = (\log 2)/2\alpha$. Then

$$\sum_{I_{m+s,j} \subset I_{m,k}} |I_{m+s,j}| \leq e^{-(1+2\delta)\alpha} |I_{m,k}|,$$

and (E.5) follows upon iteration.

This argument is due to P. Jones. There are several interesting consequences.

(b) If $\psi \in \mathrm{BMO}$ and if

(E.9)
$$\sup_I \frac{1}{|I|} \int_I e^{A/\psi - \psi_I|} \, dx < \infty,$$

then for some $\varepsilon > 0$

$$\sup_I \frac{1}{|I|} \int e^{(A+\varepsilon)|\psi - \psi_I|} \, dx < \infty.$$

Thus the set $\{A : (E.9) \text{ holds}\}$ does not contain its supremum.

   (c)   If w satisfies (A_p), then w satisfies $(A_{p-\varepsilon})$ for some $\varepsilon > 0$. (Use part (a) and Lemma 6.5 with $\psi = \log((1/w)^{1/(p-1)}$.)

16.   Let H denote the Hilbert transform and let B denote the operator of multiplication by a fixed function $b(x)$, $Bg(x) = b(x)g(x)$. The commutator $[B, H]$ is defined by

$$[B, H]g = B(Hg) - H(Bg) = b(x)(Hg)(x) - H(bg)(x).$$

Then $[B, H]$ is bounded on L^2 if and only if $b(x) \in \text{BMO}$, and

$$C_1 \|b\|_* \leq \|[B, H]\| \leq C_2 \|b\|_*.$$

(Hint: Use duality and Riesz factorization; see Coifman, Rochberg, and Weiss [1976].) The following proof that $\|[B, H]\| \leq C \|b\|_*$ is due to Rochberg. By Section 6 there is $\delta > 0$ such that whenever $\|b\|_* \leq \delta$, e^{2b} has condition (A_2). Hence by Theorem 6.2.

$$Tf(x) = \int \frac{e^{(b(x)-b(y))}}{x - y} f(y) dy$$

satisfies $\|Tf\|_2 \leq C \|f\|_2$ when $\|b\|_* \leq \delta$, since

$$\|Tf\|_2^2 = \int e^{2b(x)} |H(e^{-b}f)(x)|^2 \, dx \leq C \int e^{2b(x)} e^{-2b(x)} |f(x)|^2 \, dx.$$

For $|z| = 1$ the same holds for the operator T_z defined with zb in place of b. However

$$\frac{1}{2\pi i} \int_{|z|=1} T_z f \frac{dz}{z^2} = \int \frac{b(x) - b(y)}{x - y} f(y) dy = [B, H](f).$$

17.   If μ is a probability measure and if $f \geq 0$ is μ-measurable, then

$$\overline{\lim_{n \to \infty}} \frac{1}{n} \left(\int |f|^n d\mu \right)^{1/n} = \frac{1}{eA(f)},$$

where

$$A(f) = \sup \left\{ A : \int e^{Af} d\mu < \infty \right\}.$$

(Expand e^{Af} into a power series and use the root test and Stirling's formula.)

18.   (a)   Suppose $u \in L^\infty(T)$, $\|u\|_\infty \leq \pi/2$, and set $f = e^{-i(u+i\tilde{u})}$. Then $|f(e^{i\theta})|$ is weak L^1, since $\text{Re } f \geq 0$. Hence

$$|\{\theta : |\tilde{u}(e^{i\theta})| > \lambda\}| \leq Ce^{-\lambda}.$$

By conformal invariance, that means

$$\omega_z(\{\theta : |\tilde{u}(e^{i\theta}) - \tilde{u}(z| > \lambda\}) \leq Ce^{-\lambda}$$

for all $z \in D$, where $\omega_z(E) = (1/2\pi) \int_E P_z d\theta$. Consequently,

$$|\{\theta \in I : |\tilde{u}(e^{i\theta}) - \tilde{u}(I) > \lambda\})| \leq Ce^{-\lambda}|I|$$

for every arc I.

(b)   In a similar fashion, the Helson–Szegö condition implies the condition (A_2).

19.   (a)   Let μ be a locally finite positive Borel measure on $\mathbb{R}$ for which the maximal function

$$M\mu(x) = \sup_{x \in I} \mu(I)/|I|$$

is finite Lebesgue almost everywhere. Then $\log M\mu \in$ BMO (Coifman and Rochberg [1980]).

(b)   Let E be a subset of $[0, 1]$ with $|E| \leq 4^{-1/\varepsilon}$. Then there exists $\varphi \in$ BMO such that

$$0 \leq \varphi \leq 1.$$

$$\varphi = 1 \quad \text{on } E, \qquad \varphi = 0 \quad \text{off} [-1, 2],$$

$$\|\varphi\|_* \leq c\varepsilon,$$

where c is an absolute constant. (Hint: Take $\varphi = (\alpha + \beta \log M(\chi_E))^+$. A different proof is given by Garnett and Jones [1978].)

★20.   (a)   When $f \in L^1_{\text{loc}}(\mathbb{R})$, define

$$f^\#(x) = \sup_{x \in I} \frac{1}{|I|} \int_I |f - f_I|\, dt.$$

Thus $f \in$ BMO if and only if $f^\# \in L^\infty$. By the maximal theorem, $\|f^\#\|_p \leq C_p\|f\|_p$, $1 < p < \infty$. Prove the converse:

$$\|f\|_p \leq C'_p\|f^\#\|_p, \qquad 1 < p < \infty$$

(Fefferman and Stein [1972]).

(b)   Suppose T is a mapping from $L^2(\mathbb{R}) \cap L^\infty(\mathbb{R})$ into measurable functions on $\mathbb{R}$ such that

$$\|Tf\|_2 \leq A_0\|f\|_2, \qquad \|Tf\|_* \leq A_1\|f\|_*,$$

$f \in L^2 \cap L^\infty$. Then $\|Tf\|_p \leq A_p\|f\|_p, 2 \leq p < \infty$. (Use $(Tf)^\#$ and the Marcinkiewicz interpolation theorem.) Consequently the M. Riesz theorem follows from Theorem 1.5.

21. A positive locally integrable weight function $w(x)$ is said to satisfy condition (A_1) if

$$\sup_I \left\{ \left(\frac{1}{|I|} \int_I w \, dx \right) \operatorname*{ess\,sup}_{x \in I} \frac{1}{w(x)} \right\} < \infty.$$

(a) $w(x)$ satisfies $(A_1,)$ if and only if

$$Mw(x) \le Cw(x),$$

where M is the Hardy–Littlewood maximal function.

(b)   If $w(x)$ satisfies (A_1), then $w(x)$ satisfies (A_p) for all $p > 1$.

(c) $w(x)$ satisfies (A_1) if and only if the maximal function operator or the Hilbert transform is weak-type 1–1 on $L^1(w \, dx)$ (Muckenhoupt [1972]; Hunt, Muckenhoupt, and Wheeden [1973]).

(d)   Let $\varphi = \log w$. Then w has (A_1) if and only if

$$\sup_I \frac{1}{|I|} \int_I e^{\varphi - \varphi_I} \, dx < \infty \quad \text{and} \quad \sup_I \left(\varphi_I - \operatorname*{ess\,inf}_{x \in I} \varphi(x) \right) < \infty.$$

The space of functions satisfying the latter condition is called BLO, for bounded lower oscillation. If $\varphi \in$ BLO then $\varphi \in$ BMO and so $e^{\varepsilon \varphi}$ has (A_1) for some $\varepsilon > 0$.

★★★(e)   If w_1 and w_2 have (A_1), then by Hölder's inequality $w = w_1 w_2^{1-p}$ has (A_p). The converse is also true but quite difficult. See Jones [1980c]. With the Helson–Szegö theorem, the converse implies that if $\|v\|_\infty < 1$, then

$$Hv = u - Hu_1 + Hu_2,$$

where $u \in L^\infty$, $\|u_j\|_\infty \le 1$, and $Hu_j \in$ BLO, $j = 1, 2$.

(f)   If w satisfies A_1, then $w^{1+\delta}$ satisfies A_1 for some $\delta > 0$. (Use part (b) and Corollary 6.10.)

(g)   If μ is a positive Borel measure finite on compact sets such that $M(d\mu) < \infty$ almost everywhere and if $0 < \alpha < 1$, then $(M(d\mu))^\alpha$ has A_1 (Coifman and Rochberg [1980]).

★★(h)   For any function $w(x) \ge 0$ and for any $s > 1$, define

$$A_s(w) = M(Mw^s)^{1/s}.$$

Then for $1 < p < \infty$

$$\int |Hf|^p w \, dx \le C_{p,s} \int |f|^p A_s(w) \, dx$$

(see Córdoba and Fefferman [1976]). It then follows easily from part (a) that

$$\int |Hf|^p w \, dx \le C_p \int |f|^p w \, dx,$$

$1 < p < \infty$ if w satisfies (A_1).

22. On the unit circle let BMOA $= H^2 \cap$ BMO. If $f \in$ BMOA then

$$\inf_{g \in I^\infty} \|f - g\|_* \le C \inf_{g \in L^\infty} \|f - g\|_*.$$

★**23.** (a) Let $f(z)$ be a univalent function on the unit disc. If $f(z)$ has no zeros then $\log f(e^{i\theta}) \in$ BMO. Moreover, if $0 < p < \frac{1}{2}$, then $|f(e^{i\theta})|^p$ satisfies (A_2) (Baernstein [1976]; see also Cima and Petersen [1976], Cima and Schober [1976]).

(b) Let $f(z)$ be analytic on D. Then $f \in$ BMOA if and only if $f = \alpha \log g'(z)$, where α is a constant and g is a conformal mapping from D onto a region bounded by a rectifiable Jordan curve Γ satisfying

$$l(w_1, w_2) \le c|w_1 - w_2|,$$

$w_1, w_2 \in \Gamma$, where $l(w_1, w_2)$ is the shorter are on Γ joining w_1 to w_2 (Pommerenke [1977]). See also Pommerenke [1978] for a similar description of VMOA.

★**24.** Let E be a closed set on the Riemann sphere, $\infty \in E$. Then every analytic function on D having values in $\mathbb{C} \backslash E$ is in BMOA if and only if there is $r > 0$ and $\delta > 0$ such that

$$\text{cap}\,(E \cap \{|z - z_0| < r\}) > \delta$$

for all $z_0 \in \mathbb{C} \backslash E$. Here cap($S$) denotes the logarithmic capacity of S (Hayman and Pommerenke [1978]; Stegenga [1979]; Baernstein [1980] has yet another proof).

25. The Bloch class B is the set of analytic functions $f(z)$ on D for which

$$\sup_z (1 - |z|^2)|f'(z)| < \infty.$$

(a) If $f(z)$ is analytic on D, then $f \in B$ if and only if

$$\left\{ f\left(\frac{z - w}{1 - \bar{w}z}\right) : w \in D \right\}$$

is a normal family.

(b) BMOA $\subset B$.

(c) The function

$$\sum_{n=1}^{\infty} z^{2^n}$$

is in B but not in BMOA.

★(d) Let $f(z)$ be analytic on D, and let $F'(z) = f(z)$. Then $f \in B$ if and only if F is in the Zygmund class Λ^*:

$$F(e^{i(\theta+h)}) + F(e^{i(\theta-h)}) - 2F(e^{i\theta}) = O(h)$$

(see Duren [1970], Zygmund [1968]). Since Λ^* contains singular functions (Kahane [1969]; Piranian [1966]; Duren, Shapiro, and Shields [1966]), we have another proof that $B \neq \text{BMOA}$.

(e) $f(z) \in B$ if and only if $f(z) = \alpha \log g'(z)$, where α is constant and $g(z)$ is univalent on D. (See Duren, Shapiro, and Shields [1966]; Pommerenke [1970]. Compare with 23(b).)

(f)   On the other hand, B does coincide with the analytic functions in BMO of the unit disc defined by

$$\frac{1}{\pi r^2} \int_{D \cap \{|z-z_0| < r\}} |f(z) - f(z_0)| \, dx \, dy \leq C,$$

$|z_0| < 1$ (Coifman, Rochberg, and Weiss [1976]).

★(g)   When $f(z)$ is analytic in D, let $n(w)$ be the number of solutions of $f(z) = w, z \in D$. Suppose

$$\sup_{w_0 \in \mathbb{C}} \iint_{|w-w_0| < 1} n(w) \, du \, dv < \infty,$$

$w = u + iv$. Then $f \in \text{BMOA}$ if and only if $f \in B$, and $f \in \text{VMOA} = H^2 \cap \text{VMO}$ if and only if $f \in B_0$, defined by

$$\lim_{|z| \to 1} (1 - |z|^2)|f'(z)| = 0$$

(see Pommerenke [1977]).

Anderson, Clunie, and Pommerenke [1974] give an excellent overview of the theory of Bloch functions.

VII

Interpolating Sequences

A sequence $\{z_j\}$ in the disc or upper half plane is an *interpolating sequence* if every interpolation problem

$$f(z_j) = a_j, \quad j = 1, 2, \ldots,$$

with $\{a_j\}$ bounded, has solution $f(z) \in H^\infty$. Interpolating sequences are very interesting in their own right and they will play crucial roles in the analysis of H^∞ in the succeeding chapters. For example, they will be used in Chapter IX to characterize the closed algebras between H^∞ and L^∞ and they will be surprisingly important in the discussion of the maximal ideal space in Chapter X.

The notion of generations is introduced in this chapter. Similar to the stopping times in the proof of the John–Nirenberg theorem, generations arise naturally in several of the deeper proofs in this subject. They will be used frequently in the next chapter. Some other important techniques are also introduced. These include

(i) solving an extremal problem by a variational argument (this is done in Section 2) and

(ii) using certain ideas borrowed from harmonic analysis, such as the averaging process in the proof of Theorem 2.2 and the use of Khinchin's inequality in Section 4.

Two proofs of the interpolation theorem are given. Carleson's original proof by duality is in Section 1, because it sheds light on the geometry of interpolating sequences. Earl's elementary proof, reminiscent of the Pick–Nevanlinna theorem, is in Section 5.

1. Carleson's Interpolation Theorem

Let $\{z_j\}$ be a sequence in the upper half plane. We want to determine when every interpolation problem

(1.1) $$f(z_j) = a_j, \quad j = 1, 2, \ldots,$$

with $\{a_j\}$ bounded, has a solution $f(z) \in H^\infty$. The sequence $\{z_j\}$ is called an *interpolating sequence* if (1.1) has solution in H^∞ for every $\{a_j\} \in l^\infty$. If $\{z_j\}$ is an interpolating sequence, then the linear operator $T : H^\infty \to l^\infty$ defined by $Tf(j) = f(z_j)$ is a bounded linear mapping of H^∞ onto l^∞. The open mapping theorem then gives a constant M such that (1.1) has a solution $f(z)$ with

$$\|f\|_\infty \le M \sup_j |a_j| = M\|a_j\|_\infty.$$

The smallest such constant M is called the *constant of interpolation*

$$M = \sup_{\|a_j\|_\infty \le 1} \inf\{\|f\|_\infty : f \in H^\infty, f(z_j) = a_j, j = 1, 2, \ldots\}.$$

Let z_j and z_k be distinct points in the interpolating sequence. Then there exists $f \in H^\infty$ such that

$$f(z_j) = 0, \quad f(z_k) = 1, \quad \text{and} \quad \|f\|_\infty \le M.$$

By Schwarz's lemma this means

$$\left|\frac{z_k - z_j}{z_k - \bar{z}_j}\right| = \rho(z_j, z_k) \ge \frac{|f(z_j)/M - f(z_k)/M|}{|1 - \overline{f(z_k)}f(z_j)/M^2|} = \frac{1}{M},$$

and so

(1.2)
$$\left|\frac{z_k - z_j}{z_k - \bar{z}_j}\right| \ge a > 0, \quad j \ne k,$$

with $a = 1/M$. A sequence is said to be *separated* if (1.2) holds with constant $a > 0$ not depending on j and k. We have just proved that an interpolating sequence is separated.

The above reasoning can be carried further to yield a necessary condition for interpolation that will also be a sufficient condition. Fix z_k, let $f \in H^\infty, \|f\|_\infty \le M$, interpolate the values

$$f(z_k) = 1, \quad f(z_j) = 0, \quad j \ne k.$$

Let $B^{(k)}$ be the Blaschke product with zeros $\{z_j, j \ne k\}$. Since $f \ne 0$ this product exists. Then $f = B^{(k)}g$, where $g \in H^\infty$, and $\|g\|_\infty \le M$, so that

$$1 = |f(z_k)| = |B^{(k)}(z_k)||g(z_k)| \le M|B^{(k)}(z_k)|$$

and $|B^{(k)}(z_k)| \ge 1/M$. Since z_k is arbitrary, we conclude that

(1.3)
$$\inf_k \prod_{j, j \ne k} \left|\frac{z_k - z_j}{z_k - \bar{z}_j}\right| \ge \delta > 0$$

holds for an interpolating sequence $\{z_j\}$ with constant $\delta = 1/M$. Carleson's theorem asserts that the necessary condition (1.3) conversely implies that $\{z_j\}$ is an interpolating sequence.

Before stating the theorem in full, let us interpret (1.3) geometrically. In the unit disc (1.3) becomes

$$(1.4) \qquad \inf_k \prod_{j, j \neq k} \left| \frac{z_k - z_j}{1 - \bar{z}_j z_k} \right| \geq \delta > 0.$$

Let $B(z)$ be the full Blaschke product

$$B(z) = \prod_{j=1}^{\infty} \frac{-\bar{z}_j}{|z_j|} \left(\frac{z - z_j}{1 - \bar{z}_j z} \right).$$

If we view z_k as the origin by taking $w = (z - z_k)/(1 - \bar{z}_k z)$ as the coordinate function on the disc, then the zeros of B are

$$w_j = \frac{z_j - z_k}{1 - \bar{z}_k z_j}, \quad j = 1, 2, \ldots,$$

and (1.4) holds if and only if for all k

$$\prod_{j, j \neq k} |w_j| \geq \delta.$$

Since $1 - |w| \leq \log 1/|w|$, this gives

$$(1.5) \qquad \sum_j (1 - |w_j|) \leq 1 + \log 1/\delta,$$

and (1.4) holds if and only if the Blaschke sum (1.5) has a bound that does not depend on which point z_k is regarded as the origin. This fact, of course, reflects the conformal invariance of the interpolation problem (1.1). With the identity

$$1 - \left| \frac{z_j - z_k}{1 - \bar{z}_k z_j} \right|^2 = \frac{(1 - |z_k|^2)(1 - |z_j|^2)}{|1 - \bar{z}_k z_j|^2},$$

from Section 1 of Chapter I, (1.5) gives

$$\sup_k \sum_j \frac{(1 - |z_k|^2)}{|1 - \bar{z}_k z_j|^2} (1 - |z_j|^2) \leq C(\delta).$$

If this supremum were taken over all points in the disc instead of only over sequence points, we should have

$$(1.6) \qquad \sup_{z_0 \in D} \sum_j \frac{(1 - |z_0|^2)}{|1 - \bar{z}_0 z_j|^2} (1 - |z_j|^2) \leq C'(\delta).$$

By the conformally invariant description of Carleson measures (Chapter VI, Lemma 3.3), (1.6) holds if and only if the measure

$$\sum (1 - |z_j|) \delta_{z_j}$$

is a Carleson measure on the disc. Now it is not very hard to see that (1.4) does imply (1.6), and that conversely, if the sequence is separated, (1.6) implies (1.4). (The complete arguments are given below during the proof of the theorem.) This discussion does not prove the theorem, but it should clarify the connection between Carleson measures and interpolation. Historically it was with the interpolation theorem that Carleson measures first arose.

To state the theorem we return to the half plane.

Theorem 1.1. *If $\{z_j\}$ is a sequence in the upper half plane, then the following conditions are equivalent:*

(a) *The sequence is an interpolating sequence: Every interpolation problem*

$$f(z_j) = a_j, \quad j = 1, 2, \ldots,$$

with $\{a_j\} \in l^\infty$ has solution $f \in H^\infty$.

(b) *There is $\delta > 0$ such that*

$$(1.3) \qquad \prod_{j, j \neq k} \left| \frac{z_k - z_j}{z_k - \bar{z}_j} \right| \geq \delta, \quad k = 1, 2, \ldots.$$

(c) *The points z_j are separated,*

$$\rho(z_j, z_k) = \left| \frac{z_j - z_k}{z_j - \bar{z}_k} \right| \geq a > 0, \quad j \neq k,$$

and there is a constant A such that for every square $Q = \{x_0 \leq x \leq x_0 + \ell(Q), 0 < y \leq \ell(Q)\}$,

$$(1.7) \qquad \sum_{z_j \in Q} y_j \leq A\ell(Q).$$

The constant δ in (1.3) and the constant of interpolation

$$M = \sup_{\|a_j\|_\infty \leq 1} \inf\{\|f\|_\infty : f(z_j) = a_j, j = 1, 2, \ldots, f \in H^\infty\}$$

are related by the inequalities

$$(1.8) \qquad \frac{1}{\delta} \leq M \leq \frac{c}{\delta}\left(1 + \log\frac{1}{\delta}\right),$$

in which c is some absolute constant.

Except for the value of the numerical constant c, the upper bound given for M in (1.8) is sharp. An example illustrating this will be given after the proof.

Of course, (1.7) says that $\sum y_j \delta_{z_j}$ is a Carleson measure. Condition (c), being more geometric, is in some ways more useful than (b). Before turning to the proof we consider two examples. First, suppose the points z_j lie on a horizontal line $\{y_j = y > 0\}$. Then (1.7) holds as soon as the points are separated. So a horizontal sequence is an interpolating sequence if and only if it is separated. This fact can also be derived from (1.3) without much difficulty.

For the second example, suppose the points z_j lie on the vertical line $x = 0$. Then (1.7) holds if and only if

$$\sum_{y_j \le y_k} y_j \le A y_k, \quad k = 1, 2, \ldots,$$

This condition is satisfied if the points are separated. If the y_j are bounded above and if the points are indexed so that $y_{j+1} < y_j$, then this condition holds if and only if the points tend to the boundary exponentially:

$$y_{j+1}/y_j \le \alpha < 1.$$

Thus a vertical sequence is an interpolating sequence if and only if it is separated. Of course, not every separated sequence satisfies (1.7). A sequence having only (1.2) need not be a Blaschke sequence; it could have subsequences converging nontangentially to each point on the line.

Proof of Theorem 1.1. We have already seen that (a) implies (b), along with the estimate $M \ge 1/\delta$.

There are two remaining steps in the proof. First we show that (b) and (c) are equivalent. This is really only a matter of comparing infinite products to infinite sums. Second, we must show that (b) and (c) together imply (a). This will be done with a dual extremal problem.

To show that (b) and (c) are equivalent we need an elementary lemma.

Lemma 1.2. *Let $B(z)$ be the Blaschke product in the upper half plane with zeros $\{z_j\}$. Then*

$$(1.9) \qquad -\log|B(z)|^2 \ge \sum_j \frac{4yy_j}{|z - \bar{z}_j|^2}, \quad z = x + iy.$$

Conversely, if

$$\inf_j \rho(z, z_j) = \inf_j \left| \frac{z - z_j}{z - \bar{z}_j} \right| = a > 0,$$

then

$$(1.10) \qquad -\log|B(z)|^2 \le \left(1 + 2\log\frac{1}{a}\right) \sum \frac{4yy_j}{|z - \bar{z}_j|^2}.$$

Proof. The inequality $-\log t \ge 1 - t, t > 0$, gives

$$-\log\left| \frac{z - z_j}{z - \bar{z}_j} \right|^2 \ge 1 - \left| \frac{z - z_j}{z - \bar{z}_j} \right|^2 = \frac{4yy_j}{|z - \bar{z}_j|^2}.$$

Summing now gives (1.9). The reverse inequality,

$$-\log t \le \frac{-2\log a}{1 - a^2}(1 - t) \le \left(1 + 2\log\frac{1}{a}\right)(1 - t),$$

is valid for $a^2 < t < 1$, and in a similar fashion it gives (1.10). $\square$

Now suppose (c) holds. Then (1.7) and (3.8) of Chapter VI give

(1.11) $$\sum_j \frac{4y_k y_j}{|z_k - \bar{z}_j|^2} = \sum_j \frac{4y_k y_j}{|z_j - \bar{z}_k|^2} \le C', \quad k = 1, 2, \ldots.$$

For convenience we directly derive (1.11) from (1.7), essentially repeating the proof of (3.8) of Chapter VI. Fix $z_k = x_k + iy_k$ and let $S_n = \{z \in H : |z - x_k| \le 2^n y_k\}$, $n = 0, 1, 2, \ldots$. By (1.7), $\sum_{S_n} y_j \le c2^{n+1} y_k$. When $z_j \in S_0$, we have $|z_j - \bar{z}_k|^2 \ge y_k^2$, and when $z_j \in S_n \backslash S_{n-1}$, $n \ge 1$, we have $|z_j - \bar{z}_k|^2 \ge 2^{2n-2} y_k^2$. Consequently,

$$\sum_j \frac{4y_k y_j}{|z_j - \bar{z}_k|^2} \le 4 \sum_{z_j \in S_0} \frac{y_j}{y_k} + 16 \sum_{n=1}^{\infty} \left(\sum_{z_j \in S_n \backslash S_{n-1}} \frac{y_j}{2^{2n} y_k} \right)$$

$$\le 8A + 32A \sum_{1}^{\infty} 2^{-n} = A'.$$

Since $\inf_{j, j \ne k} |(z_k - z_j)/(z_k - \bar{z}_j)| \ge a$, we can now use (1.10) on the Blaschke product $B^{(k)}$ with zeros $\{z_j : j \ne k\}$ to obtain

$$\prod_{j, j \ne k} \left| \frac{z_k - z_j}{z_k - \bar{z}_j} \right| \ge \delta = \delta(a, A).$$

Hence (c) implies (b).

Now suppose (b) holds; that is, suppose

$$\inf_k \prod_{j, j \ne k} \left| \frac{z_k - z_j}{z_k - \bar{z}_j} \right| \ge \delta.$$

Then trivially

$$\left| \frac{z_k - z_j}{z_k - \bar{z}_j} \right| \ge \delta, \quad j \ne k,$$

and the points are separated. Using (1.9) with the Blaschke product $B^{(k)}(z)$ formed by deleting one zero z_k, we obtain

$$\sum_{j, j \ne k} \frac{4y_j y_k}{|z_k - \bar{z}_j|^2} \le 2 \log \frac{1}{\delta}.$$

Consider a square

$$Q = \{x_0 \le x \le x_0 + \ell(Q), 0 < y \le \ell(Q)\}.$$

We first treat the special case in which the top half

$$T(Q) = \{z \in Q : y > \ell(Q)/2\}$$

contains a sequence point z_k. Then we have $|z_k - \bar{z}_j|^2 \leq 5(\ell(Q))^2$, $z_j \in Q$, so that

$$y_j \leq \frac{5(\ell(Q))^2}{4y_k} \frac{4y_j y_k}{|z_k - \bar{z}_j|^2} \leq \frac{5\ell(Q)}{2} \frac{4y_j y_k}{|z_k - \bar{z}_j|^2}.$$

Hence

$$\sum_{z_j \in Q} y_j \leq \frac{5\ell(Q)}{2} \sum_j \frac{4y_j y_k}{|z_k - \bar{z}_j|^2} \leq \frac{5\ell(Q)}{2} \left(1 + 2\log \frac{1}{\delta}\right),$$

and (1.7) holds with $A = c(1 + \log(1/\delta))$ for squares Q such that $T(Q) \cap \{z_k\} \neq \varnothing$.

To obtain (1.7) for all squares Q we use a stopping time argument. Let $Q = Q_0 = \{x_0 \leq x \leq x_0 + \ell(Q), 0 < y \leq \ell(Q)\}$. Partition $Q \setminus T(Q)$ into two squares Q_1, of side $\ell(Q)/2$. Partition each $Q_1 \setminus T(Q_1)$ into two squares Q_2 of side $\ell(Q_1)/2$ and continue. At stage n we have 2^n squares Q_n of side $2^{-n}\ell(Q)$ whose top halves $T(Q_n)$ are all congruent to $T(Q)$ in the hyperbolic metric. These squares Q_n have pairwise disjoint interiors and they cover $\{z \in Q : 0 < y \leq 2^{-n}\ell(Q)\}$. Let $Q^1, Q^2, \ldots$ be those squares Q_n such that

(i) $T(Q_n) \cap \{z_j\} \neq \varnothing$ and
(ii) Q_n is contained in no larger square satisfying (i)

Then $Q \cap \{z_j\} \subset Q^1 \cup Q^2 \cup \cdots$, and the projections of the selected squares Q^k onto the axis $\{y = 0\}$ have pairwise disjoint interiors, so that

$$\sum \ell(Q^k) \leq \ell(Q).$$

See Figure VII. 1. We have already seen that (1.7) holds for each of the selected squares Q^k, with constant $A = c(1 + \log 1/\delta)$. Summing over the Q^k, we

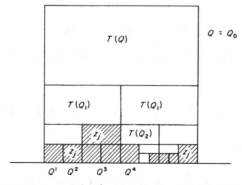

Figure VII.1. The shaded squares $Q^1, Q^2, \ldots$ are maximal Q_n for which $T(Q_n) \cap \{z_j\} \neq 0$.

obtain

$$\sum_{z_j \in Q} y_j \le c \left(1 + \log \frac{1}{\delta} \right) \ell(Q),$$

which proves (1.7) in general. Hence (b) implies (c).

In order to get the sharp inequality (1.8) we must make use of the fact that the constant A in (1.7) has the form

$$(1.12) \qquad A \le c \left(1 + \log \frac{1}{\delta} \right).$$

Now we must show that (b) and (c) imply (a). Let $\{a_j\} \in l^\infty$, $|a_j| \le 1$, and consider the finite problem

$$(1.13) \qquad f(z_j) = a_j, \quad 1 \le j \le n.$$

Since the points are distinct it is trivial that the finite problem (1.13) has a solution $f(z) \in H^\infty$. For example, take $f(z) = p(z)/(z+i)^n$, where $p(z)$ is a polynomial of degree n. Let

$$M_n(\{a_j\}) = \inf\{\| f \|_\infty : f \in H^\infty, f(z_j) = a_j, 1 \le j \le n\},$$

and let

$$M_n = \sup_{\|a_j\|_\infty \le 1} M_n(\{a_j\}).$$

By normal families the theorem will be proved and inequality (1.8) will be established if we show that

$$\lim_n M_n \le \frac{c}{\delta} \left(1 + \log \frac{1}{\delta} \right).$$

Let

$$B_n(z) = \prod_{j=1}^n \frac{z - z_j}{z - \bar{z}_j}.$$

For fixed $\{a_j\}$ let $f_0 \in H^\infty$ be a solution of (1.13). Then

$$M_n(\{a_j\}) = \inf\{\| f_0 + B_n g \|_\infty : g \in H^\infty\} = \inf\{\| f_0 \bar{B}_n + g \|_\infty : g \in H^\infty\}.$$

Since

$$H^\infty = \left\{ g \in L^\infty : \int g G \, dx = 0 \text{ for all } G \in H^1 \right\},$$

duality now gives

$$M_n(\{a_j\}) = \sup \left\{ \left| \int f_0 \bar{B}_n G \, dx \right| : G \in H^1, \|G\|_1 \le 1 \right\}.$$

Now by Cauchy's theorem we have, for $G \in H^1$,

$$\int_{\mathbb{R}} f_0 \bar{B}_n G \, dx = \int_{\mathbb{R}} \frac{f_0(x) G(x)}{B_n(x)} dx = 2\pi i \sum_{j=1}^{n} \frac{f_0(z_j) G(z_j)}{B_n'(z_j)}.$$

(In the contour integral the large semicircles tending to infinity are disregarded because we can take $G(z)$ in the H^1 norm dense set of functions $G_1(z)$ with $|G_1(z)| = O(|z|^{-2})$). Since $f_0(z_j) = a_j$, $j = 1, 2, \ldots, n$, this gives

$$M_n = \sup_{\|a_j\|_\infty \leq 1} \sup \left\{ 2\pi \left| \sum_{j=1}^{n} \frac{a_j G(z_j)}{B_n'(z_j)} \right| : G \in H^1, \|G\|_1 \leq 1 \right\}.$$

For fixed $G(z)$, the $\{a_j\}$ can be chosen so that $|a_j| = 1$ and so that

$$a_j G(z_j)/B_n'(z_j) \geq 0.$$

We therefore have

(1.14) $\qquad M_n = \sup \left\{ 2\pi \sum_{j=1}^{n} \frac{|G(z_j)|}{|B_n'(z_j)|} : G \in H^1, \|G\|_1 \leq 1 \right\}.$

Now

$$B_n'(z_j) = \frac{-i}{2y_j} \prod_{\substack{k=1 \\ k \neq j}}^{n} \left(\frac{z_j - z_k}{z_j - \bar{z}_k} \right),$$

and so (b) implies that

$$|B_n'(z_j)| \geq \delta/2y_j.$$

Hence

$$M_n \leq \frac{4\pi}{\delta} \sup \left\{ \sum_{j=1}^{n} y_j |G(z_j)| : G \subset H^1, \|G\|_1 \leq 1 \right\}.$$

By condition (c), the measure $\sum y_j \delta_{z_j}$ is a Carleson measure, and so by Theorem II.3.9

$$\sup \left\{ \sum_{j=1}^{n} y_j |G(z_j)| : G \in H^1, \|G\|_1 \leq 1 \right\} \leq CA,$$

where A is the constant from (1.7). Consequently

$$\lim_{n \to \infty} M_n \leq 4\pi CA/\delta$$

and (a) is proved.

We have now shown that (a)–(c) are equivalent. Moreover, using the estimate (1.12) of the constant A in (1.7), we get

$$M = \lim_n M_n \le \frac{c}{\delta} \left(1 + \log \frac{1}{\delta} \right)$$

and (1.8) is proved.   □

Here is an example to show that (1.8) is sharp. Let $\omega = e^{2\pi i/N}$ be a primitive Nth root of unity. In the disc take the finite sequence

$$z_j = r\omega^j, \quad j = 1, 2, \ldots, N.$$

The two parameters N and $r < 1$ will be fixed later. The Blaschke product with zeros z_j is

$$B(z) = \frac{z^N - r^N}{1 - r^N z^N}.$$

From (1.4) we have

$$\delta = \inf(1 - |z_j|^2)|B'(z_j)| = Nr^{N-1}\frac{1 - r^2}{1 - r^{2N}}.$$

Consider the interpolation problem

$$f(z_j) = a_j = \omega^{-j}.$$

By Theorem I.2.4 or by IV.1.8 this finite problem has a unique interpolating function $f(z)$ of minimal norm and

$$f(z) = mB_1(z),$$

where B_1 is a Blaschke product with at most $N - 1$ zeros. Since

$$z_{j+1} = \omega z_j, \quad a_{j+1} = \omega^{-1} a_j,$$

the uniqueness implies that

$$f(z) = \omega f(\omega z).$$

Hence the zero set of $B_1(z)$ is invariant under multiplication by ω. As there are at most $N - 1$ zeros, all the zeros are at $z = 0$, and

$$f(z) = mz^p$$

with $p \le N - 1$. A simple calculation then gives

$$f(z) = r^{1-N}z^{N-1}.$$

Let $r \to 1$ and $N \to \infty$ in such a way that $m = r^{1-N}$ is fixed. Then

$$\delta = Nm\frac{1 - r^2}{m^2 - r^2} \quad \text{and} \quad \lim_{N\to\infty} N(1 - r^2) = 2\log m,$$

so that for large N

$$\delta \geq (\log m)/m.$$

Consequently there exist finite interpolating sequences such that

$$M \geq m \geq \frac{1}{\delta} \log \frac{1}{\delta}.$$

2. The Linear Operator of Interpolation

Let $\{z_j\}$ be an interpolating sequence in the upper half plane. In this section we prove that there exist interpolating functions that depend linearly on the interpolated values $\{a_j\}$. This useful result is obtained through a nonlinear extremal problem.

Theorem 2.1. *Let $\{z_j\}$ be an interpolating sequence in the upper half plane, and let*

$$M = \sup_{\|a_j\| \leq 1} \inf\{\| f \|_\infty : f \in H^\infty, f(z_j) = a_j, j = 1, 2, \ldots\}$$

be the constant of interpolation. Then there are functions $f_j(z) \in H^\infty$ such that

(2.1) $$f_j(z_j) = 1, \quad f_j(z_k) = 0, \quad k \neq j,$$

and

(2.2) $$\sum_j |f_j(z)| \leq M.$$

Before proving this theorem let us give two applications. Suppose $\{a_j\} \in l^\infty$. By (2.2) the function

(2.3) $$f(z) = \sum_j a_j f_j(z)$$

is in H^∞, and by (2.1) this function interpolates

$$f(z_j) = a_j, \quad j = 1, 2, \ldots.$$

Thus (2.3) produces interpolating functions that depend linearly on $\{a_j\}$. With (2.3) we have defined a linear operator $S : l^\infty \to H^\infty$ by $S(\{a_j\}) = \sum a_j f_j$, and S is a linear operator of interpolation, which simply means that S is linear and that

(2.4) $$S(\{a_j\})(z_k) = a_k, \quad k = 1, 2, 3, \ldots.$$

By (2.2) the operator S is bounded and $\|S\| \leq M$. Since the constant M in (2.2) is the constant of interpolation, we actually have $\|S\| = M$. Now any operator, linear or not, that satisfies (2.4) must have

$$\sup_{\|a_j\|_\infty \leq 1} \|S(\{a_j\})\|_\infty \geq M$$

because M is the constant of interpolation. Thus (2.3) solves all interpolation problems (1.1) simultaneously with a linear operator whose norm is as small as possible. The inequalities

$$\|a_j\|_\infty \leq \|S(\{a_j\})\|_\infty \leq M\|a_j\|_\infty$$

show that the range $S(l^\infty)$ is a closed subspace of H^∞ and that, as a Banach space, $S(l^\infty)$ is isomorphic to l^∞. The linear operator $P : H^\infty \to S(l^\infty)$ defined by

$$Pg = \sum_j g(z_j) f_j$$

is a bounded operator from H^∞ onto $S(l^\infty)$ such that $P^2 = P$. By definition, this means that P is a projection and that $P(H^\infty) = S(l^\infty)$ is a complemented subspace of H^∞. (The complement is the kernel of P.) Hence Theorem 2.1 shows that H^∞ contains a closed complemented subspace isomorphic to l^∞.

The second application concerns interpolation by bounded analytic functions having values in a Banach space. A function $f(z)$ from an open set in the plane to a Banach space Y is analytic if $f(z)$ can be locally represented as a sum of a power series that has coefficients in Y and that is absolutely convergent. Equivalently, for each $y^* \in Y^*$, the complex valued function $z \to \langle y^*, f(z)\rangle$ is analytic. To quote Hoffman [1962a], "Any two reasonable-sounding definitions of an analytic function with values in a Banach space are equivalent."

Let $\{z_j\}$ be an interpolating sequence in the upper half plane. Let Y be a Banach space and let $\{a_j\}$ be a bounded sequence in Y,

$$\sup_j \|a_j\|_Y < \infty.$$

If $\{f_j\}$ is the sequence of scalar-valued functions given by Theorem 2.1, then

$$f(z) = \sum f_j(z) a_j$$

is an analytic function on the upper half plane with values in Y. By (2.1), $f(z)$ solves the interpolation problem

$$f(z_j) = a_j, \quad j = 1, 2, \ldots,$$

and by (2.2), $f(z)$ is bounded:

$$(2.5) \quad \|f\| = \sup_z \|f(z)\|_Y \leq \sup_z \sum |f_j(z)| \|a_j\|_Y \leq M \sup_j \|a_j\|_Y.$$

We conclude that $\{z_j\}$ is also an interpolating sequence for the Y-valued bounded analytic functions. Conversely, it is trivial that an interpolating sequence for Y-valued H^∞ functions is an interpolating sequence for scalar

functions. Just interpolate scalar multiples of a fixed vector in Y. Inequality
(2.5) shows that the constant of interpolation is the same in the Banach space
case as it is in the scalar case.

Proof of Theorem 2.1.   For $\lambda_1, \lambda_2, \ldots, \lambda_n \geq 0$, define the functional

$$\varphi(G) = \sum_{j=1}^{n} \lambda_j |G(z_j)|, \quad G \in H^1,$$

and consider the extremal problem

(2.6) $m_n = m_n(\{\lambda_j\}) = \sup\{\varphi(G) : G \in H^1, \|G\|_1 \leq 1\}.$

This is a nonlinear problem to which the method of Chapter IV does not apply.
But we can use an older method, usually referred to as a variational argument.
There clearly exists an extremal function G_0 for (2.6), and G_0 must be an outer
function,

$$G_0 = e^{U_0 + i\tilde{U}_0}.$$

Let $u(x)$ be a real compactly supported continuous function. Then for $t \in \mathbb{R}$

$$G_t = G_0 e^{tu + it\tilde{u}}$$

is another H^1 function and

$$m_n \|G_t\|_1 \geq \varphi(G_t),$$

with equality at $t = 0$. Write this inequality as $m_n \|G_t\|_1 - \varphi(G_t) \geq 0$ and
differentiate with respect to t at $t = 0$. We obtain

$$m_n \int |G_0(x)| u(x)\, dx = \sum_{j=1}^{n} \lambda_j |G_0(z_j)| u(z_j)$$

$$= \int \sum \frac{\lambda_j |G_0(z_j)|}{\pi} \frac{y_j}{(x_j - x)^2 + y_j^2} u(x)\, dx.$$

Because $u(x)$ is arbitrary, this means

$$m_n |G_0(x)| = \frac{1}{\pi} \sum_j \frac{\lambda_j |G_0(z_j)| y_j}{(x_j - x)^2 + y_j^2}$$

and we have almost everywhere

(2.7) $$m_n = \frac{1}{\pi} \sum_{j=1}^{n} \frac{|G_0(z_j)|}{|G_0(x)|} \frac{\lambda_j y_j}{(x - x_j)^2 + y_j^2}.$$

Now let

$$B_n(z) = \prod_{j=1}^{n} \frac{z - z_j}{z - \bar{z}_j}$$

and put $\lambda_j = 2\pi |B'_n(z_j)|^{-1}$ in (2.6). Then (2.6) is the same as the extremal problem (1.14) and

$$m_n = M_n = \sup_{|a_j| \leq 1} \inf\{\|f\|_\infty : f \in H^\infty, f(z_j) = a_j, 1 \leq j \leq n\}.$$

From (2.7) we now have

(2.8) $$M_n = \sum_{j=1}^n \frac{|G_0(z_j)|}{|G_0(x)|} \frac{1}{|B'_n(z_j)|} \frac{2y_j}{|x - z_j|^2}$$

almost everywhere. Now set

$$f_j^{(n)}(z) = \frac{G_0(z_j)}{G_0(z)} \frac{B_n(z)}{B'_n(z_j)} \frac{2iy_j}{(z - z_j)(z - \bar{z}_j)}, \quad j = 1, 2, \ldots, n.$$

Then by (2.8),

$$\sum |f_j^{(n)}(x)| = M_n$$

almost everywhere. The functions $f_j^{(n)}$ are in N^+, because G_0 is outer, and so

$$\sum |f_j^{(n)}(z)| \leq \sum \int |f_j^{(n)}(t)| P_z(t) \, dt \leq M_n.$$

Clearly $f_j^{(n)}(z_k) = 0, k \neq j, 1 \leq k \leq n$. Calculating $B'_n(z_j)$ explicitly shows that $f_j^{(n)}(z_j) = 1$. Taking a limit as $n \to \infty$, we obtain functions f_j in H^∞, $j = 1, 2, \ldots$ that satisfy (2.1) and (2.2). $\square$

Generally, it is not always the case that a linear operator of extension exists. However, there is an elegant result from uniform algebra theory that implies Theorem 2.1 with a poorer bound on $\sum |f_j(z)|$. The idea comes from harmonic analysis.

Theorem 2.2. *Let A be a uniform algebra on a compact space X. Let $\{p_1, p_2, \ldots, p_n\}$ be a finite set of points in X and let*

$$M = \sup_{\|a_j\|_\infty \leq 1} \inf\{\|g\| : g \in A, g(p_j) = a_j, j = 1, 2, \ldots, n\}.$$

For any $\varepsilon > 0$ there are functions $f_1, f_2, \ldots, f_n$ in A such that

(2.9) $$f_j(p_j) = 1, \quad f_j(p_k) = 0, \quad k \neq j$$

and such that

$$\sup_{x \in X} \sum_{j=1}^n |f_j(x)| \leq M^2 + \varepsilon.$$

By normal families, Theorem 2.2 implies Theorem 2.1 except that the sharp inequality (2.2) is replaced by the weaker

$$\sum_{j=1}^{\infty} |f_j(z)| \leq M^2.$$

For a general uniform algebra it is not possible to extend Theorem 2.2 to infinite interpolating sequences.

Proof. Let $\omega = e^{2\pi i/n}$ be a primitive nth root of unity. Let $g_j(x) \in A$, $\|g_j\| \leq M + \delta$, where $\delta > 0$, interpolate

$$g_j(p_k) = \omega^{jk}, \quad k = 1, 2, \ldots, n.$$

Set

$$f_j(x) = \left(\frac{1}{n} \sum_{k=1}^{n} \omega^{-jk} g_k(x) \right)^2.$$

Then $f_j \in A$ and $f_j(p_j) = 1$. Since $\sum_{k=1}^{n} \omega^{(l-j)k} = 0$ for $l \neq j$, $f_j(p_l) = 0$ if $l \neq j$. Moreover

$$\sum_{j=1}^{n} |f_j(x)| = \frac{1}{n^2} \sum_{j=1}^{n} \left(\sum_{k=1}^{n} \omega^{-jk} g_k(x) \sum_{l=1}^{n} \omega^{jl} \overline{g_l(x)} \right)$$

$$= \frac{1}{n^2} \sum_{k=1}^{n} \sum_{l=1}^{n} g_k(x) \overline{g_l(x)} \sum_{j=1}^{n} \omega^{(l-k)j}$$

$$= \frac{1}{n^2} \sum_{k=1}^{n} n |g_k(x)|^2 \leq (M + \delta)^2 \leq M^2 + \varepsilon$$

if δ is small. $\square$

3. Generations

Let $\{z_j\}$ be a sequence in the upper half plane. We assume that $\{z_j\}$ is separated:

(3.1) $$|z_j - z_k| \geq by_j, \quad k \neq j,$$

with $b > 0$. This condition (3.1) is clearly equivalent to (1.2), but (3.1) is slightly more convenient for our present purpose. Then the condition

(3.2) $$\sum_{z_j \in Q} y_j \leq A\ell(Q)$$

holds for every square $Q = \{x_0 \leq x \leq x_0 + \ell(Q), 0 < y \leq \ell(Q)\}$, with constant A independent of Q, if and only if $\{z_j\}$ is an interpolating sequence.

Conditions like (3.2) will arise quite often in the rest of this book, and so we pause here to analyze (3.2) carefully. Understanding the geometry of (3.2) enables one to construct interpolating sequences very easily.

Let Q be any square with base on $\{y = 0\}$ and let $T(Q) = \{z \in Q : \ell(Q)/2 < y \le \ell(Q)\}$ be the top half of Q. Partition $Q \backslash T(Q)$ into two squares Q_1 of side $\ell(Q)/2$. Continue, just as in the proof of Theorem 1.1. At stage n there are 2^n squares Q_n of side $2^{-n}\ell(Q)$ (see Fig. VII.1).

In the hyperbolic geometry each top half $T(Q_n)$ is congruent to $T(Q)$. When the sequence is separated, each top half $T(Q)$ or $T(Q_n)$ can contain at most $C(b)$ points z_j, where b is the constant in (3.1). Indeed, if $T(Q_n)$ is partitioned into $C(b)$ squares of side $2^{-(p+n)}\ell(Q)$ with $2^{-p} < b/2\sqrt{2}$, then by (3.1) each of these little squares can contain at most one point z_j, because $y_j > \ell(Q_n)/2$. (See Figure VII.2.)

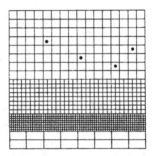

Figure VII.2. Each little square in $T(Q_n)$ contains at most one point z_j.

The *first generation* $G_1(Q)$ consists of those $Q_n \subset Q$ such that

  (i) $Q_n \ne Q$,
  (ii) $T(Q_n) \cap \{z_j\} \ne \varnothing$,
  (iii) Q_n is contained in no larger square satisfying (i) and (ii).

The squares $Q^1, Q^2, \ldots$ in $G_1(Q)$ have pairwise disjoint interiors that have pairwise disjoint projections onto $\{y = 0\}$. Hence

$$\sum_{G_1(Q)} l(Q^k) \le \ell(Q).$$

Moreover

$$\{z_j : z_j \in Q\} \subset T(Q) \cup \bigcup_{G_1(Q)} Q^k.$$

The squares in $G_1(Q)$ are shaded in Figure VII.1. Now for each $Q^k \in G_1(Q)$ we define the first generation $G_1(Q^k)$ in the same way. $G_1(Q^k)$ consists of those Q_n properly contained in Q^k such that $T(Q_n) \cap \{z_j\} \ne \varnothing$ and such that

Q_n is maximal. The *second generation is*

$$G_2(Q) = \bigcup_{G_1(Q)} G_1(Q^k).$$

The later generations $G_3(Q), G_4(Q), \dots$ are defined recursively:

$$G_{p+1} = \bigcup_{G_p(Q)} G_1(Q^k).$$

If $z_j \in Q$, then either $z_j \in T(Q)$, or $z_j \in T(Q^k)$ for some square Q^k in some generation $G_p(Q)$. In the second case y_j is comparable to $\ell(Q^k)$. Write

$$t_p = \sum_{G_p(Q)} \ell(Q^k).$$

From the definitions it is clear that

$$t_{p+1} \leq t_p \leq \ell(Q).$$

Theorem 3.1. *Assume $\{z_j\}$ is a sequence of points in the upper half plane satisfying the separation condition* (3.1). *Then $\{z_j\}$ is an interpolating sequence if and only if, for any $\varepsilon > 0$, there is q such that for any square Q*

$$(3.3) \qquad t_q = \sum_{G_p(Q)} \ell(Q^k) \leq \varepsilon \ell(Q).$$

The smallest constant q such that (3.3) *holds is related to the constant A in* (3.2) *by*

$$q \leq 1 + 2A/\varepsilon \quad and \quad A \leq C(b, \varepsilon, q),$$

where b is the constant in (3.1).

Proof. For $z_j \in T(Q_n)$, we have

$$y_j \leq \ell(Q_n) \leq 2y_j.$$

Each $T(Q_n)$ or $T(Q)$ contains at most $C(b)$ points z_j and if $Q^k \in G_p(Q)$, then $T(Q^k)$ contains at least one point z_j. Hence

$$\sum_{p=1}^{\infty} t_p \leq 2 \sum_{z_j \in Q} y_j \leq 2C(b) \left(\ell(Q) + \sum_{p=1}^{\infty} t_p \right).$$

If interpolation holds, then we have (3.2), and so

$$q t_q \leq \sum_{p=1}^{q} t_p \leq 2A\ell(Q).$$

This gives (3.3) if $q \geq 2A/\varepsilon$. Conversely, if (3.3) holds for all squares, then the recursive definition of the generations gives

$$t_{p+q} \leq \varepsilon t_p, \quad p = 1, 2, \ldots,$$

so that

$$\sum_{Nq+1}^{Nq+q} t_p \leq \varepsilon^N \sum_{1}^{q} t_p \leq q \varepsilon^N \ell(Q).$$

Therefore

$$\sum_{z_j \in Q} y_j \leq C(b) \left(\ell(Q) + \sum_{N=0}^{\infty} q \varepsilon^N \ell(Q) \right)$$

$$\leq C(b) \left(1 + \frac{q}{1-\varepsilon} \right) \ell(Q),$$

and we have (3.2). $\square$

4. Harmonic Interpolation

The proof we shall give of the next theorem requires a randomization technique that has been very useful in many other areas of analysis. Thus we pause briefly to discuss Khinchin's inequality before turning to the theorem on harmonic interpolating sequences.

Given finitely many complex numbers $\alpha_1, \alpha_2, \ldots, \alpha_n$, consider the 2^n possible sums

$$\sum_{j=1}^{n} \pm \alpha_j$$

obtained as the plus–minus signs vary in the 2^n possible ways. Let $p > 0$. Khinchin's inequality is an estimate on the expectation

$$E \left(\left| \sum_{j=1}^{n} \pm \alpha_j \right|^p \right).$$

The expectation is the average value of $|\sum \pm \alpha_j|^p$ over the 2^n choices of sign. To be precise, let Ω be the set of 2^n points

$$\omega = (\omega_1, \omega_2, \ldots, \omega_n),$$

where $\omega_j = \pm 1$. Define the probability μ on Ω so that each point ω has probability 2^{-n}. Also define

$$X(\omega) = \sum_{j=1}^{n} \alpha_j \omega_j.$$

Then $X(\omega)$ is a more rigorous expression for $\sum \pm \alpha_j$, and by definition

$$\mathrm{E}\left(\left|\sum_{j=1}^n \pm \alpha_j\right|^p\right) = \frac{1}{2^n} \sum_{\omega \in \Omega} |X(\omega)|^p = \int_\Omega |X(\omega)|^p d\mu.$$

The following lemma is called Khinchin's inequality.

Lemma 4.1. *If* $0 < p < \infty$, *then*

(4.1)
$$\left(\mathrm{E}\left(\left|\sum_{j=1}^n \pm \alpha_j\right|^p\right)\right)^{1/p} \le C_p \left(\sum |\alpha_j|^2\right)^{1/2},$$

where C_p *is a constant that does not depend on* n.

The important thing in (4.1) is that C_p does not increase as n increases. We prove Lemma 4.1 only in the easy case $p \le 2$ because we need only use that case here. (See Zygmund [1968] for the complete proof and for other applications.)

Proof for $p \le 2$. The case $p \le 2$ is easier because it is only Hölder's inequality in disguise. Let $X_j(\omega) = \omega_j$, $j = 1, 2, \ldots, n$. Then $|X_j^2(\omega)| = 1$, and for $j \ne k$, $\mathrm{E}(X_j X_k) = 0$ because $X_j X_k$ takes each value ± 1 with probability $\frac{1}{2}$. This means that $\{X_1, X_2, \ldots, X_n\}$ are orthonormal in $L^2(\mu)$. Since $X = \alpha_1 X_1 + \alpha_2 X_2 + \cdots + \alpha_n X_n$ and since $p \le 2$, Hölder's inequality gives

$$\left(\mathrm{E}\left|\sum_{j=1}^n \pm \alpha_j\right|^p\right)^{1/p} = \left(\int |X(\omega)|^p \, d\mu\right)^{1/p}$$

$$\le \left(\int |X(\omega)|^2 \, d\mu\right)^{1/2} = \left(\sum_{j=1}^n |\alpha_j|^2\right)^{1/2}.$$

This proves (4.1) with $C_p = 1$ when $p \le 2$. □

Now let $\{z_j\}$ be a sequence in the upper half plane. We say that $\{z_j\}$ is a *harmonic interpolating sequence* if every interpolation problem

(4.2)
$$u(z_j) = a_j, \quad j = 1, 2, \ldots, \quad \{a_j\} \in l^\infty,$$

can be solved with a bounded harmonic function $u(z)$. Obviously every H^∞ interpolating sequence is a harmonic interpolating sequence. Our theorem is the converse.

Theorem 4.2. *If* $\{z_j\}$ *is a harmonic interpolating sequence, then* $\{z_j\}$ *is an interpolating sequence for* H^∞.

Proof. Since every bounded harmonic function is the Poisson integral of an L^∞ function, (4.2) holds if and only if there is an L^∞ solution to every moment problem

$$(4.3) \qquad \int u(t) P_j(t)\, dt = a_j, \quad j = 1, 2, \ldots, \quad \{a_j\} \in l^\infty,$$

where

$$P_j(t) = \frac{1}{\pi y_j} \frac{1}{1 + ((t - x_j)/y_j)^2}$$

is the Poisson kernel for z_j. We show that the geometric conditions (1.2) and (1.7) hold if every moment problem (4.3) can be solved with $u \in L^\infty$. By Theorem 1.1 this will mean that $\{z_j\}$ is an H^∞ interpolating sequence.

To get started we need an inequality. Consider the linear operator $T : L^\infty \to l^\infty$ defined by

$$Tu(j) = u(z_j) = \int u(t) P_j(t)\, dt.$$

Since $\|P_j\|_1 = 1$, this operator is bounded. We are assuming T maps L^∞ onto l^∞. By the open mapping theorem, there is a constant M such that every problem (4.3) has solution u such that

$$(4.4) \qquad \|u\|_\infty \le M \sup_j |a_j|.$$

The inequality we need is

$$(4.5) \qquad \sum |\lambda_j| \le M \| \sum \lambda_j P_j \|_1,$$

which is the dual formulation of (4.4). The equivalence of (4.4) and (4.5) follows from the fact that a linear operator has closed range if and only if its adjoint has closed range (Dunford and Schwartz [1958]). (T is the adjoint of an operator from l^1 to L^1.) But (4.5) is also easily derived directly from (4.4). Given $\lambda_1, \lambda_2, \ldots, \lambda_n$, pick $u \in L^\infty$, $\|u\|_\infty \le M$, solving

$$\int u(t) P_j(t)\, dt = \bar{\lambda}_j/|\lambda_j|, \quad j = 1, 2, \ldots, n.$$

Then

$$\sum_{j=1}^{n} |\lambda_j| = \left| \int u \sum_{j=1}^{n} \lambda_j P_j\, dt \right| \le \|u\|_\infty \left\| \sum_{j=1}^{n} \lambda_j P_j \right\|_1,$$

and (4.5) is proved.

If $j \ne k$, then (4.5) gives $\|P_j - P_k\|_1 \ge 2/M$. By Harnack's inequality, for example, this means that

$$|z_j - z_k|/y_j \ge b(M) > 0, \quad j \ne k,$$

and (1.2) holds for the sequence $\{z_j\}$.

The proof of (1.7) uses Khinchin's inequality. Let Q be a square $\{x \in I, 0 < y \le |I|\}$. Reindexing, let $z_1, z_2, \ldots, z_n$ be finitely many points in the sequence and in the square Q. Set $\lambda_j = \pm y_j$, $j = 1, 2, \ldots, n$. Then by (4.5)

$$\sum_{j=1}^{n} y_j \le M \int \left| \sum_{j=1}^{n} \pm y_i P_j(t) \right| dt.$$

Taking the expectation for each $t \in \mathbb{R}$, we get from (4.1)

$$(4.6) \qquad \sum_{j=1}^{n} y_j \le M \int \left(\sum_{j=1}^{n} y_j^2 P_j^2(t) \right)^{1/2} dt.$$

Let $\tilde{I}$ be that interval concentric with I but having length $|\tilde{I}| = 3|I|$. The right side of (4.6) is

$$M \int_{\tilde{I}} \left(\sum_{j=1}^{n} y_j^2 P_j^2 \right)^{1/2} dt + M \int_{\mathbb{R} \setminus \tilde{I}} \left(\sum_{j=1}^{n} y_j^2 P_j^2 \right)^{1/2} dt.$$

For the first integral the Cauchy–Schwarz inequality gives the bound

$$M |\tilde{I}|^{1/2} \left(\int_{\tilde{I}} \sum y_j^2 P_j^2 \, dt \right)^{1/2} \le c M |I|^{1/2} \left(\sum_{j=1}^{n} y_j \right)^{1/2}$$

because

$$\int P_j^2(t) \, dt = \frac{1}{y_j} \int P^2(t) \, dt,$$

where $P(t) = 1/\pi(1 + t^2)$. For the outside integral we have the inequality

$$P_j^2(t) \le \frac{c y_j^2}{(t - x_0)^4}, \quad t \notin \tilde{I},$$

where x_0 is the center of I, because $z_j \in Q$. Then for $t \notin \tilde{I}$,

$$\left(\sum y_j^2 P_j^2(t) \right)^{1/2} \le c \left(\sum y_j^4 \right)^{1/2} \frac{1}{(t - x_0)^2}$$

$$\le c |I|^{3/2} \left(\sum y_j \right)^{1/2} \frac{1}{(t - x_0)^2}.$$

Hence the second integral is bounded by

$$c M |I|^{3/2} \left(\sum y_j \right)^{1/2} \int_{\mathbb{R} \setminus \tilde{I}} \frac{dt}{(t - x_0)^2} \le c M |I|^{1/2} \left(\sum_{j=1}^{n} y_j \right)^{1/2}.$$

From (4.6) we now have

$$\left(\sum_{j=1}^{n} y_j\right)^{1/2} \leq 2cM|I|^{1/2},$$

and this gives (1.7). $\square$

In the proof of Theorem 4.2, we were able to conclude that there was interpolation by bounded analytic functions only because the sequence satisfied the geometric conditions (1.2) and (1.7). Harmonic functions were only used to get the inequality (4.5). There is a generalization of Theorem 4.2, having essentially the same proof, which makes no mention of harmonicity. Suppose $P(t) \in L^1(\mathbb{R})$. We treat $P(t)$ as a kernel by writing

$$P_z(t) = \frac{1}{y} P\left(\frac{x-t}{y}\right), \quad z = x + iy, \quad y > 0.$$

Since $\|P_z\|_1 = \|P\|_1$, the operator

$$Tu(z) = \int u(t) P_z(t) \, dt$$

is a bounded linear mapping from L^∞ into the space of bounded (and continuous, in fact) functions on the upper half plane. In the special case $P(t) = 1/\pi(1 + t^2)$, P_z is of course the Poisson kernel and the operator T solves the Dirichlet problem.

Theorem 4.3. *Let $P(t) \in L^1$. Let $\{z_j\}$ be a sequence in the upper half plane. If every interpolation problem*

$$(4.7) \qquad Tu(z_j) = \int u(t) P_{z_j}(t) \, dt = a_j, \quad j = 1, 2, \ldots,$$

for $\{a_j\} \in l^\infty$ has solution $u(t) \in L^\infty$, then $\{z_j\}$ is an interpolating sequence for H^∞.

Proof. We shall use (4.7) to show that the distribution of the points $\{z_j\}$ satisfies (1.2) and (1.7). Theorem 1.1 then permits us to conclude that $\{z_j\}$ is an H^∞ interpolating sequence.

By the open mapping theorem, every moment problem (4.7) has solution $u(t)$ with $\|u\|_\infty \leq M \sup_j |a_j|$, where the constant M does not depend on $\{a_j\}$. Then a duality argument, as in the proof of Theorem 4.2, gives us

$$(4.8) \qquad \sum_{j=1}^{n} |\lambda_j| \leq M \left\| \sum_{j=1}^{n} \lambda_j P_{z_j} \right\|_1$$

for any $\lambda_1, \lambda_2, \ldots, \lambda_n$. To simplify the estimates that need be made at the end of the proof, choose a compactly supported continuous function $K(t)$ such that

$$\|P - K\|_1 < 1/2M.$$

Writing

$$K_{z_j}(t) = \frac{1}{y_j} K\left(\frac{x_j - t}{y_j}\right),$$

we have $\|P_{z_j} - K_{z_j}\|_1 < 1/2M$ by a change of scale. Hence (4.8) gives

(4.9) $$\sum_{j=1}^n |\lambda_j| \le 2M \left\|\sum_{j=1}^n \lambda_j K_{z_j}\right\|_1.$$

For $j \ne k$, (4.9) yields $\|K_{z_k} - K_{z_j}\| \ge \frac{1}{M}$. Let $x_0 = (x_k - x_j)/y_j$, $y_0 = y_k/y_j$. Then a change of variables gives

$$\|K_{z_k} - K_{z_j}\|_1 = \int \left|K(s) - \frac{1}{y_0} K\left(\frac{s + x_0}{y_0}\right)\right| ds$$

$$\le \|K(s) - K(s + x_0)\|_1 + \left\|K(s + x_0) - \frac{1}{y_0} K\left(\frac{s + x_0}{y_0}\right)\right\|_1$$

$$= \|K(s) - K(s + x_0)\|_1 + \left\|K(s) - \frac{1}{y_0} K\left(\frac{s}{y_0}\right)\right\|_1.$$

Since K is continuous with compact support, there is $\delta > 0$ such that $\|K(s) - K(s + x_0)\|_1 \le 1/4M$ if $|x_0| < \delta$ and such that

$$\|K(s) - (1/y_0)K(s/y_0)\|_1 < 1/4M$$

if $|1 - y_0| < \delta$. We conclude that

$$\max\left(\frac{|x_j - x_k|}{y_j}, \frac{|y_j - y_k|}{y_j}\right) \ge \delta,$$

so that $|z_j - z_k| \ge \delta y_j$ and the points $\{z_j\}$ are separated.

The proof that the points $\{z_j\}$ satisfy (1.7) is now simpler than the argument in Theorem 4.2 because the kernel $K(t)$ has compact support. Khinchin's inequality and (4.9) give

(4.10) $$\sum_{j=1}^n y_j \le 2M \int \left(\sum_{j=1}^n y_j^2 K_{z_j}^2(t)\right)^{1/2} dt$$

whenever $z_1, z_2, \ldots, z_n$ are points from the sequence lying in a square $Q = \{x \in I, 0 < y \le |I|\}$. Let $A > 0$ be such that $K(t) = 0$, $|t| > A$. Since $z_j \in Q$, this means that $K_{z_j}(t) = 0$ if $t \notin J$, where

$$J = \{t : \text{dist}(t, I) < A|I|\}.$$

The Cauchy–Schwarz inequality now yields

$$\int \left(\sum_{j=1}^{n} y_j^2 K_{z_j}^2(t) \right)^{1/2} dt \leq |J|^{1/2} \left(\int \sum y_j^2 K_{z_j}^2(t) dt \right)^{1/2}$$

$$\leq |J|^{1/2} \left(\sum y_j \right)^{1/2} \| K \|_2$$

because $\int K_{z_j}^2(t)\, dt = (1/y_j) \int K^2(t) dt$. From (4.10) we now obtain

$$\left(\sum_{j=1}^{n} y_j \right)^{1/2} \leq (2A + 1)^{1/2} 2M \| K \|_2 |I|^{1/2},$$

and this gives (1.7). $\quad\square$

In the proof just completed, condition (4.9) means that every interpolation (4.7), with P_{z_i} replaced by K_{z_i}, has solution in L^∞. This trick of replacing one kernel, even the Poisson kernel, by another simpler kernel is often helpful.

The converse of Theorem 4.3 is not true generally. If $\mathrm{P}(t) = \chi_{(-1,1)}(t)$, then interpolation is not possible on the finite set $\{i, (1 + i)/2, (-1 + i)/2\}$ because the kernels for these points are linearly dependent.

The proof above of Theorem 4.2 is due to Varopoulos [1972], who has also found another elegant proof of the same theorem. His second proof is based on the roots of unity argument used in Theorem 2.2. We give the proof for the more general Theorem 4.3 but with the extra hypothesis $\mathrm{P} \in L^2$. We know that every bounded interpolation problem

$$u(z_j) = a_j, \quad j = 1, 2, \ldots,$$

has solution with

$$\| u \|_\infty \leq M \sup_j |a_j|,$$

where M is a constant. From this it is trivial to verify (1.2) and our real task is to establish (1.7). Fix a square $Q = \{x \in I, 0 < y \leq |I|\}$ and let $z_1, z_2, \ldots, z_n$ be finitely many sequence points in the square Q. Let $u_j \in L^\infty$, $\| u_j \|_\infty \leq M$, interpolate

$$u_j(z_k) = \omega^{jk}, \quad k = 1, 2, \ldots, n,$$

where $\omega = e^{2\pi i/n}$. The functions

$$U_j(z) = \frac{1}{n} \sum_{l=1}^{n} \omega^{-jl} u_l(z)$$

satisfy $U_j(z_j) = 1$, $j = 1, 2, \ldots, n$. The proof of Theorem 2.2 shows that

$$(4.11) \qquad \sum_{j=1}^{n} |U_j(t)|^2 \leq M^2 + \varepsilon$$

almost everywhere on $\mathbb{R}$. Let J be the interval

$$J = \{t : \text{dist}(t, I) \leq cM|I|\},$$

where c is chosen so that

$$\int_{\mathbb{R} \setminus J} \mathrm{P}_z(t)\, dt < 1/2M$$

for all $z \in Q$. Then we have

$$1 \leq \int |U_j(t)| \mathrm{P}_{z_j}(t)\, dt,$$

while by (4.11),

$$\int_{\mathbb{R} \setminus J} |U_j(t)| \mathrm{P}_{z_j}(t)\, dt \leq M/2M = \tfrac{1}{2}.$$

Consequently, we have

$$\tfrac{1}{2} \leq \int_J |U_j(t)| \mathrm{P}_{z_j}(t)\, dt \leq \left(\int_J |U_j(t)|^2\, dt \right)^{1/2} \| \mathrm{P}_{z_j} \|_2$$

$$\leq \frac{c'}{y_j^{1/2}} \left(\int_J |U_j(t)|^2\, dt \right)^{1/2},$$

and

$$y_j \leq (4c')^2 \int_J |U_j(t)|^2\, dt.$$

Summing, we obtain by (4.11),

$$\sum_{j=1}^{n} y_j \leq (4c')^2 M^2 |J| \leq (4c')^2 M^3 c |I|,$$

and (1.7) is proved. The same reasoning also yields a proof of Theorem 4.3.

A refinement of Theorem 4.2 will be proved by a different method in Chapter X.

5. Earl's Elementary Proof

There is another proof of the main theorem (Theorem 1.1) that does not use duality. This constructive proof, due to J. P. Earl, shows that when $\{z_j\}$ is an interpolating sequence there are interpolating functions of the form $CB(z)$, where $B(z)$ is a Blaschke product and C is a constant. The Blaschke product $B(z)$ has simple zeros $\{\zeta_j\}$ which are hyperbolically very close to the $\{z_j\}$. It follows that $\{\zeta_j\}$ is also an interpolating sequence.

Theorem 5.1. *Let $\{z_j\}$ be a sequence in the upper half plane such that*

$$(5.1) \qquad \prod_{j,\,j\neq k} \left| \frac{z_k - z_j}{z_k - \bar{z}_j} \right| \geq \delta > 0, \quad k = 1, 2, \ldots .$$

Then there is a constant K such that whenever $\{a_j\} \in l^\infty$, there exists $f(z) \in H^\infty$ such that

$$(5.2) \qquad f(z_j) = a_j, \quad j = 1, 2, \ldots ,$$

and such that

$$(5.3) \qquad f(z) = K \left(\sup_j |a_j| \right) B(z),$$

where $B(z)$ is a Blaschke product. The zeros $\{\zeta_j\}$ of $B(z)$ satisfy

$$\rho(\zeta_j, z_j) = \left| \frac{\zeta_j - z_j}{\zeta_j - \bar{z}_j} \right| \leq \frac{\delta}{3}$$

and

$$(5.4) \qquad \prod_{j,\,j\neq k} \left| \frac{\zeta_k - z_j}{\zeta_k - \bar{z}_j} \right| \geq \frac{\delta}{3},$$

so that $\{\zeta_j\}$ is also an interpolating sequence.

The constant K obtained in the proof of Theorem 5.1 will not be the minimal constant of interpolation. We get $K = O(\delta^{-2})$, while the constant of interpolation has the bound $M = O(\delta^{-1} \log(\delta^{-1}))$.

Lemma 5.2. *If $0 < \alpha < \beta_n < 1$, then*

$$(5.5) \qquad \prod_{n=1}^{\infty} \frac{\beta_n - \alpha}{1 - \alpha\beta_n} \geq \frac{(\Pi\beta_n) - \alpha}{1 - \alpha\Pi\beta_n}.$$

Proof. If $B(z)$ is the Blaschke product with zeros β_n, then the left side of (5.5) is $B(\alpha)$. By Schwarz's lemma

$$\rho(B(\alpha), B(0)), \leq \alpha,$$

and the euclidean description of the disc $\{\rho(w, B(0)) \leq \alpha\}$ from Chapter I, Section 1, then gives

$$B(\alpha) \geq \frac{|B(0)| - \alpha}{1 - \alpha|B(0)|},$$

which is the right side of (5.5). $\square$

Lemma 5.3. *Let*

$$0 < \lambda < \frac{2\lambda}{1 + \lambda^2} < \delta < 1$$

and let $\{z_j\}$ be a sequence in the upper half plane such that (5.1) holds. If $\{\zeta_j\}$ satisfies

$$\rho(\zeta_j, z_j) \leq \lambda, \quad j = 1, 2, \ldots,$$

then

(5.6)
$$\prod_{j, j \neq k} \left| \frac{\zeta_k - \zeta_j}{\zeta_k - \bar{\zeta}_j} \right| \geq \frac{\delta - 2\lambda/(1 + \lambda^2)}{1 - 2\lambda\delta/(1 + \lambda^2)}.$$

In particular, if $\rho(\zeta_j, z_j) \leq \delta/3, j = 1/2, \ldots$, then

$$\prod_{j, j \neq k} \left| \frac{\zeta_k - \zeta_j}{\zeta_k - \bar{\zeta}_j} \right| \geq \frac{\delta}{3}.$$

Proof. For $j \neq k$, Lemma 1.4 of Chapter I gives

$$\rho(\zeta_j, \zeta_k) \geq \frac{\rho(\zeta_j, z_k) - \rho(z_k, \zeta_k)}{1 - \rho(z_k, \zeta_k)\rho(\zeta_j, z_k)} \geq \frac{\rho(\zeta_j, z_k) - \lambda}{1 - \lambda\rho(\zeta_j, z_k)}$$

and

$$\rho(\zeta_j, z_k) \geq \frac{\rho(z_j, z_k) - \lambda}{1 - \lambda\rho(z_j, z_k)}.$$

Writing $\alpha = 2\lambda/(1 + \lambda^2)$, we now have

$$\rho(\zeta_j, \zeta_k) \geq \left(\frac{\rho(z_j, z_k) - \lambda}{1 - \lambda\rho(z_j, z_k)} - \lambda \right) \bigg/ \left(1 - \lambda \frac{\rho(z_j, z_k) - \lambda}{1 - \lambda\rho(z_j, z_k)} \right) = \frac{\rho(z_j, z_k) - \alpha}{1 - \alpha\rho(z_j, z_k)}$$

when $j \neq k$. Lemma 5.2 then yields

$$\prod_{j, j \neq k} \left| \frac{\zeta_k - \zeta_j}{\zeta_k - \bar{\zeta}_j} \right| = \prod_{j, j \neq k} \rho(\zeta_j, \zeta_k) \geq \frac{\delta - \alpha}{1 - \alpha\delta},$$

which is (5.6). When $\lambda = \delta/3$ a calculation shows $(\delta - \alpha)/(1 - \alpha\delta) > \delta/3$, so that the final assertion of the lemma is true. □

Fix $\lambda = \delta/3$, and let Δ_j be the closed disc defined by

$$\Delta_j = \{\zeta_j : \rho(\zeta_j, z_j) \leq \delta/3\}.$$

For $\zeta_j \in \Delta_j, j = 1, 2, \ldots, n$, write

(5.7)
$$B_{(\zeta_1, \zeta_2, \ldots, \zeta_n)}(z) = \prod_{j=1}^{n} \left(\frac{z - \zeta_j}{z - \bar{\zeta}_j} \right) \frac{\bar{\zeta}_j}{\zeta_j}.$$

Then $B_{(\zeta_1,\dots,\zeta_n)}$ is a finite Blaschke product normalized so that $B_{(\zeta_1,\dots,\zeta_n)}(0) = 1$ and so that

$$\overline{B_{(\zeta_1,\dots,\zeta_n)}(\bar{z})} = (B_{(\zeta_1,\dots,\zeta_n)}(z))^{-1}.$$

Consequently $(\zeta_1, \zeta_2, \dots, \zeta_n) = (\zeta_1', \zeta_2', \dots, \zeta_n')$ if

$$B_{(\zeta_1,\zeta_2\dots,\zeta_n)}(z_j) = B_{(\zeta_1',\zeta_2'\dots,\zeta_n')}(z_j), \quad j = 1, 2, \dots, n,$$

because the difference of these two Blaschke products is a rational function of degree at most $2n$ which vanishes at the $2n + 1$ points $\{0, z_1, z_2, \dots, z_n, \bar{z}_1, \bar{z}_2, \dots, \bar{z}_n\}$.

The main step of the proof is the following lemma, in which it is shown that if $|a_j|$ is small, every finite interpolation problem (5.2) can be solved with a Blaschke product of the form (5.7).

Lemma 5.4. *Suppose that for* $j = 1, 2, \dots, n,$

$$(5.8) \qquad |a_j| < \frac{\delta}{3} \inf_{\zeta_k \in \Delta_k} \prod_{\substack{k=1 \\ k \ne j}}^{n} \left| \frac{z_j - \zeta_k}{z_j - \bar{\zeta}_k} \right|.$$

Then there are $\zeta_j \in \Delta_j, j = 1, 2, \dots, n$ *such that*

$$B_{(\zeta_1,\dots,\zeta_n)}(z_j) = a_j, \quad j = 1, 2, \dots, n.$$

Proof. We use induction on n. For $n = 1$ we have the mapping

$$w = w(z_1) = B_{(\zeta_1)}(z_1) = \left(\frac{z_1 - \zeta_1}{z_1 - \bar{\zeta}_1} \right) \frac{\bar{\zeta}_1}{\zeta_1}$$

from Δ_1 into $|w| \le \delta/3$. This mapping is one-to-one and $\delta\Delta_1 = \{\zeta_1 : \rho(\zeta_1, z_1) = \delta/3\}$ is mapped into the circle $|w| = \delta/3$. Even though w is not analytic in ζ_1, the argument principle can be used to show $w(\Delta_1)$ covers the disc $|w| < \delta/3$. Since $w(z_1) = 0$, the curve $w(\partial\Delta_1)$ has nonzero winding number relative to $w = 0$. If $|a_1| < \delta/3$, then the same curve also has nonzero winding number about a_1. If $a_1 \notin w(\Delta_1)$, then each curve $w(\{\rho(\zeta_1, z_1) = r\}), 0 < r < \delta/3$ has the same nonzero index relative to a_1. This is impossible for r small. Hence the lemma is true for $n = 1$.

Suppose the lemma is true for $n - 1$. For each $\zeta_n \in \Delta_n$, we can, by the induction hypothesis, find

$$\zeta_1 = \zeta_1(\zeta_n), \quad \zeta_2 = \zeta_2(\zeta_n), \quad \dots, \quad \zeta_{n-1} = \zeta_{n-1}(\zeta_n),$$

with $\zeta_j \in \Delta_j$ such that

$$B_{(\zeta_1,\zeta_n,\dots,\zeta_{n-1})}(z_j) = a_j \left(\frac{z_j - \bar{\zeta}_n}{z_j - \zeta_n} \right) \frac{\zeta_n}{\bar{\zeta}_n},$$

because these values satisfy (5.8). The points $\zeta_1, \ldots, \zeta_{n-1}$ are unique and they depend continuously on ζ_n. Now consider the mapping

$$w = w(\zeta_n) = \left(\frac{z_n - \zeta_n}{z_n - \bar{\zeta}_n} \right) \frac{\bar{\zeta}_n}{\zeta_n} B_{(\zeta_1, \ldots, \zeta_{n-1})}(z_n),$$

which is continuous in ζ_n. The right side is the value at z_n of a Blaschke product of the form (5.7) which interpolates a_j of z_j, $j \leq n - 1$, no matter which $\zeta_n \in \Delta_n$ is chosen. We want to find $\zeta_n \in \Delta_n$ for which $w(z_n) = a_n$. By (5.8) we have $|a_n| < \inf\{|w(\zeta_n)| : \zeta_n \in \partial\Delta_n\}$. Since $w(z_n) = 0$, the curve $w(\partial\Delta_n)$ has nonzero index relative to $w = 0$, and hence it also has nonzero index relative to $w = a_n$. As in the case $n = 1$, this means there is $\zeta_n \in \Delta_n$ for which $w(\zeta_n) = a_n$. The lemma is proved. $\square$

Proof of Theorem 5.1. It is enough to show that if

$$|a_j| \leq \frac{\delta}{3} \frac{\delta - \delta/3}{1 - \delta^2/3} = \frac{2\delta^2}{3(3 - \delta^2)},$$

then there are $\zeta_j \in \Delta_j$, $j = 1, 2, \ldots$, such that if B is the Blaschke product with zeros ζ_j, then

$$e^{i\theta} B(z_k) = a_k, \quad k = 1, 2, 3, \ldots,$$

for some constant $e^{i\theta}$. By (5.7) and (5.8), and by Lemma 5.4, there are $\zeta_1^{(n)}, \zeta_2^{(n)}, \ldots, \zeta_n^{(n)}$ such that $\zeta_j^{(n)} \in \Delta_j$ and

$$B_{(\zeta_1^{(n)}, \zeta_2^{(n)}, \ldots, \zeta_n^{(n)})}(z_k) = a_k, \quad k = 1, 2, \ldots, n.$$

By a change of scale we can assume $i = \sqrt{-1} \notin \cup \Delta_j$. Write

$$B_{(\zeta_1^{(n)}, \ldots, \zeta_n^{(n)})}(z) = e^{i\theta_n} B_n(z),$$

where $B_n(i) > 0$. Take a subsequence n_k so that

$$e^{i\theta_{n_\kappa}} \to e^{i\theta}, \quad \zeta_j^{(n_\kappa)} \to \zeta_j \in \Delta_j,$$

for all $j = 1, 2, \ldots$. Let $B(z)$ be the Blaschke product with zeros ζ_j, normalized so that every subproduct is positive at $z = i$. We claim that

(5.9) $e^{i\theta} B(z_j) = a_j, \quad j = 1, 2, \ldots.$

The proof of Lemma 5.3 shows that

$$\lim_{N \to \infty} \inf_{\zeta_j \in \Delta_j} \prod_{j \geq N} \left| \frac{z - \zeta_j}{z - \bar{\zeta}_j} \right| = 1$$

uniformly on compact sets. Since the products are normalized to be positive at $z = i$, this means that

$$B_{n_\kappa}(z) \to B(z) \quad (k \to \infty),$$

and so (5.9) holds. This gives (5.3) with $K = 3(3 - \delta^2)/2\delta^2$. By Lemma 5.3, the zeros $\{\zeta_j\}$ satisfy (5.4). $\square$

Notes

The proof in Section 1 follows Carleson's original paper [1958] except that the Carleson measure is treated differently. Hörmander's paper [1967b] clarifies the geometry of the problem. Shapiro and Shields [1961] and later Amar [1977a] have approached interpolating sequences using Hilbert space. The argument in Section 1 yields results when $\{z_j\}$ is not an interpolating sequence (see Exercise 9 and Garnett [1977]). Interpolating sequences can also be characterized in terms of H^p (see Duren [1970] and Exercise 11). The example showing $M \geq (1/\delta) \log(1/\delta)$ is due to A. M. Gleason. Other recent expositions of interpolation can be found in Havin and Vinogradov [1974] and Sarason [1979].

Theorem 2.1 is due to Pehr Beurling (see Carleson [1962b]). A general discussion of linear operators of interpolation is given by Davie [1972]. For Theorem 2.2, see Varopoulos [1971a] and Bernard [1971]. The source of the idea is in harmonic analysis (see Drury [1970] and Varopoulos [1970]).

Another proof of Theorem 4.2 is in Garnett [1971b]. See Garnett [1978] and Exercise 12 for extensions of Theorem 4.2 to L^p, $p > 1$, and to BMO.

Theorem 5.1 is due to Earl [1970], who has also found an elementary approach to Theorem 2.1 (Earl [1976]). P. Jones has sharpened Earl's method to obtain interpolating functions whose norms have the minimal order of magnitude (see Exercise 10).

There is an interesting open problem on harmonic interpolation in higher dimensions. Consider the upper half plane $\mathbb{R}_+^{n+1} = \{(x, y) : x \in \mathbb{R}^n, y > 0\}$. Each bounded harmonic function on $\mathbb{R}_+^{n+1}$ is the Poisson integral of a function in $L^\infty(\mathbb{R}^n)$. If a sequence $\{p_j\} = \{(x_j, y_j)\}$ in $\mathbb{R}_+^{n+1}$ is an interpolating sequence for the bounded harmonic functions, then the analog of condition (c), Theorem 1.1, holds for $\{p_j\}$,

(i) $|p_j - p_k|/y_j \geq a > 0$, $j \neq k$,

(ii) $\sum_{p_j \in Q} y_j^n \leq C\ell(Q)^n$

for every cube

$$Q = \{(x, y) : |x_i - x_i^0| < \ell(Q)/2, i = 1, 2, \ldots, n, 0 < y < \ell(Q)\}.$$

This follows from the proof of Theorem 4.2. The unsolved problem is the converse. Do (i) and (ii) characterize bounded harmonic interpolating sequences in $\mathbb{R}_+^{n+1}$? (See Carleson and Garnett [1975].)

Exercises and Further Results

1. Let $B(z)$ be the Blaschke product with zeros $\{z_j\}$ in H . Then $\{z_j\}$ is an interpolating sequence if and only if

$$\inf_j y_j |B'(z_j)| > 0.$$

2. If S and T are disjoint interpolating sequences, then $S \cup T$ is an interpolating sequence if and only if

$$\rho(S, T) = \inf\{\rho(z, w) : z \in S, w \in T\} > 0.$$

3. If $\{z_j\}$ is an interpolating sequence in H and $\{a_j^{(n)}\}$, $n = 0, 1, 2, \ldots, N$, are finitely many sequences such that $y_j^n |a_j^{(n)}| \le 1$, then there is $f \in H^\infty$ such that

$$f^{(n)}(z_j) = a_j^{(n)}, \quad j = 1, 2, \ldots, \quad n = 0, 1, 2, \ldots, N,$$

where $f^{(n)}$ denotes the nth derivative.

4. Let $B(z)$ be a Blaschke product with distinct zeros $\{z_j\}$ on an interpolating sequence. If $m \in \mathfrak{M}_{H^\infty}$ is such that $B(m) = 0$, then m is in the closure of the zeros $\{z_j\}$ in the topology of $\mathfrak{M}_{H^\infty}$.

5. (Naftalevitch). If $|z_j| < 1$ and if $\sum(1 - |z_j|) < \infty$, then there is an interpolating sequence $\{w_j\}$ with $|w_j| = |z_j|$.

6. If $\{z_j\}$ is a sequence in the upper half plane and if $\sum y_j \delta_{z_j}$ is a Carleson measure, then $\{z_j\}$ is the union of finitely many interpolating sequences.

7. If $\{z_j\}$ is an interpolating sequence and if $\{w_j\}$ is a separated sequence— that is, if $\{w_j\}$ satisfies (1.2)—and if

$$\rho(z_j, w_j) < \lambda < 1,$$

then $\{w_j\}$ is an interpolating sequence.

8. Let X be a Banach space and let $\{z_j\}$ be a sequence of linear functionals on X, $\|z_j\| = 1$. Suppose that for every $\{a_j\} \in l^\infty$ there is $x \in X$ such that

$$|z_j(x) - a_j| \le \tfrac{1}{2} \|a_j\|_\infty$$

and $\|x\| \le K \|a_j\|_\infty$. Prove that $\{z_j\}$ is an interpolating sequence: That is, prove that whenever $\{a_j\} \in l^\infty$ there is $x \in X$ such that

$$z_j(x) = a_j, \quad j = 1, 2, \ldots.$$

Now suppose that interpolation is possible whenever $\{a_j\}$ is an idempotent sequence: For each j either $a_j = 0$ or $a_j = 1$. Then $\{z_j\}$ is again an interpolating sequence. (Use Baire category to show all idempotents can be interpolated by a uniformly bounded set in X.)

9. Let $\{z_j\}$ be a sequence in the upper half plane, and let

$$\delta_k = \prod_{j, j \neq k} \left| \frac{z_k - z_j}{z_k - \bar{z}_j} \right|, \quad k = 1, 2, \ldots.$$

Suppose $\delta_k > 0$ but $\inf_k \delta_k = 0$. If $|a_j| \leq \delta_j (1 + \log 1/\delta_j)^{-2}$, then there is $f \in H^\infty$ such that

$$f(z_j) = a_j, \quad j = 1, 2, \ldots.$$

More generally, if $h(t)$ is a positive decreasing function on $[0, \infty)$, if

$$\int_0^\infty h(t)\, dt < \infty,$$

and if

$$|a_j| \leq \delta_j h(1 + \log 1/\delta_j),$$

then the interpolation $f(z_j) = a_j, j = 1, 2, \ldots$, has solution $f \in H^\infty$. To prove this show that $\sum (|a_j| y_j / \delta_j) \delta_{z_j}$ is a Carleson measure.
   The result cited above is sharp. If $h(t)$ is positive and decreasing on $[0, \infty)$ and if $\int_0^\infty h(t) dt = \infty$, then there is a separated sequence $\{z_j\}$ and values a_j such that

$$|a_j| = \delta_j h(1 + \log 1/\delta_j)$$

but such that interpolation is impossible. (See Garnett [1977] for details.)

10. If $\{z_j\}$ is an interpolating sequence with

$$\inf_k \prod_{j, j \neq k} \left| \frac{z_k - z_j}{z_k - \bar{z}_j} \right| = \delta > 0,$$

then $\{z_j\}$ can be partitioned into $K \log 1/\delta$ subsequences such that for each subsequence $\{w_j\}$,

$$\inf_k \prod_{j, j \neq k} \left| \frac{w_k - w_j}{w_k - \bar{w}_j} \right| = \frac{1}{2},$$

where K is an absolute constant. If $Y_1, \ldots, Y_N$ are these subsequences, if B_k is the Blaschke product with zeros $\bigcup_{j \neq k} Y_j$, and if f_k interpolates $a_j / B_k(z_j)$ on Y_k, then $\sum B_k f_k$ is an interpolating function with norm less than

$$C\delta^{-1}(\log 1/\delta) \sup_j |a_j|.$$

11. Let $\{z_j\}$ be a sequence in the upper half plane. For $0 < p < \infty$, let T_p be the linear operator defined on H^p by

$$T_p f(j) = y_j^{1/p} f(z_j).$$

Then $\|T_p f\|_\infty \le C\|f\|_p, f \in H^p$.

(a) If T_p is bounded from H^p to l^p, then $\sum y_j \delta_{z_j}$ is a Carleson measure. Using the closed graph theorem, obtain the same conclusion if $T_p(H^p) \subset l^p$.

(b) If $T_p(H^p) = l^p$, then $\{z_j\}$ is an H^∞ interpolating sequence.

(c) If $\{z_j\}$ is an H^∞ interpolating sequence, then $T_p(H^p) = l^p, 0 < p < \infty$. For $p > 1$, use a duality argument. For $p \le 1$ let $B_k(z)$ be the Blaschke product with zeros $\{z_j, j \ne k\}$. Then since $\sum y_k |a_k|^p < \infty$,

$$f(z) = \sum_{k=1}^{\infty} \frac{B_k(z)}{B_k(z_k)} \left(\frac{2iy_k}{(z - \bar{z}_k)} \right)^{2/p} a_k$$

is in H^p and $f(z_k) = a_k, k = 1, 2, \ldots$.

(See Shapiro and Shields [1961].)

12. If $p \ge 1$, the operator T_p of Exercise 11 has a natural extension to L^p defined using the Poisson kernel.

(a) If $T_p(L^p) = l^p$, then T_p is bounded from L^p to l^p and $\{z_j\}$ is an H^∞ interpolating sequence.

(b) If $1 < p < \infty$ and if $T_p(L^p) \supset l^p$, then $\{z_j\}$ is an H^∞ interpolating sequence. If $T_\infty(BMO) \supset l^\infty$, then $\{z_j\}$ is an interpolating sequence. This result for $p > 2$ or for BMO follows from a modification of the proof of Theorem 4.2. Another argument is needed for $p < 2$. (See Garnett [1978] for details and extensions.)

(c) If $T_1(L^1) \supset l^1$, it need not follow that $\{z_j\}$ is an interpolating sequence.

13. (a) Prove Khinchin's inequality for $p = 4$:

$$\left(E\left(\left| \sum_{j-1}^{n} \pm \alpha_j \right|^4 \right) \right)^{1/4} \le C_4 \left(\sum |\alpha_j|^2 \right)^{1/2}$$

where C_4 does not depend on n.

(b) Prove Khinchin's inequality for all finite $p > 2$ by first considering the case when p is an even integer.

14. A sequence $\{z_j\}$ in the disc is *nontangentially dense* if almost every $e^{i\theta} \in T$ is the nontangential limit of a subsequence of $\{z_j\}$. For a discrete sequence $\{z_j\}$ in the disc, the following are equivalent:

(i) $\{z_j\}$ is nontangentially dense.

(ii) For each point $z \in D$ there are positive weights λ_j such that

$$u(z) = \sum_j \lambda_j u(z_j)$$

for each bounded harmonic function $u(z)$.

(iii) For some point $z_0 \notin \{z_j\}$ there are complex weights β_j such that $\sum |\beta_j| < \infty$ and

$$f(z_0) = \sum_j \beta_j f(z_j)$$

for all $f \in H^\infty$.

(See Brown, Shields, and Zeller [1960] and Hoffman and Rossi [1967].)

15. Let $\{z_j\}$ be an interpolating sequence in the disc and let $\{a_j\} \in \ell^\infty$. Let $f \in H^\infty$ interpolate

$$f(z_j) = a_j, \quad j = 1, 2, \ldots,$$

with $\|f\|_\infty$ minimal.

(a) If $\lim a_j = 0$, then f is the unique interpolating function of minimal norm. If also $\lim z_j = 1$ nontangentially, then f is a constant times a Blaschke product.

(b) For some choices of $\{z_j\}$ and $\{a_j\}$, there is no unique minimal norm interpolating function.

(See Øyma [1977].)

VIII

The Corona Construction

This chapter is an extensive discussion of Carleson's corona theorem. Several proofs of the theorem will be presented, because the ideas behind each proof have proved useful for other problems. We first give T. Wolff's recent, very elegant proof, which is based on Littlewood–Paley integrals and which employs analyticity in a decisive way. Then we take up Carleson's original proof. It consists of a geometric construction that has led to many of the deeper results in this theory and that applies to harmonic functions and to more general situations.

We begin with two theorems bounding solutions of certain inhomogeneous Cauchy–Riemann equations. One of these theorems is then used in Section 2 to prove the corona theorem and a generalization.

Section 3 contains two theorems on minimum modulus. A simplified version of the main construction is then used to establish a separation theorem about Blaschke products. That theorem will have an important application in the next chapter.

Carleson's original proof is discussed in Section 5 and a less function-theoretic alternate approach to the construction is given in Section 6. In Section 7 we circumvent the duality argument used in the original proof, thereby making the corona proof quite constructive. At that point interpolating sequences reappear to play a decisive role.

1. Inhomogeneous Cauchy–Riemann Equations

Define

$$\frac{\partial}{\partial \bar{z}} = \frac{1}{2}\left(\frac{\partial}{\partial x} + i\frac{\partial}{\partial y}\right).$$

Thus a function $h(z)$ is analytic if and only if $\partial h / \partial \bar{z} = 0$. Let $G(\zeta)$ be C^1 and bounded on the open disc D. We want to solve the *inhomogeneous Cauchy–Riemann equation*

(1.1) $$\frac{\partial F}{\partial \bar{z}} = G(z), \quad |z| < 1,$$

with a good bound on $\|F\|_\infty = \sup_{|z|=1}|F(z)|$. For $\zeta = \xi + i\eta$, write $d\zeta = d\xi + i\,d\eta, d\bar{\zeta} = d\xi - i\,d\eta$. If $\varphi \in C^\infty$ has compact support contained in D, then by Green's theorem

$$\frac{1}{2\pi i}\iint \frac{\partial\varphi}{\partial\bar\zeta}\frac{1}{\zeta - z}d\zeta \wedge d\bar\zeta = -\frac{1}{2\pi i}\iint \frac{\partial}{\partial\bar\zeta}\left(\frac{\varphi(\zeta)}{\zeta - z}\right)d\bar\zeta \wedge d\zeta$$
$$= \lim_{\varepsilon \to 0}\frac{1}{2\pi i}\int_{|\zeta-z|=\varepsilon}\frac{\varphi(\zeta)d\zeta}{\zeta - z}$$
$$= \varphi(z).$$

So if (1.1) has solutions $F(z)$ on $|z| < 1$, then one solution should be given by

(1.2)
$$F(z) = \frac{1}{2\pi i}\iint_{|\zeta|<1} G(\zeta)\frac{1}{\zeta - z}d\zeta \wedge d\bar\zeta.$$

It is easy to see that the convolution $F(z)$ defined by (1.2) is continuous on the complex plane and that $F(z)$ is C^1 on the open disc. Moreover, $F(z)$ is a solution of (1.1). Indeed, if $\varphi \in C^\infty$ has compact support contained in the unit disc, then

$$\iint F\frac{\partial\varphi}{\partial\bar z}dz \wedge d\bar z + \iint \varphi\frac{\partial F}{\partial\bar z}dz \wedge d\bar z = -\iint \frac{\partial F\varphi}{\partial\bar z}\,d\bar z \wedge dz$$
$$= -\int_{|z|=1} F\varphi\,dz = 0,$$

and hence

$$\iint F\frac{\partial\varphi}{\partial\bar z}dz \wedge d\bar z = -\iint \varphi\frac{\partial F}{\partial\bar z}dz \wedge d\bar z.$$

If we also show that

$$\iint F\frac{\partial\varphi}{\partial\bar z}dz \wedge d\bar z = -\iint \varphi G\,dz \wedge d\bar z,$$

then we can obtain (1.1) by letting φ run through the translates of an approximate identity. But by (1.2) and Fubini's theorem,

$$\iint F\frac{\partial\varphi}{\partial\bar z}dz \wedge d\bar z = \iint_{|\zeta|<1} G(\zeta)\left(\frac{-1}{2\pi i}\iint \frac{\partial\varphi}{\partial\bar z}\frac{1}{z - \zeta}dz \wedge d\bar z\right)d\zeta \wedge d\bar\zeta$$
$$= -\iint_{|\zeta|<1} G(\zeta)\varphi(\zeta)d\zeta \wedge d\bar\zeta,$$

and so (1.1) holds when $F(z)$ is defined by (1.2).

Now (1.2) does not give a unique solution of (1.1) on $D = \{|z| < 1\}$. But any function $b(z)$ continuous on $\bar{D}$ and C^1 on D that solves

$$(1.3) \qquad \frac{\partial b}{\partial \bar{z}} = G(z), \quad |z| < 1,$$

does have the form $b(z) = F(z) + h(z)$, where $h(z)$ is in the disc algebra $A_o = H^\infty \cap C(\bar{D})$, because $\partial h / \partial \bar{z} = 0$ on D. We want an estimate on the minimal norm of such solutions $b(z)$ of (1.3). Because we shall ultimately study functions analytic on D, the norm of interest here is the supremum on ∂D,

$$\|b\|_\infty = \sup_{|z|=1} |b(z)|.$$

We use duality and the theorem on Carleson measures to make two different estimates on the minimal norm of solutions of (1.3).

Theorem 1.1. *Assume that $G(z)$ is bounded and C^1 on the disc D and that $|G| dx\, dy$ is a Carleson measure on D,*

$$(1.4) \qquad \iint_S |G| dx\, dy \le A\ell(S)$$

for every sector

$$S = \{re^{i\theta} : 1 - \ell(S) < r < 1, |\theta - \theta_0| < \ell(S)\}.$$

Then there is $b(z)$ continuous on $\bar{D}$ and C^1 on D such that

$$\partial b / \partial \bar{z} = G(z), \quad |z| < 1,$$

and such that

$$\|b\|_\infty = \sup_{|z|=1} |b(z)| \le CA,$$

with C an absolute constant.

In this theorem (and in the next theorem) it is not important that G be bounded on D and the upper bound for $|G(z)|$ does not occur in the estimate of $\|b\|_\infty$. We have assumed G is bounded only to ensure that (1.1) has at least one bounded solution (see Exercise 1).

Proof. Let $F(z)$ be the solution of (1.3) defined by (1.2). Then every solution of (1.3) has the form

$$b(z) = F(z) + h(z), \quad h \in A_o.$$

The minimal norm of such solutions is

$$\inf_{h \in A_o} \|F + h\|_\infty,$$

the norm being the essential supremum ∂D. By duality,

$$\inf_{h \in A_o} \|F + h\|_\infty = \sup \left\{ \left| \frac{1}{2\pi} \int_0^{2\pi} Fk\, d\theta \right| : k \in H_0^1, \|k\|_1 \le 1 \right\}$$

$$= \sup \left\{ \left| \frac{1}{2\pi i} \int_{|z|=1} F(z)k(z)dz \right| : k \in H^1, \|k\|_1 \le 1 \right\}.$$

(See Chapter IV, Theorem 1.3 and Lemma 1.6.)

By Green's theorem and by continuity,

$$\frac{1}{2\pi i} \int_{|z|=1} F(z)k(z)dz = \lim_{r \to 1} \frac{1}{2\pi i} \int_{|z|=r} F(z)k(z)dz$$

$$= -\frac{1}{2\pi i} \iint_{|z| \le 1} \frac{\partial F}{\partial \bar{z}} k(z)dz \wedge d\bar{z},$$

since $\partial k/\partial \bar{z} = 0$. Consequently,

$$\inf_{h \in A_o} \|F + h\|_\infty \le \sup \left\{ \frac{1}{\pi} \iint_{|z|<1} |G(z)||k(z)|dx\, dy : k \in H^1, \|k\|_1 \le 1 \right\}.$$

By the disc version of the theorem on Carleson measures, Theorem 3.8 of Chapter II, there is a constant C_1 so that

$$\frac{1}{\pi} \iint |G(z)||k(z)|dx\, dy \le C_1 A \|k\|_1$$

whenever $k \in H^1$. Taking $C > C_1$ we see (1.3) has solution $b(z) = F(z) + h(z), h \in A_o$, such that $\|b\|_\infty \le CA$. $\quad\square$

The second estimate involves ideas from the proof of the H^1–BMO duality. Write

$$\frac{\partial}{\partial z} = \frac{1}{2} \left(\frac{\partial}{\partial x} - i \frac{\partial}{\partial y} \right).$$

Theorem 1.2 (Wolff). *Assume that $G(z)$ is bounded and C^1 on the disc D and assume that the two measures*

$$|G|^2 \log(1/|z|)dx\, dy \quad \text{and} \quad |\partial G/\partial z| \log(1/|z|)dx\, dy$$

are Carleson measures,

$$(1.5) \qquad \iint_S |G|^2 \log \frac{1}{|z|} dx\, dy \le B_1 \ell(S)$$

and

(1.6)
$$\iint_S \left|\frac{\partial G}{\partial z}\right| \log \frac{1}{|z|} dx \, dy \leq B_2 \ell(S)$$

for every sector

$$S = \{re^{i\theta} : 1 - \ell(S) < r < 1, |\theta - \theta_0| < \ell(S)\}.$$

Then there is $b(z)$ continuous on $\bar{D}$ and C^1 on D such that

$$\partial b/\partial \bar{z} = G(z), \quad |z| < 1,$$

and such that

$$\|b\|_\infty = \sup_{|z|=1} |b(z)| \leq C_1 \sqrt{B_1} + C_2 B_2;$$

with C_1 and C_2 absolute constants.

Proof. As before, we have

$$\inf\left\{\|b\|_\infty : \frac{\partial b}{\partial \bar{z}} = G\right\} = \sup\left\{\left|\frac{1}{2\pi}\int_0^{2\pi} Fk \, d\theta\right| : k \in H_0^1, \|k\|_1 \leq 1\right\},$$

where $F(z)$ is defined by (1.2). Since $G \in C^1$, F is C^1 on D, and since G is bounded, F is continuous on $\bar{D}$. We may suppose $k(z) \in H_0^1$ is smooth across ∂D. Then by Green's theorem

$$\frac{1}{2\pi}\int_0^{2\pi} F(e^{i\theta})k(e^{i\theta})d\theta = \frac{1}{2\pi}\iint_D \Delta(F(z)k(z)) \log\frac{1}{|z|}dx \, dy$$

$$= \frac{2}{\pi}\iint_D k(z)\frac{\partial G}{\partial z} \log\frac{1}{|z|}dx \, dy$$

$$+ \frac{2}{\pi}\iint_D k'(z)G(z) \log\frac{1}{|z|}dx \, dy$$

$$= I_1 + I_2,$$

because $\Delta k = 0$, $\Delta F = 4(\partial/\partial z)\partial F/\partial \bar{z} = 4 \, \partial G/\partial z$, and $\nabla F \cdot \nabla k = F_x k_x + F_y k_y = 4(\partial F/\partial \bar{z})\partial k/\partial z = 4G(z)k'(z)$. By (1.6) and the theorem on Carleson measures,

$$|I_1| \leq C_2 B_2 \|k\|_1 \leq C_2 B_2.$$

To estimate I_2 we write $k = (k_1 + k_2)/2$ where $k_j \in H^1$ is zero free and $\|k_j\|_1 \leq 2$. (See the proof of Theorem VI.4.4.) Then $k_j(z) = g_j^2(z)$, $g_j \in$

H^2, $\|g_j\|_2^2 \le 2$, and

$$\left| \frac{2}{\pi} \iint_D k'_j(z) G(z) \log \frac{1}{|z|} dx\, dy \right| = \left| \frac{4}{\pi} \iint_D g_j(z) g'_j(z) G(z) \log \frac{1}{|z|} dx\, dy \right|$$

$$\le \left(\frac{4}{\pi} \iint_D |g'_j(z)|^2 \log \frac{1}{|z|} dx\, dy \right)^{1/2}$$

$$\times \left(\frac{4}{\pi} \iint_D |g_j(z)|^2 |G(z)|^2 \log \frac{1}{|z|} dx\, dy \right)^{1/2}.$$

By (1.5) the second factor has bound $(CB_1\|g_j\|_2^2)^{1/2} \le C\sqrt{B_1}\|g_j\|_2$, and by the Littlewood–Paley identity (Chapter VI, Lemma 3.1), the first factor is

$$\left(\frac{2}{\pi} \iint_D |\nabla g_j(z)|^2 \log \frac{1}{|z|} dx\, dy \right)^{1/2} = \left(\frac{1}{\pi} \int |g_j(e^{i\theta}) - g_j(0)|^2 \, d\theta \right)^{1/2}$$

$$\le \sqrt{2}\|g_j\|_2.$$

Consequently $I_2 \le C_1\sqrt{B_1}$ and Theorem 1.2 is proved. $\square$

Although Theorem 1.2 is less straightforward than Theorem 1.1, we shall see that hypothesis (1.5) and (1.6) are sometimes easier to verify than (1.4). On the other hand Theorem 1.1 is more powerful because (1.4) depends only on $|G|$ whereas (1.6) may hold for $G(z)$ and not $\bar{G}(z)$.

2. The Corona Theorem

The unit disc D is homeomorphically embedded in the maximal ideal space $\mathfrak{M}$ of H^∞. Carleson's famous corona theorem asserts that D is dense in $\mathfrak{M}$. In other words, the "corona" $\mathfrak{M} \setminus \bar{D}$ is the empty set. Because of the topology of $\mathfrak{M}$, the theorem can be formulated in this way: *If $f_1 f_2, \ldots, f_n$ are functions in H^∞ such that*

(2.1) $$\|f_j\|_\infty \le 1$$

and

(2.2) $$\max_j |f_j(z)| \ge \delta > 0, \quad z \in D,$$

then $f_1, f_2, \ldots, f_n$ lie in no maximal ideal of H^∞. This means that the ideal generated by $\{f_1, f_2, \ldots, f_n\}$ contains the constant function 1, and there exist

$g_1, g_2, \ldots, g_n$ in H^∞ such that

$$(2.3) \qquad\qquad\qquad f_1 g_1 + \cdots + f_n g_n = 1.$$

Something formally stronger is true. There exist solutions $g_1, g_2, \ldots, g_n$ of (2.3) that have bounds depending only on the number n of functions and on the constant δ in (2.2). This ostensibly stronger statement is actually equivalent to the corona theorem itself (see Exercise 2 below).

Theorem 2.1.   *There is a constant $C(n, \delta)$ such that if $f_1, f_2, \ldots, f_n$ are H^∞ functions satisfying (2.1) and (2.2), then there are H^∞ functions $g_1, g_2, \ldots, g_n$ such that (2.3) holds and such that*

$$(2.4) \qquad\qquad\qquad \|g_j\|_\infty \le C(n, \delta), \quad 1 \le j \le n.$$

We refer to $g_1, g_2, \ldots, g_n$ as *corona solutions* and to $f_1, f_2, \ldots, f_n$ as *corona data.* By normal families it is enough to find solutions satisfying (2.4) when $f_1, f_2, \ldots, f_n$ are analytic on a neighborhood of the closed disc.

Let us now reduce the corona theorem to the problem of solving certain inhomogeneous Cauchy–Riemann equations. Assume $f_1, f_2, \ldots, f_n$ are corona data; that is, assume (2.1) and (2.2). Furthermore, assume that each $f_j(z)$ is analytic on some neighborhood of $\bar{D}$, $j = 1, 2, \ldots, n$. Choose functions $\varphi_1, \varphi_2, \ldots, \varphi_n$ of class C^1 on $\bar{D}$ such that

$$f_1 \varphi_1 + \cdots + f_n \varphi_n = 1, \quad z \in D,$$

and such that

$$(2.5) \qquad\qquad\qquad |\varphi_j(z)| \le C_1(n, \delta), \quad j = 1, 2, \ldots, n.$$

These can be easily accomplished using (2.1) and (2.2). For example, take

$$(2.6) \qquad\qquad \varphi_j(z) = \bar{f}_j(z) / \sum |f_k(z)|^2, \quad j = 1, 2, \ldots, n.$$

The difficulty, of course, is that $\varphi_j(z)$ may not be analytic on D. To rectify that, we write

$$(2.7) \qquad\qquad g_j(z) = \varphi_j(z) + \sum_{k=1}^{n} a_{j,k}(z) f_k(z),$$

with the functions $a_{j,k}(z)$ to be determined. We require that

$$(2.8) \qquad\qquad\qquad a_{j,k}(z) = -a_{k,j}(z),$$

which implies

$$f_1 g_1 + f_2 g_2 + \cdots + f_n g_n = 1 \quad \text{on} \quad \bar{D}.$$

The alternating condition (2.8) will hold if

$$a_{j,k}(z) = b_{j,k}(z) - b_{k,j}(z).$$

If we also require that

(2.9) $$\frac{\partial b_{j,k}}{\partial \bar{z}} = \varphi_j \frac{\partial \varphi_k}{\partial \bar{z}} = G_{j,k}(z), \quad |z| < 1,$$

then we get

$$\frac{\partial g_j}{\partial \bar{z}} = \frac{\partial \varphi_j}{\partial \bar{z}} + \sum_{k=1}^{n} f_k \left(\varphi_j \frac{\partial \varphi_k}{\partial \bar{z}} - \varphi_k \frac{\partial \varphi_j}{\partial \bar{z}} \right)$$

$$= \frac{\partial \varphi_j}{\partial \bar{z}} + \varphi_j \frac{\partial}{\partial \bar{z}} \sum_{k=1}^{n} f_k \varphi_k - \frac{\partial \varphi_j}{\partial \bar{z}} \sum_{k=1}^{n} f_k \varphi_k$$

$$= 0, \quad |z| < 1, \qquad \bullet$$

because $\partial f_k / \partial \bar{z} = 0$, $\sum f_k \varphi_k = 1$, and $(\partial / \partial \bar{z}) \sum f_k \varphi_k = 0$. Therefore the functions $g_j(z)$ defined by (2.6) and (2.8) are analytic solutions of (2.3). (See the appendix to this chapter for a more systematic explanation of the passage from φ_j to g_j.)

But we also need the bounds $\|g_j\|_\infty \le C(n, \delta)$ (not just to obtain Theorem 2.1 but also to be able to invoke normal families in proving the corona theorem itself). Since each φ_j is bounded, a look at (2.7) shows we only have to estimate $|a_{j,k}|$, or better yet, $|b_{j,k}|$. Thus the proof of Theorem 2.1 has been reduced to the problem of finding solutions $b_{j,k}(z)$ of (2.9) that obey the estimate

(2.10) $$|b_{j,k}(z)| \le C_2(n, \delta), \quad |z| < 1.$$

We are going to solve this problem four different ways. In each case it is crucial that the smooth solutions φ_j be chosen adroitly. The first solution will be given momentarily, the others occur in Sections 5–7.

Proof of Theorem 2.1 (Wolff). Set

$$\varphi_j(z) = \bar{f}_j(z) \Big/ \sum_{l=1}^{n} |f_l(z)|^2, \quad j = 1, 2, \ldots, n.$$

By (2.2) the denominator is bounded below, so that φ_j is C^2 on $\bar{D}$, $|\varphi_j(z)| \le \delta^{-2}$, and $f_1 \varphi_1 + f_2 \varphi_2 + \cdots + f_n \varphi_n = 1$ on $\bar{D}$. By the above discussion, Theorem 2.1 will be proved when we show the equations

$$\frac{\partial b_{j,k}}{\partial \bar{z}} = \varphi_j \frac{\partial \varphi_k}{\partial \bar{z}} = G_{j,k}(z), \quad |z| < 1, \quad 1 \le j, k \le n,$$

have solutions satisfying (2.10).

We use Theorem 1.2. It is clear that $G_{j,k}$ is bounded and C^1 on D. Since $|\varphi_j| \le \delta^{-1}$, we have

$$|G_{j,k}|^2 \log \frac{1}{|z|} \le \delta^{-2} \left| \frac{\partial \varphi_k}{\partial \bar{z}} \right| \log \frac{1}{|z|},$$

while since $\partial \bar{f}_l / \partial \bar{z} = \bar{f}_l'$, $\partial f_l / \partial \bar{z} = 0$,

$$\frac{\partial \varphi_k}{\partial \bar{z}} = \frac{\bar{f}_k'}{\sum |f_l|^2} - \frac{\bar{f}_k \sum f_l \bar{f}_l'}{(\sum |f_l|^2)^2} = \frac{\sum f_l (\bar{f}_l \bar{f}_k' - \bar{f}_k \bar{f}_l')}{(\sum |f_l|^2)^2}.$$

Thus by (2.2),

$$\left| \frac{\partial \varphi_k}{\partial \bar{z}} \right|^2 \le \frac{2(\sum |f_l|^2)^2 \sum |f_l'|^2}{(\sum |f_l|^2)^4} \le 2\delta^{-4} \sum |f_l'|^2$$

and

$$|G_{j,k}|^2 \le 2\delta^{-6} \sum |f_l'|^2.$$

By Theorem 3.4 of Chapter VI, $d\lambda_l = |f_l'|^2 \log(1/|z|) dx\, dy$ is a Carleson measure with bounded constant $N(\lambda_l) \le C \|f_l\|_\infty^2$. Hence

$$|G_{j,k}|^2 \log \frac{1}{|z|} dx\, dy$$

is a Carleson measure and (1.5) holds with $B_1 \le Cn\delta^{-6}$.

Also, because $\partial \bar{f}_l / \partial z = \overline{(\partial f_l / \partial \bar{z})} = 0$, we have

$$\frac{\partial G_{j,k}}{\partial z} = \frac{\partial \varphi_j}{\partial z} \frac{\partial \varphi_k}{\partial \bar{z}} + \varphi_j \frac{\partial^2 \varphi_k}{\partial z \partial \bar{z}}$$

$$= \left(\frac{-\bar{f}_j \sum \bar{f}_l f_l'}{(\sum |f_l|^2)^2} \right) \left(\frac{\sum f_l (\bar{f}_l \bar{f}_k' - \bar{f}_k \bar{f}_l')}{(\sum |f_l|^2)^2} \right)$$

$$+ \frac{\bar{f}_j}{\sum |f_l|^2} \left(\frac{\sum f_l' (\bar{f}_l \bar{f}_k' - \bar{f}_k \bar{f}_l')}{(\sum |f_l|^2)^2} - \frac{2(\sum f_l' f_l) \sum f_l (\bar{f}_l \bar{f}_k' - \bar{f}_k \bar{f}_l')}{(\sum |f_l|^2)^3} \right).$$

All terms look roughly the same and we have

$$\left| \frac{\partial G_{j,k}}{\partial z} \right| \le C \frac{\sum_{p,q} |f_p'| |f_q'|}{(\sum |f_l|^2)^2} \le Cn\delta^{-4} \sum |f_l'|^2.$$

Again using Theorem 3.4 of Chapter VI, we obtain (1.6) with $B_2 \le Cn\delta^{-4}$. By Theorem 1.2, we have (2.10) with $C_2(n, \delta) \le Cn^{1/2}\delta^{-3} + Cn\delta^{-4}$. That gives (2.4) with

$$C(n, \delta) \le C(n^{3/2}\delta^{-3} + n^2\delta^{-4}). \quad \square$$

Repeating the proof, but with Theorem 3.4 of Chapter VI replaced by the following lemma, we can get the sharper estimate

$$C(n, \delta) \le C(n^{3/2}\delta^{-2} + n^2\delta^{-3})$$

for the constant in (2.4). The lemma will also be used for Theorem 2.3 below.

Lemma 2.2. *If $f \in H^2$ and if $f(e^{i\theta}) \in$ BMO, then*

$$\frac{|f'(z)|^2}{|f(z)|} \log \frac{1}{|z|} dx \, dy$$

is a Carleson measure on D with Carleson norm at most $K\|f\|_*$, where K is some absolute constant.

Proof. By an approximation we can suppose $f(z)$ is analytic on $\bar{D}$. Then $f(z)$ has finitely many nonzero zeros $z_1, z_2, \ldots, z_N$ in D. When $f(z) \neq 0$, a calculation yields

$$\Delta(|f(z)|) = |f'(z)|^2 / |f(z)|.$$

For small $\varepsilon > 0$ let Ω_ε be the domain

$$\Omega_\varepsilon = D \backslash \bigcup_{j=0}^{N} \Delta_j,$$

where $\Delta_0 = \{|z| \leq \varepsilon\}, \Delta_j = \{|z - z_j| \leq \varepsilon\}, j = 1, 2, \ldots, N$. Then by Green's theorem (see Section 3 of Chapter VI),

$$\iint_{\Omega_\varepsilon} \frac{|f'(z)|^2}{|f(z)|} \log \frac{1}{|z|} dx \, dy = \int |f(e^{i\theta})| d\theta$$

$$- \sum_{j=0}^{N} \iint_{\partial \Delta_j} \left(\left(\frac{\partial}{\partial n} |f| \right) \log \frac{1}{|z|} - |f| \frac{\partial}{\partial n} \log \frac{1}{|z|} \right) ds,$$

where $\partial / \partial n$ is the normal derivative outward from Δ_j. Let ε tend to zero. For $j > 0$ the boundary integrand remains bounded and the arc length tends to zero. For $j = 0$ we get

$$- \lim_{\varepsilon \to 0} \int_0^{2\pi} |f(\varepsilon e^{i\theta})| d\theta = -2\pi |f(0)|,$$

because the other integral over $\partial \Delta_0$ has limit zero. At $z_j, |f'(z)|^2 / |f(z)|$ has singularity at worst $O(|z - z_j|^{-1})$, which is area integrable. Therefore we obtain

$$(2.11) \qquad \iint_D \frac{|f'(z)|^2}{|f(z)|} \log \frac{1}{|z|} dx \, dy \leq \int |f(e^{i\theta})| d\theta - 2\pi |f(0)|$$

$$\leq \int |f(e^{i\theta})| - f(0)| d\theta.$$

Now the lemma is nothing but the conformally invariant formulation of (2.11). Let

$$S = \{re^{i\theta} : 1 - h \leq r < 1, |\theta - \theta_0| < h\}.$$

By (2.11) we can suppose $h < \frac{1}{2}$. Set $z_0 = (1 - h)e^{i\theta_0}$ and $w = (z - z_0)/(1 - \bar{z}_0 z)$. For $z \in S$ we have

$$\frac{|1 - \bar{z}_0 z|^2}{1 - |z_0|^2} \le ch,$$

and by (1.5) of Chapter I

$$\log \frac{1}{|z|} \le c(1 - |z|^2) = c\frac{(1 - |w|^2)|1 - \bar{z}_0 z|^2}{1 - |z_0|^2}$$

$$\le ch(1 - |w|^2) \le ch \log \frac{1}{|w|}.$$

Since $|f'(z)|^2 dx\, dy = |g'(w)|^2 du\, dv$, $g(w) = f(z)$, $w = u + iv$, we obtain

$$\iint_S \frac{|f'(z)|^2}{|f(z)|} \log \frac{1}{|z|} dx\, dy \le ch \iint_D \frac{|g'(w)|^2}{|g(w)|} \log \frac{1}{|w|} du\, dv$$

$$\le 2\pi ch \int |g - g(0)| d\theta \le C\|f\|_* h,$$

and the lemma is proved. $\square$

Now let $f_1, f_2, \ldots, f_n$ be H^∞ functions, and suppose $g \in H^\infty$ satisfies

(2.12) $$|g(z)| \le |f_1(z)| + \cdots + |f_n(z)|.$$

In light of the corona theorem it is natural to ask if (2.12) implies that $g \in J(f_1, f_2, \ldots, f_n)$, the ideal generated by $\{f_1, f_2, \ldots, f_n\}$. In other words, does it follow that

$$g = g_1 f_1 + \cdots + g_n f_n$$

with $g_j \in H^\infty$? Unfortunately, the answer is no (see Exercise 3). However, T. Wolff has proved that

$$g^3 \in J(f_1, f_2, \ldots, f_n).$$

At this writing the question for g^2 remains unresolved (see Exercise 4, but also page vii).

Theorem 2.3. *Suppose $f_1, f_2, \ldots, f_n$ and g are H^∞ functions for which* (2.12) *holds. Then there are $g_1, g_2, \ldots, g_n$ in H^∞ such that*

(2.13) $$g^3 = g_1 f_1 + \cdots + g_n f_n.$$

Proof. As in the proof of Theorem 2.1, we convert smooth solutions of (2.13) into H^∞ solutions, using Theorem 1.2 to control the norms of the correcting functions. We assume $\|f_j\| \le 1$, $\|g\| \le 1$, and, by normal families, we can suppose g and $f_1, f_2, \ldots, f_n$ are analytic on $\bar{D}$ because we shall obtain a priori bounds on the solutions g_j. Set

$$\psi_j = g\bar{f}_j / \sum |f_i|^2 = g\varphi_j, \quad j = 1, 2, \ldots, n.$$

Then ψ_j is bounded, $|\psi_j| \leq 1$, and C^∞ on $\bar{D}$ (at a common zero of $f_1, f_2, \ldots, f_n$ examine the power series expansions) and

$$\psi_1 f_1 + \cdots + \psi_n f_n = g.$$

Suppose we can solve

$$(2.14) \qquad \frac{\partial b_{j,k}}{\partial \bar{z}} = g\psi_j \frac{\partial \psi_{k'}}{\partial \bar{z}} = g^3 G_{j,k}, \quad 1 \leq j, k \leq n,$$

with

$$(2.15) \qquad\qquad |b_{j,k}| \leq M.$$

Then

$$g_j = g^2 \psi_j + \sum_{k=1}^{n} (b_{j,k} - b_{k,j}) f_k$$

satisfies

$$\sum g_j f_j = g^2 \sum \psi_j f_j = g^3$$

and

$$\frac{\partial g_j}{\partial \bar{z}} = g^2 \cdot \frac{\partial \psi_j}{\partial \bar{z}} + \sum_{k=1}^{n} g f_k \left(\psi_j \frac{\partial \psi_k}{\partial \bar{z}} - \psi_k \frac{\partial \psi_j}{\partial \bar{z}} \right)$$

$$= g^2 \frac{\partial \psi_j}{\partial \bar{z}} - g\left(\sum f_k \psi_k\right) \frac{\partial \psi_j}{\partial \bar{z}} + g\psi_j \frac{\partial}{\partial \bar{z}} \sum f_k \psi_k = 0.$$

Moreover, $|g_j| \leq 1 + 2Mn$, so that $g_j \in H^\infty$.

Looking back at the proof of Theorem 2.1, we see that

$$|g^3 G_{j,k}|^2 \leq \frac{|g|^6 |\bar{f}_j|^2}{(\sum |f_i|^2)^2} \frac{|\sum f_i(\bar{f}_i f_k' - \bar{f}_k \bar{f}_i')|^2}{(\sum |f_i|^2)^4} \leq C \sum |f_i'|^2,$$

and so $|g^3 G_{j,k}|^2 \log(1/|z|) dx\, dy$ is a Carleson measure. Also,

$$\frac{\partial}{\partial z}(g^3 G_{j,k}) = 3g^2 g' G_{j,k} + g^3 \frac{\partial G_{j,k}}{\partial z}$$

and

$$|g^2 g' G_{j,k}| \le \frac{|g^2||g'||\bar{f}_j|}{\sum |f_i|^2} \left| \frac{\sum f_l(\bar{f}_i \bar{f}'_k - \bar{f}_k \bar{f}'_l)}{(\sum |f_i|^2)^2} \right|$$

$$\le \frac{C(|g'|^2 + \sum |f'_l|^2)}{(\sum |f_i|^2)^{1/2}}$$

$$\le C \frac{|g'|^2}{|g|} + C \sum \frac{|f'_l|^2}{|f_l|},$$

while

$$\left| g^3 \frac{\partial G_{j,k}}{\partial z} \right| \le \frac{C|g|^3 \sum_{p,q} |f'_p||f'_q|}{(\sum |f_i|^2)^2}$$

$$\le Cn \sum \frac{|f'_l|^2}{|f_l|},$$

by the calculations in the proof of Theorem 2.1. Hence by Lemma 2.2, $\partial/\partial z(g^3 G_{j,k})$ is a Carleson measure with constant depending only on n, and by Theorem 1.2, (2.14) has solutions satisfying (2.15). $\square$

Recently Gamelin [1981], and independently A. M. Davie, refined Wolff's proof still more, even removing the notion of Carleson measure (see Exercise 5).

There is a close connection between the corona theorem and the H^1–BMO duality. The theorem was reduced to the question of finding functions $b_{j,k}$ such that

(2.16) $$\frac{\partial b_{j,k}}{\partial \bar{z}} = \varphi_j \frac{\partial \varphi_k}{\partial \bar{z}} = G_{j,k}, \quad |z| < 1,$$

and such that

$$\|b_{j,k}\|_\infty = \sup_{|z|=1} |b_{j,k}(z)| \le C_2(n, \delta).$$

Let

(2.17) $$F_{j,k}(z) = \frac{1}{2\pi i} \iint_{|\zeta|<1} G_{j,k}(\zeta) \frac{1}{\zeta - z} d\zeta \wedge d\bar{\zeta}$$

be the solution of (2.16) given by the integral formula (1.2). Any solution of (2.16) has the form $b_{j,k} = F_{j,k} - h_{j,k}$, $h_{j,k} \in H^\infty$, and Corollary 4.6 of Chapter VI gives

$$c_1 \|F_{j,k} - i\tilde{F}_{j,k}\|_* \le \inf\{\|b_{j,k}\|_\infty : b_{j,k} \text{ satisfies (2.16)}\}$$

$$\le c_2 \|F_{j,k} - i\tilde{F}_{j,k}\|_*,$$

where $\tilde{F}_{j,k}$ denotes the conjugate function and where c_1 and c_2 are constants. By (2.17), $F_{j,k}(z)$ is analytic on $|z| > 1$, so that $\bar{F}_{j,k} \in H^\infty$, when $\bar{F}_{j,k}$ is viewed as a function only on the circle $\{|z| = 1\}$. Also, $\int F_{j,k}d\theta = 0$ by (2.17). Hence $\tilde{F}_{j,k} = iF_{j,k}$ and the minimal norm of solutions of (2.16) is comparable to 2 $\|F_{j,k}\|_*$. When φ_j is defined by (2.6), the proofs of Theorems 2.1 and 1.2 show by duality that $\|F_{j,k}\|_*$ is bounded. We will show later that φ_j can be chosen so that $|G_{j,k}|dx\,dy$ is a Carleson measure, and in that case one can directly verify that $\|F_{j,k}\|_* \leq C(n, \delta)$.

The corona problem is equivalent to the problem of finding bounds on solutions of equations like (2.16). To prove this, consider the problem of two functions f_1 and f_2, analytic on a neighborhood of $\bar{D}$, that satisfy

$$\max(|f_1(z)|, |f_2(z)|) > \delta > 0, \quad |z| \leq 1,$$
$$\|f_1\| \leq 1, \quad \|f_2\| \leq 1.$$

Let φ_1 and φ_2 be C^∞ solutions of

$$\varphi_1 f_1 + \varphi_2 f_2 = 1$$

with $\|\varphi\|_\infty \leq C_1(2, \delta)$, $j = 1, 2$. If g_1 and g_2 are analytic solutions, then since $g_1 f_1 + g_2 f_2 = 1$, we have

$$g_1 = \varphi_1 + f_2\left(\frac{\varphi_2 - g_2}{f_1}\right), \quad g_2 = \varphi_2 - f_1\left(\frac{g_1 - \varphi_1}{f_2}\right).$$

The function

$$R = \frac{\varphi_2 - g_2}{f_1} = \frac{g_1 - \varphi_1}{f_2}$$

has the bound $\|R\|_\infty \leq C_2(2, \delta)$ if and only if $\|g_j\| \leq C(2, \delta)$, $j = 1, 2$. Now

$$\frac{\partial R}{\partial \bar{z}} = \frac{1}{f_1}\frac{\partial \varphi_2}{\partial \bar{z}} = \frac{(f_1\varphi_1 + f_2\varphi_2)}{f_1}\frac{\partial \varphi_2}{\partial \bar{z}}$$

$$= \varphi_1\frac{\partial \varphi_2}{\partial \bar{z}} + \frac{\varphi_2}{f_1}\frac{\partial}{\partial \bar{z}}(1 - f_1\varphi_1)$$

$$= \varphi_1\frac{\partial \varphi_2}{\partial \bar{z}} - \varphi_2\frac{\partial \varphi_1}{\partial \bar{z}}.$$

Thus bounded analytic solutions g_1, g_2 exist if and only if

$$\frac{\partial R}{\partial \bar{z}} = \varphi\frac{\partial \varphi_2}{\partial \bar{z}} - \varphi_2\frac{\partial \varphi_1}{\partial \bar{z}}$$

can be solved with $\|R\|_\infty \leq C_2(2, \delta)$ whenever φ_1 and φ_2 are smooth bounded corona solutions.

3. Two Theorems on Minimum Modulus

Let $u(z)$ be a bounded, complex-valued, harmonic function on the upper half plane. Assume $\|u\|_\infty \leq 1$. Let Q be a square with base on $\{y = 0\}$. For $0 < \alpha < 1$, set

$$E_\alpha = \{z \in Q : |u(z)| \leq \alpha\}$$

and let

$$E_\alpha^* = \{x : x + iy \in E_\alpha \text{ for some } y > 0\}$$

be the orthogonal projection of E_α onto $\{y = 0\}$. Denote by $T(Q)$ the top half of Q. Fix $\beta > 0$ and suppose there is $z_0 \in T(Q)$ such that $|u(z_0)| > \beta$. See Figure VIII.1. Since $\|u\|_\infty \leq 1$, the Poisson integral representation shows that for $t \in Q^* = \{x : x + iy \in Q \text{ for some } y > 0\}$, $|u(t) - u(z_0)|$ can be large only when t is in a set of small measure, provided of course that $1 - \beta$ is sufficiently small. This in turn implies that $|E_\alpha^*|/|Q^*|$ is small. This reasoning is made precise in the proof of the following theorem.

Theorem 3.1. *Let $u(z)$ be harmonic in the upper half plane. Assume $\|u\|_\infty \leq 1$. If $0 < \alpha < 1$ and if $0 < \varepsilon < 1$, then there is $\beta = \beta(\alpha, \varepsilon), 0 < \beta < 1$, such that if Q is any square with base on $\{y = 0\}$, then*

$$\sup_{T(Q)} |u(z_0)| \geq \beta$$

implies

(3.1) $$|E_\alpha^*| \leq \varepsilon \ell(Q).$$

where $\ell(Q) = |Q^*|$ is the edge length of Q.

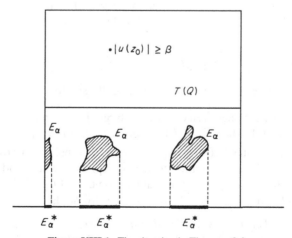

Figure VIII.1. The situation in Theorem 3.1.

Proof. The appropriate tool for this proof is the vertical maximal function

$$f^+(x) = \sup_{y>0} |f(x+iy)|,$$

where $f(z)$ is a harmonic function on the upper half plane.

We may assume $Q = \{0 \le x \le 1, 0 < y \le 1\}$. Suppose $z_0 \in T(Q)$ has $|u(z_0)| \ge \beta$. Then

$$(3.2) \qquad \left(\int |u(t) - u(z_0)| P_{z_0}(t) dt \right)^2 \le \int |u(t) - u(z_0)|^2 P_{z_0}(t) dt$$

$$= \int |u(t)|^2 P_{z_0}(t) dt - |u(z_0)|^2$$

$$\le 1 - \beta^2.$$

Let $I = [-1, 2]$ be the triple of the base of Q and let

$$f(t) = (u(t) - u(z_0))\chi_I(t).$$

Since $|P_{z_0}| \ge c_1$ on I, we have $\|f\|_1 \le c_1^{-1}(1 - \beta^2)^{1/2}$ by (3.2). Also, for $z \in Q$ but for $t \notin I$, $P_z(t)/P_{z_0}(t) \le c_2$, so that when $z \in E_\alpha \subset Q$,

$$|f(z)| = \left| \int f(t) P_z(t) dt \right|$$

$$\ge \left| \int (u(t) - u(z_0)) P_z(t) dt \right| - \int_{\mathbb{R} \setminus I} |u(t) - u(z_0)| P_z(t) dt$$

$$\ge (\beta - \alpha) - c_2 \int_{\mathbb{R} \setminus I} |u(t) - u(z_0)| P_{z_0}(t) dt$$

$$\ge (\beta - \alpha) - c_2(1 - \beta^2)^{1/2} = \gamma,$$

again by (3.2). Since α is fixed, we can make $\gamma > 0$ by taking β sufficiently close to one. Then $E_\alpha^* \subset \{t : f^+(t) > \gamma\}$ and the weak-type estimate for f^+ gives

$$|E_\alpha^*| \le Cc_1^{-1}(1 - \beta^2)^{1/2}/\gamma.$$

We conclude that $|E_\alpha^*| < \varepsilon$ if $1 - \beta$ is sufficiently small. $\square$

In Theorem 3.1 it is necessary that β be large. For example, if $u(t) = 1 - \chi_I(t)$, with $I = [0, 1]$, then $E_\alpha^* = I$ for every $\alpha > 0$, while $\sup_{T(Q)} |u(z_0)| > \frac{1}{2}$ when Q is the unit square. Thus we must have $\beta > \frac{1}{2}$ no matter how small we take α. However, if the harmonic function is replaced by a bounded analytic function $f(z)$, then by exploiting the subharmonicity of $\log |f(z)|$, we can fix any $\beta > 0$ and find $\alpha = \alpha(\beta) > 0$ so that (3.1) still holds.

Theorem 3.2. *Let $f(z)$ be a bounded analytic function on the upper half plane. Assume $\|f\|_\infty \le 1$. For $0 < \beta < 1$ and for $0 < \varepsilon < 1$, there exists*

$\alpha = \alpha(\beta, \varepsilon), 0 < \alpha < 1$, *such that for any square Q with base on* $\{y = 0\}$

$$\sup_{T(Q)} |f(z)| \geq \beta$$

implies

$$|E_\alpha^*| < \varepsilon \ell(Q),$$

where E_α^* is the vertical projection onto $\{y = 0\}$ of

$$E_\alpha = \{z \in Q : |f(z)| \leq \alpha\}.$$

Proof. We again suppose that Q is the unit square $\{0 \leq x \leq 1, 0 < y \leq 1\}$. Write $f = Bg$, where B is a Blaschke product with zeros $\{z_j\}$, and where $g(z)$ has no zeros, $\|g\|_\infty \leq 1$. Then

$$E_\alpha \subset F_\alpha \cup G_\alpha,$$

where

$$F_\alpha = \{z \in Q : |B(z)| < \sqrt{\alpha}\} \quad \text{and} \quad G_\alpha = \{z \in Q : |g(z)| < \sqrt{\alpha}\}.$$

Clearly, $E_\alpha^* \subset F_\alpha^* \cup G_\alpha^*$. We estimate $|F_\alpha^*|$ and $|G_\alpha^*|$ separately.

To bound $|G_\alpha^*|$ we use Theorem 3.1 on the function $u(z) = g(z)^p$, where $p > 0$ will be determined in a moment. Fix $\alpha_1 > 0$ and take $\beta_1 = \beta_1'(\alpha_1, \varepsilon/3)$ so that Theorem 3.1 holds. Let $p > 0$ satisfy $\beta^p = \beta_1$. Applying Theorem 3.1 to g^p, we obtain $|G_\alpha^*| < \varepsilon/3$, provided that $\alpha^{p/2} \leq \alpha_1$.

The estimate of $|F_\alpha^*|$ is based on the fact that $|B(z_0)| > \beta$ for some $z_0 = x_0 + iy_0 \in T(Q)$. By Lemma 1.2 of Chapter VII, this means

$$(3.3) \qquad \sum \frac{4y_0 y_j}{|z_0 - \bar{z}_j|^2} \leq 2 \log \frac{1}{\beta}.$$

Let $S = \{z_j : \text{dist}(z_j, Q) < 1\}$. Then $|z_0 - \bar{z}_j|^2/4y_0 \leq 7$ when $z_j \in S$, and hence

$$(3.4) \qquad \sum_{z_j \in S} y_j \leq 14 \log 1/\beta.$$

We excise the discs

$$\Delta_j = \{z : |z - z_j| < (\varepsilon/6)(14 \log 1/\beta)^{-1} y_j\}.$$

Taking ε small enough, we can assume $\Delta_j \cap F_\alpha = \varnothing$ when $z_j \notin S$, so that $F_\alpha \backslash \bigcup \Delta_j = F_\alpha \backslash \bigcup_S \Delta_j$. By (3.4) we have

$$\sum_{z_j \in S} |\Delta_j^*| < \varepsilon/3,$$

and so we only have to estimate the size of the projection of $F_\alpha \setminus \bigcup_S \Delta_j$. But when $z \in F_\alpha \setminus \bigcup \Delta_j$, Lemma 1.2 of Chapter VII also gives

$$\log \frac{1}{\alpha} \leq \log \frac{1}{|B(z)|^2} \leq c(\varepsilon, \beta) \sum \frac{4yy_j}{|z - \bar{z}_j|^2}.$$

In this sum the main contribution comes from the $z_j \in S$. Indeed, when $z_0 \in T(Q)$ and $z_j \notin S$ (so that $\text{dist}(z_j, Q) > 1$),

$$\sup_{z \in Q} \frac{yy_j}{|z - \bar{z}_j|^2} \leq c_1 \frac{y_0 y_j}{|z_0 - \bar{z}_j|^2},$$

and hence by (3.3)

$$\sup_{z \in Q} \sum_{z_j \notin S} \frac{4yy_j}{|z - \bar{z}_j|^2} \leq c_1 \sum \frac{4y_0 y_j}{|z_0 - \bar{z}_j|^2} \leq 2c_1 \log \frac{1}{\beta}.$$

Taking $\alpha \leq \alpha_0(\beta, \varepsilon)$, we conclude that

$$(3.5) \qquad \frac{2}{c(\varepsilon, \beta)} \log \frac{1}{\alpha} \leq \sum_{z_j \in S} \frac{4yy_j}{|z - \bar{z}_j|^2}$$

holds when $z \in F_\alpha \setminus \bigcup \Delta_j$.

Using (3.5) we can estimate $|(F_\alpha \setminus \bigcup \Delta_j)^*|$ in terms of a maximal function in the same way as before. Consider the positive discrete measure

$$\mu = \sum_{z_j \in S} 4y_j \delta_{z_j}.$$

Then (3.5) gives

$$(3.6) \qquad \int \frac{y}{|z - \bar{\zeta}|^2} d\mu(\zeta) \geq \frac{2}{c(\varepsilon, \beta)} \log \frac{1}{\alpha}$$

for $z \in F_\alpha \setminus \bigcup \Delta_j$, while on the other hand, (3.4) gives

$$\int d\mu \leq 56 \log \frac{1}{\beta}.$$

Project the mass μ vertically onto $\{y = 0\}$. This increases the integral in (3.6). We obtain a positive measure σ on $\mathbb{R}$ such that $\int d\sigma \leq 56 \log 1/\beta$ and such that

$$\int P_z(t) d\sigma(t) \geq \frac{2}{\pi c(\varepsilon, \beta)} \log \frac{1}{\alpha}$$

when $z \in F_\alpha \setminus \bigcup \Delta_j$. Consequently,

$$(F_\alpha \setminus \bigcup \Delta_j)^* \subset \{x : M(d\sigma) > (2/\pi c(\varepsilon, \beta)) \log 1/\alpha\},$$

where $M(d\sigma)$ denotes the Hardy–Littlewood maximal function of σ. The weak-type estimate for $M(d\sigma)$ now yields

$$\left|\left(F_\alpha \backslash \bigcup \Delta_j\right)^*\right| \le \frac{C\pi c(\varepsilon, \beta)}{2} \frac{\log 1/\beta}{\log 1/\alpha} < \frac{\varepsilon}{3},$$

provided α is sufficiently small.

We conclude that

$$|E_\alpha^*| \le |G_\alpha^*| + \sum_j |\Delta_j^*| + \left|\left(F_\alpha \backslash \bigcup \Delta_j\right)^*\right| \le 3\frac{\varepsilon}{3} = \varepsilon$$

if α is small enough. □

4. Interpolating Blaschke Products

A Blaschke product is called an *interpolating Blaschke product* if it has distinct zeros and if these zeros form an interpolating sequence. We do not know if the set of interpolating Blaschke products span a dense subspace of H^∞, as does the set of all Blaschke products (see page vii). However, the interpolating Blaschke products do separate the points of the upper half plane in a very strong way. The precise result is this.

Theorem 4.1.   *Let $u(z)$ be a bounded harmonic function on the upper half plane such that*

$$|u(t)| = 1 \quad almost\ everywhere\ on\ \mathbb{R}.$$

Let $\delta > 0$ and let $0 < \alpha < 1$. Then there exist $\beta = \beta(\alpha), 0 < \beta < 1$, and an interpolating Blaschke product $B(z)$ such that

(4.1) $|B(z)| \le \delta \quad if \quad |u(z)| \le \alpha$

and

(4.2) $|u(z)| \le \beta \quad if \quad B(z) = 0.$

Moreover if $B(z)$ has zeros $\{z_k\}$, then

(4.3) $$\delta(B) = \inf_{B(z_n)=0} \prod_{k, k\neq n} \left| \frac{z_n - z_k}{z_n - \bar{z}_k} \right|$$

$$\ge \delta_0(\alpha, \beta, \delta) > 0.$$

The reader may find Theorem 4.1 acutely specialized. We have put the theorem here for two reasons. First it will be an essential step in the description of the closed subalgebras of L^∞ containing H^∞ given in the next chapter. The second reason is pedagogical. Its proof using Theorem 3.1 above and generations like those introduced in the last chapter is an easier version of the construction originally used to prove the corona theorem.

Before proving Theorem 4.1 we use it to derive Ziskind's refinement of Newman's characterization of the Šilov boundary of H^∞. (See Theorem 2.2 of Chapter V.)

Theorem 4.2. *Let m be a complex homomorphism of H^∞. Then m is in the Šilov boundary of H^∞ if and only if $|m(B)| = 1$ for every interpolating Blaschke product $B(z)$.*

Proof. If m is in the Šilov boundary, then by Newman's theorem $|m(B)| = 1$ for every Blaschke product $B(z)$. What requires proof is the reverse implication.

If m is not in the Šilov boundary, then by Newman's theorem there is a Blaschke product $B_0(z)$ such that $m(B_0) = 0$. Using Theorem 4.1 with $u(z) = B_0(z)$ and with $\alpha = \delta = \frac{1}{2}$, we obtain an interpolating Blaschke product $B(z)$ such that $|B(z)| \le \frac{1}{2}$ if $|B_0(z)| \le \frac{1}{2}$. By the corona theorem there exists a net (z_j) in the upper half plane that converges to m in the topology of $\mathfrak{M}$

$$\lim_j f(z_j) = m(f), \quad f \in H^\infty.$$

Because $m(B_0) = 0$, we have $|B_0(z_j)| \le \frac{1}{2}$ when the index j is sufficiently large. Hence $|B(z_j)| \le \frac{1}{2}$ for large j and $|m(B)| = \lim_j |B(z_j)| \le \frac{1}{2}$. $\square$

For some points $m \in \mathfrak{M}$ not in the Šilov boundary it is not possible to find an interpolating Blaschke product $B(z)$ such that $m(B) = 0$. This is one of the mysterious things about the maximal ideal space we shall take up later (see Exercise 2(c) of Chapter X).

Proof of Theorem 4.1. Let Q be any closed square with base on $\{y = 0\}$ and, as before, let $T(Q)$ denote the top half of Q. A simple comparison of Poisson kernels shows that there is $\alpha' = \alpha'(\alpha) < 1$ such that whenever $u(z)$ is harmonic on the upper half plane and $|u(z)| \le 1$,

$$(4.4) \qquad \inf_{T|(Q)} |u(z)| < \alpha \quad \Rightarrow \quad \sup_{T(Q)} |u(z)| < \alpha'.$$

This constant $\alpha'(\alpha)$ does not depend on the function $u(z)$, or on the square Q, because (4.4) is conformally invariant. Set

$$\beta = \beta(\alpha', \frac{1}{2}),$$

so that the conclusion of Theorem 3.1 holds with α' and with $\varepsilon = \frac{1}{2}$.

For $n = 1, 2, \ldots$, we form the 2^n closed squares Q_n contained in Q that have side $\ell(Q_n) = 2^{-n}\ell(Q)$ and base on $\{y = 0\}$. The projections Q_n^* are chosen to be a partition of Q^*. For special squares Q we have to single out certain of the subsquares Q_n using a stopping time argument. There are two cases.

Case I:
$$\sup_{T(Q)} |u(z)| \geq \beta.$$

Define the first generation $G_1(Q)$ as the set of those $Q_j \subset Q$ such that

$$\sup_{T(Q_j)} |u(z)| < \alpha'$$

and Q_j is maximal. Call these red squares. See Figure VIII.2. The squares in $G_1(Q)$ have pairwise disjoint interiors. By Theorem 3.1 and our choice of β,

(4.5)
$$\sum_{G_1(Q)} \ell(Q_j) \leq \frac{1}{2}\ell(Q).$$

By (4.4),

$$\{z \in Q : |u(z)| < \alpha\} \subset \bigcup_{G_1(Q)} Q_j.$$

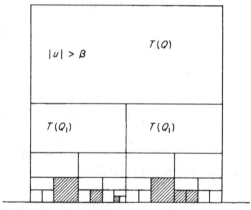

Figure VIII.2. Case I: The six shaded squares form $G_1(Q)$. They are red squares.

Case II:
$$\sup_{T(Q)} |u(z)| < \beta.$$

In this case the first generation $G_1(Q)$ consists of those $Q_j \subset Q$ such that

$$\sup_{T(Q_j)} |u(z)| \geq \beta$$

and Q_j is maximal. Call these blue squares. See Figure VIII.3. The squares in $G_1(Q)$ have pairwise disjoint interiors. Let

$$R(Q) = (\text{int } Q) \setminus \bigcup_{G_1(Q)} Q_j.$$

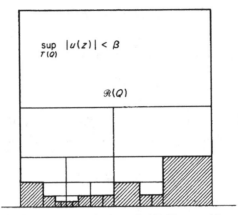

Figure VIII.3. Case II: The shaded squares form $G_1(Q)$. They are blue squares. In the region R (Q) above the blue squares, $|u(z)| < \beta$.

Then R (Q) has rectifiable boundary and

$$(4.6) \qquad\qquad \text{length}(\partial \text{R} \,(Q)) \le 6\ell(Q)$$

On R (Q) we have $|u(z)| < \beta$. Since $u(z)$ has nontangential limits of absolute value 1 almost everywhere, and since $\beta < 1$, we also have

$$(4.7) \qquad\qquad |\partial \text{R} \,(Q)| \cap \{y = 0\}| = 0.$$

Begin with the unit square Q^0. Apply Case I or II to Q^0, obtaining the first generation $G_1 = \{Q_1^1, Q_2^1, \dots\}$. To each $Q_j^1 \in G_1$ we apply Case I or II and get a new family $G_1(Q_j^1)$ of generation squares. Define the second generation to be $G_2 = \bigcup\{G_1(Q_j^1) : Q_j^1 \in G_1\} = \{Q_1^2, Q_2^2, \dots\}$. Repeating the process with G_2 and continuing inductively, we obtain later generations $G_p = \{Q_1^p, Q_2^p, \dots\}$. Since $\alpha' < \beta$, we alternate between Case I and Case II as we move from one generation to the next, and we apply the same case to all squares in a given generation. In other words, each generation consists entirely of red squares or blue squares, and the next generation consists only of squares of the other color. By the construction,

$$(4.8) \quad \{z \in Q^0 : |u(z)| < \alpha\} \subset \bigcup_{\text{Case II}} \text{R} \,(Q_j^p) \subset \{z \in Q^0 : |u(z)| < \beta\}.$$

Set

$$\Gamma = Q^0 \cap \bigcup_{\text{Case II}} \partial \text{R} \,(Q_j^p).$$

The important results of this construction are

(a) *the arc length measure on Γ is a Carleson measure, and*
(b) *if $B(z)$ is a bounded harmonic function on $\{y > 0\}$ and if $|B(z)| \le \delta$ on Γ, then $|B(z)| \le \delta$ on $\bigcup_{\text{Case II}} \text{R} \,(Q_j^p)$.*

To prove (a) it is enough to check that length $(\Gamma \cap Q) \leq C\ell(Q)$ when $Q = Q_n$ is a square from the decomposition of Q^0, because any $Q \subset Q^0$ can be covered by two such Q_n with $\ell(Q_n) \leq 2\ell(Q)$. So consider a square $Q = Q_n$. There is a smallest index p such that Q contains squares from G_p. By (4.5) and (4.6), squares from $G_p \cup G_{p+1} \cup \cdots$ contribute no more than

$$\sum_{k=0}^{\infty} \frac{6\ell(Q)}{2^k} = 12\ell(Q)$$

to length$(\Gamma \cap Q)$. The rest of $\Gamma \cap Q$ comes from squares in G_{p-1} or G_{p-2}, but not both. Their contribution does not exceed $6\ell(Q)$. Altogether we have

$$\text{length}(\Gamma \cap Q) \leq 18\ell(Q)$$

and (a) holds.

To prove (b) it is enough to consider one set R (Q) and to show that $|B(z)| \leq \delta$ on R (Q) if $|B(z)| \leq \delta$ on $\{y > 0\} \cap \partial R\,(Q)$. This follows from (4.7) and a Phragmén–Lindelöf argument which we now outline. Because of (4.7), there are positive harmonic functions $V_n(z)$ on the upper half plane such that

$$\lim_{z \to t} V_n(z) = +\infty, \quad t \in \mathbb{R} \cap \partial R\,(Q)$$

and such that

$$\lim_{n \to \infty} V_n(z) = 0, \quad z \in \mathbb{H}$$

(see Exercise 9 of Chapter I). When $\zeta \in \partial R\,(Q)$ we then have

$$\overline{\lim_{R\,(Q) \ni z \to \zeta}} |B(z) + V_n(z)| \leq \delta + \lim_{R\,(Q) \ni z \to \zeta} V_n(z).$$

Indeed, if Im $\zeta > 0$, this inequality follows from the continuity of $B(z) + V_n(z)$, whereas if Im $\zeta = 0$ the inequality is obvious since $\lim_{z \to \zeta} V_n(z) = +\infty$. The maximum principle for subharmonic functions now gives

$$|B(z) + V_n(z)| \leq \delta + V_n(z), \quad z \in \mathrm{R}\,(Q).$$

Sending $n \to \infty$, we obtain (b).

It is now quite easy to find the interpolating Blaschke product $B(z)$. Let us first construct an interpolating Blaschke product $B_1(z)$ that has (4.2) and that satisfies (4.1) only for points in the unit square Q^0. Choose points $\{z_j\}$ in $\bigcup \partial T(Q_n)$, where Q_n ranges through all squares in the decomposition of Q^0, including Q^0 itself, so that

(4.9) $$\inf_j \rho(z_j, z) < \delta, \quad z \in \bigcup \partial T(Q_n)$$

and so that

(4.10) $$\rho(z_j, z_k) \geq \eta > 0, \quad j \neq k.$$

That can be done as follows. Along each $\partial T(Q_n)$, mark off equally spaced points, $2^{-N}\ell(Q_n)$ units apart, including the corners of $T(Q_n)$. Let $\{z_j\}$ be the union, over $\{Q_n\}$, of the sets of marked points. Then (4.10) holds, and if $N = N(\delta)$ is large enough, (4.9) also holds. We let $B_1(z)$ be that Blaschke product with zeros $\{z_j : z_j \in \Gamma\}$ (see Figure VIII.4).

By (4.10) and condition (a), the zeros of $B_1(z)$ satisfy the geometric condition (c) of Theorem 1.1 of Chapter VII. Hence these zeros form an interpolating sequence, and because the estimates depend only on α, β and δ, (4.3) holds for B_1. By (4.8), (4.9), and condition (b), we have $|B_1(z)| < \delta$ when $z \in Q^0$ and when $|u(z)| \leq \alpha$. By (4.8) the zeros of $B_1(z)$ lie in $\{|u(z)| \leq \beta\}$.

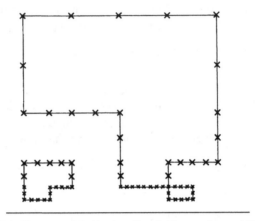

Figure VIII.4. The zeros of $B_1(z)$ when $N = 2$. Three regions R (Q) are shown.

We have found an interpolating Blaschke product B_1 that satisfies (4.2) and that satisfies (4.1) for points in Q^0. To obtain (4.1) for all points in the upper half plane, choose a second interpolating Blaschke product $B_2(z)$ with zeros $\{w_k\}$ such that B_2 satisfies (4.2) and such that $B_2(z)$ satisfies (4.1), with $\delta/2$ instead of δ, for points $z \notin Q^0$. After a conformal mapping, the construction of B_2 is the same as the construction of B_1. The product $B = B_1 B_2$ then satisfies (4.1) and (4.2). The zeros of $B = B_1 B_2$ form an interpolating sequence if

$$\inf_{k, j} \rho(w_k, z_j) > 0,$$

where $\{z_j\}$ denotes the zeros of B_1. Let $\varepsilon > 0$. Remove from B_2 any zero w_k for which

$$\inf_{j} \rho(w_k, z_j) < \varepsilon.$$

The remaining product $B = B_1 B_2$ is then an interpolating Blaschke product satisfying (4.2) and (4.3). Since the original product B_2 satisfied (4.1) for $z \notin Q^0$ with $\delta/2$ in place of δ, the new product $B = B_1 B_2$ will still have (4.1) if ε is sufficiently small. $\square$

5. Carleson's Construction

Theorem 5.1. *Let $\delta > 0$. If $f(z)$ is analytic on D and if $|f(z)| \leq 1$, then there is $\psi(z) \in C^\infty(D)$ such that*

(a) $0 \leq \psi(z) \leq 1$,
(b) $\psi(z) = 1$ if $|f(z)| \geq \delta$,
(c) $\psi(z) = 0$ if $|f(z)| < \varepsilon = \varepsilon(\delta)$, and
(d) $\iint_s |\partial\psi/\partial\bar{z}|\, dx\, dy \leq A(\delta)\ell$, for every sector

$$S = \{z = re^{i\theta} : \theta_0 < \theta < \theta_0 + \ell, 1 - \ell < r < 1\}.$$

The constants $\varepsilon(\delta) > 0$ and $A(\delta)$ depend only on δ.

Notice that Theorem 2.1 follows easily from this result. Suppose $f_1, f_2, \ldots, f_n \in H^\infty$ satisfy $\|f_j\|_\infty \leq 1$ and

$$\max_j |f_j(z)| \geq \delta > 0, \qquad z \in D.$$

For this δ and for each f_j, Theorem 5.1 gives us a function ψ_j. Set

$$\varphi_j = \psi_j / f_j \sum \psi_l.$$

Then $|\varphi_j(z)| \leq 1/\varepsilon, z \in D$, and

$$\varphi_1 f_1 + \cdots + \varphi_n f_n = 1.$$

To replace the φ_1 by H^∞ functions we boundedly solve the equations

(5.1) $$\frac{\partial b_{j,k}}{\partial \bar{z}} = \varphi_j \frac{\partial \varphi_k}{\partial \bar{z}}, \qquad 1 \leq j, k \leq n.$$

However, since

$$\frac{\partial \varphi_k}{\partial \bar{z}} = \sum \left(\psi_l \frac{\partial \psi_k}{\partial \bar{z}} - \psi_k \frac{\partial \psi_l}{\partial \bar{z}} \right) \Big/ f_k \left(\sum \psi_l \right)^2,$$

condition (d) and Theorem 1.1 ensure that (5.1) has bounded solutions.

Proof of Theorem 5.1. We do the construction in the upper half plane. In fact, we only construct ψ in the unit square $Q^0 = \{0 \leq x \leq 1, 0 < y \leq 1\}$. Simple conformal mappings and a partition of unity on D can then be used to produce ψ on the disc.

For each dyadic square $Q = \{j2^{-k} \leq x \leq (j+1)2^{-k}, 0 \leq y \leq 2^{-k}\}$ contained in Q^0, we again let $T(Q)$ denote the top half of Q, and we form the 2^n dyadic squares Q_n contained in Q, having base on $\{y = 0\}$, and having side $\ell(Q_n) = 2^{-n-k} = 2^{-n}\ell(Q)$. Thus $Q\setminus T(Q)$ is the union of the two squares Q_1.

Let $N = N(\delta)$ be a positive integer. For each dyadic square Q of side $\ell(Q)$ with base on $\{y = 0\}$, partition $T(Q)$ into 2^{2N-1} dyadic squares S_j of side

$\ell(S_j) = 2^{\sim N}\ell(Q)$. We call the S_j small squares. With respect to the hyperbolic, metric small squares S_j from different $T(Q)$ are roughly the same size. Let $\tilde{S}_j$ be the open square concentric with S_j having side $\ell(\tilde{S}_j) = 3\ell(S_j)$. By Schwarz's lemma we can choose $N = N(\delta)$ such that

$$(5.2) \qquad \sup_{z, w \in \tilde{S}_j} |f(z) - f(w)| < 6 \cdot 2^{-N} < \delta/10$$

for each S_j. Taking N larger, we can also require that whenever Q is a square with

$$\sup_{T(Q)} |f(z)| \geq \delta/2,$$

the vertical projection E^* of

$$E = \{z \in Q : |f(z)| < 2^{-N+2}\}$$

has

$$(5.3) \qquad |E^*| \leq \ell(Q)/2$$

This can be done using Theorem 3.2.

In Q^0 we will define a region R as the union of certain small squares. This region R will have the following two properties:

  (i) If

$$\inf_{S_k} |f(z)| \leq 2^{-N}$$

  and if $S_k \subset Q^0$, then $S_k \subset R$.
 (ii) If, on the other hand, $S_k \subset R$, then

$$\sup_{\tilde{S}_k} |f(z)| < \delta.$$

By (5.2) conditions (i) and (ii) are consistent. Setting $\varepsilon = \varepsilon(\delta) = 2^{-N}$, we see that ∂R separates $\{z \in Q : |f(z)| \geq \delta\}$ from $\{z \in Q : |f(z)| \leq 2^{-N} = \varepsilon\}$. Condition (5.3) will show that arc length on ∂R is a Carleson measure with constant $A_1(\delta)$. This means that except for smoothness, the function $\psi_0 = \chi_{Q_0 \setminus R}$ has the desired properties (a)–(d) on Q_0. The final function ψ is a mollification of ψ_0 with respect to the hyperbolic metric.[†] To define R we consider two cases. Let Q be any square in the decomposition of the unit square Q^0.

Case I: $\qquad\qquad \sup_{T(Q)} |f(z)| \geq \delta/2.$

[†] The smoothness of ψ is of no real importance. It is the price paid for avoiding distribution derivatives. One can work directly with ψ_0 using Cauchy's theorem instead of Green's theorem in Theorem 1.1.

Consider all $S_j \subset Q$ for which

$$\inf_{S_j} |f(z)| \leq 2^{-N}$$

and such that S_j lies below no other small square with the same property. We let $A(Q)$ be the family of such S_j. By (5.2) and (5.3), we have

(5.4)
$$\sum_{A(Q)} \ell(S_j) \leq \frac{1}{2}\ell(Q),$$

because the projections S_j^* of the squares S_j have pairwise disjoint interiors. For each $S_j \in A(Q)$ let $Q_j^{(1)}$ be the square with base S_j^*. Each $Q_j^{(1)}$ is a dyadic square with base on $\{y = 0\}$, and the interiors of the squares $Q_j^{(1)}$ are pairwise disjoint. Define the first generation

$$G_1(Q) = \{Q_j^{(1)} : S_j \in A(Q)\}.$$

Then by (5.4), we have

(5.5)
$$\sum_{G_1(Q)} \ell(Q_j^{(1)}) \leq \frac{1}{2}\ell(Q).$$

Also let $B(S_j)$ be the set of S_k such that

$$S_k^* \subset S_j^* \qquad \text{and} \qquad \ell(S_j) \leq \inf_{S_k} y \leq \inf_{S_j} y.$$

Thus $B(S_j)$ consists of S_j and all S_k below S_j except those S_k contained in $Q_j^{(1)}$. (If we think of S_j as an elevator car on the top floor, then $Q_j^{(1)}$ is the elevator car on the bottom floor, and $B(S_j)$ is a partition of the elevator shaft with the bottom floor excluded. See Figure VIII.5.) Notice that

(5.6)
$$\sum_{S_k \in B(S_j)} \ell(S_k) \leq 2^{2N}\ell(S_j)$$

because the very small S_k contained in $Q_j^{(1)}$ have been excluded. For $z \in Q$ we have $|f(z)| > 2^{-N}$ unless $z \in \bigcup Q_j^{(1)}$ or $z \in S_k \in B(S_j)$, for some $S_j \in A(Q)$. Define

$$R(Q) = \bigcup_{S_j \in A(Q)} \bigcup \left\{ S_k \in B(S_j) : \inf_{S_k} |f(z)| \leq 2^{-N} \right\}.$$

If $S_k \subset Q \setminus \bigcup Q_j^{(1)}$, and if $\inf_{S_k} |f(z)| \leq 2^{-N}$ then $S_k \subset R(Q)$. On the other hand, by (5.2) $\sup_{\tilde{S}_k} |f(z)| < \delta$ if $S_k \subset R(Q)$. Thus (i) and (ii) hold for $S_k \subset Q \setminus \bigcup Q_j^{(1)}$. By (5.4) and (5.6), $\partial R(Q)$ satisfies

(5.7)
$$\text{length}(\partial R(Q)) \leq 2^{2N+1}\ell(Q).$$

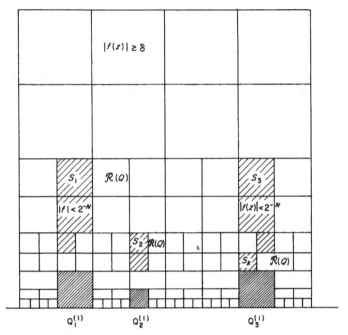

Figure VIII.5. A case I square Q when $N = 2$. The first generation $G_1(Q)$ consists of the darkly shaded squares. The squares S_k on which $\inf |f(z)| < 2^{-N}$ in R (Q) are lightly shaded.

Case II:
$$\sup_{T(Q)} |f(z)| < \delta/2.$$

In this case the first generation $G_1(Q)$ consists of those $Q_n \subset Q$ such that

$$\sup_{Q_n} |f(z)| \geq \delta/2$$

and Q_n is maximal. The squares $Q_j^{(1)}$ in $G_1(Q)$ have pairwise disjoint interiors. Let

$$R(Q) = Q \backslash \bigcup_{G_1(Q)} Q_j^{(1)}.$$

In this case the picture is like Fig. VIII.3. On R (Q) we have $|f(z)| \leq \delta/2$, and if $S_k \subset R(Q)$, then by (5.2), $\sup_{\bar{S}_k} |f(z)| < \delta$. Also, we have

(5.8) $\ell(\partial R(Q)) \leq 6\ell(Q) \leq 2^{2N+1}\ell(Q).$

Starting with the unit square Q^0, we apply either Case I or II, and form the region R (Q^0). Then apply either Case I or II to each square $Q_j^{(1)}$ in the first generation $G_1 = G_1(Q^0)$. We obtain new regions R $(Q_j^{(1)})$ and a second generation

$$G_2 = \bigcup_{G_1} G_1(Q_j^{(1)}),$$

which consists of all first generation descendants of squares from G_1. Continue this process indefinitely, obtaining succeeding generations $G_3, G_4, \ldots$. Define

$$\mathrm{R} = \mathrm{R}(Q^0) \cup \bigcup_{p=1}^{\infty} \bigcup \{ \mathrm{R}(Q_j^{(p)}) : Q_j^{(p)} \in G_p \}.$$

By the construction, (i) and (ii) hold for $S_k \subset Q^0$.

The same case need not apply to all the squares in a given generation G_p, but each first generation descendant of a Case II square is a Case I square. This means that we never use Case II two times in succession. By (5.7), (5.8), and especially (5.5), we see that arc length on $\partial \mathrm{R}$ is a Carleson measure. We have

$$\ell(Q \cap \partial \mathrm{R}) \leq 2^{2N+1} \ell(Q) + 2^{2N+1} \sum \{ \ell(Q_j^{(p)}) : Q_j^{(p)} \subset Q \},$$

and, because generation squares are nested, (5.5) gives

$$\sum \{ \ell(Q_j^{(p)}) : Q_j^{(p)} \subset Q \} \leq 2 \sum_{q=1}^{\infty} 2^{-q} \ell(Q) = 2\ell(Q).$$

Consequently we obtain

(5.9) $$\ell(Q \cap \partial \mathrm{R}) \leq 3 \cdot 2^{2N+1} \ell(Q).$$

which shows that arc length on $\partial \mathrm{R}(Q)$ is a Carleson measure with constant $A = 3 \cdot 2^{2N+1}$ depending only on δ. Because we want a smooth function $\psi(z)$ we need a slightly different formulation of (5.9). For any Q let

$$\mathrm{E}(Q) = \{ S_k : \tilde{S}_k \cap Q \cap \partial \mathrm{R} \neq \varnothing \}.$$

If $S_k \in \mathrm{E}(Q)$ and if $\tilde{\tilde{S}}_k$ is concentric with S_k but nine times as large, then

$$\ell(S_k) \leq c\ell(\tilde{\tilde{S}}_k \cap Q \cap \partial \mathrm{R}),$$

because the only contribution to $\partial \mathrm{R}$ within $\tilde{\tilde{S}}_k$ comes from squares S_j with edge length $\ell(S_j) \geq \ell(S_k)/2$. No point lies in more than nine squares $\tilde{S}_k$, so that (5.9) yields

(5.10) $$\sum_{\mathrm{E}(Q)} \ell(S_k) \leq c 2^{2N+1} \ell(Q).$$

For each small square S_j, let $\psi_j \in C^\infty(\{y > 0\})$ satisfy

$$\begin{aligned}
\psi_j &= 0 \quad \text{on} \quad S_j, \\
\psi_j &\equiv 1 \quad \text{on} \quad \{y > 0\} \backslash \widetilde{S}_j, \\
0 \leq \psi_j &\leq 1, \quad |\nabla \psi_j| \leq c/\ell(S_j).
\end{aligned}$$

Since no point lies in more than nine squares $\widetilde{S}_j$, the function

$$\psi(z) = \prod_{S_j \subset R} \psi_j(z)$$

is C^∞ on the upper half plane, and

(5.11) $$|\nabla \psi(z)| \leq c/\ell(S_k), \qquad z \in S_k.$$

Moreover $\nabla \psi(z) = 0$ except on those squares $\widetilde{S}_k$ with $\widetilde{S}_k \cap \partial R \neq \varnothing$. By (5.10) and (5.11) this means that

$$\iint_Q |\nabla_\psi| dx \, dy \leq C2^{N+1} \ell(Q)$$

for any square Q on $\{y = 0\}$. Hence $|\nabla \psi| dx \, dy$ is a Carleson measure with constant $A(\delta)$. Clearly $0 \leq \psi \leq 1$, and (a) holds. By (i) and (ii), conditions (b) and (c) hold for points in Q^0. $\square$

6. Gradients of Bounded Harmonic Functions

Much of the difficulty with the corona theorem rests in the fact that when $f \in H^\infty$, $|f'(z)| dx \, dy$ need not be a Carleson measure. See Chapter VI, Exercise 9, for an example. The theorem in this section provides a detour around that obstruction.

Theorem 6.1. *Let $u(z)$ be a bounded harmonic function on the upper half plane $\mathbb{H}$ and let $\varepsilon > 0$. There exists a C^∞ function $\varphi(z)$ on $\mathbb{H}$ such that*

(6.1) $$|\varphi(z) - u(z)| < \varepsilon$$

and such that $|\nabla \varphi| dx \, dy$ is a Carleson measure,

(6.2) $$\iint_Q |\nabla \varphi| dx \, dy \leq C\varepsilon^{-6} \|u\|_\infty^5 \ell(Q),$$

whenever $Q = \{a \leq x \leq a + \ell(Q), 0 < y < \ell(Q)\}$. The constant C in (6.2) is independent of ε and $u(z)$.

This theorem is a compromise between something true, that $y|\nabla u|^2 dx\, dy$ is a Carleson measure, and something more desirable but false, that $|\nabla u| dx\, dy$ should be a Carleson measure. Theorem 5.1 is an immediate corollary. Let $f \in H^\infty$, $\|f\|_\infty \leq 1$, and let $\delta > 0$. Take $h \in C^\infty(\mathbb{R})$ with $h(x) = 1$, $x > 3\delta/4$, $h(x) = 0$, $x < \delta/2$ and with $0 \leq h(x) \leq 1$. If φ is the C^∞ function given by Theorem 6.1 with $\varepsilon = \delta/4$, then $\psi = h \circ \varphi$ has the properties asserted in Theorem 5.1, with $\varepsilon(\delta) = \delta/4$.

The theorem holds more generally when $u(z)$ is the Poisson integral of a BMO function, and the BMO result, even with ε large, gives yet another proof of the H^1–BMO duality (see Exercise 11). The proof of duality in Chapter VI used the disc analog of the inequality

$$(6.3) \qquad \iint\limits_Q y|\nabla u|^2 \, dx\, dy \leq C\|u\|_*^2 \ell(Q),$$

which is much more accessible than (6.2). (See Chapter VI, Theorem 3.4, and Exercise 5 of that chapter.)

Examples exist for which the function $\varphi(z)$ in Theorem 6.1 cannot be a harmonic function (see Exercise 12).

Theorem 6.1 includes a quantative formulation of Fatou's theorem. Let $u(z)$ be a bounded harmonic function on $\{y > 0\}$ and let $\varepsilon > 0$. For $x \in \mathbb{R}$, let $N_\varepsilon(x)$ denote the number of times $u(x + iy)$ oscillates by ε units in the segment $0 < y < 1$. To be precise, say $N_\varepsilon(x) \geq n$ if there are

$$0 < y_0 < y_1 < y_2 < \cdots < y_n \leq 1$$

such that $|u(x + iy_j) - u(x + iy_{j+1})| \geq \varepsilon$. Fatou's theorem asserts that for each $\varepsilon > 0$, $N_\varepsilon(x) < \infty$ almost everywhere.

Corollary 6.2.   *If $\varepsilon > 0$, if $u(z)$ is harmonic on $\{y > 0\}$, and if $\|u\|_\infty \leq 1$, then*

$$\int_I N_\varepsilon(x)dx \leq C\varepsilon^{-7}$$

whenever I is an interval of unit length.

Proof.   Let $\varphi \in C^\infty$ satisfy (6.1) and (6.2) with $\varepsilon/2$ in place of ε. If $N_\varepsilon(x) \geq n$, then there are $y_0 < y_1 < y_2 < \cdots < y_n \leq 1$ such that $|\varphi(x + iy_j) - \varphi(x + iy_{j+1})| \geq \varepsilon/3$. Hence

$$\int_0^1 \left|\frac{\partial \varphi}{\partial y}(x + iy)\right| dy \geq \sum_{j=0}^{n-1} \left|\int_{y_j}^{y_{j+1}} \frac{\partial \varphi}{\partial y}(x + iy)dy\right|,$$

$$\geq n\varepsilon/3,$$

and so $\int_0^1 |\partial\varphi/\partial y| dy \geq N_\varepsilon(x)\varepsilon/3$. Then if $|I| = 1$, (6.2) gives

$$\int_I N_\varepsilon(x)dx \leq 3/\varepsilon \int_I \int_0^1 |\nabla\varphi| dy\, dx \leq c\varepsilon^{-7},$$

as desired. $\square$

This corollary is suggestive of the theorem on harmonic interpolation sequences, Theorem 4.2 of Chapter VII, because both results restrict the oscillations of a bounded harmonic function, and one can easily derive the harmonic interpolation theorem from Corollary 6.2. The connection between Theorem 6.1 and harmonic interpolations will be discussed more fully in Chapter X. Corollary 6.2 does not obviously follow from inequality (6.3). However, the analogous result for averages over dyadic intervals, Lemma 6.4, is an easy consequence of the martingale version of (6.3). See the remarks after the proof of Lemma 6.4.

The constants ε^{-7} in Corollary 6.2 and ε^{-6} in (6.2) are not sharp. Dahlberg [1980] recently obtained ε^{-1} in (6.2). His local theorem contains Theorem 6.1 and a similar result for Poisson integrals of L^p functions, $p \geq 2$. Dahlberg uses the Lusin area integral for Lipschitz domains where we shall compare $u(z)$ to a simpler martingale.

Here is the strategy of the proof of Theorem 6.1. We know that $y|\nabla u|^2 dx\, dy$ is a Carleson measure. So at points where $u(z)$ has large oscillation, that is, where $y|\nabla u| \geq \delta(\varepsilon)$, we have $|\nabla u(z)| \leq \delta^{-1}(\varepsilon)y|\nabla u|^2$. Thus the restriction of $|\nabla u(z)| dx\, dy$ to $\{z : y|\nabla u(z)| \geq \delta(\varepsilon)\}$ is already a Carleson measure, and we can take $\varphi(z) = u(z)$ on that set. We are left with the set where $u(z)$ has small oscillation. Temporarily relaxing the requirement that $\varphi \in C^\infty$, we choose a piecewise constant function $\varphi(z)$ such that $|\varphi(z) - u(z)| < \varepsilon$ at almost every point where the oscillation is small, and such that $\varphi(z)$ jumps by about ε units when z crosses the edges of certain dyadic squares. These squares are determined by a stopping time argument and they satisfy the nesting condition

$$\sum_{Q_j \subset Q} \ell(Q_j) \leq C(\varepsilon)\ell(Q).$$

As a distribution, $|\nabla\varphi|$ resembles arc length on the boundaries of the Q_j, and hence $|\nabla\varphi|$ is a Carleson measure. Thus the idea is simply to flatten out the small oscillations as much as possible.

It will be convenient to replace Poisson integrals by the averages of $u(t)$ over dyadic intervals.

Lemma 6.3. *Let $u(z) = P_y * u(x)$ be a bounded harmonic function on the upper half plane and let I be an interval on $\mathbb{R}$. Then for $0 < \delta < \frac{1}{2}$,*

$$(6.4) \qquad \left| \frac{1}{|I|} \int_I u(t)dt - \frac{1}{|I|} \int_I u(x + i\delta|I|)dx \right| \leq c\delta \log\frac{1}{\delta} \|u\|_\infty,$$

where the constant c does not depend on $u(z)$, on δ, or on I.

Proof. By a change of scale we can take $I = [0, 1]$. We can also assume $\|u\|_\infty = 1$. Then the left side of (6.4) has supremum $\|F\|_1$, where

$$F(t) = \chi_I(t) - \int_I P_\delta(x - t)dx.$$

If dist $(t, I) > \delta$, then

$$|F(t)| \leq \frac{\delta}{\pi} \int_I \frac{dx}{(x - t)^2} = \frac{\delta}{\pi} \left(\frac{1}{\text{dist}(t, I)} - \frac{1}{1 + \text{dist}(t, I)} \right).$$

Hence

$$\int_{\text{dist}(t,I)>\delta} |F(t)|dt \leq \frac{2\delta}{\pi} \int_\delta^\infty \left(\frac{1}{s} - \frac{1}{s+1} \right) = \frac{2\delta}{\pi} \log \left(1 + \frac{1}{\delta} \right).$$

Since $\|F\|_\infty \leq 1$ we also have

$$\int_{|t|<\delta} |F(t)|dt + \int_{|1-t|<\delta} |F(t)|dt \leq 4\delta.$$

What remains is the interval $J = (\delta, 1 - \delta)$. For $t \in J$,

$$-F(t) = \int_{-\infty}^0 P_\delta(x - t)dx + \int_1^\infty P_\delta(x - t)dx = G_1(t) + G_2(t)$$

and

$$\int_J |G_1(t)|dt = \int_J |G_2(t)|dt \leq \frac{\delta}{\pi} \int_\delta^{1-\delta} \int_{-\infty}^0 \frac{dx}{(x - t)^2} dt$$

$$< \frac{\delta}{\pi} \int_\delta^{1-\delta} \frac{dt}{t} \leq \frac{\delta}{\pi} \log \frac{1}{\delta}.$$

That establishes (6.4). □

Now fix a dyadic interval I and fix a positive integer N. For $k = 1, 2, \ldots$, partition I into 2^{Nk} closed dyadic intervals I_k of length $|I_k| = 2^{-Nk}|I|$. Fix $\varepsilon > 0$ and let $u(t)$ be an L^∞ function defined on I. Define the first generation $G_1 = G_1(I)$ to be the set of maximal $I_k \subset I$ for which

$$|u_{I_k} - u_I| \geq \varepsilon.$$

The intervals in $G_1(I)$ have pairwise disjoint interiors. For $I_k \in G_1(I)$ define $G_1(I_k)$ in the same way and set

$$G_2 = G_2(I) = \bigcup \{G_1(I_k) : I_k \in G_1\}.$$

Later generations $G_3, G_4, \ldots$ are defined inductively, so each $I_k \in G_{p+1}$ is contained in a unique $I_j \in G_p$ and $|u_{I_k} - u_{I_j}| \geq \varepsilon$. Lebesgue's theorem asserts that almost every point lies in a finite number of generation intervals. We need a quantitative formulation of that theorem.

Lemma 6.4. *For every $\varepsilon > 0$ and for every positive integer N,*

$$\sum_{p=1}^{\infty} \sum_{I_j \in G_p} |I_j| \leq \frac{\|u\|_{\infty}^2}{\varepsilon^2} |I|.$$

Proof. Set $G_0 = \{I\}$ and set $E_p = I \setminus \bigcup_{G_p} I_j$, $p = 0, 1, 2, \dots$. Define

$$Y_p(t) = u(t)\chi_{E_p}(t) + \sum_{G_p} u_{I_j} \chi_{I_j}(t).$$

Then $|Y_p - Y_{p-1}| \geq \varepsilon$ on $\bigcup_{G_p} I_j$, and hence

$$\sum_{p=1}^{\infty} \sum_{G_p} |I_j| \leq \frac{1}{\varepsilon^2} \int_I \sum_{p=1}^{\infty} |Y_p - Y_{p-1}|^2 \, dt.$$

When $I_k \in G_{p-1}$ we have

$$\int_{I_k} Y_p \, dt = \int_{I_k \cap E_p} u(t) \, dt + \sum_{\substack{I_j \subset I_k \\ I_j \in G_p}} \int_{I_j} u_{I_j} \, dt$$

$$= \int_{I_k} u(t) \, dt = |I_k| u_{I_k},$$

and, since $Y_{p-1} = u_{I_k}$ on I_k,

$$\int_{I_k} Y_p Y_{p-1} \, dt = |I_k| u_{I_k}^2 = \int_{I_k} Y_{p-1}^2 \, dt.$$

Similarly, because $E_p \supset E_{p-1}$ we have

$$\int_{E_{p-1}} Y_p Y_{p-1} \, dt = \int_{E_{p-1}} u^2(t) \, dt = \int_{E_{p-1}} Y_{p-1}^2 \, dt.$$

Consequently,

$$\int_I Y_p Y_{p-1} \, dt = \int_I Y_{p-1}^2 \, dt$$

and

$$\int_I |Y_p - Y_{p-1}|^2 \, dt = \int_I Y_p^2 \, dt - 2 \int_I Y_p Y_{p-1} \, dt + \int_I Y_{p-1}^2 \, dt$$

$$= \int_I Y_p^2 \, dt - \int_I Y_{p-1}^2 \, dt.$$

Hence

$$\int_I \sum_{p=1}^{\infty} |Y_p - Y_{p-1}|^2 \, dt = \lim_{p \to \infty} \int_I (|Y_p|^2 - |Y_0|^2) \, dt = \int_I u^2 \, dt - u_I^2 |I|$$

$$\leq \|u\|_\infty^2 |I|,$$

and the lemma is proved. □

Lemma 6.4 is really a theorem about martingales. One can view $\{Y_p\}$ as a martingale restricted to a sequence of stopping times. In the proof above, the dominant expression $\sum |Y_p - Y_{p-1}|^2$ is the square of the *martingale S-function*. Curiously, the S-function is the analog for martingales of the Littlewood–Paley expression $y|\nabla u|^2 \, dx \, dy$.

Proof of Theorem 6.1.   We first find a discontinuous function $\varphi_2(z)$ satisfying (6.1) almost everywhere such that the distribution $|\nabla \varphi_2|$ is a Carleson measure. This function will only be constructed in the unit square $Q_0 = \{0 \leq x \leq 1, 0 < y \leq 1\}$. A partition of unity can be used to obtain a similar function on the upper half plane. At the end we mollify the latter function into a C^∞ function satisfying (6.1) and (6.2).

Choose $\delta = 2^{-N}$ so that we have

$$\frac{\varepsilon}{8} < c\delta \log \frac{1}{\delta} < \frac{\varepsilon}{4}$$

in (6.4). For $k = 1, 2, \ldots$, consider the 2^{Nk} dyadic squares Q_k of the form $Q_k = \{j2^{-Nk} \leq x \leq (j+1)2^{-Nk}, 0 < y \leq 2^{-Nk}\}$. Set $S(Q_k) = Q_k \backslash \bigcup Q_{k+1} = Q_k \cap \{y > 2^{-N}\ell(Q_k)\}$, and let $I_k = Q_k^*$ be the vertical projection of Q_k. Write

$$a_{Q_k}(u) = \frac{1}{|I_k|} \int_{I_k} u(x + i2^{-N}|I_k|) dx$$

for the average of $u(z)$ over the bottom edge of $S(Q_k)$. By Lemma 6.3 and by the choice of N,

(6.5) $|a_{Q_k}(u) - u_{I_k}| \leq (\varepsilon/4)\|u\|_\infty$

holds for every harmonic function $u(z)$.

We say $S(Q_k)$ is a *blue rectangle* if

(6.6) $\sup_{S(Q_k)} |u(z) - u(w)| \leq \varepsilon/4.$

A rectangle $S(Q_k)$ on which (6.6) fails is called a *red rectangle*. There are two steps in the proof. In step I we approximate $u(z)$ on the blue rectangles by a piecewise constant function. In step II we correct the approximation on the red rectangles.

We assume that $\|u\|_\infty = 1$ and that $\varepsilon < 1$.

Step I: The zero generation G_0 consists of the unit square Q_0. The first generation $G_1 = G_1(Q_0)$ consists of the maximal $Q_k \subset Q_0$ for which

$$(6.7) \qquad |u_{I_k} - u_{I_0}| \geq \varepsilon/4.$$

When Q_j is in the first generation, $G_1(Q_j)$ is defined in the same manner. The second generation $G_2 = G_2(Q_0)$ is $\bigcup\{G_1(Q_j) : Q_j \in G_1\}$, and the later generations $G_3, G_4, \ldots$ are defined inductively. Except in that we are using $\varepsilon/4$ instead of ε, the generations of squares here correspond naturally to the generations of intervals I_k defined before Lemma 6.4. For each generation square Q_j we form the region

$$R(Q_j) = Q_j \setminus \bigcup_{G_1(Q_j)} Q_k.$$

Then by (6.7), $R(Q_j)$ is a union of rectangles $S(Q_n) \subset Q_j$ such that $|u_{I_n} - u_{I_j}| < \varepsilon/4$. By (6.5) this means that

$$(6.8) \qquad |a_{Q_n}(u) - a_{Q_j}(u)| < 3\varepsilon/4,$$

when $S(Q_n) \subset R(Q_j)$. Each $S(Q_n)$ is contained in a unique $R(Q_j)$. When two generation squares Q_j and Q_k are distinct, their regions $R(Q_j)$ and $R(Q_k)$ have disjoint interiors. Relative to the open upper half plane, $\partial R(Q_j)$ consists of horizontal and vertical segments. The intersections of these segments from one $\partial R(Q_j)$ with any square Q have lengths summing to no more than $6\ell(Q)$. See Figure VIII.6.

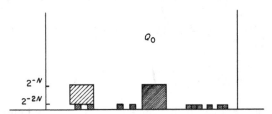

Figure VIII.6. The squares Q_k if $N = 3$. The rectangle $S(Q_0)$ is blue. On the left a red rectangle $S(Q_1)$ has been lightly shaded to indicate that $\sup_{S(Q_1)} |u(z) - u(w)| > \varepsilon/4$. The squares in the first generation $G_1(Q_0)$ are darkly shaded. For a point in the unshaded region, $|u(z) - a_{Q_0}(u)| < \varepsilon$. The region $R(Q_0)$ is the union of the unshaded region and the red rectangle.

For each generation square Q_j, including the unit square Q_0, define $\varphi_1(z) = a_{Q_j}(u)$ on the interior $R^0(Q_j)$ of $R(Q_j)$. Thus

$$\varphi_1(z) = \sum_{p=0}^{\infty} \sum_{Q_j \in G_p} a_{Q_j}(u) \chi_{R^0(Q_j)}(z).$$

The set $\bigcup \partial R(Q_j)$, on which φ_1 has not been defined, has area zero and can be ignored because for now we are only trying to find $\varphi(z)$ with $|\varphi - u| < \varepsilon$ almost everywhere.

Suppose $S(Q_n)$ is a blue rectangle. Then by (6.6),

$$\sup_{S(Q_n)} |u(z) - a_{Q_n}(u)| \le \varepsilon/4 \,.$$

There is a unique generation square Q_j such that $S(Q_n) \subset \mathrm{R}\,(Q_j)$, and by (6.8), $|a_{Q_n}(u) - a_{Q_j}(u)| < 3\varepsilon/4$, so that

$$(6.9) \qquad\qquad |\varphi_1(z) - u(z)| < \varepsilon$$

on $S(Q_n) \cap \mathrm{R}\,^0(Q_j)$. Therefore (6.9) holds almost everywhere on each blue rectangle.

We are interested in $|\nabla\varphi_1|$ only as a distribution on the upper half plane, where

$$\nabla\varphi_1 = \left(\frac{\partial\varphi_1}{\partial x}, \frac{\partial\varphi_1}{\partial y}\right)$$

$$= \sum_{p=0}^{\infty} \sum_{Q_j \in G_p} a_{Q_j}(u) \left(\frac{\partial\chi_{\mathrm{R}\,^0(Q_j)}}{\partial x}, \frac{\partial\chi_{\mathrm{R}\,^0(Q_j)}}{\partial y}\right).$$

As a distribution on $\{y > 0\}$, $\partial\chi_{\mathrm{R}\,^0(Q_j)}/\partial y$ is the measure $-dx$ along the top edge of Q_j, plus the sum of the measures dx along the other horizontal segments in $\{y > 0\} \cap \partial\mathrm{R}\,(Q_j)$. Similarly, $\partial\chi_{\mathrm{R}\,^0(Q_j)}/\partial x$ is a signed sum of the measures dy on the vertical segments in $\partial\mathrm{R}\,(Q_j)$. Hence $|\nabla\chi_{\mathrm{R}\,^0(Q_j)}|$ is the arc length measure on $\Gamma_j = \{y > 0\} \cap \partial\mathrm{R}\,(Q_j)$. Since $|a_{Q_j}(u)| \le \|u\|_\infty = 1$,

$$|\nabla\varphi_1| \le \sum_{p=0}^{\infty} \sum_{Q_j \in G_p} |\nabla\chi_{\mathrm{R}\,(Q_j)}|$$

and

$$(6.10) \qquad\qquad \int_Q |\nabla\varphi_1| \le \sum \text{length}\,(Q \cap \Gamma_j),$$

the sum being taken over all generation squares Q_j.

We claim that

$$(6.11) \qquad\qquad \sum \text{length}(Q \cap \Gamma_j) \le C\varepsilon^{-2}\ell(Q)$$

for every square Q resting on $y = 0$. With (6.10), that proves $\nabla\varphi_1$ is a Carleson measure. In proving (6.11) we can assume that Q is a dyadic square $Q = \{j2^{-n} \le x \le (j+1)2^{-n}, 0 < y \le 2^{-n}\}$. Consider first those Q_j such that $Q \cap \Gamma_j = Q \cap \partial\mathrm{R}\,(Q_j) \cap \{y > 0\} \ne \varnothing$, but such that $Q_j \not\subset Q$. If, in this case, $\ell(Q_j) \le \ell(Q)$, then $Q \cap \Gamma_j$ is a segment along one vertical edge of Q. These segments are pairwise disjoint, so that these Q_j contribute at most $2\ell(Q)$ to the sum (6.11). There can be at most two squares Q_j such that $\ell(Q_j) > \ell(Q)$ and such that $Q \cap \Gamma_j \ne \varnothing$. For each of these Q_j we have length $(Q \cap \Gamma_j) \le 6\ell(Q)$, so these squares contribute no more than $12\ell(Q)$ in (6.11).

Now consider the generation squares Q_j such that $Q_j \subset Q$. By Lemma 6.4,

$$\sum_{Q_j \subset Q} \text{length}(Q \cap \Gamma_j) \le 6 \sum_{Q_j \subset Q} \ell(Q_j) \le 6\ell(Q) + \frac{16}{\varepsilon^2}\ell(Q).$$

Since $\varepsilon < 1$, we have inequality (6.11) and $|\nabla\varphi_1|$ is a Carleson measure.

Step II: We have found a function $\varphi_1(z)$ such that $|\nabla\varphi_1|$ is a Carleson measure, such that $\varphi_1(z)$ is constant on the interior of each rectangle $S(Q_k)$ and such that $|\varphi_1 - u| < \varepsilon$ almost everywhere on the blue rectangles $S(Q_k)$. Now we approximate $u(z)$ on the red rectangles, which are the $S(Q_k) = \{x \in I_k, 2^{-N}|I_k| < y \le |I_k|\}$ such that

$$\sup_{S(Q_k)} |u(z) - u(w)| > \varepsilon/4.$$

Write

$$R = \bigcup_{\text{red}}\{S(Q_k)\}$$

for the union of the red rectangles.

Let $S(Q_k)$ be a red rectangle and take $z_1, z_2 \in S(Q_k)$ such that $|u(z_1) - u(z_2)| > \varepsilon/4$. At some point z_0 on the segment joining z_1 to z_2 we have

$$|z_1 - z_2||\nabla u(z_0)| > \varepsilon/4.$$

Since $|z_1 - z_2| < \sqrt{2}\ell(Q_k) < 2^{N+1} \text{Im } z_0 = 2^{N+1}y_0$, this gives

$$2^{N+3}y_0|\nabla u(z_0)| > \varepsilon,$$

so that by the subharmonicity of $|\nabla u|^2$,

$$\iint_{|z-z_0|<y_0/2} y|\nabla u|^2 dx\, dy \ge \frac{1}{2}y_0 \iint_{|z-z_0|<y_0/2} |\nabla u|^2 dx\, dy$$
$$\ge c\varepsilon^2 2^{-2N}y_0.$$

Letting $\tilde{S}(Q_k) = \bigcup_{z_0 \in S(Q_k)}\{z : |z - z_0| < \text{Im } z_0/2\}$, we obtain

(6.12) $$\iint_{\tilde{S}(Q_k)} y|\nabla u|^2 dx\, dy \ge c\varepsilon^2 2^{-3N}\ell(Q_k).$$

No point lies in more than four regions $\tilde{S}(Q_k)$. Because $y|\nabla u|^2$ is a Carleson measure, (6.12) then yields

(6.13) $$\sum\{\ell(Q_k): Q_k \subset Q, S(Q_k) \text{ red}\} \le c\varepsilon^{-2}2^{3N}\ell(Q)$$

for every square Q. In particular, the arc length ∂R is a Carleson measure with constant $c\varepsilon^{-2}2^{3N}$.

On the other hand, the inequality $y|\nabla u| \leq c$ yields

(6.14)
$$\iint_{S(Q_k)} |\nabla u| dx\, dy \leq c \iint_{S(Q_k)} \frac{dx\, dy}{y} = cN\ell(Q_k)$$

for any rectangle $S(Q_k)$. When $S(Q_k)$ is red, (6.12) and (6.14) give

(6.15)
$$\iint_{S(Q_k)} |\nabla u| dx\, dy \leq cN\varepsilon^{-2} 2^{3N} \iint_{\tilde{S}(Q_k)} y|\nabla u|^2 dx\, dy,$$

so that $|\nabla u(z)|\chi_R(z) dx\, dy$ is a Carleson measure with constant $cN\varepsilon^{-2} 2^{3N}$.
Now define

$$\varphi_2(z) = \begin{cases} \varphi_1(z), & z \notin R, \\ u(z), & z \in R. \end{cases}$$

Then $|\varphi_2(z) - u(z)| < \varepsilon$ almost everywhere on H . As a distribution

$$\nabla\varphi_2 = \chi_{Q_0 \setminus R} \nabla\varphi_1 + \chi_R \nabla u + J,$$

where the remainder term J accounts for the jumps in $\varphi_2(z)$ as z crosses ∂R .
Since $|\varphi_2| < 1 + \varepsilon$, J is a measure and $|J|$ is dominated by $1 + \varepsilon$ times arc
length on ∂R. Thus by (6.13), $|J|$ is a Carleson measure with constant $C2^{3N}\varepsilon^{-2}$.
By (6.15), $\chi_R |\nabla u| dx\, dy$ is also a Carleson measure, with constant $CN2^{3N}\varepsilon^{-2}$.
In the discussion of step I we showed $|\nabla\varphi_1|$ is a Carleson measure with constant
$C\varepsilon^{-2}$. The integer N was chosen so that $N2^{-N} \sim \varepsilon$ and the worst constant in
the three estimates comes from $\chi_R |\nabla u| dx\, dy$. We obtain

(6.16)
$$\int_Q |\nabla\varphi_2| dx\, dy \leq C\varepsilon^{-6}\ell(Q)$$

for every square Q.

We have built a function φ_2 on the unit square Q_0 such that $|\nabla\varphi_2|$ satisfies
(6.16) and such that $|\varphi_2 - u| < \varepsilon$ almost everywhere on Q_0. Using a partition
of unity, it is now easy to get a function φ_3 on H such that $|\varphi_3 - u| < \varepsilon$ almost
everywhere on H and such that $|\nabla\varphi_3|$ satisfies (6.16). A C^∞ function satisfying
(6.1) and (6.2) can now be obtained by mollifying φ_3. Let $h(z) \in C^\infty(\mathbb{R}^2)$ be
such that $h(z) \geq 0$, $\int h(z) dx\, dy = 1$ and $h(z) = 0$ if $|z| \geq 1$. Assume also
that $h(z)$ is radial, $h(z) = h(|z|)$. For $\delta > 0$, set $h_\delta(z) = (1/\delta^2) h(z/\delta)$. Then
$\varphi_3 * h_\delta$ is C^∞. If $\delta < \operatorname{Im} z$, then $|\varphi_3 * h_\delta(z) - u(z)| < \varepsilon$, because $u * h_\delta(z) = u(z)$ since u is harmonic and h_δ is radial. To keep $\delta < \operatorname{Im} z$, partition $\{y > 0\}$
into strips $T_n = \{2^{-n} \leq y \leq 2^{-n+1}\}$ and set $\tilde{T}_n = T_n \cup T_{n+1} \cup T_{n-1}$. Choose
$g_n = g_n(y) \in C^\infty$ such that g_n is supported on $\tilde{T}_n$, such that $0 \leq g_n \leq 1$ and
$|\nabla g_n| = |\partial g_n/\partial y| \leq c2^n$, and such that $\sum_{-\infty}^{\infty} g_n = 1$ on $\{y > 0\}$.

$$\varphi(z) = \sum_{-\infty}^{\infty} g_n(y)(\varphi_3 * h_{2-n-2})(z)$$

is C^∞ and $|\varphi(z) - u(z)| < \varepsilon$. For $z \in \tilde{T}_n$, $\varphi_3 * h_{2-n-2}(z) = \varphi_3(z)$ if

$$\text{dist}(z, \partial R \cup \bigcup \Gamma_j) \geq 2^{-n-2}$$

because then φ_3 is harmonic on $|w - z| < 2^{-n-2}$. Thus we only have to estimate $|\nabla \varphi| dx\, dy$ on

$$V = \bigcup_{-\infty}^{\infty} \tilde{T}_n \cap \{z : \text{dist}(z, \partial R \cup \bigcup \Gamma_j) < 2^{-n-2}\}.$$

Fix a dyadic square $Q = \{j2^{-p} \leq x \leq (j+1)2^{-p}, 0 < y \leq 2^{-p}\}$, p an integer. Then because

$$\nabla \varphi(z) = \sum_{-\infty}^{\infty}(\varphi_3 * h_{2-n-2}(z))\nabla g_n(z) + \sum_{-\infty}^{\infty} g_n(z)(\nabla \varphi_3 * h_{2-n-2})(z),$$

we have

$$\iint_{Q \cap V} |\nabla \varphi(z)| dx\, dy \leq c \sum_{n=p-1}^{\infty} 2^n |Q \cap V \cap \tilde{T}_n|$$

$$+ \sum_{n=p-1}^{\infty} \int_{Q \cap \tilde{T}_n} |\nabla \varphi_3| * h_{2-n-2}(z)\, dx\, dy.$$

Since $|Q \cap V \cap \tilde{T}_n| \leq c2^{-n}$ length $(Q \cap \tilde{T}_n \cap (\partial R \cup \bigcup \Gamma_j))$, the first sum is bounded by $c\varepsilon^{-6}\ell(Q)$. The second sum does not exceed

$$\sum_{n=p-1}^{\infty} \int_{W_n} |\nabla \varphi_3|,$$

where $W_n = \{z : \text{dist}(z, Q \cap \tilde{T}_n) \leq 2^{-n-2}\}$. These sets W_n are locally finite, no point lies in more than five W_n. The union $\bigcup_{n \geq p-1} W_n$ is contained in a square 16 times as large as Q. Hence

$$\sum_{n=p-1}^{\infty} \int_{W_n} |\nabla \varphi| \leq C\varepsilon^{-6}2^{-p}.$$

Since $\ell(Q) = 2^{-p}$, we see that $\varphi(z)$ satisfies (6.2). That completes the proof of the theorem. □

7. A Constructive Solution of $\partial b / \partial \bar{z} = \mu$

To prove the corona theorem we used a duality argument to get the crucial estimate

$$(7.1) \qquad \|b\|_\infty \leq C \sup_S \frac{1}{\ell(S)} \int_S |G| dx\, dy$$

for some solution of the equation

$$\partial b / \partial \bar{z} = G(z), \qquad |z| < 1.$$

Recently Peter Jones found a direct way to obtain (7.1). His method simultaneously yields a constructive proof of the basic decomposition

$$(7.2) \qquad \varphi = u + Hv, \qquad u, v \in L^\infty,$$

of a BMO function. The construction is quite explicit and it should have further applications.

Let μ be a Carleson measure on the upper half plane. Assume μ is positive and normalized, so that

$$N(\mu) = \sup_Q \mu(Q)/\ell(Q) \leq 1.$$

Jones solved

$$(7.3) \qquad \partial b / \partial \bar{z} = \mu, \qquad \text{Im } z > 0,$$

with $|b(t)| \leq C$ almost everywhere on $\mathbb{R}$, by exploiting the relation between Carleson measures and interpolating sequences. First consider a very special case.

Case I: $\mu = \sum \alpha_j y_j \delta_{z_j}$, where $\{z_j\}$ is a finite sequence of points such that

$$(7.4) \qquad \prod_{j:j\neq k} \left| \frac{z_k - z_j}{z_k - \bar{z}_j} \right| \geq \delta > 0$$

with δ a fixed constant, and where $0 \leq \alpha_j \leq 1$. Let $B_1(z)$ be the Blaschke product with zeros $\{z_j\}$. By Green's theorem the distribution $(\partial/\partial\bar{z})(1/B_1(z))$ equals

$$\sum_j \frac{\pi}{B_1'(z_j)} \delta_{z_j} = \sum_j \beta_j y_j \delta_{z_j},$$

where $1 \leq |\beta_j| \leq 1/\delta$. By Earl's proof of the interpolation theorem (Theorem 5.1 of Chapter VII) there is a second Blaschke product $B_2(z)$ with zeros ζ_j satisfying

$$\rho(z_j, \zeta_j) \leq \delta/3$$

such that

$$K\delta^{-3}B_2(z_j) = \alpha_j/\beta_j,$$

where K is an absolute constant. The rational function

$$b(z) = K\delta^{-3}(B_2(z)/B_1(z))$$

solves (7.3) and $|b(t)| \le K\delta^{-3}$ on $\mathbb{R}$.

Case II: Again μ is a positive discrete measure with finite support, $\mu = \sum \alpha_j y_j \delta_{z_j}$, but $N(\mu) \le 1$ and the coefficients α_j are rational, $\alpha_j = k_j/N$. The main difference between Cases II and I is that here the constant δ in (7.4) may be very small. Since $N(\mu) \le 1$, we have $1 \le k_j \le N$. Relabeling the points so that z_j occurs with multiplicity k_j, we have

$$\mu = \frac{1}{N} \sum y_j \delta_{z_j}.$$

The top half of any square Q resting on $\{y = 0\}$ contains at most $2N$ points z_j, counting multiplicities, because $N(\mu) \le 1$. We are going to partition $\{z_j\}$ into $4N$ interpolating sequences having uniformly bounded constants. Write

$$S_n = \{z_j : 2^{-n-1} < y_j \le 2^{-n}\}$$

and order the $z_j \in S_n$ according to increasing real parts $S_n = \{x_{k,n} + iy_{k,n}\}$ with

$$x_{k-1,n} \le x_{k,n} \le x_{k+1,n}.$$

Split $\{z_j\}$ into $2N$ subsequences $Y_1, Y_2, \dots, Y_{2N}$, evenly distributing the points in each S_n. That is, put $z_j = x_{k,n} + iy_{k,n}$ in Y_p if and only if $k \equiv p \mod(2N)$. Fix a dyadic square Q and let $M_n(Q)$ be the number of points z_j in $S_n \cap Q$. The subsequences have been chosen so that each set $Y_p \cap S_n \cap Q$ contains at most

$$1 + M_n(Q)/2N$$

points z_j. Consequently

$$\sum_{Y_p \cap Q} y_j \le \sum_{2^{-n} \le \ell(Q)} \left(1 + \frac{M_n(Q)}{2N}\right) 2^{-n}$$

$$\le \sum_{2^{-n} \le \ell(Q)} 2^{-n} + \frac{1}{2N} \sum_{z_j \in Q} 2y_j$$

$$\le 2\ell(Q) + \mu(Q) \le 3\ell(Q).$$

Also, the points in $Y_p \cap \bigcup_{n \text{ even}} S_n$ are very well separated. Two points from one of these subsequences satisfy

$$|z_j - z_k| \ge y_j/2.$$

Thus each Y_p is the union of two interpolating sequences having large constants δ in (7.4). By the very special case treated above, we have a rational function $b_p(z)$ such that

$$\frac{\partial b_p}{\partial \bar{z}} = \sum_{Y_p} y_j \delta_{z_j}$$

and $|b_p(t)| \leq C', t \in \mathbb{R}$. Then

$$b(z) = \frac{1}{N} \sum_{p=1}^{2N} b_p(z)$$

is a solution of (7.3) and

$$|b(t)| \leq C, \qquad t \in \mathbb{R},$$

with C independent of N.

Case III: Now let μ be any positive measure on H such that $N(\mu) \leq 1$. There is a sequence $\{\mu_n\}$ of measures of the type treated in Case II such that

$$\int \varphi \, d\mu_n \to \int \varphi \, d\mu, \qquad \varphi \in C_0^\infty(\text{H}),$$

and such that

$$\int |f| d\mu_n \to \int |f| d\mu, \qquad f \in H^1.$$

(First restrict μ to a compact subset of H. Then partition H into hyperbolically small squares and concentrate the mass of each square at its center.) Let $b_n(z)$ be the solution of $\partial b_n / \partial \bar{z} = \mu_n$ obtained in Case II. Then $\{b_n(t)\}$ converges weak-star to $b(t) \in L^\infty$ and $\|b\|_\infty \leq C$. Moreover, $\{b_n(z)\}$ converges to a distribution solution of (7.3) on the upper half plane. For applications this distribution must be reconciled with its boundary function $b(t)$, but that is not a serious difficulty. To avoid unimportant technicalities, let us move to the unit disc. Then $b_n(z)$ is a rational function with no poles on ∂D, and by residues,

$$\frac{1}{2\pi i} \int_{|z|=1} f(z) b_n(z) dz = \int_D f(z) d\mu_n(z), \qquad f \in H^1.$$

Consequently we have

(7.5) $$\frac{1}{2\pi i} \int_{|z|=1} f(z) b(z) dz = \int_D f(z) d\mu(z), \qquad f \in H^1.$$

Now (7.5) is an interpretation of $\partial b / \partial \bar{z} = \mu$ sufficient for our applications. $\qquad \square$

In the corona problem we had $d\mu = (1/\pi) G \, dx \, dy$, with $G \in C^\infty(\bar{D})$, and by (1.2) we had a function $F \in C(\bar{D}) \cap C^\infty(D)$, with $\partial F / \partial \bar{z} = G(z)$.

Then by Green's theorem

$$\frac{1}{2\pi i}\int_{|z|=1} f(z)F(z)dz = \frac{1}{\pi}\iint_D f(z)G(z)dx\,dy.$$

By (7.5) this means $h(e^{i\theta}) = F(e^{i\theta}) - b(e^{i\theta}) \in H^\infty$, and so

$$b(z) = F(z) - h(z)$$

is a smooth solution bounded on the disc that satisfies $|b(e^{i\theta})| \leq C$ almost everywhere. If we put such functions $b(z)$ into (2.7) we obtain a proof of the corona theorem without recourse to duality.

Jones's approach is even more transparent in the notation of distribution derivatives. Then our differential equation is

$$\partial b/\partial\bar{z} = \mu,$$

where μ can be taken absolutely continuous to arc length on a contour Γ of the type constructed in Section 4 or 5. On Γ arc length is a Carleson measure and the density defining μ is bounded above and below. In this case the interpolating Blaschke products are easy to visualize; they resemble the Blaschke product constructed in Section 4. The solution $b(z)$ can be recognized as an average of interpolating Blaschke products. (See the example at the end of this section.)

To prove the decomposition (7.2) let φ be a real function in BMO(T), $\|\varphi\|_* \leq 1$. By Exercise 13 of Chapter VI, there exist $\psi \in L^\infty$, $\|\psi\|_\infty \leq A$, and $F(z) \in C^\infty(D)$ and $g \in L^1$ such that

$$|F(re^{i\theta})| \leq g(e^{i\theta}),$$

such that

$$\varphi(\theta) = \psi(\theta) + \lim_{r\to 1} F(re^{i\theta}) = \psi(\theta) + F(\theta),$$

and such that $|\nabla F|dx\,dy$ is a Carleson measure with $N(|\nabla F|dx\,dy) \leq A$. Let $b(z)$ be the solution of

$$\frac{\partial b}{\partial\bar{z}} = \frac{1}{\pi}\frac{\partial F}{\partial\bar{z}}dx\,dy$$

given by Jones's procedure. Then $|b(\theta)| \leq CA$ and

$$\varphi = \psi + b + (F - b)$$

almost everywhere on T. Arguing formally for a moment, we have $(F - b)^\sim = -i(F - b)$, because $F - b$ is analytic, and so

$$\varphi = \psi + b + i\tilde{F} - i\tilde{b}.$$

Taking real parts, we obtain

$$\varphi = (\psi + \operatorname{Re} b) + (\operatorname{Im} b)^\sim,$$

which is (7.2). To make this reasoning precise, notice that by (7.5)

$$\frac{1}{2\pi i} \int z^n b(z) dz = \frac{1}{\pi} \iint_D z^n \frac{\partial F}{\partial \bar{z}} dx\, dy.$$

On the other hand, since $|F(re^{i\theta})| \le g(\theta) \in L^1$, dominated convergence gives

$$\frac{1}{2\pi i} \int z^n F(z) dz = \frac{1}{\pi} \iint_D z^n \frac{\partial F}{\partial \bar{z}} dx\, dy.$$

Consequently $F(e^{i\theta}) - b(e^{i\theta}) \in H^1$, so that we really do have

$$(F - b)^{\sim} = -i(F - b).$$

An attractive aspect of this approach to (7.2) is that the Varopoulos construction of $F(z)$ is very explicit. The method in this section also gives a constructive way of finding $g \in H^\infty$ such that

$$\|f - g\|_\infty \le C \operatorname{dist}(f, H^\infty)$$

when $f \in L^\infty$.

For an example, write $x \in [0, 1]$ in base 5, $x = \sum \alpha_k 5^{-k}$, and let $\varphi(x) = \sum \chi_{E_k}(x)$, $E_k = \{x : \alpha_k = 1 \text{ or } \alpha_k = 3\}$. Then $\varphi \in \mathrm{BMO}$ (see Figure VIII.7).

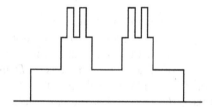

Figure VIII.7. The function $\varphi(x)$.

If $F(x, y)$ is constructed with pentadic squares instead of dyadic squares, then

$$F(x, y) = \sum_k \chi_{E_k}(x) \chi_{(0, 5^{-k})}(y).$$

Then $\varphi(x) = \lim_{y \to 0} F(x, y)$ and $|\nabla F|$ is bounded by arc length on the curves in H pictured in Figure VIII.8. Cut these curves into segments of bounded hyperbolic length, uniformly parametrized by t, $0 < t < 1$. For each t, let $B_1(t, z)$ be the Blaschke product with one zero on each segment at position t. The zeros of $B_1(t, z)$, form an interpolating sequence, with constant δ independent of t. They are marked by crosses in Figure VIII.8, for $t = 0.8$, say. The bounded solution of $\partial b/\partial \bar{z} = \partial F/\partial \bar{z}$ is

$$b(x) = c \int_0^1 \frac{B_2(t, x)}{B_1(t, x)} dt,$$

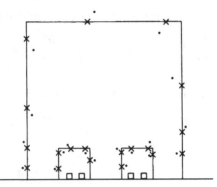

Figure VIII.8. The zeros of $B_1(t, z)$ are marked by crosses; the zeros of $B_2(t, z)$ are marked by dots.

where $B_2(t, x)$ is another interpolating Blaschke product. Its zeros are marked with dots in the figure. The decomposition (7.2) is

$$\varphi(x) = \operatorname{Re} b(x) + (\operatorname{Im} b(x))^{\sim}.$$

Appendix: The Koszul Complex

The Koszul complex is a general algebraic mechanism that, in the case of the corona problem, leads from smooth corona solutions $\varphi_1, \ldots, \varphi_n$ to the appropriate differential equations (2.9). In our setting the complex works as follows. Let $\mathfrak{A}$ be the ring of all analytic functions on D and let E be the ring of all C^∞ functions on D. Then $\mathfrak{A}$ is a subring of E. Also let $\mathrm{E}_{(0,1)}$ be the E module of $(0, 1)$ forms $g \, d\bar{z}$, $g \in \mathrm{E}$. For the case we are considering $\mathrm{E}_{(0,1)}$ is isomorphic to E, but it will be convenient to distinguish between functions $h(z)$ and differential forms $g \, d\bar{z}$ of type $(0, 1)$. Define

$$\bar{\partial} : \mathrm{E} \to \mathrm{E}_{(0,1)}$$

by $\bar{\partial} h = (\partial h / \partial \bar{z}) d\bar{z}$. Then $\mathfrak{A} \subset \mathrm{E}$ is the kernel of $\bar{\partial}$.

Let R denote either $\mathfrak{A}$, or E, or $\mathrm{E}_{(0,1)}$. Let $\Lambda^0(\mathrm{R}) = \mathrm{R}$ and let $\Lambda^1(\mathrm{R})$ be the module of all expressions of the form

$$\sum_{j=1}^{n} h_j e_j, \qquad h_j \in \mathrm{R},$$

where the e_j are place markers. Thus $\Lambda^1(\mathrm{R})$ is the direct sum of n copies of R. Also let $\Lambda^2(\mathrm{R})$ be the module of all expressions of the form

$$\sum_{j,k=1}^{n} h_{j,k} e_j \wedge e_k, \qquad h_{j,k} \in \mathrm{R},$$

where we require that

(A.1) $$e_j \wedge e_k = -e_k \wedge e_j,$$

a familiar condition in alternating linear algebra. Thus $\Lambda^2(R)$ has dimension $n(n-1)/2$ over R. We define

$$\bar{\partial}: \Lambda^p(E) \to \Lambda^p(E_{(0,1)}), \qquad p = 0, 1, 2,$$

by differentiating the coefficients of the place markers e_j or $e_j \wedge e_k$. That is,

$$\bar{\partial}\left(\sum h_j e_j\right) = \sum (\bar{\partial} h_j) e_j \quad \text{and} \quad \bar{\partial}\left(\sum h_{j,k} e_j \wedge e_k\right) = \sum \bar{\partial} h_{j,k} e_j \wedge e_k.$$

Then $\Lambda^p(\mathfrak{A}) \subset \Lambda^p(E)$ is the kernel of $\bar{\partial}$. Using (1.2), and a partition of unity if our functions are not bounded, we see that for each $p = 0, 1, 2$ the $\bar{\partial}$ mapping is surjective. This means the sequence

$$0 \to \Lambda^p(\mathfrak{A}) \to \Lambda^p(E) \xrightarrow{\bar{\partial}} \Lambda^p(E_{(0,1)}) \to 0$$

is exact; at each step the kernel of the outgoing map is the range of the incoming map. Let $\tau \in \Lambda^p(E)$, $p = 0, 1$ and let $\omega \in \Lambda^q(R)$, $q = 0, 1$, where R is either of the E modules E or $E_{(0,1)}$. The wedge product $\tau \wedge \omega \in \Lambda^{p+q}(R)$ is defined by setting

$$f \wedge g = fg,$$
$$f \wedge g e_j = f e_j \wedge g = f g e_j,$$
$$f e_j \wedge g e_k = f g e_j \wedge e_k,$$

and by requiring that $\tau \wedge \omega$ be E-bilinear; that is, $\tau \wedge \omega$ is to be E-linear in each variable.

Now let $f_1, f_2, \ldots, f_n$ be corona data. For $p = 1, 2$, define

$$J : \Lambda^p(R) \to \Lambda^{p-1}(R)$$

by

$$J(h e_{j_1} \wedge e_{j_2} \wedge \cdots \wedge e_{j_p}) = \sum_{k=1}^{p} (-1)^{k+1} f_{j_k} h e_{j_1} \wedge \cdots \wedge \widehat{e_{j_k}} \wedge \cdots \wedge e_{j_p},$$

where the circumflex over a marker e_{j_k} indicates that e_{j_k} is deleted. The mapping J is then defined on the full module $\Lambda^p(R)$ by requiring that J be R-linear. The factors $(-1)^{k+1}$ make the definition of J consistent with the alternating condition (A.1). In particular

$$J\left(\sum_{j=1}^{n} g_j e_j\right) = \sum_{j=1}^{n} f_j g_j.$$

A simple calculation shows that $J^2 = 0$, and we have another sequence

$$\Lambda^2(R\ ) \overset{J}{\to} \Lambda^1(R\ ) \overset{J}{\to} R\ \to 0.$$

Now let $\varphi_1, \varphi_2, \ldots, \varphi_n$ be any set of C^∞ corona solutions. Then $\varphi = \sum \varphi_j e_j \in \Lambda^1(E\ )$ and $J(\varphi) = \sum f_j \varphi_j = 1$. If $\lambda \in \Lambda^p(R\ )$ and if $J(\lambda) = 0$, then

$$J(\varphi \wedge \lambda) = J(\varphi) \wedge \lambda - \varphi \wedge J(\lambda) = \lambda.$$

Consequently the J-sequence is exact when $R\ = E$ or $R\ = E_{(0,1)}$. Finally, since the f_j are analytic,

$$J\bar\partial = \bar\partial J,$$

and we have the commutative diagram

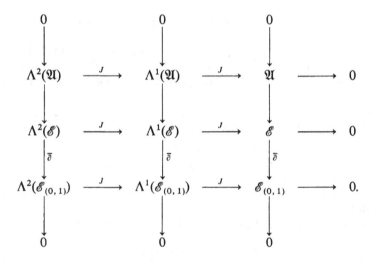

In the diagram each column is exact and, except for the $\mathfrak{A}$-row, each row is exact. We are seeking $g = \sum g_j e_j \in \Lambda^1(\mathfrak{A})$ such that $J(g) = \sum f_j g_j = 1$. We do have $\varphi = \sum \varphi_j e_j \in \Lambda^1(E\ )$ with $J(\varphi) = 1$, and a diagram chase will now yield g and the differential equations (2.9). Since $J\,\bar\partial\varphi = \bar\partial J\varphi = 0$, $\bar\partial\varphi = J\omega$ for $\omega = \varphi \wedge \bar\partial\varphi \in \Lambda^2(E_{(0,1)})$. There is $b = \sum b_{j,k} e_j \wedge e_k \in \Lambda^2(E\ )$ such that $\bar\partial b = \omega$. This is precisely the system (2.9). Then $\bar\partial J b = J\,\bar\partial b = J\omega = \bar\partial\varphi$, so that $\bar\partial(\varphi - Jb) = 0$ and $g = \varphi - Jb \in \Lambda^1(\mathfrak{A})$. Since $J^2 = 0$, we also have

$$J(g) = J(\varphi - Jb) = 1.$$

The components of this vector $g \in \Lambda^1(\mathfrak{A})$ are exactly the corona solutions given by (2.7) and (2.8).

For the corona problem on a domain in $\mathbb{C}^n$, $n > 1$, the Koszul complex has longer rows and columns, but it does reduce the problem to a system of differential equations.

Notes

Theorem 1.1 is in Carleson [1962a], while Theorem 1.2 is a recent idea of Wolff's [1980]. Theorem 2.1 is due to Carleson [1962a], but the line of reasoning in Section 2 comes from Wolff [1980].

The connection between the corona theorem and differential equations and the Koszul complex was noticed by Hörmander [1967a] (see also Carleson [1970]), but the basic difficulties have remained the same as they were in Carleson [1962a].

Theorem 3.2 is from Carleson's exposition [1970]. The theorems in Section 3 can also be derived from a theorem on harmonic measures known as *Hall's lemma*. Let E be a compact subset of the upper half plane such that $\Omega = H \setminus E$ is connected, and let $\omega(z)$, $z \in \Omega$, be the harmonic measure of E. Hall's lemma is the estimate

$$\omega(z) \geq \frac{2}{3} \frac{1}{\pi} \int_{\hat{E}} \frac{y}{(|x| + t)^2 + y^2} \, dt,$$

where $\hat{E} = \{|z| : z \in E\}$ is the angular projection of E onto the positive real axis and where the Poisson kernel represents the image $-|x| + iy$ of $z = x + iy$ under a folding. For the proof see Duren [1970] or Hall [1937]. By means of elementary conformal mappings this inequality gives bounds on the lengths of vertical projections. Hall's lemma has certain advantages because it gives simple relations between the numbers α and β of Theorems 3.1 and 3.2 (see Exercise 7). On the other hand, the proof in the text illustrates the power of maximal functions. Incidentally, the best constant to replace $\frac{2}{3}$ in Hall's lemma is not known (Hayman [1974]).

Theorem 4.1 is due to Marshall [1976b]. Ziskind [1976] had obtained a slightly weaker result in the course of proving Theorem 2.2. The construction of the contour Γ originates in Carleson's [1962a] fundamental paper. If Γ is taken to be a level set, $|u(z)| = v$ will not work and it is necessary to approximate the level sets by shorter curves (see Exercise 8 below).

Theorem 6.1 is motivated by the work of Varopoulos [1977a, b]. The philosophy that harmonic functions mimic very simple martingales is delightfully expressed in Fefferman's lecture [1974]. Dahlberg [1980] has improved upon Theorem 6.1 substantially.

The results in Section 7 come from Jones [1980b]. The extension $F(z)$, so effective in the proof of (7.2), is due to Varopoulos [1977a]. Finding a constructive proof of the euclidean space analog of (7.2) is an important open problem in real analysis. As is often the case with higher-dimensional generalizations, the real question here is how to eliminate Blaschke products from the construction. See page vii.

Exercises and Further Results

1. Let $G(z)$ be continuous on the open unit disc. Suppose that on $|z| < 1 - 1/n$ there is $b_n(z) \in C^1$ such that

$$\partial b_n / \partial \bar{z} = G$$

and $|b_n| \le K$, with K independent of n. Then by a normal families argument there is $b(z)$ of class C^1 such that $\partial b / \partial \bar{z} = G$ on $|z| < 1$.

2. (a)   If G is a bounded function of compact support in the plane and if

$$F(z) = \frac{1}{2\pi i} \iint \frac{G(\zeta)}{\zeta - z} d\zeta \wedge d\bar{\zeta},$$

then F is bounded and continuous and $\partial F / \partial \bar{z} = G$ in the distribution sense.

   (b)   Let $f(z)$ be a bounded analytic function on an open set W in the complex plane and let $\lambda \in \partial W$. Declare $f = 0$ on $\mathbb{C} \setminus W$. Let $\chi(z) \in C_0^\infty$ have support $\{|z - \lambda| < \varepsilon\}$. Suppose $0 \le \chi \le 1$, $\chi = 1$ on $\{|z - \lambda| < \varepsilon/2\}$, and $|\partial \chi / \partial \bar{z}| \le c/\varepsilon$. Then

$$F(z) = \frac{-1}{2\pi i} \iint \frac{f(\zeta) - f(z)}{\zeta - z} \frac{\partial \chi}{\partial \bar{\zeta}} d\zeta \wedge d\bar{\zeta},$$

has the following properties:

   (i) $\partial F / \partial \bar{z} = \chi \, \partial f / \partial \bar{z}$, as distributions,
   (ii) $F(z)$ is analytic on $W \cup \{|z - \lambda| > \varepsilon\}$,
   (iii) $|F(z)| \le C \sup \{|f(\zeta)| : |\zeta - \lambda| < \varepsilon\}$, and
   (iv) $F(z) - f(z)$ is continuous on $\{|z - \lambda| < \varepsilon/2\}$.

   (c)   Let Ω be an open set in the plane and let $f_1, f_2, \ldots, f_n \in H^\infty(\Omega)$, the bounded analytic functions on Ω. If there is a finite open cover $\{U_k\}_{k=1}^N$ of Ω and if there are $g_j^{(k)} \in H^\infty(\Omega \cap U_k)$ such that $\sum_j f_j g_j^{(k)} = 1$ on $\Omega \cap U_k$, then there are $g_1, g_2, \ldots, g_n \in H^\infty(\Omega)$ such that $\sum f_j g_j = 1$. Let $\chi_k \in C_0^\infty(U_k)$, $\sum \chi_k = 1$ on Ω. Perturb the C^∞ solutions $\varphi_j = \sum_k \chi_k g_j^{(k)}$ to get analytic solutions g_j. Use part (a) to solve the Cauchy–Riemann equations (2.9).

   (d)   From the assumption that the disc is dense in $\mathfrak{M}$ it follows that there exists a constant $C(n, \delta)$ such that every corona problem (2.1), (2.2) in the disc has solutions satisfying $\|g_j\| \le C(n, \delta)$. For the proof it is enough to establish the corona theorem for the open set $W = \bigcup_{k=1}^\infty \{-1 < x < 1, 2^{-k-1} < y < 2^{-k}\}$, which is conformally equivalent to an infinite disjoint union of discs. So let $f_1, f_2, \ldots, f_n \in H^\infty(W)$ satisfy $\|\cdot f\|_\infty \leq 1$, $\max_j |f_j(z)| \geq \delta, z \in W$. By the localization theorem in part (c) we only have to find solutions on $W \cap \{|z - \lambda| < \eta\}$, for some $\eta > 0$, for $-1 \leq \lambda \leq 1$. By part (b) there is a bounded simply connected domain Ω_λ containing $W \cap \{|z - \lambda| < \varepsilon\}$ and there are $F_1, F_2, \ldots, F_n \in H^\infty(\Omega_\lambda)$ such that $\lim_{z \to \lambda} F_j(z) - f_j(z) = 0$. On Ω_λ, $F_0(z) = z - \lambda$ and $F_1, \ldots, F_n$ are corona data. From the solutions $G_0, G_1, \ldots, G_n$ of $\sum F_j G_j = 1$, one can obtain solutions to the original problem on $W \cap \{|z - \lambda| < \eta\}$.

★(e)   There are constants $C(n, \delta, m)$ such that every corona problem (2.1), (2.2) on every plane domain of connectivity m has solutions with $\|g_j\| \leq C(n, \delta, m)$.

Parts (c)–(e) are from Gamelin [1970], who also observed that the corona theorem will hold for all plane domains if and only if the constants in (e) have bounds not depending on m; see Behrens [1971] as well. Part (e) above is not true for general uniform algebras (see Rosay [1968]).

3. Let $f_1, f_2, \ldots, f_n \in H^\infty$, $\|f_j\| \leq 1$, and let $g \in H^\infty$ satisfy

$$|g(z)| \leq \max_j |f_j(z)|.$$

It does not follow that g is in the ideal generated by $f_1, f_2, \ldots, f_n$. If B_1 and B_2 are two Blaschke products with distinct zeros such that $\inf(|B_1(z)| + |B_2(z)|) = 0$, then $|B_1 B_2| \leq \max(|B_1|^2, |B_2|^2)$ but B_1, B_2 is not of the form $g_1 B_1^2 + g_2 B_2^2$. This example is due to Rao [1967].

4. Suppose g, f_1, and f_2 are H^∞ functions, $\|g\|_\infty \leq 1, \|f_j\|_\infty \leq 1$, and suppose

$$|g(z)| \leq \max(|f_1(z)|, |f_2(z)|).$$

Then $g^2 = g_1 f_1 + g_2 f_2$ with $g_j \in H^\infty$ if and only if

$$\frac{\partial b}{\partial \bar{z}} = \frac{g^2(\bar{f}_1 f_2' - \bar{f}_1' \bar{f}_2)}{(|f_1|^2 + |f_2|^2)^2}$$

has a solution bounded on ∂D.

5. Let $f_1, f_2, \ldots, f_n$ be corona data. Then from (2.5),

$$\varphi_j \frac{\partial \varphi_k}{\partial \bar{z}} = \frac{\bar{f}_j}{\psi^3} \sum_{l \neq j} f_l(\bar{f}_l \bar{f}_k' - \bar{f}_k \bar{f}_k'),$$

where $\psi = \sum |f|^2$, and the analyticity of f_l can be exploited to solve

$$\frac{\partial b}{\partial \bar{z}} = \varphi_j \frac{\partial \varphi_k}{\partial \bar{z}}$$

boundedly without recourse to Carleson measures. By duality and Green's theorem, as in the proof of Theorem 1.2, it is enough to bound

$$J_1 = \left| \iint_D k'(z)\varphi_j(z)\frac{\partial \varphi_k}{\partial \bar{z}} \log \frac{1}{|z|} dx\, dy \right|$$

and

$$J_2 = \left| \iint_D k(z)\frac{\partial}{\partial z}\left(\varphi_j \frac{\partial \varphi_k}{\partial \bar{z}}\right) \log \frac{1}{|z|} dx\, dy \right|,$$

where $k \in H_0^1$, $\|k\|_1 = 1$. Writing $k = (k_1 + k_2)/2$ with k_j zero free, we can replace $k(z)$ by $g^2(z)$ where $g \in H^2$, $\|g\|_2^2 \leq 2$. Then J_1 is dominated by a sum of $2(n-1)$ terms of the form

$$2\delta^{-3} \int |gg'\bar{f}_l'| \log \frac{1}{|z|} dx\, dy \leq 2\delta^{-3} \left(\iint |g'|^2 \log \frac{1}{|z|} dx\, dy \right)^{1/2}$$

$$\times \left(\iint |gf_l'|^2 \log \frac{1}{|z|} dx\, dy \right)^{1/2}.$$

By analyticity,

$$gf_l' = (gf_l)' - g'f_l,$$

the second factor is bounded by

$$\left(\iint |(gf_l)'|^2 \log \frac{1}{|z|} dx\, dy \right)^{1/2} + \left(\iint |g'|^2 \log \frac{1}{|z|} dx\, dy \right)^{1/2},$$

and $J_1 \leq 2\delta^{-3} \|g\|_2 (\|gf_l\|_2 + \|g\|_2) \leq 8\delta^{-3}$. The form of $(\partial/\partial z)(\varphi_j \partial \varphi_k / \partial \bar{z})$ computed in the proof of Theorem 2.1 shows that J_2 is dominated by a sum of cn^2 terms of the form

$$\delta^{-4} \iint |g^2 f_q' f_p'| \log \frac{1}{|z|} dx\, dy \leq \delta^{-4} \left(\iint |gf_p'|^2 \log \frac{1}{|z|} dx\, dy \right)^{1/2}$$

$$\times \left(\iint |gf_q'|^2 \log \frac{1}{|z|} dx\, dy \right)^{1/2}$$

Twice repeating the trick from above therefore yields $J_2 \leq cn^2\delta^{-4}$ (see Gamelin [1981]).

6. (a) In Theorem 3.1 the vertical projection can be replaced by a nontangential projection. Let $E_\alpha^{(a)} = \{t : \Gamma_a(t) \cap E_\alpha \neq \varnothing\}$, where $\Gamma_a(t)$ is the cone $\{z : |x - t| < ay\}$, $a > 0$. If $\|u\|_\infty \leq 1$ and if $\sup_{T(Q)} |u(z)| \geq \beta = \beta(\alpha, \varepsilon, a)$, then $|E_\alpha^{(a)}| < \varepsilon$.

(b)   For nontangential projections the situation with Theorem 3.2 is a little more delicate. Let N be a fixed positive integer and let $\delta > 0$. Let $B(z)$ be the Blaschke product with zeros

$$\{j\delta/N + i\delta, -\infty < j < \infty\}.$$

Then $B(z)$ has a zero in each cone $\Gamma_{1/N}(t) = \{z : |x - t| < y/N\}, t \in \mathbb{R}$. So whatever $\alpha > 0$ we choose, $E_\alpha = \{|B(z)| < \alpha\}$ meets every cone $\Gamma_{1/N}(t)$. However, when N is fixed

$$\lim_{\delta \to 0} |B(i)| = e^{-2\pi N}.$$

Thus if $\beta < e^{-2\pi N}$, the conclusion of Theorem 3.2 does not hold for cones with angle $2 \tan^{-1}(1/N)$.

(c)   However, if $\beta > 0$ is fixed, there exist $N = N(\beta, \varepsilon)$ and $\alpha = \alpha(\beta, \varepsilon)$ such that if $f \in H^\infty$, $\|f\|_\infty \leq 1$ and $\sup_{T(Q)} |f(z)| \geq \beta$ then

$$|\{t : \Gamma_{1/N}(t) \cap E_\alpha \neq \varnothing\}| < \varepsilon.$$

7.   Derive the following variant of Theorem 3.2 from Hall's lemma. Let R be the rectangle $\{0 \leq x \leq A, 0 < y \leq 1\}$, and let $0 < \beta < 1$. If $f(z) \in H^\infty$, $\|f\| \leq 1$, and if $|f(z_0)| \geq \beta$ at some point z_0 in the top half of R, then, provided A is sufficiently large, the vertical projection of

$$\{z \in R : |f(z)| \leq \beta^7\}$$

has measure not exceeding $0.9A$. Obtain a similar result when $\log |f(z)|$ is replaced by any negative subharmonic function.

8.   Let $u(z)$ be the bounded harmonic function on $\mathbb{H}$ with boundary values

$$u(t) = \begin{cases} +1, & 2n \leq t < 2n + 1, \\ -1, & 2n - 1 \leq t < 2n, \end{cases}$$

n an integer. Arc length on the set $\{u = 0\}$ is not a Carleson measure because this set contains each vertical line $\{x = n\}$. So there is an outer function whose modulus has a large level set. Let $B(z)$ be the Blaschke product with zeros $\{n + i : -\infty < n < \infty\}$. On the level set $|B(z)| = e^{-2\pi}$ arc length is not a Carleson measure. Moving these functions to the disc, one obtains level sets of infinite length. (See Piranian and Weitsman [1978] and Belna and Piranian [1981] for similar examples.) Finding a function whose every level set has infinite length is considerably more difficult. (See the paper of Jones [1980a].)

9.   Use Corollary 6.2 to prove that a harmonic interpolating sequence is an H^∞ interpolating sequence.

10. Let $1/\varepsilon$ be a positive integer and let

$$u(e^{i\theta}) = \prod_{n=1}^{1/\varepsilon^2}(1 + i(10\varepsilon \cos 4^{4^n}\theta)).$$

Then $|u(e^{i\theta})| \leq C_1$, with C_1 independent of ε. Let $N_\varepsilon(\theta)$ denote the number of times the harmonic extension of $u(e^{i\theta})$ oscillates by ε units on the radius $re^{i\theta}, 0 < r < 1$. Then

$$\int_0^{2\pi} N_\varepsilon(\theta)d\theta \geq c_2/\varepsilon^2,$$

with c_2 independent of ε.

11. (a) If $u(z) = P_y * u(x)$ with $u(x) \in$ BMO and if $0 < \delta < \frac{1}{2}$, then for every interval I,

$$\left| u_I - \frac{1}{|I|}\int_I u(x + i\delta|I|)dx \right| \leq C\delta \log \frac{1}{\delta}\|u\|_*,$$

where C is independent of $u(z)$ and I.

(b) If $\|u\|_* \leq 1$ and if $0 < \varepsilon < 1$, then there is $\varphi(z) \in C^\infty(y > 0)$ such that

$$|\varphi(z) - u(z)| < \varepsilon \quad \text{and} \quad \iint_Q |\nabla\varphi|dx\,dy \leq C(\varepsilon)\ell(Q).$$

The proof of Theorem 6.1 can be followed except that the BMO condition must be used to control the jumps of φ_1 and φ_2.

(c) Use (b) with $\varepsilon = 1$ to show BMO is the dual of $H_{\mathbb{R}}^1$.

12. Let $0 < \varepsilon < 1$. Suppose that whenever $u(z)$ is a bounded harmonic function in the disc, $\|u\|_\infty \leq 1$, there exists a second harmonic function $\varphi(z)$ such that

$$|\varphi(z) - u(z)| \leq \varepsilon \quad \text{and} \quad \iint |\nabla\varphi|dx\,dy \leq C(\varepsilon).$$

Then if $f \in L^1$ has mean value zero,

$$\|f\|_1 \leq C' \sup_{|z|<1}(1 - |z|)|\nabla f(z)|,$$

where $\nabla f(z)$ denotes the gradient of the harmonic extension of f. The latter inequality fails when $f(z) = \sum_{k \leq N} z^{2k}$. This means the function $\varphi(z)$ in Theorem 6.1 cannot always be a harmonic function.

13. If $G(z)$ is continuous on the closed upper half plane and if $G(z)$ has Compact support, then the methods of Section 7 yield a continuous function $b(z)$ such that $\partial b/\partial \bar{z} = G$ and $\|b\|_\infty \leq CN(|G|dx\,dy)$, where $N(\mu)$ is the Carleson norm of a measure μ.

14. Given $f \in L^\infty$ use the methods of Section 7 to construct $g \in H^\infty$ such that

$$\|f - g\| \le C \operatorname{dist}(f, H^\infty),$$

with the constant C not dependent on f.

15. Let B be the smallest closed subalgebra of H^∞ containing the functions z and $(1 - z)^i$. Then D is an open subset of $\mathfrak{M}_B$, and $|(1 - z)^i|$ is bounded below on D but $(1 - z)^{-i} \notin B$. Thus D is not dense in $\mathfrak{M}_B$ (Dawson [1975]).

IX

Douglas Algebras

We come to the beautiful theory, due to D. Sarason, S.-Y. Chang, and D. E. Marshall, of the uniform algebras between H^∞ and L^∞. The results themselves are very pleasing esthetically, and the proofs present an interesting blend of the concrete and the abstract. The corona construction and the BMO duality proof from Chapter VI provide the hard techniques, but the theory of maximal ideals holds the proof together.

The local Fatou theorem, a fundamental result on harmonic functions, is discussed in Section 5. This theorem could actually have been treated in the first chapter. It occurs here, somewhat misplaced, for reasons of logistics.

1. The Douglas Problem

Let A be a uniformly closed subalgebra of L^∞ containing H^∞. For an example, let B be any set of inner functions in H^∞ and take $A = [H^\infty, \bar{\text{B}}]$, the closed algebra generated by $H^\infty \cup \bar{\text{B}}$. Because $f_1 \bar{b}_1 + f_2 \bar{b}_2 = (f_1 b_2 + f_2 b_1) \bar{b}_1 \bar{b}_2$ for $f_j \in H^\infty, b_j \in \text{B}$, A is simply the norm closure of

$$\mathfrak{A} = \{ f \bar{b}_1^{n_1} \bar{b}_2^{n_2} \cdots \bar{b}_k^{n_k} : f \in H^\infty, b_1, \ldots, b_k \in \text{B} \}.$$

Algebras of the form $[H^\infty, \bar{\text{B}}]$ are called *Douglas algebras*. The simplest example $[H^\infty, \bar{z}]$ will be analyzed in Section 2. It is a beautiful theorem that every closed algebra between H^∞ and L^∞ actually is a Douglas algebra. This was conjectured by R. G. Douglas and proved by S.-Y. Chang and D. E. Marshall, following influential earlier work of D. Sarason.

Given a closed algebra A with $H^\infty \subset A \subset L^\infty$, write

$$\text{B}_A = \{ b : b \in H^\infty, b \text{ inner}, b^{-1} \in A \}.$$

(In this chapter we use the lowercase b to denote an inner function. The more usual symbol B will be reserved for an algebra between H^∞ and L^∞.) Then B $_A$ is the largest set B for which we can have $A = [H^\infty, \bar{\text{B}}]$, and A is a Douglas algebra if and only if $A = [H^\infty, \bar{\text{B}}_A]$. When $A = H^\infty$, B $_A$ is the set of unimodular constants and Douglas' question is trivial. When $A = L^\infty$, B $_A$

is the set of all inner functions and $L^\infty = [H^\infty, \bar{B}_A]$ by the Douglas–Rudin theorem (Theorem V.2.1). Define $A^{-1} = \{f \in A : f^{-1} \in A\}$ and

$$U_A = \{u \in A^{-1} : |u| = 1 \text{ almost everywhere}\}.$$

Theorem 1.1. *If A is a closed subalgebra of L^∞ containing H^∞, then A is generated by H^∞ and U_A. That is, $A = [H^\infty, U_A]$.*

Proof. Let $f \in A^{-1}$. Then $\log |f| \in L^\infty$ and there is $g \in (H^\infty)^{-1}$ such that $|g| = |f|$ almost everywhere. Then $u = g^{-1}f$ is a unimodular function invertible in A such that $f = gu \in [H^\infty, U_A]$. Since every function A is the sum of a constant and an invertible function, the theorem is proved. $\square$

Since $A \subset L^\infty, b^{-1} = \bar{b}$ when $b \in B_A$, and so $\bar{B}_A \subset U_A$. Solving the Douglas problem therefore amounts to showing that when U_A is cut down to $\bar{B}_A$, the algebra A will still be generated as an H^∞ module.

Maximal ideal spaces will play a pivotal role in the proof of the Douglas conjecture. Recall our notations $\mathfrak{M} = \mathfrak{M}_{H^\infty}$ for the spectrum, or maximal ideal space, of H^∞ and $X = \mathfrak{M}_{L^\infty}$ for the spectrum of L^∞, which is also the Šilov boundary of H^∞.

Theorem 1.2. *If A is a closed subalgebra of L^∞ containing H^∞, then its maximal ideal space $\mathfrak{M}_A$ can be identified with a closed subset of $\mathfrak{M}$ which contains X, and X is the Šilov boundary of A.*

Proof. We can identify X with a closed subset of $\mathfrak{M}_A$ because A is a closed subalgebra of $L^\infty = C(X)$ and, since $A \supset H^\infty$, A separates the points of X. This means X is a closed boundary for A. But since A is a logmodular subalgebra of $C(X)$ (Chapter V, Section 4), X is its Šilov boundary.

The natural restriction mapping $\pi : \mathfrak{M}_A \to \mathfrak{M}$ is the identity mapping on X. What we need to show is that π is one-to-one. The reason for this is that each $m \in \mathfrak{M}_{H^\infty}$ has a unique representing measure on X, by Theorem V.4.2. Consider $m_1, m_2 \in \mathfrak{M}_A$ and let μ_1, μ_2 be their representing measures on X with respect to the larger algebra A. If $\pi(m_1) = \pi(m_2)$, then

$$\int f d\mu_1 = \int f d\mu_2, \qquad f \in H^\infty.$$

So by the uniqueness of H^∞ representing measures, $\mu_1 = \mu_2$ and

$$\int_X g \, d\mu_1 = \int_X g \, d\mu_2, \qquad g \in A.$$

Hence $m_1 = m_2$ and π is 1–1. Because π is continuous and $\mathfrak{M}_A$ is compact, π is a homeomorphism of $\mathfrak{M}_A$ onto a subset of $\mathfrak{M}$. $\square$

For $m \in \mathfrak{M}$, we write μ_m for its unique representing measure on X. If $m \in \mathfrak{M}_A$, then of course μ_m is also its unique representing measure with respect to A. By a compactness argument, the mapping $m \to \mu_m$ is a homeomorphism between $\mathfrak{M}$ and a weak-star compact set of probability measures on X. When

$x \in X$ is naturally identified with the point mass δ_x, this homeomorphism is the identity on X. By duality there is an isometry from $L^\infty = C(X)$ into $C(\mathfrak{M})$, and we can identify L^∞ with its image in $C(\mathfrak{M})$ through the definition

$$(1.1) \qquad\qquad f(m) = \int f \, d\mu_m, \qquad f \in L^\infty.$$

When $m \in D$, (1.1) is only the Poisson integral formula in disguise. The mapping (1.1) is not surjective and it is not multiplicative. In fact, Theorem 1.2 says that

$$(1.2) \qquad \mathfrak{M}_A = \{m \in \mathfrak{M} : f(m)g(m) = (fg)(m), \text{ all } f, g \in A\}$$

whenever A is an algebra between H^∞ and L^∞. Consequently, if A and B are algebras such that $H^\infty \subset B \subset A \subset L^\infty$, then $\mathfrak{M}_A \subset \mathfrak{M}_B$.

Theorem 1.3. *Let A be a closed subalgebra of L^∞ containing H^∞ and let $U \subset U_A$ be a set of functions in A^{-1}, unimodular on X, such that $A = [H^\infty, U]$. Then*

$$\mathfrak{M}_A = \bigcap_{u \in U} \{m \in \mathfrak{M} : |u(m)| = 1\}.$$

Proof. Because $A \subset L^\infty$, we have $u^{-1} = \bar{u}$ when $u \in U_A$. We also have $\bar{u}(m) = \overline{u(m)}$ whenever $u \in L^\infty$ and $m \in \mathfrak{M}$, because μ_m is real. Therefore, if $m \in \mathfrak{M}$ and $u \in U_A$, then $1 = u(m)\bar{u}(m) = |u(m)|^2$. Conversely, if $|u(m)| = \|u\| = 1$, then $u = u(m)$ on the closed support of μ_m, because μ_m is a probability measure. Thus if m satisfies $|u(m)| = 1$ for all $u \in U$, then the restriction of $A = [H^\infty, U]$ to the closed support of μ_m coincides with the same restriction of H^∞. That means m is multiplicative on A. $\qquad \square$

By Theorem 1.3, distinct Douglas algebras have distinct maximal ideal spaces, because the spectrum $\mathfrak{M}_A$ determines which inner functions are in A^{-1}. Thus a corollary of the Chang–Marshall theorem is that every closed algebra between H^∞ and L^∞ is uniquely determined by its spectrum. A similar situation exists with Wermer's maximality theorem: An algebra between the disc algebra A_o and $C(T)$ is either the disc algebra itself or $C(T)$. The choice there depends on whether or not the inner function z is invertible in the given algebra; that is, on whether or not the maximal ideal space contains zero. Hoffman and Singer based a proof of Wermer's theorem on just that distinction. In our setting their argument yields the following result.

Theorem 1.4. *If A is a closed subalgebra of L^∞ containing H^∞, then either $A = H^\infty$ or $A \supset [H^\infty, \bar{z}]$.*

Proof. If $z \in A^{-1}$, then $\bar{z} \in A$ and $[H^\infty, \bar{z}] \subset A$. If $z \notin A^{-1}$ then z lies in some maximal ideal of A, and there is $m \in \mathfrak{M}_A$ with $z(m) = 0$. Since $\mathfrak{M}_A \subset \mathfrak{M}$, the only possible such m is evaluation at the origin. Then by the uniqueness of representing measures, $d\theta/2\pi$ is multiplicative on A and

$$\frac{1}{2\pi} \int e^{in\theta} f(\theta)d\theta = 0, \qquad n = 1, 2, \ldots ,$$

for all $f \in A$. Hence $A \subset H^\infty$. $\square$

Now there are many algebras between H^∞ and L^∞, but each such algebra is determined by its invertible inner functions (by the Chang–Marshall theorem) or equivalently, by its maximal ideal space. The Chang–Marshall proof consists of two steps.

(i) If A is a Douglas algebra, and if B is another algebra having the same spectrum, then $B = A$.

(ii) Given an algebra B, $H^\infty \subset B \subset L^\infty$, there is a Douglas algebra having the same spectrum as B.

The proof of Theorem 1.4 consists of two similar steps and, in hindsight, it can be said that the strategy underlying the Chang–Marshall proof originates in the Hoffman–Singer argument about maximal ideals.

There are two C^*-algebras associated with an algebra A such that $H^\infty \subset A \subset L^\infty$. The first is

$$Q_A = A \cap \bar{A},$$

the largest self-adjoint subalgebra of A. The second is

$$C_A = [\mathrm{B}_A, \bar{\mathrm{B}}_A],$$

the self-adjoint algebra (or C^*-algebra) generated by B_A, those inner functions invertible in A. For $A = H^\infty$, $Q_A = C_A = \mathbb{C}$, the space of complex numbers. For $A = L^\infty$, $Q_A = L^\infty$, and by the Douglas–Rudin theorem, $C_A = L^\infty$ also. However, in general Q_A and C_A do not coincide. Along with solving the Douglas problem, we want to understand the C^*-algebras Q_A and C_A. Let us first turn to the simplest special case, so that we may know what to expect in the general case.

2. $H^\infty + C$

Let $C = C(T)$ denote the continuous functions on the unit circle and let $H^\infty + C = \{f + g : f \in H^\infty, g \in C\}$, which is a linear subspace of L^∞.

Lemma 2.1. $H^\infty + C$ *is uniformly closed.*

Proof. Recall from Theorem IV.1.6 that when $g \in C$

$$\mathrm{dist}(g, H^\infty) = \mathrm{dist}(g, A_o),$$

where A_o is the disc algebra. If $h \in L^\infty$ lies in the closure of $H^\infty + C$, there are $f_n \in H^\infty$ and $g_n \in C$ such that $\|h - (f_n + g_n)\| < 2^{-n}$. Then $\mathrm{dist}((g_n - g_{n+1}), H^\infty) < 2^{-n+1}$ and there are $k_n \in A_o$ such that $\|(g_n - g_{n+1}) - k_n\|_\infty < 2^{-n+1}$. Write $K_1 = 0$ and $K_n = K_1 + \cdot 0 \cdot + k_{n-1}$ for $n > 1$. Then

$G_n = g_n + K_n \in C$ and $\|G_n - G_{n+1}\|_\infty < 2^{-n+1}$. Hence $\{G_n\}$ has a uniform limit $g \in C$. But then $f_n = F_n - K_n = (f_n + g_n) - G_n$ is in H^∞ and $\{F_n\}$ converges in norm to $h - g$. Since H^∞ is closed, $h - g \in H^\infty$ and $h \in H^\infty + C$. $\square$

Theorem 2.2. $H^\infty + C$ *is a closed subalgebra of* L^∞. *In fact,*

$$H^\infty + C = [H^\infty, \bar{z}].$$

The maximal ideal space of $H^\infty + C$ *is* $\mathfrak{M} \backslash D$, *the complement of the unit disc in* $\mathfrak{M}$.

Proof. The set of functions

$$f(z) + \sum_1^N a_k \bar{z}^k, \quad f \in H^\infty,$$

is, by definition, dense in $[H^\infty, \bar{z}]$. By the Weierstrass theorem, this set is also dense in $H^\infty + C$. Because $H^\infty + C$ is closed, that means $H^\infty + C = [H^\infty, \bar{z}]$. Theorem 1.3 then implies that $\mathfrak{M} \backslash D$ is the spectrum of $H^\infty + C$. $\square$

Consequently $H^\infty + C$ is a Douglas algebra. An inner function $b(z)$ is invertible in $H^\infty + C$ if and only if $|b(m)| > 0$ on $\mathfrak{M} \backslash D$, which happens if and only if $b(z)$ is a finite Blaschke product. Hence $B_{H^\infty + C}$ is the set of finite Blaschke products, and the corresponding self-adjoint algebra $C_{H^\infty + C}$ coincides with C.

Now let us determine $Q_{H^\infty + C}$, which is called QC, the space of *quasicontinuous functions*. The conformal mapping $f(z)$ of D onto $\{0 < x < 1, -2 < y < \sin(1/x)\}$ is an H^∞ function with continuous real part but discontinuous imaginary part. Then Im $f \in H^\infty + C$, and because it is real valued, Im $f \in QC$. Thus $QC \neq C$.

Theorem 2.3. $QC = L^\infty \cap \text{VMO}$.

Proof. If $f \in L^\infty \cap \text{VMO}$, then by Chapter VI, Theorem 5.2 there are $\phi, \psi \in C$ such that $f = \phi + H\psi$. But then $H\psi \in L^\infty$ and $\psi + iH\psi \in H^\infty$. Thus

$$f = -i(\psi + iH\psi) + (\varphi + i\psi) \in H^\infty + C.$$

The same holds for $\bar{f}$, so that $f \in (H^\infty + C) \cap \overline{(H^\infty + C)} = QC$.

Conversely suppose $f \in H^\infty + C$ is real valued. Then

$$f = (u + iHu) + (v + iw),$$

with $u + iHu \in H^\infty$ and $v + iw \in C$. Since f is real, $Hu = -w \in C$ and $u = Hw$. Therefore $f = v + Hw, v \in C, w \in C$, and $f \in \text{VMO}$. Because QC is spanned by the real functions in $H^\infty + C$ we conclude that $QC \subset L^\infty \cap \text{VMO}$. $\square$

Corollary 2.4. *Let* $f \in L^\infty$ *and let* $f(z)$ *be its Poisson integral. If* $|f(z)|$ *extends continuously to* $\bar{D}$ *then* $f \in QC$.

Proof. If $1 - |z|$ is small enough, and if $e^{i\theta}$ is close to $z/|z|$, then by hypothesis $|f(e^{i\theta})|$ is close to $|f(z)|$. Thus

$$\frac{1}{2\pi} \int |f(e^{i\theta}) - f(z)|^2 P_z(\theta) d\theta = \frac{1}{2\pi} \int |f(e^{i\theta})|^2 P_z(\theta) d\theta - |f(z)|^2 < \varepsilon$$

because $P_z(\theta) d\theta / 2\pi$ has most of its mass near $z/|z|$. This means $f \in \text{VMO}$, and so $f \in QC$ by Theorem 2.3. $\square$

Corollary 2.5. *Let A be a closed subalgebra of L^∞ containing H^∞. If $\mathfrak{M}_A = \mathfrak{M} \setminus D$, then $A = H^\infty + C$.*

Proof. It is clear that $H^\infty + C \subset A$ because $z \in A^{-1}$. Now if $f, g \in A$, then

$$d(m) = (fg)(m) - f(m)g(m)$$

is continuous on $\mathfrak{M}$ and $d(m) = 0$ on $\mathfrak{M}_A = \mathfrak{M} \setminus D$. By continuity there is $\delta > 0$ such that

$$(2.1) \qquad |(fg)(z) - f(z)g(z)| < \varepsilon, \qquad 1 - |z| < \delta.$$

Letting $f = u, g = \bar{u}, u \in \mathrm{U}_A$, we obtain $|1 - |u(z)|^2| < \varepsilon$ for $1 - |z| < \delta$. Then by Corollary 2.4, $\mathrm{U}_A \subset H^\infty + C$, and by Theorem 1.1, $A \subset H^\infty + C$. $\square$

Notice how the abstract condition $\mathfrak{M}_A = \mathfrak{M} \setminus D$ was brought into the last proof. It was converted into the more manageable condition (2.1) that the Poisson integral is "asymptotically multiplicative" on A. Since by the corona theorem, (2.1) is equivalent to the hypothesis $\mathfrak{M}_A \supset \mathfrak{M} \setminus D$, we see that if $H^\infty \subset A \subset L^\infty$ and if the Poisson integral is asymptotically multiplicative on A, then either $A = H^\infty$ or $A = H^\infty + C$.

Much of the proof of the Chang–Marshall theorem will amount to generalizing Theorem 2.3 and Corollary 2.5 to arbitrary Douglas algebras.

3. The Chang–Marshall Theorem

Theorem 3.1. *If B is a closed subalgebra of L^∞ containing H^∞, then there is a set B of inner functions in H^∞ such that*

$$B = [H^\infty, \bar{\mathrm{B}}].$$

In other words, every uniform algebra between H^∞ and L^∞ is a Douglas algebra. The proof of this theorem breaks up into two pieces. We reverse the historical order, giving Marshall's part first.

Theorem 3.2. *If B is a uniform algebra with $H^\infty \subset B \subset L^\infty$, then there is a set B of interpolating Blaschke products such that*

$$\mathfrak{M}_B = \mathfrak{M}_{[H^\infty, \bar{\mathrm{B}}]}.$$

The theorem says there is a Douglas algebra having the same spectrum as B.

Lemma 3.3. *If $b(z)$ is an interpolating Blaschke product having zeros $\{z_n\}$ in D and if $m \in \mathfrak{M}$ is such that $b(m) = 0$, then m lies in the closure of $\{z_n\}$ with respect to the topology of $\mathfrak{M}$.*

Proof. Assuming the contrary, we have $f_1, f_2, \ldots, f_k \in H^\infty$, with $f_k(m) = 0$, such that $\{z_n\}$ is disjoint from

$$\bigcap_{j=1}^{k} \{z : |f_j(z)| < 1\}.$$

Then $\max_j |f_j(z_n)| \geq 1, n = 1, 2, \ldots$, and because $\{z_n\}$ is an interpolating sequence, there are $g_1, g_2, \ldots, g_k \in H^\infty$ such that $G = f_1 g_1 + \cdots + f_k g_k$ satisfies $G(z_n) = 1, n = 1, 2, \ldots$ Hence $1 = G + bh$ for some $h \in H^\infty$. But since $G(m) = 0$, this is a contradiction. $\square$

Proof of Theorem 3.2. We have already done the hard work back in the proof of Theorem VIII.4.1. For each $u \in U_B$ and for each $\alpha, 0 < \alpha < 1$, Theorem VIII.4.1 gives us an interpolating Blaschke product $b_{\alpha,u}$ such that when $z \in D$

$$(3.1) \qquad\qquad |b_{\alpha,u}(z)| \leq \tfrac{1}{2} \quad \text{if} \quad |u(z)| \leq \alpha$$

and

$$(3.2) \qquad\qquad |u(z)| \leq \beta(\alpha) < 1 \qquad \text{if} \quad b_{\alpha,u}(z) = 0.$$

Set $B = \{b_{\alpha,u} : u \in U_B, 0 < \alpha < 1\}$. We claim that $[H^\infty, \bar{B}]$ has the same spectrum as B.

If $b_{\alpha,u}(m) = 0$, then by (3.2) and Lemma 3.3 $|u(m)| \leq \beta(\alpha) < 1$. This means $m \notin \mathfrak{M}_B$ by Theorem 1.3. Hence each $b_{\alpha,u}$ is invertible in B, so that $[H^\infty, \bar{B}] \subset B$ and $\mathfrak{M}_B \subset \mathfrak{M}_{[H^\infty, \bar{B}]}$, by (1.2).

Now suppose $m \in \mathfrak{M}_{[H^\infty, \bar{B}]}$. By Theorem 1.3, $|b_{\alpha,u}(m)| = 1$ for all $b_{\alpha,u} \in B$. By the corona theorem there is a net (z_j) in D that converges to m. Consequently $\lim_j |b_{\alpha,u}(z_j)| = 1$ for each $b_{\alpha,u} \in B$, and (3.1) yields $|u(m)| = 1$ for all $u \in U_B$. Thus $m \in \mathfrak{M}_B$, again by Theorem 1.3, and $\mathfrak{M}_{[H^\infty, \bar{B}]} \subset \mathfrak{M}_B$. $\square$

We now turn to Chang's half of Theorem 3.1.

Theorem 3.4. *Let A and B be two subalgebras of L^∞ containing H^∞. Assume that $\mathfrak{M}_A = \mathfrak{M}_B$ and that A is a Douglas algebra. Then $A = B$.*

Clearly, Theorems 3.2 and 3.4 prove the main result, Theorem 3.1. In fact, something stronger is true, because Theorem 3.2 gives a family of interpolating Blaschke products: *Every uniform algebra A such that $H^\infty \subset A \subset L^\infty$ is generated by H^∞ and the complex conjugates of a set of interpolating Blaschke products.*

The proof of Theorem 3.4 depends on a characterization of a Douglas algebra $A = [H^\infty, \bar{B}\,]$ in terms of Poisson integrals. For $b(z)$ an inner function and for $0 < \delta < 1$, we define the region

$$G_\delta(b) = \{z : |b(z)| > 1 - \delta\}.$$

For example, $G_\delta(z)$ is the annulus $1 - \delta < |z| < 1$, while if $b = \exp((z + 1)/(z - 1))$, $G_\delta(b)$ is the region between T and a disc in D tangent to T at $z = 1$. When $f \in L^1(T)$, let

$$dv_f = \left|\frac{\partial f}{\partial \bar{z}}\right|^2 \log \frac{1}{|z|} dx\, dy.$$

A related measure, with $|\partial f/\partial \bar{z}|^2$ replaced by $|\nabla f|^2$, was denoted by λ_f in Chapter VI, Section 3. Since

$$|\nabla f|^2 = 2(|\partial f/\partial z|^2 + |\partial f/\partial \bar{z}|^2)$$

and since $\partial f/\partial z = \partial \hat{f}/\partial \bar{z}$, we have $\lambda_f = 2(v_f + v_{\hat{f}})$. In particular, when f is real-valued, λ_f and v_f are equivalent. From Chapter VI, Exercise 14, we know that $f \in \text{VMO}$ if and only if to each $\varepsilon > 0$ there corresponds $\delta, 0 < \delta < 1$, such that when $0 < h < \delta$

$$\lambda_f(S(\theta_0, h)) < \varepsilon h,$$

where $S(\theta_0, h)$ is the sector $\{re^{i\theta} : |\theta - \theta_0| \le h, 1 - h \le r < 1\}$. Therefore, by Theorem 2.3, $QC = (H^\infty + C) \cap \overline{(H^\infty + C)}$ consists of those $f \in L^\infty$ such that

$$\lambda_f(S(\theta_0, h)) < \varepsilon h$$

when $h < \delta = \delta(\varepsilon, f)$. An arbitrary Douglas algebra is characterized by a similar condition with $\{1 - \delta < |z| < 1\}$ replaced by some region $G_\delta(b), b \in B_A$. So that we may characterize A instead of the self-adjoint algebra $Q_A = A \cap \bar{A}$, we must also replace λ_f by v_f.

Theorem 3.5. *Let $A = [H^\infty, \bar{B}_A]$ be a Douglas algebra. When $f \in L^\infty$ the following conditions are equivalent.*

(i) *$f \in A$.*

(ii) *For any $\varepsilon > 0$, there are $b \in B_A$ and $\delta, 0 < \delta < 1$, such that for all $z \in G_\delta(b)$*

$$(3.3) \qquad \inf_{g \in H^2} \frac{1}{2\pi} \int |f - g|^2 P_z(\theta) d\theta < \varepsilon.$$

(iii) *For any $\varepsilon > 0$, there are $b \in B_A$ and $\delta, 0 < \delta < 1$, such that*

$$(3.4) \qquad v_f(G_\delta(b) \cap S(\theta_0, h)) < \varepsilon h$$

for every sector

$$S(\theta_0, h) = \{re^{i\theta} : |\theta - \theta_0| \le h, 1 - h \le r < 1\}.$$

Before proving this theorem we use it to derive Theorem 3.4.

Proof of Theorem 3.4. Let A be a Douglas algebra and let B be another algebra, $H^\infty \subset B \subset L^\infty$, such that $\mathfrak{M}_A = \mathfrak{M}_B$. By Theorem 1.3, we have $\overline{B}_A \subset B$, so that $A = [H^\infty, \overline{B}_A] \subset B$.

To prove that $B \subset A$ recall that B is generated by H^∞ and U_B, the unimodular functions in B^{-1}. We show $u \in \mathsf{U}_B$ satisfies (3.3) for every $\varepsilon > 0$. Since $\mathfrak{M}_A = \mathfrak{M}_B$, Theorem 1.3 implies $|u(m)| = 1$ on

$$\mathfrak{M}_A = \bigcap_{b \in \mathsf{B}_A} \{m : |b(m)| = 1\}.$$

By compactness, this means $|u| > 1 - \varepsilon/2$ on some finite intersection of sets $\{|b(m)| =\}, b \in \mathsf{B}_A$. Taking a product, we obtain a single $b \in \mathsf{B}_A$ such that $|u(m)| > 1 - \varepsilon/2$ when $|b(m)| = 1$. Consequently, there is $\delta > 0$ such that $|u(z)| > 1 - \varepsilon/2$ when $z \notin G_\delta(b)$. But then

$$\frac{1}{2\pi} \int |u(\theta) - u(z)|^2 P_z(\theta) d\theta = \frac{1}{2\pi} \int |u(\theta)|^2 P_z(\theta) d\theta - |u(z)|^2$$
$$= 1 - |u(z)|^2 < \varepsilon,$$

when $z \in G_\delta(b)$. Thus (3.3) holds with g the constant function $u(z)$. By Theorem 3.5, $B \subset A$ and Theorem 3.4 is proved. $\square$

Proof of Theorem 3.5. We show (i) $\Rightarrow$ (ii) $\Rightarrow$ (iii) $\Rightarrow$ (i).

Assume (i) holds. Then there are $b \in \mathsf{B}_A$ and $h \in H^\infty$ such that $\|f - \bar{b}h\|_\infty < \varepsilon$. Fix $z_0 \in G_\delta(b)$ and let $g = \overline{b(z_0)}h$. Then $g \in H^2$ and

$$\frac{1}{2\pi} \int |\bar{b}h - g|^2 P_{z_0}(\theta) d\theta \le \frac{\|h\|_\infty^2}{2\pi} \int |b(\theta) - b(z_0)|^2 P_{z_0}(\theta) d\theta$$
$$= \|h\|_\infty^2 (1 - |b(z_0)|^2) \le 2\delta \|h\|_\infty^2.$$

Consequently,

$$\frac{1}{2\pi} \int |f - g|^2 P_{z_0}(\theta) \, d\theta \le 2\varepsilon^2 + 4\delta \|h\|_\infty^2,$$

which gives (3.3) for $\varepsilon < \frac{1}{2}$ if δ is small enough.

Now assume (ii) and choose $b \in \mathsf{B}_A$ and $\delta, 0 < \delta < 1$, so that (3.3) holds. For the moment suppose that $G_\delta(b) \cap S(\theta_0, h) \subset \{|z| > \frac{1}{4}\}$. By a stopping time argument, which should by now be quite familiar, it is enough to prove (3.4) for a sector whose inside half $\{|\theta - \theta_0| \le h, 1 - h \le r < 1 - h/2\}$ contains a point $z_1 \in G_\delta(b)$. (Otherwise partition the outside half of $S(\theta_0, h)$ into two sectors $S(\theta_1, h/2)$ and continue, stopping at maximal sectors whose inside halves meet $G_\delta(b)$.) Let $k(z) = f(z) - g(z)$, where $g \in H^2$ is chosen to attain the infimum (3.3) with respect to $P_{z_1}(\theta) d\theta$. Then $k(z)$ is conjugate analytic and $k(z_1) = 0$, because in the Hilbert space $L^2(P_{z_1} d\theta)$, g is the orthogonal projection of f onto H^2. Consequently,

$$|\nabla k(z)|^2 = 2|\partial k/\partial \bar{z}|^2 = 2|\partial f/\partial \bar{z}|^2,$$

and the Littlewood–Paley identity (3.3) of Chapter VI gives

$$(3.5) \qquad \frac{1}{2\pi} \int |f - g|^2 P_{z_1}(\theta) d\theta = \frac{2}{\pi} \iint \left|\frac{\partial f}{\partial \bar{z}}\right|^2 \log \left|\frac{1 - \bar{z}_1 z}{z - z_1}\right| dx\, dy.$$

On $G_\delta(b) \cap S(\theta_0, h) \subset \{|z| > \frac{1}{4}\}$ we have the familiar inequalities

$$\log \frac{1}{|z|} \leq C(1 - |z|^2) \leq Ch \frac{(1 - |z|^2)(1 - |z_1|^2)}{|1 - \bar{z}_1 z|^2}$$

$$= Ch \left(1 - \frac{|z - z_1|^2}{|1 - \bar{z}_1 z|^2}\right) \leq Ch \log \left|\frac{z - \bar{z}_1 z}{z - z_1}\right|,$$

so that by (3.5)

$$\nu_f(G_\delta(b) \cap S(\theta_{0,h})) = \iint\limits_{G_\delta(b) \cap S(\theta_0, h)} \left|\frac{\partial f}{\partial \bar{z}}\right|^2 \log \frac{1}{|z|} dx\, dy$$

$$\leq Ch \int |f - g|^2 P_{z_1}(\theta)\, d\theta \leq C\varepsilon h$$

by (3.3). That is (3.4), the inequality we wanted to prove.

There remains the rather uninteresting case $G_\delta(b) \cap S(\theta, h) \cap \{|z| \leq \frac{1}{4}\} \neq \emptyset$. If $b(z)$ is not constant, it can be replaced by $b^N(z)$, and $G_\delta(b^N) \subset \{|z| > \frac{1}{4}\}$ if N is large. If $b(z)$ is constant, then $0 \in G_\delta(b)$ and (3.3) at $z = 0$ gives (3.4) for sectors meeting $\{|z| \leq \frac{1}{4}\}$. Thus (iii) follows from (ii).

We come to the main step, (iii) $\Rightarrow$ (i). Let $\varepsilon > 0$, and fix $b \in B_A$ and $\delta, 0 < \delta < 1$, so that we have (3.4). We estimate

$$\text{dist}(f, A) \leq \text{dist}(f, \bar{b}^n H^\infty) = \text{dist}(b^n f, H^\infty) = \sup_{\substack{F \in H_0^1 \\ \|F\|_1 = 1}} \frac{1}{2\pi} \int f b^n F\, d\theta,$$

as we did in the proof of the H^1–BMO duality in Chapter VI, Section 4. Note that when $F \in H^1$

$$(\nabla f) \cdot (\nabla b^n F) \equiv f_x (b^n F)_x + f_y (b^n F)_y = 2\left(\frac{\partial f}{\partial \bar{z}}\right) \frac{\partial}{\partial z}(b^n F).$$

Since $F(0) = 0$, the polarized Littlewood–Paley identity now yields

$$(3.6) \qquad \frac{1}{2\pi} \int f b^n F\, d\theta = \frac{2}{\pi} \iint \frac{\partial f}{\partial \bar{z}} b^n \frac{\partial F}{\partial z} \log \frac{1}{|z|} dx\, dy$$

$$+ \frac{2}{\pi} \iint \frac{\partial f}{\partial \bar{z}} n b^{n-1} \frac{\partial b}{\partial z} F(z) \log \frac{1}{|z|} dx\, dy.$$

Writing $F = b_0 H$, where b_0 is a Blaschke product and H is zero free, we have $F = ((b_0 - 1)H + (b_0 + 1)H)/2 = (G_1^2 + G_2^2)/2$, where $G_j \in H^2$, $\|G_j\|_2^2 \leq 2\|F\|_1$. Thus we can make the additional assumption $F = G^2$, $G \in H^2$, when we bound the right side of (3.6), which then becomes

$$\frac{4}{\pi} \iint\limits_{G_\delta(b)} \frac{\partial f}{\partial \bar{z}} b^n G(z) \frac{\partial G}{\partial z} \log \frac{1}{|z|} dx\, dy + \frac{4}{\pi} \iint\limits_{D \backslash G_\delta(b)} \frac{\partial f}{\partial \bar{z}} b^n G(z) \frac{\partial G}{\partial z} \log \frac{1}{|z|} dx\, dy$$

$$+ \frac{2}{\pi} \iint\limits_{G_\delta(b)} \frac{\partial f}{\partial \bar{z}} \frac{\partial b^n}{\partial z} G^2(z) \log \frac{1}{|z|} dx\, dy$$

$$+ \frac{2}{\pi} \iint\limits_{D \backslash G_\delta(b)} \frac{\partial f}{\partial \bar{z}} nb^{n-1} \frac{\partial b}{\partial z} G^2(z) \log \frac{1}{|z|} dx\, dy$$

$$= I_1 + I_2 + I_3 + I_4.$$

Use the Cauchy–Schwarz inequality on each of I_1, I_2, I_3, and I_4. We obtain

$$I_1 \leq C \left(\iint\limits_{G_\delta(b)} \left| \frac{\partial f}{\partial \bar{z}} \right|^2 |b^n G|^2 \log \frac{1}{|z|} dx\, dy \right)^{1/2} \left(\iint \left| \frac{\partial G}{\partial z} \right|^2 \log \frac{1}{|z|} dx\, dy \right)^{1/2}$$

$$\leq C(\varepsilon \|b^n G\|^2)^{1/2} \|G\|_2 = C\varepsilon^{1/2} \|F\|_1$$

by (3.4), the theorem on Carleson measures, and the Littlewood–Paley identity. Similarly, we get

$$I_2 \leq C(1 - \delta)^n \left(\iint\limits_{D} \left| \frac{\partial f}{\partial \bar{z}} \right|^2 |G|^2 \log \frac{1}{|z|} dx\, dy \right)^{1/2}$$

$$\times \left(\iint\limits_{D} \left| \frac{\partial G}{\partial \bar{z}} \right|^2 \log \frac{1}{|z|} dx\, dy \right)^{1/2}$$

$$\leq C(1 - \delta)^n (\|f\|_* \|G\|_2)(\|G\|_2)$$

$$\leq C(1 - \delta)^n \|f\|_\infty \|F\|_1$$

by using Theorem 3.4 of Chapter VI instead of condition (3.4).

To estimate I_3 and I_4, let $|G|^2 \log(1/|z|) dx\, dy$ be the measure in the Cauchy–Schwarz inequality. Then we have

$$I_3 \leq C \left(\iint\limits_{G_\delta(b)} \left| \frac{\partial f}{\partial \bar{z}} \right|^2 |G|^2 \log \frac{1}{|z|} dx\, dy \right)^{1/2} \left(\iint \left| \frac{\partial b^n}{\partial z} \right|^2 |G|^2 \log \frac{1}{|z|} dx\, dy \right)^{1/2}$$

$$\leq C(\varepsilon \|G\|_2^2)^{1/2} (\|b^n\|_*^2 \|G\|_2^2)^{1/2}$$

$$\leq C_\varepsilon^{1/2} \|F\|_1,$$

using both (3.4) and Theorem 3.4 of Chapter VI. As with I_2, we have

$$
I_4 \leq Cn(1-\delta)^{n-1} \left(\iint\limits_D \left| \frac{\partial f}{\partial \bar{z}} \right|^2 |G|^2 \log \frac{1}{|z|} dx\, dy \right)^{1/2}
$$

$$
\times \left(\iint\limits_D \left| \frac{\partial b}{\partial z} \right|^2 |G|^2 \log \frac{1}{|z|} dx\, dy \right)^{1/2}
$$

$$
\leq Cn(1-\delta)^{n-1}(\|f\|_*\|G\|_2)(\|b\|_*\|G\|_2)
$$

$$
\leq Cn(1-\delta)^{n-1}\|f\|_\infty\|F\|_1,
$$

this time twice using Chapter VI, Theorem 3.4.

Taken together, the estimates yield

$$
\text{dist}(f, A) \leq \lim_{n\to\infty} (c\varepsilon^{1/2} + C(1-\delta)^{n-1}\|f\|_\infty + Cn(1-\delta)^{n-1}\|f\|_\infty) = C\varepsilon^{1/2}
$$

and $f \in A$. $\square$

It is worth reflecting on the role played by the maximal ideal space in the proof of Theorem 3.1. It is possible to merge the two parts of the proof and to write down a proof that is almost free of Banach algebra theory (see Exercise 13). At present the only place where maximal ideal spaces are still needed is when Lemma 3.3 is applied, and it is quite possible that there will be a future proof containing no reference to Banach algebras. Such a constructive proof is certainly worth finding, if only for the new ideas it should yield. However, it is difficult to imagine a first proof of this theorem not using maximal ideals. Banach algebra theory sets the framework on which the proof can be built, it points out which inequalities one should aim for, and it provides the economy of thought necessary for a long and difficult proof.

4. The Structure of Douglas Algebras

Many of Sarason's results in Section 2 have generalizations to an arbitrary Douglas algebra A. Because it is so close at hand, we begin with the analog of Theorem 2.3. Recall that $Q_A = A \cap \bar{A}$ is the largest C^*-subalgebra of A, and that $\mathrm{B}_A = \{b \in H^\infty : b \text{ inner}, \bar{b} \in A\}$. Define VMO_A to be the set of $f \in$ BMO such that to each $\varepsilon > 0$, there correspond $\delta, 0 < \delta < 1$, and $b \in \mathrm{B}_A$ such that

$$
(4.1) \quad \lambda_f(G_\delta(b) \cap S(\theta_0, h)) = \iint\limits_{G_\delta(b)\cap S(\theta_0,h)} |\nabla f|^2 \log \frac{1}{|z|} dx\, dy < \varepsilon h
$$

for every sector $S(\theta_0, h)$. When $A = H^\infty + C$, VMO_A is just our old friend VMO.

Theorem 4.1. *If A is a closed subalgebra of L^∞ containing H^∞, then*

$$Q_A = L^\infty \cap \text{VMO}_A.$$

Proof. The theorem follows directly from Theorem 3.5. Since

$$\lambda_f = 2(\nu_f + \nu_{\bar f})$$

and ν_f and $\nu_{\bar f}$ are positive, (4.1) means that both f and $\bar f$ satisfy condition (3.4), and so $f \in A \cap \bar A = Q_A$. Conversely, if $f \in Q_A$ and if $\varepsilon > 0$, then there are $b_1 \in B_A$ and $\delta_1, 0 < \delta_1 < 1$, for which f satisfies (3.4). There are also $b_2 \in B_A$ and $\delta_1, 0 < \delta_2 < 1$, for which $\bar f$ satisfies (3.4). Therefore (4.1) holds with $b = b_1 b_2$ and $\delta = \min(\delta_1, \delta_2)$. $\square$

The space VMO_A can also be described in terms of Hilbert space projections. If $f \in \text{BMO}$, then $f \in \text{VMO}_A$ if and only if for each $\varepsilon > 0$ there exist $b \in B_A$ and $\delta, 0 < \delta < 1$, such that

$$(4.2) \qquad\qquad \frac{1}{2\pi} \int |f - f(z)|^2 P_z(\theta) d\theta < \varepsilon$$

for all $z \in G_\delta(b)$. The derivation of (4.2) is left as an exercise.

Recall the notation $C_A = [B_A, \bar B_A]$ for the C^*-algebra generated by B_A.

Theorem 4.2. *If A is a closed subalgebra of L^∞ containing H^∞, then*

$$A = H^\infty + C_A.$$

We need a lemma, whose proof we postpone for a moment.

Lemma 4.3. *Let $u \in L^\infty$ with $|u| = 1$ almost everywhere. If*

$$(4.3) \qquad\qquad\qquad \text{dist}(u, H^\infty) < 1$$

but

$$(4.4) \qquad\qquad\qquad \text{dist}(u, H_0^\infty) = 1,$$

where $H_0^\infty = \{f \in H^\infty : f(0) = 0\}$, then

$$\bar u \in [H^\infty, u].$$

Proof of Theorem 4.2. When $f \in C_A$ we claim

$$(4.5) \qquad\qquad \text{dist}(f, H^\infty) = \text{dist}(f, H^\infty \cap C_A),$$

generalizing Lemma 1.6 of Chapter IV. Clearly $\text{dist}(f, H^\infty) \leq \text{dist}(f, H^\infty \cap C_A)$. Because (4.5) is trivial when $\text{dist}(f, H^\infty) = 0$, we can assume $\text{dist}(f, H^\infty) = 1 - \varepsilon$ for some small $\varepsilon > 0$. Since $f \in C_A$, there is $g = \sum_{j=1}^n \lambda_j b_j, b_j \in B_A$, and there is $b_0 \in B_A$ such that $\|f - \bar b_0 g\|_\infty < \varepsilon$. Then $\text{dist}(\bar b_0 g, H^\infty) < 1$. Looking back at the theory of dual extremal problems, in particular at Case I in Theorem 4.3 of Chapter IV, we see there exists $u \in L^\infty, |u| = 1$ almost everywhere, such that

$$(4.6) \qquad\qquad u - \bar{b}_0 g \in H^\infty,$$

which means that $\text{dist}(u, H^\infty) < 1$, and such that

$$(4.7) \qquad\qquad \text{dist}(u, H_0^\infty) = 1.$$

By (4.6), $b_0 u \in H^\infty$—in fact $b_0 u$ is an inner function—and $u \in [H^\infty, \bar{b}_0] \subset A$, so that by (4.7) and Lemma 4.3, $\bar{u} \in A$. Therefore $\bar{b}_0 \bar{u} \in A$ and $b_0 u \in \text{B}_A$. Hence $u = \bar{b}_0(b_0 u) \in C_A$ and $u - \bar{b}_0 g \in C_A$. That gives

$$\text{dist}(f, H^\infty \cap C_A) \le \| f - (\bar{b}_0 g - u) \|_\infty \le \varepsilon + \| u \|_\infty$$
$$\le \text{dist}(f, H^\infty) + 2\varepsilon,$$

and (4.5) is established.

Now (4.5) implies that $H^\infty + C_A$ is uniformly closed. This has the same proof as Lemma 2.1, but for reasons of pedagogy we phrase the argument differently. The natural injection

$$\pi : C_A \to A/H^\infty$$

has kernel $H^\infty \cap C_A$ and (4.5) shows that $\pi(C_A)$ is closed in A/H^∞. Under the quotient mapping $A \to A/H^\infty$, $\pi(C_A)$ has inverse image $H^\infty + C_A$. Thus $H^\infty + C_A$ is closed in A.

To conclude the proof of Theorem 4.2 it now suffices to show that $H^\infty + C_A$ is dense in A, and because A is a Douglas algebra it is enough to show $f = \bar{b} g \in H^\infty + C_A$ whenever $b \in \text{B}_A$ and $g \in H^\infty$. We suppose $\| f \| < 1$. Again using Case I of Theorem 4.3 of Chapter IV, we have $u \in L^\infty, |u| = 1$ almost everywhere, such that

$$(4.8) \qquad u - f = u - \bar{b} g \in H^\infty \qquad \text{and} \qquad \text{dist}(u, H_0^\infty) = 1.$$

Then bu is an inner function in H^∞, and $u \in A$. By Lemma 4.3, $\bar{u} \subset [H^\infty, u] \subset A$ so that $bu \in \text{B}_A$. But then $u = \bar{b}(bu) \in C_A = [\text{B}_A, \bar{\text{B}}_A]$, and by (4.8) $f = \bar{b} g \in H^\infty + C_A$. $\square$

Proof of Lemma 4.3. Let $f \in H^\infty$ with $\| u - f \|_\infty = \alpha < 1$. We first show $\inf_z |f(z)| \ge 1 - \alpha$, which means that f is invertible in H^∞. Now if $z_0 \in D$ and if $|f(z_0)| < 1 - \alpha$, then

$$\| u - (f - f(z_0)) \| < 1.$$

Writing $f(z) - f(z_0) = (z - z_0) g(z), g \in H^\infty$, we then have, on $|z| = 1$,

$$\left\| 1 - \bar{u} z \frac{|1 - z_0 \bar{z}|^2}{1 - \bar{z}_0 z} g \right\| = \| u - (z - z_0) g \| < 1.$$

For $(1 - |z_0|)^2/(1 + |z_0|)^2 \le t \le 1$, it follows easily that

$$\left\| 1 - t\bar{u}z \frac{|1 - z_0\bar{z}|^2}{1 - \bar{z}_0 z} g \right\| < 1,$$

because discs are convex. Setting $t = (1 - |z_0|)^2/|1 - z_0\bar{z}|^2$, we obtain

$$\left\| u - \frac{z(1 - |z_0|)^2}{1 - \bar{z}_0 z} g \right\| = \left\| 1 - t\bar{u}z \frac{|1 - \bar{z}_0 z|^2}{1 - \bar{z}_0 z} g \right\| < 1,$$

which contradicts (4.4). Hence $\inf_z |f(z)| \ge 1 - \alpha$, and $f \in (H^\infty)^{-1}$.

Since $|1 - \bar{u}f| \le \alpha$ and since the disc $\{|1 - w| < \alpha\}$ is invariant under the mapping

$$w \to \frac{1 - \alpha^2}{w},$$

we have

$$\left| 1 - \frac{(1 - \alpha^2)u}{f} \right| = \left| 1 - \frac{(1 - \alpha^2)}{\bar{u}f} \right| < \alpha.$$

Consequently, u/f is invertible in $[H^\infty, u]$, so that

$$\bar{u} = f^{-1}(u/f)^{-1} \in [H^\infty, u],$$

which is what we wanted to show. $\square$

Notice that the above proof gives more: If $|u| = 1$, if $\operatorname{dist}(u, H^\infty) = \alpha < 1$, and if $\operatorname{dist}(u, H_0^\infty) = 1$, then

$$\operatorname{dist}(\bar{u}, H^\infty) = \alpha.$$

Theorem 4.4.   *Let A be a closed subalgebra of L^∞ containing H^∞ and let $f \in BMO$. Then the following conditions are equivalent:*

(i) $f \in VMO_A$,
(ii) $f = u + \tilde{v}, u, v \in Q_A$,
(iii) $f = u + \tilde{v}, u, v \in C_A$,

where $\tilde{v}$ denotes the Hilbert transform or conjugate function of v.

Proof.   Clearly (iii) $\Rightarrow$ (ii), and because VMO_A is self-conjugate, Theorem 4.1 shows that (ii) $\Rightarrow$ (i). To complete the proof we show (i) $\Rightarrow$ (iii).

Lemma 4.5.   *If $f \in BMO$ and if f satisfies (4.1) with respect to some $b \in B_A$, then when n is large*

(4.9)
$$\sup_{\substack{F \in H_0^1 \\ \|F\|_1 \le 1}} \left| \frac{1}{2\pi} \int f(\theta)b^n(\theta)F(\theta)d\theta \right| \le C\varepsilon^{1/2}.$$

The proof of this lemma is exactly the same as the proof of Theorem 3.5. The details are left as an exercise.

Let $f \in \text{VMO}_A$. We may assume f is real valued. Write $f = u + \tilde{v}$, with $u, v \in L^\infty$. Then $g = u + iv \in L^\infty$ and, for any $F \in H_0^1 \cap L^2$,

$$\int (f - g)F \frac{d\theta}{2\pi} = \int (\tilde{v} - iv)F \frac{d\theta}{2\pi} = 0,$$

because $\tilde{v} - iv \in H^2$. By the definition of VMO_A, for every $\varepsilon > 0$ there corresponds $b \in B_A$ and $\delta, 0 < \delta < 1$, such that (4.1) holds. So by (4.9) and by the density of $H_0^1 \cap L^2$ in H_0^1, we have

$$\sup_{\substack{F \in H_0^1 \\ \|F\|_1 \leq 1}} \left| \int g b^n F \frac{d\theta}{2\pi} \right| \leq C\varepsilon^{1/2},$$

when n is large. This means

$$\text{dist}(g, [H^\infty, \bar{b}]) \leq C\varepsilon^{1/2},$$

and as ε is arbitrary, $g \in A$. But then by Theorem 4.2, $g = h + k, h \in C_A, k \in H^\infty$, and

$$f - i\tilde{f} = (u - i\tilde{u}) + (\tilde{v} + iv) = g - i\tilde{g} = h - i\tilde{h},$$

because $k - i\tilde{k} = 0$. Therefore

$$f = \text{Re}(h - i\tilde{h}),$$

and (iii) holds.   □

5. The Local Fatou Theorem and an Application

Let us leave Douglas algebras briefly to take up a basic but elementary result, the local Fatou theorem. It was first proved by Privalov. See Exercise 10 of Chapter II for a sketch of Privalov's proof. The more general real variables argument below is due to A. P. Calderón.

Let G be an open set in the upper half plane $\mathbb{H}$. We say G is *nontangentially dense* at a point $t \in \mathbb{R}$ if there exist $\alpha = \alpha(t) > 0$ and $h = h(t) > 0$ such that G contains the truncated cone

$$\Gamma_\alpha^h(t) = \{x + iy : |x - t| < \alpha y, 0 < y < h\}.$$

A function $u(z)$ is *nontangentially bounded* at t if $u(z)$ is defined and bounded in some truncated cone $\Gamma_\alpha^h(t)$. On the other hand, $u(z)$ has a *nontangential limit* at t if to every $\beta > 0$ there is $h = h(\beta, t) > 0$ such that $u(z)$ is defined on $\Gamma_\beta^h(t)$ and

$$\lim_{\Gamma_\beta^h(t) \ni z \to t} u(z) \quad \text{exists}.$$

Thus the notion of nontangential boundedness involves only one cone while nontangential convergence refers to arbitrarily wide cones.

Theorem 5.1. *Let E be a measurable subset of* $\mathbb{R}$ *and let G be a region in* H *, nontangentially dense at each point of E. Let u(z) be a harmonic function on G, nontangentially bounded at each point of E. Then u(z) has a nontangential limit at almost every point of E.*

It is implicit in the conclusion of the theorem that G contains truncated cones $\Gamma_\beta^h(t)$ with arbitrarily large β at almost every point t of E, whereas the hypotheses only assert that G contains some cone of positive angle at each $t \in E$. However, a simple geometric argument about points of density will enable us to pass from narrow cones to wide cones. The real impact of the theorem lies in the improvement to nontangential convergence over nontangential boundedness. Only because of the application to follow has the open set G been included in the theorem's statement.

Proof. We may assume that E is bounded, $E \subset [-A, A]$. Let $h_n \downarrow 0$ and $\alpha_n \downarrow 0$. Then

$$E \subset \bigcup_{n=1}^{\infty} \{t \in [-A, A] : \Gamma_{\alpha_n}^{h_n}(t) \subset G \text{ and } |u(z)| \le n \text{ on } \Gamma_{\alpha_n}^{h_n}(t)\}.$$

Since these sets increase with n, we can, after discarding a subset of small measure, assume there exist $\alpha > 0, h > 0$, and $M > 0$ such that for all $t \in E$

$$\Gamma_\alpha^h(t) \subset G \quad \text{and} \quad |u(z)| \le M, \quad z \in \Gamma_\alpha^h(t).$$

Discarding another set of small measure, we can replace E by a compact subset F.

Now let

$$\mathrm{R} = \bigcup_{t \in F} \Gamma_\alpha^h(t).$$

If t_0 is a point of density of F (i.e., if t_0 is in the Lebesgue set of χ_F) and if $\beta > 0$, then there is $h_0 = h_0(\beta, t_0)$ such that $\Gamma_\beta^{h_0}(t_0) \subset \mathrm{R}$. Indeed, let $(x, y) \in \Gamma_\beta^{h_0}(t_0)$. If $(x, y) \notin \mathrm{R}$ then the interval $J = \{t : |t - x| < \alpha y\}$ is disjoint from F (see Figure IX.1). Since J is contained in $I = \{t : |t - t_0| < (\alpha + \beta)y\}$, this means

$$\frac{|I \cap F|}{|I|} \le 1 - \frac{|J|}{|I|} = \frac{\beta}{\alpha + \beta},$$

which is a contradiction when $y < h_0$ and h_0 is small. Hence R contains cones of arbitrarily wide aperture at each Lebesgue point of F.

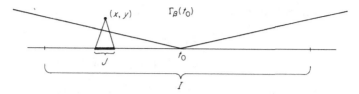

Figure IX.1. $(x, y) \in R$ if and only if $J \cap F \neq \emptyset$.

On R the harmonic function $u(z)$ is defined and bounded, $|u(z)| \leq M$. By the preceding argument, we shall be done when we show

$$(5.1) \qquad \lim_{\substack{z \to t \\ z \in R \cap \Gamma_\beta(t)}} u(z) \text{ exists}$$

for any β and for almost every $t \in F$. Define

$$\varphi_n(t) = \begin{cases} u(t + i/n), & t + i/n \in R, \\ 0, & t + i/n \notin R. \end{cases}$$

Then $\varphi_n \in L^\infty$, $\|\varphi_n\|_\infty \leq M$, and $\{\varphi_n\}$ has a weak-star accumulation point $\varphi \in L^\infty$, $\|\varphi\|_\infty \leq M$. Replace $\{\varphi_n\}$ by a convergent subsequence and write

$$\varphi(z) = \int P_z(t)\varphi(t)\, dt = \lim_{n \to \infty} \int P_z(t)\varphi_n(t)\, dt$$
$$= \lim_{n \to \infty} \varphi_n(z),$$

and

$$\psi(z) = u(z) - \varphi(z) = \lim_{n \to \infty} (u(z + i/n) - \varphi_n(z))$$
$$= \lim_{n \to \infty} \psi_n(z).$$

Then $\psi(z)$ is harmonic on R and $|\psi| \leq 2M$. By a simple majorization we will show

$$(5.2) \qquad \lim_{\substack{z \to t \\ z \in R \cap \Gamma_\beta(t)}} \psi(z) = 0$$

almost everywhere on F. Since $\varphi(z)$ has nontangential limit $\varphi(t)$ almost everywhere, (5.2) implies (5.1) with limit $\varphi(t)$. Let

$$(5.3) \qquad k(z) = c \left(y + \int P_z(t)\chi_{\mathbb{R} \setminus F}(t)\, dt \right),$$

where $c > 0$ will be determined in a moment. Then $k(z)$ has zero nontangential limit almost everywhere on F. Hence (5.2) and the theorem will follow from the estimate

$$(5.4) \qquad |\psi(z)| \leq k(z), \quad z \in R.$$

Finally, (5.4) will hold if for all large n we have

(5.5) $|\psi_n(z)| \le k(z)$

on $R_n = R \cap \{y < h - 1/n\}$, which is that part of R on which ψ_n is defined.

To ensure (5.5) we must choose the constant c in (5.3) so large that

(5.6) $k(z) \ge 2M, \quad z \in \{y > 0\} \cap \partial R_n$.

Taking $c > 3M/h$ gives $k(z) \ge 2M$ when $y \ge h - 1/n$, n large. At any other $z \in \{y > 0\} \cap \partial R_n$, we have $0 < y < h$, and the interval $J = \{t : |t - x| < \alpha y\}$ falls in $\mathbb{R} \backslash F$. Then

$$k(z) \ge \frac{c}{\pi} \int_J \frac{y}{(x - t^2) + y^2} \, dt = \frac{2c}{\pi} \arctan \alpha,$$

and (5.6) holds if c is sufficiently large.

Now if (5.5) were to fail, there would be points $z_j \in R_n$ such that

$$|\psi_n(z_j)| \ge a + k(z_j)$$

for some constant $a > 0$. By the maximum principle, $\{z_j\}$ has an accumulation point $\zeta \in \partial R_n$. It is impossible that $\text{Im } \zeta > 0$ because

$$|\psi_n(z_j)| \le 2M < a + k(\zeta)$$

by (5.6). Hence $\text{Im } \zeta = 0$ and $\zeta \in \mathbb{R} \cap \partial R = F$. But then, since $\varphi_n(t) = u(t + i/n)$ on $\{|t - \zeta| < \alpha/n\}$, our function

$$\psi_n(z) = u(z + i/n) - \int P_z(t)\varphi_n(t) dt$$

is continuous at ζ and $\psi_n(\zeta) = 0$. Therefore $\lim_{z \to \zeta} \psi_n(z) = 0$. This contradiction establishes (5.5) and the theorem is proved. $\square$

Now let $b(z)$ be an inner function on the disc and let $0 < \delta < 1$. The region $G_\delta(b)$ is nontangentially dense at almost every point of T, because $|b(e^{i\theta})| = 1$ almost everywhere. If $F(z)$ is a bounded analytic function on $G_\delta(b)$, we write $F \in H^\infty(G_\delta(b))$. By Theorem 5.1, $F \in H^\infty(G_\delta(b))$ has a non-tangential limit $F(e^{i\theta})$ almost everywhere on T.

Theorem 5.2. *Let A be a closed subalgebra of L^∞ containing H^∞ and let $f \in L^\infty$. Then $f \in A$ if and only if for each $\varepsilon > 0$, there are $b \in B_A$ and $\delta, 0 < \delta < 1$, and $F \in H^\infty(G_\delta(b))$ such that almost everywhere*

$$|f(e^{i\theta}) - F(e^{i\theta})| < \varepsilon.$$

Proof. Let us do the easy half first. If $f \in A$ then there are $b \in B_A$ and $g \in H^\infty$ such that $\|f - \bar{b}g\|_\infty < \varepsilon$. Let $F(z) = g(z)/b(z)$. For any $\delta, 0 < \delta < 1$, $F \in H^\infty(G_\delta(b))$ and $|F(e^{i\theta}) - f(e^{i\theta})| < \varepsilon$ almost everywhere.

For the converse we use the corona construction. When $b(z)$ is an inner function and $0 < \delta < 1$ there is a region $U \subset D$ with boundary Γ such that

$$(5.7) \qquad \{|b(z)| < 1 - \delta\} \subset U,$$

$$(5.8) \qquad U \subset \{|b(z)| < \eta\}, \quad 0 < \eta < 1, \quad \eta = \eta(\delta),$$

$$(5.9) \qquad \Gamma \text{ is a countable union of rectifiable Jordan curves,}$$

$$(5.10) \qquad \text{arc length on } \Gamma \cap D \text{ is a Carleson measure.}$$

See Section 4, Chapter VIII. By (5.8), $T \cap \partial U$ has zero length. We can suppose that every component of U contained in $\{|z| < r\}, 0 < r < 1$, is simply connected. Indeed by the maximum principle filling in the holes in such a component does not hurt (5.7)–(5.10). For any $r < 1$ there are finitely many components of U contained in $\{|z| < r\}$. Let $V = D \backslash \overline{U}$. Then by (5.7), $V \subset G_\delta(b)$ and $F(z) \in H^\infty(V)$. We want to use Cauchy's theorem on ∂V. To avoid cumbersome technicalities we work instead with $V_r = V \cap \{|z| < r\}$.

By duality,

$$\text{dist}(f, A) \leq \lim_n \text{dist}(f, \bar{b}^n H^\infty)$$

$$= \lim_n \sup_{\substack{k \in H_0^1 \\ \|k\|_1 = 1}} \frac{1}{2\pi} \int f(\theta) b^n(\theta) k(\theta) d(\theta),$$

and by hypothesis,

$$\left| \frac{1}{2\pi} \int f(\theta) b^n(\theta) k(\theta) d\theta \right| \leq \varepsilon + \left| \frac{1}{2\pi} \int F(\theta) b^n(\theta) k(\theta) d\theta \right|.$$

Fix $k \in H_0^1$, $\|k\|_1 = 1$. By dominated convergence and the maximal theorem,

$$\frac{1}{2\pi} \int F(\theta) b^n(\theta) k(\theta) d\theta = \lim_{r \to 1} \frac{1}{2\pi i} \int_{\{|z|=r\} \backslash U} F(z) b^n(z) k(z) \frac{dz}{z},$$

because by (5.8), $|U \cap \{|z| = r\}| \to 0 (r \to 1)$. The domain $V_r = V \cap \{|z| < r\}$ is finitely connected, and $\partial V_r = (\Gamma \cap \{|z| < r\}) \cup (\{|z| = r\} \backslash U) = \Gamma_r \cup J_r$. With suitable orientations, Cauchy's theorem now gives

$$\frac{1}{2\pi i} \int_{J_r} F(z) b^n(z) k(z) \frac{dz}{z} = \frac{1}{2\pi i} \int_{\Gamma_r} F(z) b^n(z) k(z) \frac{dz}{z}.$$

However, $k(z)/z \in H^1$ and $\|k(z)/z\|_1 = 1$, so that (5.8) and (5.10) yield

$$\left| \frac{1}{2\pi i} \int_{\Gamma_r} F(z) b^n(z) \frac{k(z)}{z} dz \right| \leq \|F\|_\infty \eta^n \int_\Gamma \left| \frac{k(z)}{z} \right| \frac{ds}{2\pi}$$

$$\leq C \|F\|_\infty \eta^n < \varepsilon$$

when n is large, and $f \in A$. $\quad\square$

With Theorems 5.2 and 3.5 we have three necessary and sufficient conditions for a function $f \in L^\infty$ to belong to a Douglas algebra A. Each condition says that, in some sense, f is almost an analytic function. It is clear what we mean by that remark in the case of Theorem 5.2. In Theorem 3.5, condition (ii) stipulates that f is close to H^2 in all the Hilbert spaces $L^2(P_2, d\theta)$, $z \in G_\delta(b)$. Condition (iii) of Theorem 3.5 says that $\partial f / \partial \bar{z}$ is small on a region $G_\delta(b)$, $b \in \mathbb{B}_A$. Thus we have three different descriptions of a Douglas algebra in terms of analyticity. Each of these conditions can be reformulated to give an upper and lower bound for dist (f, A) (see Exercise 11).

Notes

Motivated by operator theory, Douglas formulated his conjecture in [1969]. With the corona theorem and Hoffman's theorem (Chapter X), the solution of Douglas's problem is one of the major accomplishments of this theory. There is so much structure to H^∞ that very general conjectures, at first blush rather suspect, turn out pleasantly to be true. It is difficult to overestimate Sarason's influence on Douglas's problem, and we have followed his expositions ([1973, 1976, 1979]) closely.

Hoffman and Singer discuss their proof of Wermer's theorem in [1957, 1960], and its relation to the Douglas problem was noted by Sarason in [1976]. Using Chang's theorem, but anticipating Marshall's theorem, S. Axler showed that any algebra between H^∞ and L^∞ is determined by its spectrum.

The primary references for $H^\infty + C$ are by Sarason [1975, 1973]. It is noteworthy how the careful analysis of an example like $H^\infty + C$ can lead to a rich general theory.

The sources for Section 3 are Chang [1976] and Marshall [1976b]. Theorems 4.2 and 4.4 are due to Chang [1977a], and Theorem 5.2 and our proof of Theorem 4.2 are from Chang and Garnett [1978].

Further developments related to the local Fatou theorem can be found in the books of Stein [1970] and Stein and Weiss [1971].

The algebras $H^\infty \cap C_A$ have been studied by Chang and Marshall [1977]. When $A = H^\infty + C$, $H^\infty \cap C_A$ is the disc algebra A_o, and when $A = L^\infty$, $H^\infty \cap C_A$ reduces to H^∞. Thus their theorems generalize results in this chapter and the corresponding classical results for A_o. Chang and Marshall show that the disc is dense in $\mathfrak{M}_{H^\infty \cap C_A}$ and that $H^\infty \cap C_A$ is generated by its inner functions. The second result generalizes the theorems of Fisher [1968] and Marshall [1976a]. They also prove that any algebra between C_A and $H^\infty \cap C_A$ is a Douglas algebra over $H^\infty \cap C_A$, which by definition is an algebra obtained by inverting certain inner functions in $H^\infty \cap C_A$. Wermer's maximality theorem and Theorem 3.1 are special cases of this theorem. Chang and Marshall also obtain the analog of Theorem 4.2 for algebras between C_A and $H^\infty \cap C_A$.

We have ignored the operator theoretic aspects of the Douglas problem. We refer the reader to Sarason's recent survey [1979] and to its extensive bibliography.

Exercises and Further Results

1. (a) Let B be a closed subalgebra of L^∞ containing the disc algebra A_o. Assume the linear functional $\phi_0(f) = f(0)$ has a unique norm-preserving extension from A_o to B. Then either $B \supset C$ or $B \subset H^\infty$. (Hint: If $\bar{z} \notin B$, then $J = \{zf : f \in B\}$ has distance 1 from the constant 1. There is $\varphi \in B^*$ such that $\varphi(J) = 0$, $\varphi(1) = 1$, and $\|\varphi\| = 1$. From this conclude that $B \subset H^\infty$.)

★(b)   If B is a closed subalgebra of L^∞ containing A_o such that B contains at least one Riemann integrable function not in H^∞, then $B \supset C$. (Hint: φ_0 has unique norm preserving extension to B if and only if

$$\sup\{\operatorname{Re} \varphi_0(g) : g \in A_o, \operatorname{Re} g \le \operatorname{Re} f\}$$
$$= \inf\{\operatorname{Re} \varphi_0(g) : g \in A_o, \operatorname{Re} g \ge \operatorname{Re} f\}$$

for all $f \in B$.

(c)   Let K be a closed nowhere dense subset of T of positive measure, and let B be the closed subalgebra of L^∞ generated by z and χ_K. Then $\bar{z} \notin B$.

(See Lumer [1965] and Sarason [1973].) Part (a) generalizes Theorem 1.4, while part (c) shows H^∞ cannot be replaced by A_0 in that theorem.

2. (a) Let E and F be closed subspaces of a Banach space X. Then $E + F$ is closed in X if and only if there is $c > 0$ such that

$$\operatorname{dist}(x, F) \le c \operatorname{dist}(x, E \cap F)$$

for all $x \in E$.

(b) If E is a closed subspace of L^∞ containing the constants and closed under complex conjugation, then $E + H(E)$ is closed in BMO if and only if $H^\infty + E$ is closed in L^∞. Here $H(E) = \{Hu : u \in E\}$, H being the Hilbert transform or conjugate function. (See Chang [1977a].)

3. For $|\alpha| = 1$, let $\mathfrak{M}_\alpha = \{m \in \mathfrak{M} : m(z) = \alpha\}$ be the fiber of $\mathfrak{M}$ over α and let $H^\infty|\mathfrak{M}_\alpha$ be the restriction of H^∞ to $\mathfrak{M}_\alpha$. Then $H^\infty|\mathfrak{M}_\alpha$ is a closed subalgebra of $C(\mathfrak{M}_\alpha)$ with Šilov boundary $X_\alpha = X \cap \mathfrak{M}_\alpha$. If $f \in L^\infty$, then $f \in H^\infty + C$ if and only if $f|X_\alpha \in H^\infty|\mathfrak{M}_\alpha$ for every α. More generally

$$\operatorname{dist}(f, H^\infty + C) = \sup_{|\alpha|=1} \operatorname{dist}(f|X_\alpha, H^\infty|\mathfrak{M}_\alpha).$$

(Hint: An extreme point of the unit ball of $(H^\infty + C)^\perp$, a space of measures on X, has support a single X_α.)

4. Let $f \in H^\infty + C$. Then f is invertible on $H^\infty + C$ if and only if $|f(z)| \ge \delta > 0$ on some annulus $r < |z| < 1$. Consequently each $f \in (H^\infty + C)^{-1}$ has a well-defined winding number over T. Then

$$f \in (H^\infty + C)^{-1}$$

if and only if

$$f = z^n g e^{i(u+Hv)},$$

where n is the winding number of f, $g \in (H^\infty)^{-1}$, and $u, v \in C$ (see Sarason [1973]).

5. (a) Every $f \in L^\infty$ has the form $f = \bar{b}g$, where b is a Blaschke product and $g \in H^\infty + C$. Every measurable subset of T is almost everywhere the zero set of some function in $H^\infty + C$ (Axler [1977]).

(b) If $\{I_j\}$ is a sequence of arcs on T such that $\sum |I_j| < \infty$, then there is $\varphi \in \text{VMO}$, $\varphi \geq 0$, such that $\varphi_{I_j} \to \infty$.

(c) If $f \in L^\infty$, there is $h \in QA = QC \cap H^\infty$ such that $hf^n \in QC$ for all $n = 1, 2, \ldots$ Consequently every measurable subset of T is almost everywhere the zero set of some QC function, and every Blaschke sequence is the zeros of some QA function.

(d) Every unimodular function in $H^\infty + C$ is the product of an inner function and a QC function, and every inner function is a Blaschke product times a QC function. Consequently, if $u \in L^\infty$ is unimodular, then $u = (b_1/b_2)e^{i(f+\bar{g})}$, where b_1 and b_2 are Blaschke products and $f, g \in C$.

Parts (b)–(d) are from T. Wolff [1979].

★★**6.** If $f \in L^\infty$, there is a best approximation $g \in H^\infty + C$; that is, there exists $g \in H^\infty + C$ such that

$$\|f - g\| = \text{dist}(f, H^\infty + C).$$

Unless $f \in H^\infty + C$, the best approximation is not unique. As a consequence, for any $\alpha \in T$ there is $h \in H^\infty$ such that

$$\sup_{X_\alpha} |f - h| = \text{dist}(f|X_\alpha, H^\infty|\mathfrak{M}_\alpha)$$

(see Axler, Berg, Jewell, and Shields [1979]). Luecking [1980] has a proof using the theory of M-ideals. When $H^\infty + C$ is replaced by an arbitrary Douglas algebra the existence of best approximations is an interesting presently unsolved problem.

★**7.** This exercise outlines Sarason's work on the algebra B_1. His arguments anticipated the solution of the Douglas problem, and we suggest the reader work out the details below without relying on Section 3.

Let C_1 denote the space of complex functions on the unit circle which are continuous except possibly at $z = 1$ but which have one-sided limits at $z = 1$. Let

$$B_1 = [H^\infty, C_1]$$

be the closed algebra generated by H^∞ and C_1.

(a) Let $\sigma(e^{i\theta}) = e^{i\theta/2}$. Then $B_1 = [H^\infty, \sigma]$.

(b) For $\varepsilon > 0$ let G_ε be the region bounded by the unit circle, and the curve consisting of the circular arc$\{e^{i\theta}\cos\varepsilon : \varepsilon \leq |\theta| \leq \pi\}$ and the two segments $[1, e^{\pm i\varepsilon}\cos\varepsilon]$ (see Figure IX.2). If $f, g \in B_1$, then

$$\lim_{\varepsilon \to 0}\left(\sup_{G_\varepsilon} |f(z)g(z) - (fg)(z)| \right) = 0.$$

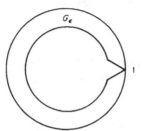

Figure IX.2. The region G_ε.

(c) There is a constant K such that when $h \in H^\infty$

$$\text{dist}(h\sigma, H^\infty) \leq K \sup_{0 < x < 1} |h(x)|.$$

(Hint: Use duality and Cauchy's theorem on $D\setminus\{0 < x < 1\}$. Actually one can take $K = 1$.)

(d) If $b(z)$ is an inner function such that $\sup_{0<x<1} |b(x)| < 1$, then $B_1 \subset [H^\infty, \bar{b}]$.

(e) Let $f(z) = -1 + \exp(2\pi i \log(1 - z))$. Then $f \in B_1^{-1}$. The inner factor of f is the Blaschke product $b(z)$ having zeros $1 - e^{-n}, n \geq 0$. Then $b \in B_1^{-1}$ and $B_1 = [H^\infty, \bar{b}]$. Thus B_1 is a Douglas algebra.

(f) If $g(z) \in H^\infty(G_\varepsilon)$ for some $\varepsilon > 0$, then $g(e^{i\theta}) \in B_1$. (Hint: Use duality and Cauchy's theorem on G_ε.)

(g) The maximal ideal space of B_1 is

$$\mathfrak{M}_1^+ \cup \mathfrak{M}_1^- \cup \bigcup_{\alpha \in T \setminus \{1\}} \mathfrak{M}_\alpha,$$

where $\mathfrak{M}_\alpha$ is the fiber $z^{-1}(\alpha)$ of $\mathfrak{M}$ at α, $\mathfrak{M}_\pm = \{m : |m(f)| \leq \sup_{X^\pm} |\hat{f}|\}$, X_1 is the fiber of $X = \mathfrak{M}_{L^\infty}$ at 1 and $X^\pm = X_1 \cap \{\hat{\sigma} = \pm 1\}$ where $\hat{\sigma}$ is the Gelfand transform of the jump function $\sigma \in L^\infty$.

(h) $H^\infty + C_1$ is closed, but it is not an algebra.

Part (h) is due to Chang [1977b]. The other parts, and more things about B_1, can be found in Sarason [1972].

8. On the real line let A_1 be the closed algebra generated by H^∞ and the bounded uniformly continuous functions. Then $A_1 = [H^\infty, e^{-ix}]$. If $f \in A_1$, then f is invertible in A_1 if and only if $|f(z)|$ is bounded away from zero in some strip $\{0 < y < \alpha\}$. If $b(z)$ is a Blaschke product whose zeros tend to ∞, then the following are equivalent:

    (i) $b \in (A_1)^{-1}$,
    (ii) $b(x)$ is uniformly continuous on $\mathbb{R}$, and
    (iii) $b'(x)$ is bounded on $\mathbb{R}$.

Assuming the Blaschke product $b(z)$ is in $(A_1)^{-1}$, prove the following are equivalent:

    (1) $A_1 = [H^\infty, \bar{b}]$,
    (2) $|b(z)|$ is bounded away from one in every half plane $\{y > a > 0\}$, and
    (3) $b'(x)$ is bounded away from zero on $\mathbb{R}$.

(See Sarason [1973].)

    9. If $f \in L^\infty$ is a nonconstant simple function, then there is an interpolating Blaschke product $b(z)$ such that

$$[H^\infty, f] = [H^\infty, \bar{b}]$$

(see Marshall [1976c]). It is not known which algebras have the form $[H^\infty, \bar{b}]$ for a single interpolating Blaschke product.

    10. Let A be a closed subalgebra of L^∞ containing H^∞ and let $f \in \mathrm{BMO}$. Show $f \in \mathrm{VMO}_A$ if and only if for each $\varepsilon > 0$ there are $b \in \mathrm{B}_A$ and $\delta, 0 < \delta < 1$, such that

$$\frac{1}{2\pi} \int |f - f(z)|^2 P_z(\theta)\, d\theta < \varepsilon$$

for all $z \in G_\delta(b)$.

    11. Let A be a closed subalgebra of L^∞ containing H^∞ and let $f \in L^\infty$.
    (a) Let ε_1 be the infimum of the set of $\varepsilon > 0$ for which there are $b \in \mathrm{B}_A$ and $\delta, 0 < \delta < 1$, such that

$$\inf_{g \in H^2} \frac{1}{2\pi} \int |f - g|^2 P_z(\theta)\,d\theta < \varepsilon$$

for all $z \in G_\delta(b)$. Then

$$\mathrm{dist}(f, A) \sim \varepsilon_1^{1/2}.$$

    (b) Let ε_2 be the infimum of the set of $\varepsilon > 0$ for which there are $b \in \mathrm{B}_A$ and $\delta, 0 < \delta < 1$, such that

$$v_f(G_\delta(b) \cap S(\theta_0, h)) < \varepsilon h.$$

Then

$$\text{dist}(f, A) \sim \varepsilon_2^{1/2}.$$

(c) Let ε_3 be the infimum of the set of $\varepsilon > 0$ for which there are $b \in$ $B_A, \delta, 0 < \delta < 1$, and $F \in H^\infty(G_\delta(b))$ such that

$$|f(e^{i\theta}) - F(e^{i\theta})| < \varepsilon$$

almost everywhere. Then

$$\text{dist}(f, A) = \varepsilon_3.$$

12. Prove Lemma 4.5.

13. The two steps of the proof of the Chang–Marshall theorem can be intertwined to yield a proof less dependent on Banach algebras. In particular, the corona theorem is not needed. The argument outlined below is due to Jewell [1976]. Let B be a closed subalgebra of L^∞ containing H^∞.

(a) For each $u \in U_B$ and each $\alpha, 0 < \alpha < 1$, there is an interpolating Blaschke product $b_{\alpha,u}$ satisfying (3.1) and (3.2).

(b) By Lemma 3.3, $b_{\alpha,u} \in B^{-1}$.

(c) Let $G_{\alpha,u} = \{z : |b_{\alpha,u}(z)| > \frac{1}{2}\}$. If $1 - \alpha$ is small, then

$$\lim_{n \to \infty} \text{dist}(u, \overline{b}_{\alpha,u}^n H^\infty) < \varepsilon.$$

Consequently $U_B \subset [H^\infty, \overline{\{b_{\alpha,u}\}}]$, which proves the Chang–Marshall theorem.

14. If A is a Douglas algebra, then ball (A) is the norm closed convex hull of $\{b_1\overline{b}_2 : b_1$ is a Blaschke product and $b_2 \in B_A$ is an interpolating Blaschke product$\}$ (see D. Marshall [1976c]).

★15.

(a) Let A be a Douglas algebra, let $u \in C_A \cap U_A$, and let $\varepsilon > 0$. Then there are $b_1, b_2 \in B_A$ such that

$$\|u^2 - b_1\overline{b}_2\|_\infty < \varepsilon$$

(see Marshall [1976c]). The generalization of the Douglas–Rudin theorem that should be true here, stating that $\|u - b_1\overline{b}_2\|_\infty < \varepsilon$, remains an open problem.

(b) However, the conjecture above is true in an interesting special case. Let E be an arbitrary subset of the circle T. Say $f \in L_E^\infty$ if $f \in L^\infty$ and if, for each $\alpha \in E$, f can be redefined on a set of measure zero so as to become continuous at α. Any function in L_E^∞ can be approximated in the L^∞ norm by functions continuous on open neighborhoods of E. Now if $u \in L_E^\infty$ is unimodular and if $\varepsilon > 0$, then there are Blaschke products b_1, b_2 analytic across E such that $\|u - b_1\bar{b}_2\|_\infty < \varepsilon$ (see Davie, Gamelin, and Garnett[1973]).

(c) $H^\infty + L_E^\infty$ is a closed algebra, and part(b) shows that $H^\infty + L_E^\infty$ is a Douglas algebra. Also, $C_{(H^\infty + L_E^\infty)} = L_E^\infty$.

X

Interpolating Sequences
and Maximal Ideals

We return to interpolating sequences and their Blaschke products, and in particular to the surprising part they play in unraveling the maximal ideal space of H^∞. In this chapter three topics are discussed.

1. Analytic structure in $\mathfrak{M}\backslash D$, and its relation to interpolating sequences. This theory, due to Kenneth Hoffman, occupies Sections 1 and 2. The theory rests on two factorization theorems for Blaschke products.

2. Two generalizations of the theorem that a harmonic interpolating sequence is an H^∞ interpolating sequence. One of these generalizations is the theorem that a sequence is an interpolating sequence if its closure in $\mathfrak{M}$ is homeomorphic to the Stone–Čech compactification of the integers. The key to these generalizations is a real-variables argument determining when one Poisson kernel can be approximated by convex combinations of other Poisson kernels. This idea is developed in Sections 3 and 4.

3. A more recent theorem, due to Peter Jones, that refines the Douglas–Rudin theorem. Here the analysis is not done on the upper half plane, but on the boundary.

The three topics have little interdependence, and they can be studied separately.

1. Analytic Discs in $\mathfrak{M}$

It will be convenient to think of two copies of the unit disc. First there is $D = \{z : |z| < 1\}$, the natural domain of H^∞ functions and an open dense subset of $\mathfrak{M} = \mathfrak{M}_{H^\infty}$. The second disc $D = \{\zeta : |\zeta| < 1\}$ will be the coordinate space for many abstract discs in $\mathfrak{M}$, including D itself. Points of D will always be denoted ζ. When $z \in D$, the mapping $L_z : D \to D$ defined by

$$L_z(\zeta) = \frac{\zeta + z}{1 + \bar{z}\zeta}$$

coordinatizes D so that z becomes the origin.

A continuous mapping $F : D \to \mathfrak{M}$ is *analytic* if $f \circ F$ is analytic on D whenever $f \in H^\infty$. An *analytic disc* in $\mathfrak{M}$ is a one-to-one analytic map $L : D \to \mathfrak{M}$. With an analytic disc we do not distinguish between the map L and its image $L(D)$. It is not required that L be a homeomorphism, and there are natural examples where it cannot be one (see Exercise 8). The mappings L_z above are examples of analytic discs. In this and the next section, we describe all the analytic maps into $\mathfrak{M}$ and we present Hoffman's fascinating theory connecting analytic discs to interpolating sequences. But first we need some general facts about possible analytic structure in $\mathfrak{M}$.

The *pseudohyperbolic distance* between $m_1 \in \mathfrak{M}$ and $m_2 \in \mathfrak{M}$ is

$$\rho(m_1, m_2) = \sup\{|f(m_2)| : f \in H^\infty, \|f\|_\infty \leq 1, f(m_1) = 0\}.$$

On D this definition of $\rho(m_1, m_2)$ coincides with the earlier one for $\rho(z_1, z_2)$ introduced in Chapter I, because if $m_j = z_j \in D$, then by Schwarz's lemma

$$\rho(m_1, m_2) = \left| \frac{z_1 - z_2}{1 - \bar{z}_2 z_1} \right|.$$

Moreover, the distance $\rho(m_1, m_2)$ on $\mathfrak{M}$ retains many of the properties of $\rho(z_1, z_2)$. If $f \in H^\infty$, $\|f\|_\infty \leq 1$, then

$$\rho(f(m_1), f(m_2)) \leq \rho(m_1, m_2).$$

because $g(z) = (f(z) - f(m_1))/(1 - \overline{f(m_1)}f(z))$ satisfies $\|g\|_\infty \leq 1$, $g(m_1) = 0$, and $|g(m_2)| = \rho(f(m_1), f(m_2))$. Choosing $\{f_n\}$ such that $\|f_n\|_\infty \leq 1$, $f_n(m_1) = 0$, and $|f_n(m_2)| \to \rho(m_1, m_2)$, we see that

$$\rho(m_1, m_2) = \sup\{\rho(f(m_1), f(m_2)) : f \in H^\infty, \|f\| \leq 1\}.$$

By Lemma 1.4 of Chapter I, we have

$$(1.1) \frac{\rho(m_0, m_2) - \rho(m_2, m_1)}{1 - \rho(m_0, m_2)\rho(m_2, m_1)} \leq \rho(m_0, m_1) \leq \frac{\rho(m_0, m_2) + \rho(m_2, m_1)}{1 + \rho(m_0, m_2)\rho(m_2, m_1)}$$

$m_j \in \mathfrak{M}$, $j = 0, 1, 2$. Indeed, the left-hand inequality follows from that lemma by taking $\rho(f(m_0), f(m_2))$ close to $\rho(m_0, m_2)$ and noting that $(s - t)/(1 - st)$ is decreasing in t when $0 \leq s, t \leq 1$; while the right-hand inequality follows by taking $\rho(f(m_0), f(m_1))$ close to $\rho(m_0, m_1)$ and noting that $(s + t)/(1 + st)$ increases in both s and t when $0 \leq s, t \leq 1$.

Clearly $\rho(m_1, m_2) \leq 1$, and by (1.1) the relation

$$m_1 \sim m_2 \quad \text{iff} \quad \rho(m_1, m_2) < 1$$

is an equivalence relation on $\mathfrak{M}$. The corresponding equivalence classes are called the *Gleason parts* of $\mathfrak{M}$. Write

$$P(m) = \{m' \in \mathfrak{M} : \rho(m, m') < 1\}$$

for the Gleason part containing m. If $m \in X$, the Šilov boundary, then $P(m) = \{m\}$. Indeed, if $m' \neq m$, then its representing measure $\mu_{m'}$ clearly satisfies $\mu_{m'}(\{m\}) \neq 1$, and by the logmodular property there is $f \in H^\infty$, $\|f\| = 1$, such that $|f(m)| = 1$ but

$$|f(m')| \leq \int |f| d\mu_{m'} < 1,$$

which means that $\rho(m, m') = 1$. The open disc D is a nontrivial Gleason part, because $|z(m)| = 1$ if $m \in \mathfrak{M} \backslash D$.

Lemma 1.1. *If $F : \mathrm{D} \to \mathfrak{M}$ is an analytic mapping, then $F(\mathrm{D})$ is contained in a single Gleason part.*

Proof. This is just Schwarz's lemma. If $F : \mathrm{D} \to \mathfrak{M}$ is analytic and if $m_j = F(\zeta_j)$, $\zeta_j \in \mathrm{D}$, $j = 1, 2$, then

$$\begin{aligned}
\rho(m_1, m_2) &= \sup\{|f \circ F(\zeta_2)| : f \in H^\infty, \|f\|_\infty \leq 1, f \circ F(\zeta_1) = 0\} \\
&\leq \sup\{|g(\zeta_2)| : g \in H^\infty(\mathrm{D}), \|g\|_\infty \leq 1, g(\zeta_1) = 0\} \\
&= \rho(\zeta_1, \zeta_2) < 1,
\end{aligned}$$

and m_1 and m_2 are in the same Gleason part. $\quad\square$

Thus an analytic disc cannot pass through a one-point Gleason part. As we have seen, each point of the Šilov boundary X comprises a single part, and the disc D is a nondegenerate Gleason part and an analytic disc. So we are searching for analytic structure in the remaining subset $\mathfrak{M} \backslash (X \cup D)$, which is a union of Gleason parts. Now, this set is not empty. Any accumulation point m of the zeros of an infinite Blaschke product is a point in $\mathfrak{M} \backslash (X \cup D)$. If the zeros form an interpolating sequence, then this set of accumulation points is homeomorphic to $\beta\mathbb{N} \backslash \mathbb{N}$. Hence $\mathfrak{M} \backslash (X \cup D)$ is very, very big.

A theorem from the general theory of logmodular algebras implies that each Gleason part of $\mathfrak{M}$ is either a one-point part or an analytic disc (see Hoffman [1962b]). Because we shall ultimately obtain a more complete result in the case of H^∞, we shall not prove that general theorem. However, let us use it to find some one-point parts in $\mathfrak{M} \backslash (X \cup D)$. Suppose $S(z)$ is a nonconstant singular inner function. Then $S(z)$ has roots $S^\alpha(z)$ for all $\alpha > 0$ and $\|S^\alpha\| = 1$. This implies that

$$K = \{m : S(m) = 0\}$$

is a union of Gleason parts. Indeed if $S(m) = 0$, then

$$\rho(m, m') \geq \lim_{\alpha \to 0} |S^\alpha(m')|,$$

so that either $\rho(m, m') = 1$ or $S(m') = 0$. Because $S^{-1} \notin H^\infty$, $K \neq \varnothing$, and because $|S| = 1$ on X, $K \subset \mathfrak{M}\backslash(X \cup D)$. The closure in $C(K)$ of $H^\infty|K$ is a uniform algebra A with maximal ideal space K. Then K contains a storng boundary point for the algebra A (Chapter V, Exercise 10), and by the maximum principle such a point cannot lie in an analytic disc for $H^\infty|K$. By the general theorem quoted above, this means that $\mathfrak{M}\backslash(X \cup D)$ contains one-point Gleason parts. Another route to this fact is given in Exercise 2.

One step in the argument above will be needed later. Recall that $P(m)$ denotes the Gleason part containing $m \in \mathfrak{M}$.

Lemma 1.2. *Let $m \in \mathfrak{M}$ and let $g \in H^\infty$ satisfy $\|g\|_\infty \leq 1$ and $g(m) = 0$. Suppose that for $n = 2, 3, \ldots$ there is a factorization*

$$g = g_1^{(n)} g_2^{(n)} \cdots g_n^{(n)}$$

with $g_j^{(n)} \in H^\infty$, $\|g_j^{(n)}\|_\infty \leq 1$ and $g_j^{(n)}(m) = 0$. Then $g \equiv 0$ on $P(m)$.

Proof. For $m' \in \mathfrak{M}$, we have

$$|g(m')| \leq \lim_{n\to\infty} \prod_{j=1}^n |g_j^{(n)}(m')| \leq \lim_{n\to\infty} (\rho(m, m'))^n,$$

so that $g(m') = 0$ if $\rho(m, m') < 1$. $\square$

The set $\mathfrak{M}^D$ of all mappings (continuous or not) from D into $\mathfrak{M}$ is a compact Hausdorff space in the product topology. In this topology a net (F_j) has limit F if and only if $F_j(\zeta) \to F(\zeta)$ for each $\zeta \in D$, that is, if and only if $f \circ F_j(\zeta) \to f \circ F(\zeta)$ for all $f \in H^\infty$ and all $\zeta \in D$. Nets are forced upon us here because $\mathfrak{M}^D$ is not a metric space. The notations (z_j) or (F_j) for a net and $\{z_n\}$ for a sequence will be used to distinguish nets from sequences. Our principal object of study here is the set $L \subset \mathfrak{M}^D$ of maximal analytic discs in $\mathfrak{M}$. Maximal means that the range $L(D)$ is contained in no larger analytic disc. It will turn out that $\{\alpha L_z : z \in D, |\alpha| = 1\}$ is dense in L. This fact can be regarded as a refinement of the corona theorem.

Nontrivial analytic maps into $\mathfrak{M}\backslash(X \cup D)$ were exhibited in Chapter V by the following reasoning. Let $B(z)$ be an interpolating Blaschke product with zeros $S = \{z_n\}$. Then

$$\inf_n (1 - |z_n|^2)|B'(z_n)| = \delta > 0.$$

Let $m \in \mathfrak{M}\backslash D$ be in the closure of S and let (z_j) be a subnet of S converging to m. Taking a finer subnet, we can be compactness suppose that (L_{z_j}) converges to a map $L_m \in \mathfrak{M}^D$. Then $L_m(0) = \lim_j L_{z_j}(0) = m$, and when $f \in H^\infty$, $f \circ L_m(\zeta) = \lim f \circ L_{z_j}(\zeta)$ is analytic on D. Moreover, L_m is not a constant mapping, because

$$|(B \circ L_m)'(0)| = \lim |(B \circ L_{z_j})'(0)| = \lim_j (1 - |z_j^2|)|B'(z_j)| \geq \delta.$$

That is as far as we went in Chapter V. Now let us study the mapping L_m more carefully.

Because S is an interpolating sequence, its closure $\bar{S}$ in $\mathfrak{M}$ is homeomorphic to $\beta\mathbb{N}$, the Stone–Čech compactification of the positive integers. Consequently any map from S into a compact Hausdorff space like $\mathfrak{M}^D$ can be uniquely extended to a continuous map from $\bar{S}$ to the same compact Hausdorff space (see Chapter V, Theorem 1.4). For the map $z_n \to L_{z_n}$ this means L_m does not depend on the choice of the subnet of $\{z_n\}$ converging to m. We have proved the following:

Lemma 1.3. *If S is an interpolating sequence and if $m \in \bar{S}$, then there is a unique nonconstant analytic map $L_m \in \mathfrak{M}^D$ such that whenever (z_j) is a net in S coverging to m,*

$$\lim_j L_{z_j} = L_m.$$

To get more precise information about L_m we need two lemmas about Blaschke products. The first lemma continues the theme of Chapter VII, Section 5.

Lemma 1.4. *Suppose $B(z)$ is an interpolating Blaschke product with zeros $S = \{z_n\}$, and suppose*

$$\inf_n (1 - |z_n|^2)|B'(z_n)| \geq \delta > 0.$$

There exist $\lambda = \lambda(\delta), 0 < \lambda < 1$ and $r = r(\delta), 0 < r < 1$ satisfying

(1.2) $$\lim_{\delta \to 1} \lambda(\delta) = 1,$$

(1.3) $$\lim_{\delta \to 1} r(\delta) = 1,$$

and having the following properties: The set $B^{-1}(\Delta(0, r)) = \{z : |B(z)| < r\}$ is the union of pairwise disjoint domains V_n, $z_n \in V_n$, and

(1.4) $$V_n \subset \{z : \rho(z, z_n) < \lambda\}.$$

$B(z)$ maps each domain V_n univalently onto $\Delta(0, r) = \{w : |w| < r\}$. If $|w| < r$, then

$$B_w(z) = \frac{B(z) - w}{1 - \bar{w}B(z)}$$

is (a unimodular constant multiple of) an interpolating Blaschke product having one zero in each V_n.

Proof. Let $h_n(\zeta) = B((\zeta + z_n)/(1 + \bar{z}_n \zeta)) = B \circ L_{z_n}(\zeta)$. Then $\|h_n\|_\infty = 1$, $h_n(0) = 0$, and $|h_n'(0)| \geq \delta$. By Schwarz's lemma,

$$\rho\left(\frac{h_n(\zeta)}{\zeta}, h_n'(0)\right) \leq |\zeta|,$$

so that when $|\zeta| = \lambda = \lambda(\delta) < \delta$,

$$\frac{|h_n(\zeta)|}{|\zeta|} \geq \frac{|h_n'(0)| - |\zeta|}{1 - |\zeta||h_n'(0)|} \geq \frac{\delta - \lambda}{1 - \lambda\delta}.$$

On $|\zeta| = \lambda$ this gives

(1.5) $$|h_n(\zeta)| \geq \frac{\delta - \lambda}{1 - \lambda\delta}\lambda = r = r(\delta),$$

and the argument principle shows that $h_n(\zeta) = w, |w| < r$, has exactly one solution in $|\zeta| < \lambda$. Hence $B(z)$ maps

$$V_n = L_{z_n}(h_n^{-1}(\Delta(0, r)))$$

univalently onto $\Delta(0, r)$, $z_n \in V_n$, and (1.4) holds.

If there exists $z \in V_n \cap V_k, n \neq k$, then by (1.4),

$$\rho(z_n, z_k) \leq \frac{\rho(z_n, z) + \rho(z, z_k)}{1 + \rho(z_n, z)\rho(z, z_k)} \leq \frac{2\lambda}{1 + \lambda^2}.$$

Choosing $\lambda = \lambda(\delta)$ so that

(1.6) $$\lambda < \frac{2\lambda}{1 + \lambda^2} < \delta$$

thus ensures that $V_n \cap V_k = \varnothing, n \neq k$. Also choosing λ so that $\lambda = \lambda(\delta) \to 1$ ($\delta \to 1$) and so that

$$\frac{\delta - \lambda}{1 - \lambda\delta} \to 1 \quad (\delta \to 1),$$

we obtain

$$\lim_{\delta \to 1} r(\delta) = 1$$

in (1.5). Thus we have (1.2) and (1.3).

If $|w| < r$, then $B_w(z) = (B(z) - w)/(1 - \bar{w}B(z))$ has one zero $z_n(w)$ in each V_n and $z_n(w)$ is a holomorphic function of w. Let $A_w(z)$ be the Blaschke product having zeros $z_n(w)$. Then $B_w = A_w g_w, \|g_w\| \leq 1$. To prove $B_w = cA_w, |c| = 1$, we show $|g_w(0)| = 1$. Now

$$|B_w(0)| = |g_w(0)| \prod_{n=1}^{\infty} |z_n(w)|.$$

write

$$H(w) = \prod_{n=1}^{\infty} \frac{\bar{z}_n}{|z_n|} z_n(w), \quad |w| < r.$$

This product converges on $|w| < r$ because its partial products are bounded and because it converges at $w = 0$. Also, since $\|g_w\| \leq 1$,

$$(1.7) \qquad |H(w)| = \prod_{n=1}^{\infty} |z_n(w)| \geq |B_w(0)| = \frac{|B(0) - w|}{|1 - \overline{B(0)}w|}.$$

Both sides of this inequality are moduli of functions analytic in w, and equality holds at $w = 0$:

$$|H(0)| = \prod |z_n| = |B(0)|.$$

We can assume $B(0) \neq 0$. Then equality holds in (1.7) for all $|w| < r$, and $|g_w(0)| = 1$ except possibly when $H(w) = 0$, that is, when $B(0) = w$. However,

$$|g_w(0)|^{-1} = \frac{|1 - \overline{B(0)}w|}{|B(0) - w|}|H(w)|$$

is the modulus of a function meromorphic on $|w| < r$. Thus $|g_w(0)| = 1$ for all w and $B_w(z)$ is unimodular constant times the Blaschke product $A_w(z)$.

Since $B_w(z)$ has the same zeros as $A_w(z)$, B_w has no zeros outside $\bigcup V_n$, and so $B^{-1}(|w| < r) = \bigcup V_n$.

The zeros of $B_w(z)$ are an interpolating sequence because, by Lemma 5.3 of Chapter VII, any sequence $\{\zeta_n\}$ with $\rho(\zeta_n, z_n) < \lambda(\delta)$ is an interpolating sequence. $\quad\square$

The second lemma employs a clever combinatorial argument due to W. Mills.

Lemma 1.5. *Let $B(z)$ be a Blaschke product with distinct zeros $\{z_n\}$. Then B admits a factorization $B = B_1 B_2$ such that*

$$(1.8) \qquad if \quad B_1(z_n) = 0, \quad then \quad (1 - |z_n|^2)|B_1'(z_n)| \geq |B_2(z_n)|;$$

$$(1.9) \qquad if \quad B_2(z_n) = 0, \quad then \quad (1 - |z_n|^2)|B_2'(z_n)| \geq |B_1(z_n)|.$$

Corollary 1.6. *If $B(z)$ is an interpolating Blaschke product with zeros $\{z_n\}$ and if*

$$\delta(B) = \inf_n (1 - |z_n|^2)|B'(z_n)|,$$

then B has a factorization $B = B_1 B_2$ such that

$$\delta(B_j) \geq (\delta(B))^{1/2}, \qquad j = 1, 2.$$

Proof. The corollary is immediate from the lemma. If $B_1(z_n) = 0$, then by (1.8),

$$(1 - |z_n|^2)|B'(z_n)| = (1 - |z_n|^2)|B_1'(z_n)||B_2(z_n)| \leq \{(1 - |z_n|^2)|B_1'(z_n)|\}^2$$

and $\delta(B_1) \geq (\delta(B))^{1/2}$. By symmetry, we also have $\delta(B_2) \geq (\delta(B))^{1/2}$. $\quad\square$

Proof of Lemma 1.5. Write

$$a_{k,n} = \log \left| \frac{z_n - z_k}{1 - \bar{z}_k z_n} \right|, \quad k \neq n; \quad a_{n,n} = 0.$$

Then $[a_{k,n}]$ is a symmetric real matrix, $a_{k,n} = a_{n,k}$, with zeros on the diagonal and with absolutely summable rows. We claim there is a subset E of the positive integers such that

(a) if $n \in E$, then

$$\sum_{k \in E} a_{k,n} \geq \sum_{k \notin E} a_{k,n};$$

(b) if $n \notin E$, then

$$\sum_{k \notin E} a_{k,n} \geq \sum_{k \in E} a_{k,n}.$$

This will prove Lemma 1.5. If B_1 has zeros $\{z_n : n \in E\}$ then by (a)

$$(1 - |z_n|^2)|B_1'(z_n)| = \prod_{\substack{k \in E \\ k \neq n}} \left| \frac{z_n - z_k}{1 - \bar{z}_k z_n} \right| \geq \prod_{k \notin E} \left| \frac{z_n - z_k}{1 - \bar{z}_k z_n} \right| = |B_2(z_n)|,$$

and (1.8) holds. Similarly, (1.9) follows from (b).

To establish (a) and (b), first suppose $a_{k,n} = 0$ if $k > N$ or $n > N$. Then we have a finite problem and there is $E \subset \{1, 2, \ldots, N\}$ that maximizes the function

$$h(E) = \sum_{\substack{k \in E \\ n \in E}} a_{k,n} + \sum_{\substack{k \notin E \\ n \notin E}} a_{k,n}.$$

Suppose $n \in E$ and let F be the subset obtained by removing n from E. Then

$$h(F) = h(E) - 2 \sum_{k \in E} a_{k,n} + 2 \sum_{k \notin E} a_{k,n}.$$

Since $h(F) \leq h(E)$, (a) holds; and since $h(E) = h(E')$, where E' is complement of E, (b) holds for the same reason.

For each N we have a subset $E_N \subset \{1, 2, \ldots, N\}$ such that (a) and (b) hold under the constraint $k, n \leq N$. The space of subsets of $\mathbb{N} = \{1, 2 \ldots\}$ is compact in the product topology, in which a sequence $\{E_j\}$ has limit E if and only if

$$n \in E \Leftrightarrow n \in E_j, \quad \text{for all} \quad j \geq j_0(n).$$

(The correspondence $E \to \chi_E$, the characteristic function of E, makes the space of subsets homeomorphic to the compact product space $\{0, 1\}^{\mathbb{N}}$.) Choose a subsequence $\{E_{N_j}\}$ of $\{E_n\}$ which tends to a limit set E. Because $\sum_k |a_{n,k}| <$

∞, we then have

$$\sum_{k \in E} a_{k,n} - \sum_{k \notin E} a_{k,n} = \lim_{j \to \infty} \left(\sum_{k \in E_{N_j}} a_{k,n} - \sum_{\substack{k \notin E_{N_j} \\ k \le N_j}} a_{k,n} \right).$$

If $n \in E$, the limit is nonnegative, while if $n \notin E$, the limit is nonpositive. Therefore (a) and (b) hold for the subset E. $\square$

We return to the analytic mapping $L_m = \lim L_{z_j}$, where (z_j) is a subnet of the interpolating sequence $S = \{z_n\}$, and $\lim z_j = m$.

Theorem 1.7.   *The mapping L_m is a one-to-one analytic mapping from D onto $P(m)$, the Gleason part containing m. For any $r < 1$ there is a Blaschke product B_r such that $B_r(m) = 0$ and $B_r \circ L_m$ is one-to-one on $|\zeta| < r$. If (w_j) is any net in D converging to m, then*

$$\lim L_{w_i} = L_m.$$

Proof.   Let $B(z)$ be the Blaschke product with zeros $\{z_n\}$, and let

$$\delta = \delta(B) = \inf_n (1 - |z_n|^2)|B'(z_n)| > 0.$$

Then $|(B \circ L_m)'(0)| = \lim_j |(B \circ L_{z_j})'(0)| \ge \delta$. By Schwarz's lemma, as in the start of the proof of Lemma 1.4, $B \circ L_m$ is univalent on $(B \circ L_m)^{-1}(|w| < r(\delta))$. Since $|B \circ L_m(\zeta)| \le |\zeta|$, this means $B \circ L_m$ and L_m are one-to-one on $|\zeta| < r(\delta)$.

Let $r < 1$. By (1.3) and Corollary 1.6, $B(z)$ has a factorization $B = B_1 B_2 \cdots B_N$ with $r(\delta(B_k)) > r$, $k = 1, 2, \ldots, N$. We can replace (z_j) by a subnet so that $B_k(z_j) = 0$ for some factor B_k not depending on z_j. The above argument then shows $B_k \circ L_m$ and L_m are one-to-one on $|\zeta| < r$. This means L_m is one-to-one on D.

By Lemma 1.1, $L_m(D) \subset P(m)$. We must show L_m maps D onto $P(m)$. Let $m' \in P(m)$ so that $\rho(m, m') < 1$. By (1.2) and by Corollary 1.6, we can assume that

$$\rho(m, m') < \lambda = \lambda(\delta(B)),$$

and by (1.3) we can assume that $|B(m')| < r(\delta(B))$. Hence by Lemma 1.4

$$\frac{B - B(m')}{1 - \overline{B(m')}B}$$

is (a constant times) an interpolating Blaschke product with zeros z_n', where $z_n' \in V_n$. Then m' is in the closure of $\{z_n'\}$ by Chapter IX, Lemma 3.3, and there is a subnet $\{z_{n(i)}'\}$ of $\{z_n'\}$ converging to m'. We claim

(1.10) $$\lim_i z_{n(i)} = m,$$

where $z_{n(i)}$ is the subnet of S determined by the rule $z'_{n(i)} \in V_{n(i)}$. Let m'' be any cluster point of $z_{n(i)}$. Then

$$\rho(m'', m') \leq \overline{\lim}\rho(z_{n(i)}, z'_{n(i)}) \leq \lambda(\delta),$$

and by (1.6)

$$(1.11) \qquad \rho(m, m'') \leq \frac{\lambda(\delta) + \rho(m, m')}{1 + \lambda(\delta)\rho(m, m')} \leq \frac{2\lambda}{1 + \lambda^2} < \delta(B).$$

On the other hand, distinct cluster points m and m'' of $S = \{z_n\}$ must satisfy

$$(1.12) \qquad \rho(m, m'') \geq \delta(B).$$

This holds because there is a subset $T \subset S$ whose closure contains m and not m''. If B_T is the Blaschke product with zeros T, then

$$\rho(m, m'') \geq \varliminf_{z_n \notin T} |B_T(z_n)| \geq \delta(B).$$

Since (1.11) and (1.12) show $m'' = m$, (1.10) holds.

To summarize, we have

$$z_{n(i)} \to m, \quad z'_{n(i)} \to m', \quad \text{and} \quad z_{n(i)} \in V_{n(i)}.$$

There is $\zeta_i, |\zeta_i| \leq \lambda(\delta)$, such that $L_{z_{n(i)}}(\zeta_i) = z'_{n(i)}$. We may suppose $\zeta_1 \to \zeta, |\zeta| \leq \lambda(\delta)$. Then $\rho(\zeta_i, \zeta) \to 0$, while by Lemma 1.3, $L_{z_{n(i)}} \to L_m$. Consequently,

$$L_m(\zeta) = \lim_i L_{z_{n(i)}}(\zeta) = \lim_i L_{z_{n(i)}}(\zeta_i) = \lim_i z'_{n(i)} = m',$$

and $L_m(D) = P(m)$.

If $\{w_i\}$ is any net in D converging to m, then $\lim B(w_i) = 0$ and $|B(w_i)| < r(\delta(B))$ for large i. By Lemma 1.4 there is $n(i)$ such that $w_i \in V_{n(i)}$, for i large. Since B is univalent on $V_{n(i)}$,

$$\rho(w_i, z_{n(i)}) \leq c|B(w_i)| \to 0.$$

Hence

$$(1.13) \qquad \rho(L_{w_i}(\zeta), L_{z_{n(i)}}(\zeta)) \to 0, \quad \zeta \in D.$$

In particular $z_{n(i)} \to m$, and by Lemma 1.3, $L_{z_{n(i)}} \to L_m$. Then (1.13) shows that $L_{w_i} \to L_m$. $\square$

Incidentally, Theorem 1.7 leads to an interesting observation about the size of $\mathfrak{M}$. If the Blaschke product corresponding to the interpolating sequence $\{z_n\}$ satisfies

$$\lim_{n \to \infty} (1 - |z_n|^2)|B'(z_n)| = 1,$$

and if $m \in \mathfrak{M} \setminus D$ is in the closure of $\{z_n\}$, then the map L_m is a homeomorphism of D onto $P(m)$ (see Exercise 8). Moreover, the mapping

$$f \to f \circ L_m(\zeta)$$

is then an algebra isomorphism from $H^\infty | P(m)$ onto H^∞. This means that the nowhere dense set $\overline{P(m)}$ is homeomorphic to $\mathfrak{M}$. Furthermore, $\overline{P(m)} \setminus P(m)$ also contains homeomorphic copies of $\mathfrak{M}$, and so on.

2. Hoffman's Theorem

Hoffman proved that all analytic structure in $\mathfrak{M}$ comes about in the manner described in Section 1. In other words, every analytic disc in $\mathfrak{M} \setminus D$ has the form $L_m = \lim L_{z_j}$, where (z_j) is a subnet of some interpolating sequence. In view of the size and intractibility of $\mathfrak{M} \setminus D$, this is a remarkable accomplishment. Write

$$G = \{m \in \mathfrak{M} : m \text{ is in the closure of some interpolating sequence}\}.$$

Exercise 2(d) provides a geometric glimpse of G. We know each $m \in G$ lies in an analytic disc. The key observation we must make is this: When $m \in \mathfrak{M} \setminus G$, the Gleason part $P(m)$ is a singleton, $P(m) = \{m\}$. It follows by Lemma 1.1 that no nonconstant analytic mapping $F : D \to \mathfrak{M}$ can include m in its range. Consequently, m lies in an analytic disc if and only if $m \in G$.

Our basic tool will be another factorization theorem for Blaschke products. Let $B(z)$ be a Blaschke product with zeros $\{z_n\}$. For $\delta > 0$, set

$$K_\delta(B) = \bigcap_{n=1}^{\infty} \{z : \rho(z, z_n) \geq \delta\}.$$

Theorem 2.1. *For $0 < \delta < 1$ there are constants $a = a(\delta)$ and $b = b(\delta)$ such that the Blaschke product $B(z)$ has a nontrivial factorization $B = B_1 B_2$ such that*

$$(2.1) \qquad a|B_1(z)|^{1/b} \leq |B_2(z)| \leq \frac{1}{a}|B_1(z)|^b$$

on $K_\delta(B)$. The factors B_1 and B_2 do not depend on δ.

Proof. We work in the upper half plane H. For $z \in K_\delta(B)$, we have

$$c(\delta) \log \left| \frac{z - \bar{z}_n}{z - z_n} \right| \leq \frac{2y y_n}{|z_n - \bar{z}|^2} \leq \log \left| \frac{z - \bar{z}_n}{z - z_n} \right|,$$

by the proof of Lemma VII.1.2. We show there are constants C_1 and C_2 such that every Blaschke sequence $\{z_n\}$ can be partitioned into two subsets S_1 and

S_2 such that for $z \in H$,

(a)
$$\sum_{S_1} \frac{2yy_n}{|z_n - \bar{z}|^2} \leq C_1 + C_2 \sum_{S_2} \frac{2yy_n}{|z_n - \bar{z}|^2},$$

(b)
$$\sum_{S_2} \frac{2yy_n}{|z_n - \bar{z}|^2} \leq C_1 + C_2 \sum_{S_1} \frac{2yy_n}{|z_n - \bar{z}|^2}.$$

Taking B_1 with zeros S_1 and B_2 with zeros S_2, we obtain (2.1) with $a = \exp C_2/C_1$ and $b = c(\delta)/C_2$, when $z \in K_\delta(B)$.

Choose $\lambda, 0 < \lambda < 1$, and form strips $T_k = \{z : \lambda^{k+1} \leq y < \lambda^k\}$, k an integer. Write the z_n in T_k as the (possibly two-sided) sequence $z_{k,j}$, j an integer, so that $x_j \leq x_l$ if $j < l$. (The Blaschke condition ensures that only finitely many z_n's from T_k can tie.) Put $z_{k,j} \in S_1$ if j is odd, $z_{k,j} \in S_2$ if j is even.

To prove (a) and (b) first consider the special case

$$y_n = \lambda^{k+1}, \quad \text{all} \quad z_n \in T_k.$$

Then we have

(2.2)
$$\left| \sum_{S_1} - \sum_{S_2} \right| \leq \frac{2(1+\lambda)}{\lambda(1-\lambda)}$$

for all $z \in H$. Indeed, when z is fixed, the terms from T_k have alternating signs and moduli

$$\frac{2y\lambda^{k+1}}{(x - x_{k,j})^2 + (y + \lambda^{k+1})^2}.$$

These moduli are monotone in $|x - x_{k,j}|$ and tend to zero at infinity. Thus the contribution from T_k to the left side of (2.2) consists of a two-sided alternating series, which is dominated by the largest modulus of its terms. Hence

$$\left| \sum_{S_1 \cap T_k} - \sum_{S_2 \cap T_k} \right| \leq \frac{2y\lambda^{k+1}}{(y + \lambda^{k+1})^2}.$$

Summation over k now gives (2.2), because if $\lambda^{n+1} < y \leq \lambda^n$, then

$$\frac{2y\lambda^{k+1}}{(y + \lambda^{k+1})^2} \leq \frac{2}{\lambda} \frac{\lambda^{k-n}}{(1 + \lambda^{k-n})^2}$$

and

$$\sum_{j=-\infty}^{\infty} \frac{\lambda^j}{(1 + \lambda^j)^2} = \sum_{j=0}^{\infty} + \sum_{j=-\infty}^{-1} \leq \frac{1}{1 - \lambda} + \frac{\lambda}{1 - \lambda}$$
$$= \frac{1 + \lambda}{1 - \lambda}.$$

Turning to the general case, we write $z'_{k,j} = x_{k,j} + i\lambda^{k+1}$ and compare the sums to those for the adjusted sequences S'_1 and S'_2. When $y > 0$ and $z_n \in T_k$,

$$\frac{2yy_n}{(x - x_n)^2 + (y + y_n)^2} \le \frac{1}{\lambda} \frac{2y\lambda^{k+1}}{(x - x_n)^2 + (y + \lambda^{k+1})^2}$$

$$\le \frac{1}{\lambda^2} \frac{2yy_n}{(x - x_n)^2 + (y + y_n)^2},$$

so that for $p = 1, 2$

$$\sum_{S_p} \le \frac{1}{\lambda} \sum_{S'_p} \le \frac{1}{\lambda^2} \sum_{S_p}.$$

With (2.2), this yields

$$\sum_{S_1} \le \frac{1}{\lambda^2} \frac{(1+\lambda)}{(1-\lambda)} + \frac{1}{\lambda^2} \sum_{S_2} \quad \text{and} \quad \sum_{S_2} \le \frac{2}{\lambda^2} \frac{(1+\lambda)}{(1-\lambda)} + \frac{1}{\lambda^2} \sum_{S_1},$$

which are (a) and (b). □

Theorem 2.2. *Suppose $m \in \mathfrak{M} \backslash G$. Let $f \in H^\infty$, $\|f\|_\infty \le 1$. If $f(m) = 0$ then $f = f_1 f_2$, $f_j \in H^\infty$, with $\|f_j\|_\infty \le 1$ and $f_j(m) = 0$.*

Proof. First let us reduce the problem to the critical case where $f(z)$ is a Blaschke product with simple zeros. Write $f = Bg$ where $g \in H^\infty$ has no zeros and $B(z)$ is a Blaschke product. If $g(m) = 0$, then $f = (Bg^{1/2})(g^{1/2})$ gives the desired factorization. So we can assume $f(z) = B(z)$. Write $B = B_1(B_2)^2$ where B_1, and B_2 are Blaschke products and B_1 has simple zeros. If $B_2(m) = 0$, then $B = (B_1 B_2)(B_2)$ and we are done. Thus we can assume $f(z) = B(z)$ is a Blaschke product with simple zeros.

Factor $B = B_1 B_2$ in accordance with Theorem 2.1. We may suppose $B_1(m) = 0$ but $B_2(m) \ne 0$. Then by (2.1) m is not in the closure of the set $K_\delta(B)$, for any $\delta > 0$. This means m is in the closure of

$$S = \{z_n\},$$

the zero sequence of B. Indeed, if (z_j) is a net in D converging to m, then by Theorem 2.1 there are $z_{n(j)}$, $B(z_{n(j)}) = 0$, such that $\rho(z_j, z_{n(j)}) \to 0$, and this implies that $z_{n(j)}$ also converges to m. The factorization $B = B_1 B_2$ splits S into $S_1 \cup S_2$, where S_j is the zero sequence for B_j. Since $B_2(m) \ne 0$, m is not in the closure of S_2. Therefore m lies in the closure of S_1.

Now factor $B_1 = B_{1,1} B_{1,2}$ using Lemma 1.5. As before we can suppose $B_{1,1}(m) = 0$, $B_{1,2}(m) \ne 0$. Fix $\varepsilon < |B_{1,2}(m)|$. Then m is not in the closure of $\{z : |B_{1,2}(z)| < \varepsilon\}$, but m is in the closure of S_1. Hence

$$T = \{z_n \in S_1 : |B_{1,2}(z_n)| \ge \varepsilon\}$$

captures m in its closure. But Lemma 1.5 shows that T is an interpolating sequence, because

$$(1 - |z_n|^2)|B_{1,1}(z_n)| \geq \varepsilon, \quad z_n \in T.$$

Therefore m lies in the closure of an interpolating sequence, contrary to the hypothesis $m \in \mathfrak{M} \backslash G$. $\square$

Corollary 2.3. *If $m \in \mathfrak{M} \backslash G$, then the Gleason part $P(m)$ reduces to the singleton $\{m\}$.*

Proof. Suppose $m' \neq m$. We show $m' \notin P(m)$. There is $g \in H^\infty$, $\|g\|_\infty = 1$, with $g(m) = 0$ but $g(m') \neq 0$. For each $n = 2, 3, \ldots,$ Theorem 2.2 gives a factorization

$$g = g_1^{(n)} g_2^{(n)} \cdots g_n^{(n)},$$

where $g_j^{(n)} \in H^\infty$, $\|g_j^{(n)}\|_\infty = 1$, and $g_j^{(n)}(m) = 0$. Hence $m' \notin P(m)$ by Lemma 1.2. $\square$

To summarize, we now have several characterizations of points in $\mathfrak{M} \backslash G$.

Theorem 2.4. *Let $m \in \mathfrak{M}$. The following are equivalent:*

(i) *The Gleason part $P(m)$ is trivial.*
(ii) *If (z_j) is a net in D converging to m, then $\lim L_{z_j}$ is a constant map $L(\zeta) = m$.*
(iii) *$m \notin G$.*
(iv) *If $f \in H^\infty$, $\|f\|_\infty \leq 1$, and $f(m) = 0$, then*

$$f = f_1 f_2,$$

with $f_j \in H^\infty$, $\|f_j\|_\infty \leq 1$, and $f_j(m) = 0$.
(v) *The ideal $J_m = \{f \in H^\infty : f(m) = 0\}$ is the closure of its own square $J_m^2 = \{\sum_{j=1}^n f_j g_j : f_j, g_j \in J_m\}$.*

Conditions (iv) and (v) are the strongest and weakest ways, respectively, of saying that J_m is equal to its own square.

Proof. Clearly (i) implies (ii), because any limit point of L_{z_j} in $\mathfrak{M}^D$ is an analytic map whose range must be contained in $P(m)$ by Lemma 1.1. Lemma 1.3 shows that (ii) imples (iii). By Theorem 2.2, (iv) follows from (iii), and by the proof of Corollary 2.3, (iv) implies (i). So except for (v), we have a logical circle.

Trivially, (v) follows from (iv). To complete the proof, we show (v) implies (iii). If (iii) fails then there is an analytic disc L_m with $L_m(0) = m$, and there is a Blaschke product $B(z) \in J_m$ such that $(B \circ L_m)'(0) \neq 0$. But then the continuous linear functional

$$f \to (f \circ L_m)'(0)$$

annihilates J_m^2, so that J_m^2 cannot be dense in J_m. $\square$

We can now determine all the analytic structure in $\mathfrak{M}$. Recall that when $m \in G$, there is an analytic disc $L_m : D \to P(m)$, $L_m(0) = m$. By Theorem 1.7 the map L_m is uniquely determined by m in the sense that whenever (w_i) is a net in D converging to m,

$$L_m = \lim L_{w_i}.$$

Theorem 2.5. *Suppose $F : D \to \mathfrak{M}$ is a nonconstant analytic map. Let $m = F(0)$. Then $m \in G$ and there is an analytic function $\tau : D \to D$, $\tau(0) = 0$, such that*

(2.3) $$F(\zeta) = L_m \circ \tau(\zeta), \quad \zeta \in D.$$

In particular, if L is an analytic disc with range $P(m)$, $m = L(0)$, then there is a constant α, $|\alpha| = 1$, such that

$$L(\zeta) = L_m(\alpha\zeta), \quad \zeta \in D.$$

Proof. By Lemma 1.1, $P(m)$ is nontrivial, because F is analytic but not constant. Hence $m \in G$ by Theorem 2.4, and we have the one-to-one analytic mapping L_m. Now use (2.3) to define the function $\tau : D \to D$, $\tau(0) = 0$. Our task is to show that τ is analytic. For any $r > 1$ there is, by Theorem 1.7, a Blaschke product $B_r(z)$ such that $h_r(\zeta) = B_r \circ L_m(\zeta)$ is univalent on $|\zeta| < r$. But then by (2.3)

$$\tau(\zeta) = h_r^{-1} \circ B_r \circ F(\zeta), \quad |\zeta| < r,$$

so that τ is analytic on $h_r(|\zeta| > r)$. Since $h_r(0) = 0$ and $\|h_r\|_\infty \leq 1$, this means τ is analytic on $|\zeta| < r$. Letting $r \to 1$, we see that τ is analytic on D.

If F maps D one-to-one onto $P(m)$, then τ is a univalent mapping from D onto D with $\tau(0) = 0$, so that by Schwarz's lemma, $\tau(\zeta) = \alpha\zeta$, a constant, $|\alpha| = 1$. $\square$

Corollary 2.6. *In the topology of $\mathfrak{M}^D$, the set of analytic maps from D into D is dense in the set of analytic maps from D into $\mathfrak{M}$. The set of maps*

$$\left\{ \alpha L_z(\zeta) = \alpha \frac{\zeta + z}{1 + \bar{z}\zeta} : z \in D, |\alpha| = 1 \right\}$$

is dense in the set L of maximal analytic discs in $\mathfrak{M}$.

Proof. Let $F \in \mathfrak{M}^D$ be an analytic map. If F is constant, the corona theorem yields a net of constants in D converging to F. If F is not constant, then F has the form (2.3) with $L_m = \lim L_{z_i}$. Then $F_i(\zeta) = L_{z_i} \circ \tau(\zeta)$ defines a net converging to F.

If L is a maximal analytic disc in $\mathfrak{M}$, then $L(D) = P(m)$, $m = L(0)$, and there is α, $|\alpha| = 1$, and a net (z_j) converging to m such that

$$L(\zeta) = L_m(\alpha\zeta) = \lim_j L_{z_j}(\alpha\zeta) = \alpha \lim_j L_{\bar{\alpha}z_j}(\zeta). \quad \square$$

An equivalent formulation of Theorem 2.5, without the language of maximal ideals, is given in Exercise 7.

3. Approximate Dependence between Kernels

Let $\{z_j\}$ be a sequence in the upper half plane. Assume $\{z_j\}$ is separated, which means there is $\alpha > 0$ such that

$$|z_j - z_k| \geq \alpha y_j, \quad k \neq j.$$

Then $\{z_j\}$ is an H^∞ interpolating sequence if and only if

$$(3.1) \qquad \sup_Q \sum_{z_j \in Q} \frac{y_j}{\ell(Q)} \leq A < \infty,$$

where Q ranges through all squares of the form $Q = \{a \leq x \leq a + \ell(Q)\}, 0 < y \leq \ell(Q)\}$. In this section and in Section 4 we prove that (3.1) holds if the points $\{z_j\}$ can be separated, in either of two ways, by bounded harmonic functions. These results generalize the theorem from Chapter VII, Section 4, that a harmonic interpolating sequence is an H^∞ interpolating sequence. The proofs of these results are of independent interest, because each proof gives a quantative formulation of Fatou's theorem. The key idea, in Lemma 3.3, is that when (3.1) fails the Poisson kernel for some point z_j can be approximated by a convex combination of the kernels representing the other points z_k.

To begin we observe that if (3.1) fails for a separated sequence $\{z_j\}$ then (3.1) also fails for a subsequence having a particularly convenient distribution.

Lemma 3.1. *Let N be a positive integer, $N \equiv 1 \pmod 3$. Let $\{z_j\}$ be a separated sequence*

$$(3.2) \qquad |z_j - z_k| > \alpha y_j, \quad k \neq j,$$

with $\alpha > 0$. Then $\{z_j\}$ can be partitioned into $n_0 = n_0(\alpha, N)$ disjoint subsequences $Y_1, Y_2, \ldots, Y_{n_0}$ having the following properties: To each z_j there corresponds an interval I_j with $|I_j| = 3N^m$, m an integer, such that

$$(3.3) \qquad 3Ny_j \leq |I_j| < 3N^2 y_j$$

and

$$(3.4) \qquad \mathrm{dist}(x_j, \partial I_j) \geq \frac{1}{3}|I_j|, \quad x_j = \mathrm{Re}\ z_j \in I_j.$$

If z_j and z_k belong to the same subsequence Y_i, and if $I_j \cap I_k \neq \emptyset$, then

$$(3.5) \qquad I_j \subset I_k \quad \text{or} \quad I_k \subset I_j,$$

and if $I_j \subset I_k$, then

$$(3.6) \qquad |I_j| \leq N^{-3}|I_k|$$

and I_j is a member of the unique partition of I_k into $|I_k|/|I_j|$ subintervals of length $|I_j|$.

Two intervals I_j and I_k from the same Y_i are illustrated in Figure X.1.

Figure X.1. Two intervals from the same Y_i.

Admittedly, this lemma is technical, and some readers may want to postpone studying its proof until after seeing its applications. Conditions (3.3) and (3.4) ensure that, for N large, $\mathbb{R}\setminus I_j$ has small harmonic measure at z_j. Condition (3.5) enables us to fit the points z_j neatly into generations, while condition (3.6) implies that when $I_j \subset I_k$, the Poisson kernel for z_k is almost constant on I_j.

Proof.   Consider all intervals of the form

$$J_{k,m} = (kN^m, (k+3)N^m),$$

with k and m integers. For fixed m the middle thirds of the $J_{k,m}$ form a paving of $\mathbb{R}$. To each z_j we let I_j be that unique $J_{k,m}$ such that $x_j \in [(k+1)N^m, (k+2)N^m)$, the middle third of $J_{k,m}$, and such that $N^{m-2} < y_j < N^{m-1}$. Then (3.3) and (3.4) hold.

Divide the family $\{J_{k,m}\}$ of intervals into three subfamilies according to whether k is congruent to 0, 1, or 2 (mod 3). Within each subfamily the intervals of each given length form a partition of $\mathbb{R}$. Moreover, if $J_{p,n}$ and $J_{q,m}$ are any two intervals from the same subfamily, and if $|J_{q,m}| < |J_{p,n}|$, then either

$$J_{q,m} \subset J_{p,n} \quad \text{or} \quad J_{q,m} \cap J_{p,n} = \varnothing.$$

This holds because $N \equiv 1$ (mod 3). Write $pN^n = kN^m, k = pN^{n-m}$. Since $n > m$ (because $|J_{p,n}| > |J_{q,m}|$), k is an integer. And since $N^{n-m} \equiv 1$ (mod 3), we have $k \equiv p$ (mod 3). Hence pN^n is an endpoint of an interval $J_{k,m}$ from the same subfamily. The same thing happens at the other endpoint $(p+3)N^n$ of $J_{p,n}$ and so $J_{q,m}$ cannot cross either end of $J_{p,n}$.

Divide the points $\{z_j\}$ into three subsequences according to whether $I_j = J_{k,m}$ with k congruent to 0, 1, or 2 (mod 3). For z_j and z_k in the same subsequence we now have

$$I_j \subset I_k, \quad I_k \subset I_j, \quad \text{or} \quad I_j \cap I_k = \varnothing,$$

and if $I_j \subset I_k$, then I_j occurs in the partition of I_k into subintervals of length $|I_j|$.

To get (3.6) we use the separation (3.2) to further partition $\{z_j\}$. In the hyperbolic metric, all rectangles of the form $R_{k,n} = \{kN^n < x < (k+$

$3)N^n, 3N^{n-5} < y \le 3N^{n-1}\}$ are congruent. By (3.2) each such rectangle contains at most $n_1 = n_1(\alpha, N)$ sequence points z_j. Partition each of our three subsequences into n_1 subsequences so that every rectangle $R_{k,n}$ contains at most one point from each subsequence. We then obtain $n_0 = 3n_1$, subsequences $Y_1, \ldots, Y_{n_0}$ for which (3.5) and (3.6) hold. $\quad\square$

By (3.5) the points z_j in each subsequence Y_i, and their corresponding intervals I_j, fit naturally into generations. When $z_j, z_k \in Y_i$, we say $z_k \in G_1(z_j)$, and $I_k \in G_1(I_j)$, if $I_k \subsetneq I_j$ and I_k is maximal. Successive generations, as always, are defined inductively:

$$G_p(I_j) = \bigcup_{G_{p-1(I_j)}} G_1(I_k).$$

Lemma 3.2. *Suppose $\{z_j\}$ is a sequence of points satisfying (3.2). Let N be given and let $Y_1, Y_2, \ldots, Y_{n_0}, n_0 = n_0(\alpha, N)$, be the subsequences given by Lemma 3.1. Let $0 < \gamma < 1$ and let r be a positive integer. Then there exists $A = A(\alpha, N, \gamma, r)$ such that if (3.1) fails with constant A, that is, if*

$$\sup_Q \sum_{z_j \in Q} \frac{y_j}{l(Q)} > A,$$

then there is a subsequence Y_i and a point $z_j \in Y_i$ such that

$$\sum_{G_r(I_j)} \frac{|I_k|}{|I_j|} > \gamma.$$

In a quantative way, this lemma says that to establish (3.1) it is enough to work with the special sequences Y_i.

Proof. If the contrary holds, then for every Y_i and for every $z_j \in Y_i$, induction gives

$$\sum_{G_q(I_j)} \frac{|I_k|}{|I_j|} \le \gamma^n,$$

when $nr \le q < (n+1)r$. Consequently,

$$\sum_{\substack{I_k \subset I_j \\ z_k \in Y_i}} \frac{|I_k|}{|I_j|} \le \frac{r}{1-\gamma}.$$

Summing over the Y_i then yields

$$\sum_{I_k \subset I} \frac{|I_k|}{|I|} \le \frac{n_0 r}{1-\gamma},$$

and since $y_k/|I_k|$ is bounded above and below, this proves that (3.1) holds with some constant $A = A(\alpha, N, \gamma, r)$. $\quad\square$

Let N be large. Then by (3.3) and (3.4) I_j contains most of the mass of the Poisson kernel P_j corresponding to z_j

$$\int_{\mathbb{R}\setminus I_j} P_j(t)dt \leq \frac{2}{\pi} \int_{t > Ny_j/3} \frac{y_j}{t^2 + y_j^2} dt \leq \frac{c_1}{N}.$$

Partition I_j into N^3 subintervals $I_{j,l}$ of length $|I_{j,l}| = N^{-3}|I_j| \leq N^{-1}y_j$, and form the step function

$$K_j(t) \sum_{l=1}^{N^3} \left(\inf_{I_{j,l}} P_j \right) \chi_{I_{j,l}}(t).$$

Then $K_j(t) \leq P_j(t)$ and K_j constant on each $I_k \in \bigcup G_q(I_j)$. If N is large, $\|P_j - K_j\|_1$ is small,

$$\|P_j - K_j\|_1 \leq \int_{\mathbb{R}\setminus I_j} P_j(t)\, dt + \sum_{l=1}^{N^3} \int_{I_{j,l}} \left(P_j(t) - \inf_{I_{j,l}} P_j \right) dt$$

$$\leq \frac{c_1}{N} + 2\int_0^\infty \left(P_j(t) - P_j\left(t + \frac{y_j}{N}\right)\right) dt \leq \frac{c^2}{N}.$$

Fix $\varepsilon > 0$ and choose $N = N(\varepsilon)$ so that

$$(3.7) \qquad\qquad \|P_j = -K_j\|_1 < \varepsilon/8$$

for each z_j. See Figure X.2.

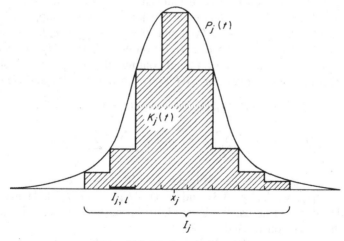

Figure X.2. The kernels P_j and K_j.

Lemma 3.3. *Suppose $\{z_j\}$ is a sequence of points satisfying (3.2). For $\varepsilon > 0$ and for $N = N(\varepsilon)$, let $Y_1, Y_2 \ldots, Y_{n_0}, n_0 = n_0(\alpha, N)$ be the subsequences* *given by Lemma 3.1. There exist $\beta = \beta(\varepsilon), 0 < \beta < 1$, and a positive integer $p = p(\varepsilon)$ such that if*

$$(3.8) \qquad \sum_{G_p(I_j)} \frac{|I_k|}{|I_j|} \geq \beta$$

for some point $z_j \in Y_i$, then there exist convex weight $\lambda_k, 0 \leq \lambda_k < 1, \sum \lambda_k = 1$ such that

$$(3.9) \qquad \|P_j - \sum \lambda_k P_k\|_1 < \varepsilon,$$
$$(3.10) \qquad \lambda_k = 0 \quad \text{if} \quad z_k \notin G_1(z_j) \cup \ldots \cup G_p(z_j),$$

and

$$(3.11) \qquad \lambda_k \leq b|I_k|/|I_j|,$$

where $b = b(N)$ is a constant.

Proof. We are going to find weights λ_k satisfying (3.10) and (3.11) such that

$$0 \leq \sum \lambda_k K_k \leq K_j$$

and such that

$$(3.12) \qquad \|K_j - \sum \lambda_k K_k\|_1 < \varepsilon/8.$$

Replacing λ_k by $\lambda_k / \sum \lambda_i$ and using (3.7), we then obtain convex weights for which we have (3.9)–(3.11).

Set $f_0(t) = K_j(t)$. Then f_0 has constant value $f_0(I_k)$ on each first generation interval $I_k \in G_1(I_j)$. These intervals are disjoint and K_k is supported on I_k. Set $\lambda_k = f_0(I_k)/\|K_k\|_\infty, z_k \in G_1(z_1)$, and $f_1 = f_0 - \sum_{G_1} \lambda_k K_k$. Then $0 \leq f_1 \leq f_0$, and f_1 has constant value $f_1(I_k)$ on each $I_k \in G_2(I_j)$. Repeat the construction using f_1 in place of f_0 and continue through p generations. At stage q, f_{q-1} has constant value $f_{q-1}(I_k)$ on each $I_k \in G_q(I_j)$, and we set

$$\lambda_k = f_{q-1}(I_k)/\|K_k\|_\infty, \qquad z_k \in G_q(z_j)$$

$$f_q = f_{q-1} - \sum_{G_q(I_j)} \lambda_k K_k = K_j - \sum_{r=1}^{q} \sum_{G_r(I_j)} \lambda_k K_k.$$

Then $0 \leq f_q \leq f_{q-1} \leq f_0$. See Figure X.3. By (3.4) we have $f_{q-1}(I_k) \leq \|f_0\|_\infty \leq c_1/|I_j|$ while by (3.3) and (3.4), $\|K_k\|_\infty \geq c_2/|I_k|$. Hence $\lambda_k \leq b|I_k|/|I_j|$ and (3.11) holds. We have (3.10) because the process is stopped when f_p has been constructed.

It remains to estimate $\|K_j - \sum \lambda_k K_k\|_1 = \int f_p \, dt$. Since $\|K_k\|_\infty \geq c_2/|I_k|$, (3.7) yields

$$\int \frac{K_k}{\|K_k\|_\infty} dt \geq \frac{(1-\varepsilon)|I_k|}{c_2} = (1-\delta)|I_k|,$$

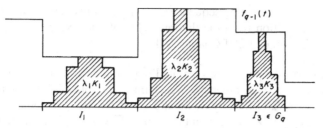

Figure X.3. For each $I_k \in G_q$, $\lambda_k K_k \le f_{q-1}(I_k)$, but $\int_{I_k} \lambda_k K_k dt \ge \delta \int_{I_k} f_{q-1} dt$.

with $0 < \delta < 1$. Consequently, when $I_k \in G_q$ we have

$$\int f_q \, dt = |I_k| f_{q-1}(I_k) - f_{q-1}(I_k) \int \frac{K_k}{\|K_k\|_\infty} \, dt$$

$$\le \delta |I_k| f_{q-1}(I_k) = \delta \int_{I_k} f_{q-1} \, dt.$$

Let $E_0 = I_j$, $E_q = \bigcup_{G_q} I_k$, $q = 1, 2, \ldots, p$. Then $E_0 \supset E_1 \supset \ldots \supset E_p$, and induction yields

$$\int_{E_p} f_p \, dt \le \delta \int_{E_p} f_{p-1} \, dt \le \delta \int_{E_{p-1}} f_{p-1} \, dt$$

$$\le \delta^p \int_{E_0} f_0 \, dt \le \delta^p.$$

Choose p so that $\delta^p < \varepsilon/16$. Recall that $\|f_p\|_\infty \le \|f_0\|_\infty \le c_1/|I_j|$, where c_1 depends only on ε. Our hypothesis is that $|E_p|/|I_j| \ge \beta$, for some $\beta = \beta(\varepsilon)$ yet to be determined. Now choose $\beta = \beta(\varepsilon)$ so that $(1 - \beta)c_1 < \varepsilon/16$. Then we obtain $\|f_0\|_\infty |I_j \backslash E_p| < \varepsilon/16$ and

$$\int f_p \, dt \le \delta^p + \int_{I_j \backslash E_p} f_p \, dt \le \delta^p + \|f_0\|_\infty |I_j \backslash E_p| < \varepsilon/8,$$

so that (3.12) holds. $\quad\square$

Theorem 3.4. *Let $\{z_j\}$ be a sequence in the upper half plane. Assume there are real valued harmonic functions $\{u_j(z)\}$ such that*

(i) $\|u_j\|_\infty \le 1$,
(ii) $u_j(z_j) \ge \delta > 0$,
(iii) $u_j(z_k) \le 0, k \ne j$,

where $\delta > 0$ is independent of j. Then $\{z_j\}$ is an H^∞ interpolating sequence.

The hypothesis of the theorem is reminiscent of the condition

$$\inf_j \prod_{k; k \ne j} \left| \frac{z_j - z_k}{z_j - \bar{z}_k} \right| = \delta > 0$$

characterizing interpolating sequences, but there seems to be no direct derivation of this condition from the theorem's hypothesis. Note that the theorem generalizes Theorem 4.2 of Chapter VII.

Proof. Clearly, the points are separated. The bounded function $\exp(u_j + i\tilde{u}_j)$ separates z_j from each other z_k. We claim (3.1) holds. That will prove the theorem. If (3.1) fails, then by Lemmas 3.2 and 3.3 there is a point z_j and there are weights $\lambda_k \geq 0$, $\lambda_j = 0$, such that $\| P_j - \sum \lambda_k P_k \|_1 < \delta$. But then

$$\delta \leq u_j(z_j) - \sum \lambda_k u_j(z_k) = \int u_j(P_j - \sum \lambda_k P_k)\, dt$$

$$\leq \|u_j\|_\infty \| P_j - \sum \lambda_k P_k \|_1 < \delta,$$

a contradiction. □

Lemma 3.3 and Theorem 3.4 are also true with approximate identities more general than the Poisson kernel (see Exercise 12).

4. Interpolating Sequences and Harmonic Separation

We continue the discussion of the preceding section. Our objective is another characterization of interpolating sequences.

Theorem 4.1. *Let $\{z_j\}$ be a sequence in the upper half plane. Then $\{z_j\}$ is an interpolating sequence if disjoint subsets of $\{z_j\}$ have disjoint closures in the maximal ideal space $\mathfrak{M}$ of H^∞.*

Before getting into its proof, let us see what this theorem means. Notice that the converse of the theorem is trivial. Is S and T are disjoint subsets of an interpolating sequence, then there is $f \in H^\infty$ such that $f(z_j) = 0$ when $z_j \in S$ and $f(z_j) = 1$ when $z_j \in T$. With respect to $\mathfrak{M}$, the closures of S and T lie in the disjoint closed sets $\{f = 0\}$ and $\{f = 1\}$ respectively. If we make the additional assumption that the closure K of $\{z_n\}$ in $\mathfrak{M}$ is a hull (i.e., if $m \in \mathfrak{M}\setminus K$, there is $f \in H^\infty$ such that $f(m) \neq 0$ but $f \equiv 0$ on K), then the theorem follows from a general result of Šilov and Exercise 8 of Chapter VII (see Hoffman [1962a]).

As we saw in Chapter V, $\mathfrak{M}$ is homeomorphic to a weak-star compact subset of the dual space $(L^\infty)^*$, and under that homeomorphism z_j corresponds to its Poisson kernel P_j. Hence the hypothesis of the theorem is that disjoint subsets of $\{P_j\}$ have disjoint weak-star closures in $(L^\infty)^*$. This hypothesis can be reformulated two ways. First, it means that the weak-star closure of $\{P_j\}$ in $(L^\infty)^*$ is homeomorphic to $\beta\mathbb{N}$, the Stone-Čech compactification of the integers. Equivalently, the closure of $\{z_j\}$ in $\mathfrak{M}$ is homeomorphic to $\beta\mathbb{N}$. Thus $\{z_j\}$ is an interpolating sequence if and only if its closure in $\mathfrak{M}$ is homeomorphic to $\beta\mathbb{N}$.

Secondly, every basic weak-star open subset V of $(L^\infty)^*$ is defined by a real number α and by finitely many functions $u_1, u_2, \ldots, u_M \in L^\infty$ in the following way:

$$V = V_\alpha = \{\varphi \in (L^\infty)^* : \varphi(u_m) < \alpha, 1 \leq m \leq M\}.$$

Let S and T be subsets of $\{P_j\}$ having disjoint weak-star closures in $(L^\infty)^*$. Each point $\varphi \in \bar{S}$, the weak-star closure of S, has a neighborhood V_α such that $V_\alpha \cap \bar{T} = \varnothing$. Since $\bar{T}$ is compact there is $\beta > \alpha$ such that $V_\beta \cap \bar{T} = \varnothing$. Since $\varphi(1) = 1$ when $\varphi \in \bar{S} \cup \bar{T}$, we can take $\alpha = -1$ and $\beta > 1$ by replacing each u_m by $au_m + b$, a and b constants. Covering $\bar{S}$ by finitely many such neighborhoods $V_{-1}^{(n)}, 1 \leq n \leq K$, we arrive at the following equivalent formulation of the hypothesis:

Whenever S and T are disjoint subsets of $\{z_j\}$, there are $\{u_m^{(n)}\} \in L^\infty, 1 \leq n \leq K, 1 \leq m \leq M$ such that

(4.1)
$$\begin{aligned}\inf_n \sup_m u_m^{(n)}(z_j) &\leq -1 \quad \text{if} \quad z_j \in S, \\ \inf_n \sup_m u_m^{(n)}(z_j) &> +1 \quad \text{if} \quad z_j \in T.\end{aligned}$$

Taking $K = M = 1$ in (4.1), we see that Theorem 4.1 also generalizes Theorem 4.2 of Chapter VII. Roughly speaking, Theorem 4.1 says that a sequence satisfies (3.1) and (3.2) if any reasonable form of separation is possible with bounded harmonic functions.

The proof of the theorem will focus on the concrete condition (4.1). It will be crucial that the numbers K, M, and

$$B = \sup_{n,m} \|u_m^{(n)}\|_\infty$$

can be chosen not to depend on the subsets S and T of $\{z_j\}$.

Lemma 4.2. *If disjoint subsets of $\{z_j\}$ have disjoint closures in $\mathfrak{M}$, then (4.1) holds with K, M, and $B = \sup_{n,m} \|u_m^{(n)}\|_\infty$ not depending on the subsets S and T of $\{z_j\}$.*

Proof. We can suppose $T = \{z_j\} \setminus S$. The set L of all subsets S of $\{z_j\}$ is a compact space in the product topology, in which a neighborhood V of a subset S_0 is determined by finitely many indices $j_1, j_2, \ldots, j_s$ by the rule

$$V = V(S_0; j_1, j_2, \ldots, j_s) = \{S \in \mathrm{L} : z_{j_l} \in S \Leftrightarrow z_{j_l} \in S_0, 1 \leq l \leq s\}.$$

For K, M, and B positive integers, let $\mathrm{E}_{K,M,B} = \{S \in \mathrm{L} : (4.1) \text{ holds with bounds } K, M, \text{ and } B\}$. Then $\mathrm{L} = \bigcup \mathrm{E}_{K,M,B}$, and by normal families each set $\mathrm{E}_{K,M,B}$ is closed in the product topology of L. By the Baire category theorem, some $\mathrm{E}_{K,M,B}$ contains an open set $V(S_0; j_1, j_2, \ldots, j_s)$. By the definition of the product topology this means that except for the finitely many points $z_{j_1}, z_{j_2}, \dots, z_{j_s}$, we have (4.1) with bounds on K, M, and B that do not depend on the subsets S and T. Adjoining finitely many additional functions $u_m^{(n)}$ to separate $z_{j_1}, z_{j_2}, \dots, z_{j_s}$, we then obtain (4.1) with uniform bounds. $\square$

We prove Theorem 4.1 using Lemma 3.3 and an iteration. Afterward we shall indicate how an alternate proof can be based on Theorem 6.1 of Chapter VIII.

Proof of Theorem 4.1. By Lemma 4.2 the points z_j are separated, and (3.2) holds with $\alpha = \alpha(B)$. The problem now is to derive the Carleson condition (3.1) from (4.1). Let $\varepsilon = (4BK)^{-1}$, and take $N = N(\varepsilon)$, $p = p(\varepsilon)$, and $\beta = \beta(\varepsilon)$ in accordance with Lemma 3.3. Here is the strategy. If (3.1) fails, then a subsequence Y_i contains a point z_0 such that

$$\sum_{G_r(z_0)} \frac{|I_k|}{|I_0|} > \gamma$$

with $r = 2(BK + 1)p$ and with $1 - \gamma$ very small. This will mean that we have (3.9) for z_j in a large portion of G_q for a large number of indices q. When the subsets S and T are chosen properly, this will contradict Lemma 4.2.

If there were $\rho < 1$ and $q \leq 2BKp$ such that for all Y_i, and all $z_0 \in Y_i$

$$(4.2) \qquad \sum_{G_q(z_0)} \left\{ \frac{|I_j|}{|I_0|} : \sum_{G_{2p(z_j)}} \frac{|I_k|}{|I_j|} < \beta \right\} > \rho,$$

then

$$\sum_{G_{q+2p}(z_0)} \frac{|I_k|}{|I_0|} < \rho\beta + (1 - \rho)$$

for all z_0 in all Y_j. By Lemma 3.2 we should then have (3.1).

Assume (4.2) fails for a fixed point $z_0 \in Y_i$, for all $q \leq 2BKp$ and for $\rho < 1$ to be determined. Choose subsets S and T of $\{z_j\}$ so that $z_0 \in T$ and so that for $sp < q \leq (s + 1)p$, $s \geq 0$, $G_p(z_0) = G_q \subset S$ if s is even and $G_q \subset T$ if s is odd. Thus we alternate between S and T after every p generations.

Note that (4.1) is unchanged if we permute the indices $\{1, 2, \dots, K\}$ or if, for any fixed n, we permute the set of functions $\{u_1^{(n)}, u_2^{(n)}, \dots, u_M^{(n)}\}$. So we can assume $u_1^{(n)}(z_0) > 1$ for $1 \leq n \leq K$. Let $v = \sum_{n=1}^{K} u_1^{(n)}$. Then $\|v\|_\infty \leq BK$, but $v(z_0) \geq K$. By Lemma 3.3 there are convex weights $\lambda_j \geq 0$ such that

$$\sum_{1}^{p} \sum_{G_q(z_0)} \lambda_j v(z_j) > K - \varepsilon BK = K - \frac{1}{4}.$$

Since $|v(z_j)| \le BK$, Chebychev's inequality (for the measure $\sum \lambda_j \delta_{z_j}$) then yields

$$\sum_1^p \sum_{G_q(z_0)} \{\lambda_j : v(z_j) > K - \tfrac{1}{2}\} \ge \frac{1}{4BK}.$$

Then by (3.11) there is q_1, $1 \le q_1 \le p$, and there is $E_1 \subset G_{q_1}$, such that $v(z_j) \ge K - \tfrac{1}{2}$ on E_1 and such that

$$\sum_{E_1} \frac{|I_j|}{|I_0|} \ge \frac{1}{bp} \frac{1}{4BK} = \delta_1,$$

where b is the constant in (3.11). Assume (4.2) fails with $1 - \rho < \delta_1/2$. Then there is $F_1 \subset E_1$ such that

$$\sum_{F_1} \frac{|I_j|}{|I_0|} \ge \tfrac{1}{2}\delta_1,$$

and for each $z_j \in F_1$ $\sum_{G_{2p(z_j)}} |I_k|/|I_j| \ge \beta$.

For $z_j \in F_1 \subset S$, we have (3.9) with weights λ_k attached only to points in $G_{p+1} \cup \cdots \cup G_{2p} \subset T$, because if $q_1 < p$ we still have (3.8) when we delete $G_1(z_j) \cup \cdots \cup G_{p-q_1}(z_j)$. (It is for this reason that $2p$ occurs inside (4.2).)

Fix $z_j \in F_1$. Then $z_j \in S$ and we can permute the indices $\{1, 2, \ldots, K\}$ so that $u_1^{(1)}(z_j) < -1$. This permutation depends on the point $z_j \in F_1$, but that does not harm what we are going to do, because distinct $z_j \in F_1$ have disjoint generations $\bigcup G_q(z_j)$. set $v' = v - u_1^{(1)} \sum_2^K u_1^{(n)}$. Then $\|v'\|_\infty < BK$ but $v'(z_j) > K + \tfrac{1}{2}$. We have convex weights λ_k such that

$$\sum_{p-q_1+1}^{2p-q_1} \sum_{G_q(z_j)} \lambda_k v'(z_k) > K + \tfrac{1}{4},$$

and Chebychev's inequality now yields

$$\sum_{p-q_1+1}^{2p-q_1} \sum_{G_q(z_j)} \{\lambda_k : v'(z_k) > K\} \ge \frac{1}{4BK}.$$

Arguing as before using (3.11), we have $q_1(z_j)$, $p < q_1(z_j) \le 2p$, and $E_1(z_j) \subset G_{q_1(z_j)}$ such that $v'(z_k) > K$ on $E_1(z_j)$ and such that

$$\sum_{E_1(z_j)} \frac{|I_k|}{|I_j|} \ge \delta_1.$$

Summing over F_1, we find there is an index q_2, $p < q_2 \geq 2p$, and a set $E_2 \subset G_{q_2}$ such that

$$\sum_{E_2} \frac{|I_k|}{|I_0|} \geq \frac{\delta_1}{p} \sum_{F_1} \frac{|I_j|}{|I_0|} \geq \frac{\delta_1^2}{2p} = \delta_2,$$

and such that for each $z_k \in E_2 v'(z_k) > K$ (after a suitable permutation of $\{1, 2, \ldots, K\}$). We assume (4.2) fails with $1 - \rho < \delta_2/2$. Then we have $F_2 \subset E_2$ such that

$$\sum_{F_2} \frac{|I_k|}{|I_0|} \geq \frac{\delta_2}{2}$$

and such that for each $z_k \in F_2$, $\sum_{G_{2p}(z_k)} |I_j|/|I_k| \geq \beta$.

Moreover, since $F_2 \subset T$, for each $z_k \in F_2$ we can permute the set $\{u_1^{(1)}, u_2^{(1)}, \ldots, u_m^{(1)}\}$ so that $u_1^{(1)}(z_k) > 1$. Then $v(z_k) = \sum_{n=1}^{k} u_1^{(n)}(z_k) > K + 1$. If (4.2) fails with $1 - \rho$ small enough, we can repeat the above argument replacing z_0 by each $z_k \in F_2$ and obtain a set $F_4 \subset G_{q_4} \subset T$ such that

$$\sum_{F_4} \frac{|I_j|}{|I_0|} \geq \frac{1}{2}\delta_4$$

for some $\delta_4 = \delta_4(\delta_2, p)$ and such that (after suitable permutations of $u_m^{(n)}$) $v(z_j) > K + 2$, $z_j \in F_4$.

Now if (4.2) fails with $1 - \rho$ extremely small, we can do the above argument $(B - 1)K$ times. We then reach a point z_k for which, after the appropriate permutations of $\{u_m^{(n)}\}$, $v(z_k) > BK$. This contradiction proves Theorem 4.1. □ ·

To derive Theorem 4.1 from Theorem 6.1 of Chapter VIII, partition the top half of each dyadic square $Q = \{k2^{-n} \leq x \leq (k + 1)2^{-n}, 0 < y \leq 2^{-n}\}$ into 2^{2N-1} small squares S_j of side 2^{-n-N}, If $B = \sup_{m,n} \|u_m^{(n)}\|_\infty$ is the bound guaranteed by Lemma 4.2, choose N so that

$$(4.3) \qquad \sup_{z,w \in S_j} |u(z) - u(w)| \leq \frac{1}{2}$$

whenever $u(z)$ is harmonic and $\|u\|_\infty \leq B$.

Fix a dyadic square Q. The first generation $G_1 = \{S_j^{(1)}\}$ consists of those small squares $S_j \subset Q$ such that $S_j^{(1)}$ contains a sequence point $z_j^{(1)}$ but such that $S_j^{(1)}$ lies below no other small square in Q having the same property. Choose one sequence point $z_j^{(1)}$ in each $S_j^{(1)}$ and let $F_1 = \{z_j^{(1)}\}$. Let $Q_j^{(1)}$ be the dyadic square having base the projection of $S_j^{(1)}$ onto $\{y = 0\}$. Then the dyadic squares $Q_j^{(1)}$ have pairwise disjoint interiors.

Inside each $Q_j^{(1)}$ select small squares $S_k^{(2)}$ in the same way. Write $G_1(S_j^{(1)})$ and

$$G_2 = \bigcup \{G_1(S_j^{(1)}) : S_j^{(1)} \in G_1\}.$$

Choose one sequence point $z_k^{(2)} \in S_k^{(2)}$ and set $F_2 = \{z_k^{(2)} : S_k^{(2)} \in G_2\}$. Form new dyadic squares $\{Q_k^{(2)}\}$ by dropping the $S_k^{(2)}$ onto the axis. Continue the process, obtaining the generations $G_3, G_4, \dots$, and the corresponding subsets $F_3, F_4, \dots$, of the original sequence $\{z_j\}$.

By Lemma 4.2, the points $\{z_j\}$ are separated. If $R_k^{(p)}$ is the rectangle, with sides parallel to the axes, joining the top edge of $Q_k^{(1)}$ to the top edge of $S_k^{(p)}$, then because the points are separated,

$$\sum_{z_j \in R_k^{(p)}} y_j \le C_1 \ell(S_k^{(p)}).$$

Consequently, $\{z_j\}$ is an interpolating sequence if and only if

$$(4.4) \qquad \sum_{p=1}^{\infty} \sum_{G_p} l(S_k^{(p)}) \le C_2 \ell(Q)$$

for each dyadic square Q lying on $\{y = 0\}$. Because $\sum_{G_1} \ell(S_j^{(1)}) \le \ell(Q)$, we only need to estimate $\sum_2^{\infty}$ in (4.4).

Replace $\{z_j\}$ by the subsequence $F_1 \cup F_2 \cup \cdots$. This does not change the sum (4.4). Partition $\{z_j\}$ by setting

$$S = \bigcup \{F_p : p \text{ odd}\}, \quad T = \bigcup \{F_p : p \text{ even}\}.$$

Then by (4.1) and (4.3) there exist finitely many bounded harmonic functions $u_m^{(n)}$ such that whenever $S_k^{(p)} \in G_1(S_j^{(p-1)})$,

$$\inf \left\{ \max_{m,n} |u_m^{(n)}(z) - u_m^{(n)}(w)| : z \in S_k^{(p)}, w \in S_j^{(p-1)} \right\} \ge 1.$$

By Theorem 6.1 of Chapter VIII, there exist smooth functions $\varphi_m^{(n)}(z)$ such that

$$(4.5) \qquad \inf \left\{ \max_{m,n} |\varphi_m^{(n)}(z) - u_m^{(n)}(w)| : z \in S_k^{(p)}, w \in S_j^{(p-1)} \right\} \ge \tfrac{1}{2},$$

and such that

$$(4.6) \qquad \iint_Q \sum_{m,n} |\nabla \varphi_m^{(n)}| \, dx \, dy \le C_3 \ell(Q)$$

for each square Q resting on $\{y = 0\}$.

Let $S_k^{(p)} \in G_p$, $p \ge 2$. Then $S_k^{(p)} \in G_1(S_j^{(p-1)})$ for some $S_j^{(p-1)} \in G_{p-1}$ and $S_k^{(p)}$ lies below $S_j^{(p-1)}$. Let $T_k^{(p)}$ be the rectangle, with sides parallel to the axes,

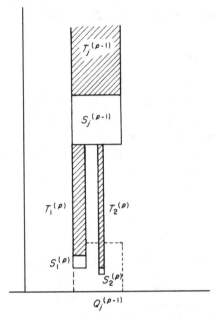

Figure X.4. Theorem VIII.6.1 implies Theorem 4.1.

joining the top edge of $S_k^{(p)}$ to the same dyadic interval on the bottom edge of $S_j^{(p-1)}$. The rectangles $\{T_k^{(p)}\}$ have pairwise disjoint interiors. See Figure X.4. By (4.5),

$$\sum_{m,n} \iint_{T_k^{(p)}} |\nabla \varphi_m^{(n)}| \, dx \, dy \geq \max_{m,n} \iint_{T_k^{(p)}} \left| \frac{\partial \varphi_m^{(n)}}{\partial y} \right| dx \, dy$$

$$\geq \tfrac{1}{2} \ell(S_k^{(p)}),$$

and since the rectangles $T_k^{(p)}$ are disjoint, (4.4) now follows from (4.6).

Of the two proofs of Theorem 4.1, the second one looks more transparent, but it is not perfectly clear which argument is stronger. For example, Theorem 3.4 does not seem to follow from Theorem VIII.6.1.

5. A Constructive Douglas–Rudin Theorem

Theorem 5.1. *Suppose* $u \in L^\infty$, $|u| = 1$ *almost everywhere. Let* $\varepsilon < 0$. *Then there exist interpolating Blaschke products* $B_1(z)$ *and* $B_2(z)$ *such that*

$$\|u - B_1/B_2\|_\infty < \varepsilon.$$

This result refines the Douglas–Rudin theorem, Chapter V, Section 2. Before proving the theorem, we mention one corollary and two related open problems.

Corollary 5.2. *The interpolating Blaschke products separate the points of the maximal ideal space $\mathfrak{M}$ of H^∞.*

Proof. Let m_1 and m_2 be distinct points of $\mathfrak{M}$. There are three cases.

Case 1. $m_1 \in X$ and $m_2 \in X$, where $X = \mathfrak{M}_{L^\infty}$ is the Šilov boundary of H^∞. By Theorem 5.1, L^∞ is the self-adjoint closed algebra generated by the interpolating Blaschke products. Consequently the interpolating Blaschke products separate the points of $X = \mathfrak{M}_{L^\infty}$.

In the remaining two cases we have $m_1 \in \mathfrak{M} \backslash X$ or $m_2 \in \mathfrak{M} \backslash X$. By symmetry we can suppose $m_1 \in \mathfrak{M} \backslash X$. Recall the notation

$$G = \{m \in \mathfrak{M} : m \text{ is in the closure of an interpolating sequence}\}.$$

Case 2. $m_1 \in G$. Then there is an interpolating Blaschke product $B(z)$ such that $B(m_1) = 0$. If also $B(m_2) = 0$, then m_1, and m_2 are in the closure of the zeros S of $B(z)$, and there are disjoint subsets T_1 and T_2 of S such that $m_1 \in \bar{T}_1, m_2 \in \bar{T}_2$. The Blaschke product with zeros T_1 then separates m_1 and m_2. (See, for instance, the conclusion of the proof of Theorem 1.7.)

Case 3. $m_1 \notin X \cup G$. Then m_1 and m_2 are in different Gleason parts. For every $\varepsilon > 0$ there is $f_\varepsilon \in H^\infty$ such that $f_\varepsilon(m_1) = 0, |f_\varepsilon(m_2)| > 1 - \varepsilon$, and $\|f_\varepsilon\| = 1$. By Corollary 2.6 of Chapter V, there is an inner function $u_c \in H^\infty$ such that $u_\varepsilon(m_1) = 0$ and $|u_\varepsilon(m_2)| > 1 - \varepsilon$. By Theorem 4.1 of Chapter VIII, there is an interpolating Blaschke product $B_\varepsilon(z)$ such that

(5.1) $|B_\varepsilon(z)| < \frac{1}{4}$ if $|u_\varepsilon(z)| < \frac{1}{4}$,

(5.2) $|u_\varepsilon(z)| < \beta$ if $B_\varepsilon(z) = 0$,

where $\beta = \beta\left(\frac{1}{4}\right)$ is some constant. Moreover,

(5.3) $\delta(B_\varepsilon) \geq \delta_0$,

where $\delta_0 > 0$ is a constant independent of ε. Now we claim there is $\eta(\varepsilon) \to 0(\varepsilon \to 0)$ such that

(5.4) $|B_\varepsilon(z)| \geq 1 - \eta(\varepsilon)$ if $|u_\varepsilon(z)| \geq 1 - \varepsilon$.

Accepting (5.4) for the moment, we choose ε so small that $1 - \eta(\varepsilon) \geq \frac{1}{2}$. Then by (5.1) and the corona theorem, $B_\varepsilon(m_1) \leq \frac{1}{4}$, while by (5.4) and the corona theorem, $|B_\varepsilon(m_2)| \geq \frac{1}{2}$. Thus the interpolating Blaschke product B_ε separates m_1 and m_2.

To prove (5.4), we use some ideas from Section 3, Chapter VIII. Suppose $|u_\varepsilon(z)| > 1 - \varepsilon$. Using a Möbius transformation, we can assume $z = 0$. Let

$$E_\beta = \left\{ \theta : \inf_{\Gamma(\theta)} |u_\varepsilon(z)| < \beta \right\},$$

where $\Gamma(\theta)$ denotes the cone

$$\Gamma(\theta) = \left\{ z : \frac{|z - e^{i\theta}|}{1 - |z|} < 2 \right\}.$$

Write $v(\theta) = |u_\varepsilon(e^{i\theta}) - u_\varepsilon(0)|^2$. Then $\int v(\theta)d\theta/2\pi = 1 - |u_\varepsilon(0)|^2 < 2\varepsilon$, whereas if $|u_\varepsilon(z)| < \beta$, then

$$\frac{1}{2\pi} \int v(\theta) P_z(\theta) d\theta \geq |u_\varepsilon(z) - u_\varepsilon(0)|^2 \geq (1 - \varepsilon - \beta)^2.$$

The weak-type estimate for the nontangential maximal function therefore yields

(5.5) $$|E_\beta| \leq \frac{2c\varepsilon}{(1 - \varepsilon - \beta)^2}.$$

Write $E_\beta = \bigcup I_j$, where the I_j are pairwise disjoint ares on ∂D and set

$$S_j = \{re^{i\theta} : \theta \in I_j, 1 - |I_j| < r < 1\}.$$

By (5.2) the zeros $\{z_n\}$ of $B_\varepsilon(z)$ lie inside $\bigcup S_j$. Hence by (5.3) and (5.5)

$$\sum (1 - |z_n|) \leq C(\delta_0) \frac{2c\varepsilon}{(1 - \varepsilon - \beta^2)}.$$

Because $\beta = \beta(\frac{1}{4})$ fixed we conclude that $|B_\varepsilon(0)| \geq 1 - \eta(\varepsilon)$, where $\eta(\varepsilon) \to 0(\varepsilon \to 0)$. That establishes (5.4). $\square$

Problem 5.3. *Do the interpolating Blaschke products generate H^∞ as a uniform algebra?*

Problem 5.4. *Can every Blaschke product be uniformly approximated by interpolating Blaschke products?*

Since the Blaschke products generate H^∞, a yes answer to Problem 5.4 would imply a yes answer to Problem 5.3. See page vii.

The proof of Theorem 5.1 is different from the constructions we have discussed above.

Lemma 5.5. *Let $\varepsilon > 0$, $\delta > 0$, and $\eta > 0$. Suppose $I_1, I_2, \ldots, I_K$ are pairwise disjoint closed bounded intervals on $\mathbb{R}$, and suppose $\alpha_1, \alpha_2, \ldots, \alpha_K$ are real numbers, $0 < \alpha_j < 2\pi$. Then there exist finite Blaschke products $B_1(z)$*

and $B_2(z)$, having simple zeros, such that

(5.6) $$\sum_j |\{x \in I_j : |\alpha_j - \operatorname{Arg} B_1(x)/B_2(x)| \ge \varepsilon/2\}| < \eta.$$

and such that

(5.7) $$0 < \operatorname{Arg} B_1(x)/B_2(x) < \delta \quad x \notin \bigcup I_j.$$

The zeros $\{z_n\}$ of $B_1(z)$ or of $B_2(z)$ satisfy

(5.8) $$\rho(z_n, z_m) \ge c\varepsilon, \quad n \ne m,$$

with c some absolute constant, and

(5.9) $$x_n = \operatorname{Re} z_n \in \bigcup_j I_j.$$

Moreover, if $0 < y < y_0(\varepsilon, \delta, K)$, then the zeros z_n may be chosen on the horizontal line

$$y_n = \operatorname{Im} z_n = y.$$

Here $\operatorname{Arg} w$ denotes the principal branch of the argument, $0 \le \operatorname{Arg} w < 2\pi$.

Proof. Because the intervals are pairwise disjoint, it is enough to prove this lemma for one interval I (with η replaced by η/K, with ε replaced by $\varepsilon/2$, and with δ replaced by $\min(\delta/K, \varepsilon/2K)$). Fix a closed interval I. After a translation we can suppose

$$I = [-\eta/4, A + \eta/4].$$

Also fix $\alpha, 0 < \alpha < 2\pi$.

Let N be a positive integer, to be determined later, and consider the points

$$x_k = k/N, \quad 0 \le k \le k_0 = [NA] - 1,$$

$[t]$ denoting the greatest integer in t. Let

$$\lambda = \alpha/2\pi N$$

and fix $y > 0$, also to be determined later. The zeros of $B_1(z)$ are

$$z_k = x_k + iy, \quad 0 \le k \le k_0,$$

and the zeros of $B_2(z)$ are

$$z_k^* = x_k + \lambda + iy, \quad 0 \le k \le k_0$$

See Figure X.5. Because $k_0/N + \lambda < A$, we have (5.9). We set

$$B_1(z) = \prod_{k=0}^{k_0} \frac{z - z_k}{z - \bar{z}_k}, \quad B_2(z) = \prod_{k=0}^{k_0} \frac{z - z_k^*}{z - \bar{z}_k^*},$$

$$\times \cdot \ \times \cdot \ \times \cdot \ \times \cdot \ \times \cdot \ \times \cdot \ \times \cdot \ \times \cdot \ \times \cdot \ \times \cdot \ \times \cdot \ \times \cdot \ \times \cdot \ \times \cdot \ \times \cdot \ \times \cdot \ \times \cdot$$

Figure X.5. Here $\alpha = 2\pi/3$. The zeros of $B_2(z)$ are slightly to the right of the zeros of $B_1(z)$.

so that, modulo 2π,

$$\text{Arg}\frac{B_1(x)}{B_2(x)} = \sum_{k=0}^{k_0}\left(\text{Arg}\left(\frac{x - z_k}{x - \bar{z}_k}\right) - \text{Arg}\left(\frac{x - z_k^*}{x - \bar{z}_k^*}\right)\right).$$

Note that

$$\text{Arg}\left(\frac{x - z_k}{x - \bar{z}_k}\right) = \pi + 2\arctan\left(\frac{x - x_k}{y}\right) = 2\int_{x_k}^{\infty}\frac{y_k}{(x - t)^2 + y_k^2}\,dt.$$

Consequently,

$$(5.10) \qquad \text{Arg}\frac{B_1(x)}{B_2(x)} = 2\sum_{k=0}^{k_0}\int_{x_k}^{x_{k+\lambda}}\frac{y_k}{(x - t)^2 + y_k^2}\,dt.$$

Because the intervals of integration are pairwise disjoint, the right side of (5.10) has value in $(0, 2\pi)$ and it is the principal branch of the argument. Setting $E = \bigcup_{k=0}^{k_0}[x_k, x_k + \lambda]$, we see that

$$\text{Arg}\frac{B_1(x)}{B_2(x)} = 2\pi\int P_y(x - t)\chi_E(t)\,dt.$$

Since dist$(E, \mathbb{R}\backslash I) \geq \eta/4$, (5.10) yields

$$\text{Arg}\frac{B_1(x)}{B_2(x)} \leq 2\int_{\eta/4}^{\infty}\frac{y\,ds}{s^2 + y^2} \leq \frac{c_1 y}{\eta}, \qquad x \notin I.$$

Therefore (5.7) will hold if

$$(5.11) \qquad\qquad\qquad y \leq c_2\,\delta\eta.$$

Now let $J = [\eta/4, A - \eta/4]$. Then $J \subset I$ and $|I\backslash J| = \eta$. We shall have (5.6) if we can get

$$(5.12) \qquad\qquad |\alpha - \text{Arg }B_1(x)/B_2(x)| < \varepsilon/2, \qquad x \in J.$$

As a preliminary approximation set $F = [0, x_{k_0} + 1/N]$ and write

$$V(x) = \frac{\alpha}{n}\int_F\frac{y}{(s - t)^2 + y^2}\,dt.$$

Taking

$$(5.13) \qquad\qquad\qquad N \geq \delta/\eta,$$

we have $x_{k_0} + 1/N \geq A - 1/N \geq A - \eta/8$, and $\mathrm{dist}(J, \mathbb{R} \backslash F) > \eta/8$. Thus as before

$$|V(x) - \alpha| \leq \frac{\delta}{\pi} \int_{\eta/8}^{\infty} \frac{y\,ds}{s^2 + y^2} \leq \frac{c_3 y}{\eta}, \quad x \in J';$$

and

$$|V(x) - \alpha| < \varepsilon/4, \quad x \in J,$$

provided that

(5.14) $$y \leq c_4 \varepsilon \eta.$$

But now for any $x \in \mathbb{R}$,

$$
\begin{aligned}
\left| V(x) - \mathrm{Arg} \frac{B_1(x)}{B_2(x)} \right| &= \left| \frac{\alpha}{\pi} \int_F \frac{y}{(x-t)^2 + y^2}\,dt - 2 \int_E \frac{y}{(x-t)^2 + y^2}\,dt \right| \\
&\leq \sum_{k=0}^{k_0} \left| \frac{\alpha}{\pi} \int_{x_k}^{x_{k+1}} \frac{y\,dt}{(x-t)^2 + y^2} - 2 \int_{x_k}^{x_{k+\lambda}} \frac{y\,dt}{(x-t)^2 + y^2} \right| \\
&= \sum_{k=0}^{k_0} \left| \int_{x_k}^{x_{k+1}} \frac{g_k(t)y}{(x-t)^2 + y^2}\,dt \right|,
\end{aligned}
$$

(5.15)

where $|g_k(t)| \leq 2$ and $\int_{x_k}^{x_{k+1}} g_k(t)\,dt = 0$. Consequently,

$$
\left| \int_{x_k}^{x_{k+1}} \frac{g_k(t)y}{(x-t)^2 + y^2}\,dt \right| \leq \frac{2}{N} \max_{x_k \leq t \leq x_{k+1}} \left(\frac{y}{(x-t)^2 + y^2} \right) \\
- \frac{2}{N} \min_{x_k \leq t \leq x_{k+1}} \left(\frac{y}{(x-t)^2 + y^2} \right),
$$

and the right side of (5.15) is bounded by a two-sided telescoping sum. Therefore

$$|V(x) - \mathrm{Arg}\,B_1(x)/B_2(x)| \leq 4/Ny, 2$$

and we have (5.12) and also (5.6) if we take

(5.16) $$Ny = c_5/\varepsilon.$$

Our restrictions (5.11), (5.13), and (5.14) on the parameters y and N are consistent with (5.16) and we can obtain (5.6) and (5.7) with arbitrarily small y. Finally, by (5.16)

$$\rho(z_n, z_m) \geq c_6 |x_n - x_m|/y \geq c\varepsilon, \quad n \neq m,$$

and so (5.8) holds. $\square$

Proof of Theorem 5.1.   First consider the critical special case in which

$$v_0 = \text{Arg } u$$

has support $(-1, 1)$.

Step 1: Set $\eta_0 = 1, \eta_1 = \frac{1}{4}$. Choose pairwise disjoint closed intervals $I_1^{(1)}, I_2^{(1)}, \dots, I_{K_1}^{(1)}$, and choose real numbers $\alpha_1^{(1)}, \dots, \alpha_{K_1}^{(1)}, 0 < \alpha_j^{(1)} < 2\pi$, such that

$$\left| \left\{ x : \left| v_0(x) - \sum_1^{K_1} \alpha_j^{(1)} \chi_{I_j^{(1)}}(x) \right| \ge \frac{\varepsilon}{4} \right\} \right| < \frac{\eta_1}{2}.$$

That can be done because simple functions are dense in L^∞ and because any bounded measurable set can be approximated in measure by a finite union of intervals. By Lemma 5.5 there are finite Blaschke products $B_1^{(1)}(z)$ and $B_2^{(1)}(z)$ such that

$$\left| \left\{ x : \left| \sum \alpha_j^{(1)} \chi_{I_j^{(1)}}(x) - \text{Arg } \frac{B_1^{(1)}(x)}{B_2^{(1)}(x)} \right| \ge \frac{\varepsilon}{2} \right\} \right| < \frac{\eta_1}{2}.$$

Hence

$$E_1 = \{ x : |v_0(x) - \text{Arg } B_1^{(1)}(x)/B_2^{(1)}(x)| > 3\varepsilon/4 \}$$

satisfies

$$|E_1| < \eta_1.$$

Moreover, the zeros $z_n^{(1)} = x_n^{(1)} + i y_n^{(1)}$ of $B_1^{(1)}$ satisfy

$$\rho(z_n^{(1)}, z_m^{(1)}) \ge c\varepsilon, \quad n \ne m, \quad \text{and} \quad x_n^{(1)} \in \bigcup I_j^{(1)}.$$

Fixing $\eta_2 < \eta_1/4$ sufficiently small, we can take

$$y_n^{(1)} = \eta_2, \qquad n = 1, 2, \dots.$$

The zeros of $B_2^{(1)}$ have the same three properties.

Step 2: Let

$$v_1(x) = \text{Arg } \left(\frac{u(x) B_2^{(1)}(x)}{B_1^{(1)}(x)} \right) \chi_{E_1}(x).$$

Choose pairwise disjoint closed intervals $I_1^{(2)}, \dots, I_{K_2}^{(2)}$ and choose real numbers $\alpha_1^{(2)}, \alpha_2^{(2)}, \dots, \alpha_{K_2}^{(2)}$ such that

$$\sum |I_j^{(2)}| \le 4|E_1| < 4\eta_1,$$

such that

$$(5.17) \quad |E_1 \triangle (\bigcup I_j^{(2)})| = |E_1 \setminus \bigcup I_j^{(2)}| + |\bigcup I_j^{(2)} \setminus E_1| < \eta_2/2,$$

and such that

(5.18)
$$\left| \left\{ x : |v_1(x) - \sum \alpha_j^{(2)} \chi_{I_j^{(2)}}(x)| > \frac{\varepsilon}{8} \right\} \right| \le \frac{\eta_2}{4}.$$

By Lemma 5.5 there are finite Blaschke products $B_1^{(2)}(z)$ and $B_2^{(2)}(z)$ such that

(5.19)
$$\left| \left\{ x : \left| \sum_j \alpha_j^{(2)} \chi_{I_j^{(2)}}(x) - \text{Arg} \frac{B_1^{(2)}(x)}{B_2^{(2)}(x)} \right| > \frac{\varepsilon}{2} \right\} \right| < \frac{\eta_2}{4},$$

and such that

(5.20)
$$0 < \text{Arg} \, B_1^{(2)}(x)/B_2^{(2)}(x) < \varepsilon/8, \quad x \notin \bigcup I_j^{(2)}.$$

Consider the set

$$E_2 = \left\{ x : \left| v_0(x) - \text{Arg} \left(\frac{B_1^{(1)}(x) B_1^{(2)}(x)}{B_2^{(1)}(x) B_2^{(2)}(x)} \right) \right| > \frac{7\varepsilon}{8} \right\}.$$

Then

$$E_2 = [E_2 \cap E_1 \cap (\bigcup I_j^{(2)})] \cup [E_2 \backslash (E_1 \cup \bigcup I_j^{(2)})]$$
$$\cup [E_2 \cap (E_1 \triangle (\bigcup I_j^{(2)}))]$$

On $E_1 \cap (\bigcup I_j^{(2)})$, (5.18) and (5.19) give us, modulo 2π,

$$\left| v_0(x) - \text{Arg} \left(\frac{B_1^{(1)}(x) B_1^{(2)}(x)}{B_2^{(1)}(x) B_2^{(2)}(x)} \right) \right| = \left| v_1 - \text{Arg} \left(\frac{B_1^{(2)}(x)}{B_2^{(2)}(x)} \right) \right|$$
$$< \frac{\varepsilon}{8} + \frac{\varepsilon}{2} = \frac{5\varepsilon}{8},$$

except on a set of measure at most $\eta_2/2$ Hence,

$$\left| E_2 \cap E_1 \cap \left(\bigcup I_j^{(2)} \right) \right| < \eta_2/2$$

By (5.20), we have

$$E_2 \backslash \left(\bigcup I_j^{(2)} \right) \subset E_1.$$

so that

$$E_2 \backslash \left(E_1 \cup \bigcup I_j^{(2)} \right) = \varnothing.$$

And by (5.17), we have

$$\left| E_2 \cap \left(E_1 \triangle \left(\bigcup I_j^{(2)} \right) \right) \right| < \eta_2/2.$$

We conclude that

$$|E_2| < \eta_2.$$

The zeros $z_n^{(2)} = x_n^{(2)} + iy_n^{(2)}$ of $B_1^{(2)}$ satisfy

$$\rho(z_n^{(2)}, z_m^{(2)}) \geq c\varepsilon, \quad n \neq m,$$
$$x_\eta^{(2)} \in \bigcup I_i^{(2)} \quad \text{and} \quad y_n^{(2)} = \eta_3,$$

where $0 < \eta_3 < \eta_2/4$ and η_3 is as small as we like. The zeros of $B_2^{(2)}$ and have the same three properties.

Step p: By induction we obtain finite Blaschke products $B_1^{(1)}(z), \dots, B_1^{(p)}(z)$ and $B_2^{(1)}(z), \dots, B_2^{(p)}(z)$ such that

$$E_p = \left\{ x : \left| v_0(x) - \mathrm{Arg}\left(\frac{B_1^{(1)}(x) \cdots B_1^{(p)}(x)}{B_2^{(1)}(x) \cdots B_2^{(p)}(x)} \right) \right| > (1 - 2^{-p-1})\varepsilon \right\}$$

satisfies

$$(5.21) \qquad |E_p| < \eta_p \leq 4^{-p}.$$

The induction step is the same as step 2 except that $\varepsilon 2^{-p-1}$ is used in the analogs of (5.18) and (5.20) but $\varepsilon/2$ is always used in the analog of (5.19). Thus Lemma 5.5 is applied at step p with $\delta = \varepsilon 2^{-p-1}$, and with $\eta = \eta_p/2$, but with the same ε.

Because we use the same ε at each stage, the zeros $z_n^{(p)} = x_n^{(p)} + iy_n^{(p)}$ of $B_1^{(p)}$ satisfy

$$(5.22) \qquad \rho(z_n^{(p)}, z_m^{(p)}) \geq c\varepsilon, \quad m \neq n,$$

with c not depending on p. These zeros also satisfy

$$(5.23) \qquad x_n^{(p)} \in \bigcup I_j^{(p)},$$

where

$$(5.24) \qquad \left| \bigcup I_j^{(p)} \right| \leq 4|E_{p-1}| \leq 4\eta_{p-1}$$

and

$$(5.25) \qquad y_n^{(p)} = \eta_{p+1} < \eta_p/4.$$

Now set

$$B_1(z) = \prod_{p=1}^{\infty} B_1^{(p)}(z), \qquad B_2(z) = \prod_{p=1}^{\infty} B_2^{(p)}(z).$$

Because the zeros remain in a bounded set, convergence factors are unnecessary and the products converge if their assigned zeros satisfy $\sum y_n < \infty$. We show more: The zeros given to $B_1(z)$ (and to $B_2(z)$) are an interpolating sequence. We consider $B_1(z)$ only. It has zeros $\bigcup_{p=1}^{\infty} \{z_n^{(p)}\}$. By (5.22) and (5.25) these zeros are separated. Let Q be a square $\{a < x < a+h, 0 < y < h\}$. Since $y_n^{(p)} \leq \eta_2$, we can assume $h < \eta_2$. Take q so that

$$\eta_{q+1} \leq h < \eta_q.$$

Then by (5.25),

$$\sum_{z_n^{(p)} \in Q} y_n^{(p)} = \sum_{p=q}^{\infty} \eta_{p+1} N_p(Q),$$

where $N_p(Q)$ is the number of $z_n^{(p)}$ inside Q. By (5.22) and (5.25),

$$|x_n^{(p)} - x_m^{(p)}| \geq c_1 \varepsilon \eta_{p+1},$$

so that by (5.23) and (5.24),

$$N_p(Q) \leq \frac{\min(|\bigcup I_j^{(p)}|, h)}{c_1 \varepsilon \eta_{p+1}} \leq \frac{c_2 \min(4\eta_{p-1}, h)}{\varepsilon \eta_{p+1}}.$$

Therefore, as $\eta_p \leq 4^{-p}$,

$$\sum_{z_n^{(p)} \in Q} y_n^{(p)} \leq \frac{c_2}{\varepsilon} \sum_{p=q+2}^{\infty} \min(4\eta_{p-1}, h) \leq \frac{2c_2}{\varepsilon} h + \frac{c_2}{\varepsilon} \sum_{p=q+2}^{\infty} 4\eta_{p-1}$$

$$\leq \frac{c_3}{\varepsilon} h \sum_{k=0}^{\infty} 4^{-k},$$

and $\bigcup \{z_n^{(p)}\}$ is an interpolating sequence. Thus $B_1(z)$ and $B_2(z)$ are interpolating Blaschke products.

In $L^2(dx/(1+x^2))$, the partial products $(B_1^{(1)} \cdots B_1^{(p)} \cdots B_1^{(p)})/(B_2^{(1)} \cdot B_2^{(p)})$ converge to B_1/B_2. So a subsequence of the partial products converges to B_1/B_2 almost everywhere. Hence by (5.21),

$$|v_0(x) - \operatorname{Arg} B_1(x)/B_2(x)| \leq \varepsilon$$

almost everywhere.

For the general case write $u = u_1 u_2$, where

$$u_1(x) = 1, \quad |x| > 1,$$
$$u_2(x) = 1, \quad |x| < 1.$$

By the special case treated above we have interpolating Blaschke products $B_{1,1}(z)$ and $B_{2,1}(z)$ such that

$$\left| u_1 - \frac{B_{1,1}(x)}{B_{2,1}(x)} \right| < \frac{\varepsilon}{2}$$

almost everywhere. Using the inversion $z \to -1/z$, we also get interpolating Blaschke products $B_{1,2}(z)$ and $B_{2,2}(z)$ such that

$$\left| u_2(x) - \frac{B_{1,2}(x)}{B_{2,2}(x)} \right| < \frac{\varepsilon}{2}$$

almost everywhere. The theorem will be proved if it can be arranged that $B_1 = B_{1,1}B_{1,2}$ and $B_2 = B_{2,1}B_{2,2}$ are interpolating Blaschke products. For that it is enough to bound below the pseudohyperbolic distance from the zeros of $B_{j,1}$ to the zeros of $B_{j,2}$, $j = 1, 2$. The zeros of $B_{1,1}$ and $B_{2,1}$ lie in $|x| < 1$ and on horizontal lines $y = \eta_p$, $p = 2, 3, \ldots$, with $\eta_{p+1} < \eta_p/4$. The zeros of $B_{1,2}$ and $B_{2,2}$ lie in $|x| > 1$ and in large circles tangent to $\mathbb{R}$ at $z = 0$. In $\{y \leq \eta_2\}$ these circles cut the lines $\{x = \pm 1\}$ at heights $y = y_q$, $q = 1, 2, \ldots$. By Lemma 5.5, we can take y_1 and y_{q+1}/y_q as small as we please. Thus we can ensure that

$$\inf_p \frac{|y_q - \eta_q|}{y_q} \geq c\varepsilon, \quad q = 1, 2, \ldots,$$

which implies that the zeros of $B_1(z)$ and of $B_2(z)$ are interpolating sequences. $\square$

Notes

Sections 1 and 2 are taken from Hoffman [1967]. Further results from his paper are given in Exercises 1–8. Lemma 3.3 is due to Carleson [1972]. Theorem 4.1 is from Carleson and Garnett [1975], who also study harmonic interpolation in $\mathbb{R}_+^{d+1}$. Theorem 5.1 is in Jones [1981]. The idea behind Lemma 5.5 is due to A. M. Davie (see Davie, Gamelin, and Garnett [1973]).

Exercises and Further Results

1. Let $m \in \mathfrak{M}$.

(a) We say m is a *nontangential* point if m is in the closure of some cone

$$\Gamma_\alpha(e^{i\theta}) = \left\{ z : \frac{|z - e^{i\theta}|}{1 - |z|} < \alpha, |z| < 1 \right\},$$

$\alpha > 1$. Every nontangential point m is in the closure of an interpolating sequence, Consequently, m lies in an analytic disc. (Hint: There are finitely many interpolating Blaschke products $B_1, \ldots, B_N$ and there is $r > 0, r < r(\delta(B_j))$

(see Lemma 1.4), such that

$$\Gamma_\alpha(e^{i\theta}) \subset \bigcup_{j=1}^{N} \{z : \rho(z, z_{j,k}) < r\},$$

where $\{z_{j,k}\}$ denotes the zeros of B_j.)

(b) We say m is an *orocycular point* if m lies in the closure of the region between two circles tangent to the unit circle at the same point. Every orocycular point lies in the closure of an interpolating sequence.

(c) Let $r(\theta)$ be continuous and decreasing on $0 \leq \theta \leq 1$, $r(0) = 1$, and let γ be the curve $\{r(\theta)e^{i\theta} : 0 < \theta \leq 1\}$, terminating at $z = 1$. If $m \in \mathfrak{M}$ is in the closure of γ, then m lies in an analytic disc.

2. Let V be a disc in D tangent to the unit circle at one point $e^{i\theta}$, and let $\bar{V}$ be the closure of V in $\mathfrak{M}$.

(a) $\bar{V}$ is disjoint from the Šilov boundary X.

(b) There exist points $m \in \bar{V}$ not in the closure of any interpolating sequence. (Otherwise, for any $\varepsilon > 0$ we could, by compactness, cover V by pseudohyperbolic discs $\{\rho(z, z_j) < \varepsilon\}$ with $\{z_j\}$ a finite union of interpolating sequences, thereby violating the geometric characterization of interpolating sequences.)

(c) Thus there are one-point Gleason parts disjoint from the Šilov boundary.

(d) Part (b) and Exercise 1 can be generalized. Let S be any subset of D, and let $\{z_j\}$ be any separated sequence in S such that $\bigcup\{z : \rho(z, z_j) < \frac{1}{2}\}$ covers S. Then every point in the closure of S lies in an analytic disc if and only if $\sum(1 - |z_j|)\delta_{z_j}$ is a Carleson measure.

In the upper half plane, let $Q_{n,j} = \{j2^{-n} \leq x \leq (j+1)2^{-n}, 2^{-n-1} \leq y \leq 2^{-n}\}$, $-\infty < n < \infty$, $-\infty < j < \infty$. Thus $\{Q_{n,j}\}$ is a paving of H by hyperbolically congruent rectangles. Let S be a subset of H and consider the sequence $\{z_k\}$ consisting of the centers of those $Q_{n,j}$ such that $Q_{n,j} \cap S \neq \emptyset$. Every point in the closure of S lies in an analytic disc if and only if $\{z_k\}$ is an interpolating sequence.

3. The mapping $z \to L_z$ from D into $\mathfrak{M}^D$ has a unique continuous extension to a mapping L from $\mathfrak{M}$ into $\mathfrak{M}^D$. If $m \in G$, the extension has value L_m. If $m \notin G$, then $L_m(\zeta) = m$, $\zeta \in D$.

4. Let $C_{\mathfrak{M}}$ denote the complex algebra of bounded continuous functions on D which admit continuous extensions to $\mathfrak{M}$.

(a) $C_{\mathfrak{M}}$ is the smallest uniformly closed algebra containing H^∞ and $\overline{H^\infty}$.

(b) $C_{\mathfrak{M}}$ is the smallest uniformly closed algebra containing the bounded harmonic functions.

(c) If $f \in H^\infty$ and if $\alpha \in D$, then

$$g(z) = f\left(\frac{\alpha + z}{1 + \bar{z}\alpha}\right) \in C_{\mathfrak{M}},$$

and for $n = 1, 2, \ldots,$

$$h_n(z) = (1 - |z|^2)^n f^{(n)}(z) \in C_{\mathfrak{M}}.$$

5. Let $S \subset D$ and suppose that for some $\varepsilon, 0 < \varepsilon < 1$, the ε pseudohyperbolic neighborhood of S, $\{z : \inf_S \rho(z, w) < \varepsilon\}$ covers D. Then the closure if S in $\mathfrak{M}$ contains every point not in the closure of any interpolating sequence. (Use the fact that $P(m) = \{m\}$ for such a point.)

6. Let $m \in \mathfrak{M}$. A necessary and sufficient condition for m to be in the closure of some interpolating sequence is the following. If E and F are subsets of D and if $m \in \bar{E} \cap \bar{F}$, then $\rho(E, F) = \inf\{\rho(z, w) : z \in E, w \in F\} = 0.$

7. Let T be an endomorphism from H^∞ into H^∞; that is, T is a (bounded) linear operator from H^∞ into H^∞ satisfying

$$T(fg) = T(f)T(g).$$

Then there exists an analytic mapping $\tau : D \to D$, $\tau(0) = 0$, and there exists a net (z_i) in D such that

(E.1) $$Tf(z) = \lim_i f\left(\frac{\tau(z) + z_i}{1 + \bar{z}_i \tau(z)}\right).$$

Tf is not constant for some $f \in H^\infty$ if and only if either

(i) (z_i) can be chosen a constant, $z_i = \zeta, \zeta \in D$, or
(ii) (z_i) can be chosen from some interpolating sequence.

Alternative (i) holds if and only if T is weak-star continuous.

Let $H^\infty(V)$ be the ring of bounded analytic functions on a Riemann surface (or analytic space) V. If $T : H^\infty(D) \to H^\infty(V)$ is a homomorphism, then there is an analytic mapping $\tau : V \to D$ and there is a net (z_i) in D such that (E.1) holds.

8. (a) Let $B(z)$ be a Blaschke product with zeros $\{z_n\}$ satisfying

$$\lim_{n \to \infty} (1 - |z_n|^2)|B'(z_n)| = 1.$$

If m is in the closure of $\{z_n\}$ then the map L_m is a homeomorphism from D onto $P(m)$ (which has the usual Gelfand topology of $\mathfrak{M}$), and L_m^{-1} is a constant multiple of B.

(b) Turning to the upper half plane, let S be the two-sided sequence $\{z_k = k + i : k \in \mathbb{Z}\}$. Then S is an interpolating sequence. Thus each point $m \in \bar{S}\backslash S$ lies in an analytic disc. Call a subsequence $\{z_{k_n}\}$ of S thin if $|k_{n+1} - k_n| \to \infty$. If m is in the closure of a thin subsequence of S then L_m is a homeomorphism.

(c) For $z_k = k + i \in H$, the coordinate map (to the upper half plane) is

$$L_{z_k}(\zeta) = k + i \left(\frac{1 + \zeta}{1 - \zeta} \right), \quad \zeta \in D.$$

If $F(z) = e^{2\pi i z}$, then

$$F \circ L_{z_k}(\zeta) = \exp \left(-2\pi \left(\frac{1 + \zeta}{1 - \zeta} \right) \right)$$

is independent of k. Thus

$$F \circ L_m(\zeta) = \exp \left(-2\pi \left(\frac{1 + \zeta}{1 - \zeta} \right) \right)$$

for all $m \in \bar{S}$, Let $m, m' \in \bar{S} \backslash S$. Then m and m' are in the same Gleason part if and only if

$$L_m(\zeta) = L_{m'}(\zeta')$$

for some $\zeta, \zeta' \in D$. This means

$$i \left(\frac{1 + \zeta'}{1 - \zeta'} \right) = n + i \left(\frac{1 + \zeta}{1 - \zeta} \right)$$

for some integer n, which is independent of ζ and ζ'.

Let $\sigma : S \to S$, $\sigma(z_k) = z_{k+1}$. Then σ extends to a homeomorphism of $\bar{S} \backslash S$ onto $\bar{S} \backslash S$, and we have $m' = \sigma^n(m)$. Conversely, if $m' = \sigma^n(m)$, then

$$m' = L_m \left(\frac{n}{n + 2i} \right)$$

and $m' \in P(m)$. Thus points m and m' of $\bar{S} \backslash S$ lie in the same Gleason part if and only if m and m' have the same orbit under the group $G = \{\sigma^n : -\infty < n < \infty\}$.

(d) Now let K be a closed subset of $\bar{S} \backslash S$ that is invariant under G and minimal among the closed invariant sets. K exists by Zorn's lemma. If $m \in K$, then m is in the closure of $\{\sigma^n(m)\}_{n=1}^{\infty}$. Thus some subnet of

$$\{L_m(n/(n + 2i))\}_{n=1}^{\infty}$$

converges to $L_m(0)$ and L_m is not a homeomorphism. In light of part (b) we see that if m is in the closure of a thin subsequence, then the closure of $\{\sigma^n(m)\}_{n=1}^{\infty}$ contains proper closed subsets invariant under G. See Hoffman [1967].

★★9. Let τ be an analytic map from D into D. Then τ extends to a continuous map from $\mathfrak{M}$ into $\mathfrak{M}$ defined by $(\tau(m)(f) = m(f \circ \tau))$.

(a) If $\tau(m) = m$ for some point $m \in G$, then $\inf_{z \in D} \rho(\tau(z), z) = 0$. For the proof, suppose $\{z_n\}$ is an interpolating sequence such that $m \in \overline{\{z_n\}}$. For $\varepsilon > 0$ small, the discs $K_n = K(z_n, \varepsilon) = \{\rho(z, z_n) < \varepsilon\}$ are pairwise disjoint, and m is not in the closure of $D \setminus \bigcup_{n=1}^{\infty} K_n$. Consequently the relation $f(z_n) \in k_j$ holds for infinitely many n, say for $n \in S$. Write $T(n) = j$ when $n \in S$ and $f(z_n) \in K_j$. The problem is to show that T has a fixed point. Define an equivalence relation in S by setting $n_1 \sim n_2$ if $T^p(n_1) = T^q(n_2)$ for some nonnegative integers p and q. Note that $n \sim T(n)$ if $n \in S$ and if $T(n) \in S$. Let n^* denote the least element of the equivalence class containing n. Let E be the set of $n \in S$ such that $p + q$ is even, where $p + q$ is the smallest such integer such that $T^p(n) = T^q(n^*)$. If T has no fixed point in S, then when $n \in S$ and $T(n) \in S$, exactly one of n and $T(n)$ lies in E.

Let $B(z)$ be the Blaschke product with zeros on S. Then $B(m) = 0$. Factor $B = B_1 B_2$, where B_1 has zeros $\{z_n : n \in E\}$ and B_2 has zeros $\{z_n : n \in S \setminus E\}$. When ε is small, the assumption that T has no fixed point now leads to a contradiction.

(b) If τ has two fixed points in G lying in different fibers, then $\tau(z) = z$. (Use part (b) and Exercise 2 of Chapter I.)

(c) Let $\mathfrak{M}_\lambda$ be the fiber of $\mathfrak{M}$ at λ, $|\lambda| = 1$. Then τ fixed a point in $G \cap \mathfrak{M}_\lambda$ if and only if the angular derivative of τ exists at λ and equals 1 there. (See Exercise 7 of Chapter I for a discussion of angular derivatives.)

(d) τ maps the Šilov boundary into itself if and only if $\tau(z)$ is an inner function.

(These results are due to Michael Behrens.)

10. Let $m \in \mathfrak{M}$ and let $f \in H^\infty$, $f(m) = 0$. Then

$$\lim_{z \to m} |f'(z)|(1 - |z|^2) = 0$$

if and only if $f = f_1 f_2$ with $f_j \in H^\infty$, $f_j(m) = 0$. (The only interesting case is that in which $f(z)$ is a Blaschke product, because both conditions hold trivially if $f^{1/2} \in H^\infty$. When $m \notin G$, both conditions follow from Theorem 2.4. When $m \in G$, each condition means that $(f \circ L_m)'(0) = 0$.)

11. Let B_0 be the class of analytic functions on D satisfying

$$\lim_{|z| \to 1} (1 - |z|^2)|f'(z)| = 0,$$

and let VMOA be the analytic functions of the form $u + \tilde{v}$, u, v harmonic on D and continuous on $\bar{D}$. Here B stands for Bloch.

(a) VMOA $\subset B_0$.

(b) $f(z) \in H^\infty \cap B_0$ if and only if $f(m)$ is constant on each Gleason part in $\mathfrak{M} \setminus D$. For this reason Behrens called this class COP, for constant on parts.

(c) Suppose $g(z)$ is in the disc algebra. Then $g(e^{i\theta})$ is in the *Zygmund class* Λ_*, defined by

$$|g(\theta + h) - g(\theta - h) - 2g(\theta)| = O(h)$$

if and only if

$$\sup_{z}(1 - |z|^2)|g''(z)| < \infty.$$

(see Zygmund [1968, Vol. I]). Similarly, $g'(z) \in B_0$ if and only if $g(e^{i\theta}) \in \lambda_*$, which is defined by

$$|g(\theta + h) + g(\theta - h) - 2g(\theta)| = o(h).$$

(d)   There exists a continuous increasing function $F(\theta)$ on $[0, 2\pi]$ such that F is singular, $F'(\theta) = 0$ almost everywhere, but such that $F \in \lambda_*$ (see Kahane [1969], Piranian [1966]). Thus $F(\theta) = \mu([0, \theta])$ for some singular measure μ, and the inner function determined by μ lies in $H^\infty \cap B_0$. It follows that $H^\infty \cap B_0$ contains an infinite Blaschke product and that VMOA $\neq B_0$.

Sarason has proposed the problem of characterizing the Blaschke products in B_0 in terms of the distribution of their zeros (see Nikolski, Havin, and Kruschev [1978]).

12.   Let $P(t) \in L^1(\mathbb{R})$ be nonnegative, $\int P(t)dt = 1$. Let $\{z_j\}$ be a sequence in the upper half plane and set

$$P_j(t) = \frac{1}{y_j} P\left(\frac{t - x_j}{y_j}\right).$$

(a)   Suppose there are real functions $u_j(t) \in L^\infty$ such that $\|u_j\|_\infty \leq 1, \int u_j(t)P_j(t)dt > \delta, \int u_j(t)P_k(t)dt \leq 0, k \neq j$, with δ independent of j. Then $\{z_j\}$ is an interpolating sequence.

(b)   If the weak-star closure of $\{P_j\}$ in $(L^\infty)^*$ homeomorphic $\beta\mathbb{N}$, the Stone–Čech compactification of the integers, then $\{z_j\}$ is an interpolating sequence.

Bibliography

Adamyan, V. M., Arov, D. Z., and Krein, M. G.
  [1968] Infinite Hankel matrices and generalized problems of Carathéodory, Fejér
    and I. Schur, *Funkcional Anal. i Priložen.* **2**, vyp 4, 1–17 (Russian); *Functional
    Anal. Appl.* **2**, 269–281.
  [1971] Infinite Hankel block matrices and related approximation problems, *Vestnik
    Akad. Nauk Armjan. SSR* **6**, 87–102 (Russian); *Amer. Math. Soc. Transl.* (2) **111**
    (1978), 133–156.

Ahern, P. R., and Clark, D. N.
  [1970a] On functions orthogonal to invariant subspaces, *Acta Math.* **124**, 191–204.
  [1970b] Radial limits and invariant subspaces, *Amer. J. Math.* **92**, 332–342.
  [1971] Radial Nth derivatives of Blaschke products, *Math. Scand.* **28**, 189–201.
  [1974] On inner functions with H^p derivatives, *Michigan Math. J.* **21**, 115–127.
  [1976] On inner functions with B^p derivatives, *Michigan Math. J.* **23**, 107–118.

Ahlfors, L. V.
  [1966] "Complex Analysis" (2nd ed.). McGraw-Hill, New York.
  [1973] "Conformal Invariants, Topics in Geometric Function Theory." McGraw-Hill,
    New York.

Alexander, H., Taylor, B. A., and Williams, D. L.
  [1971] The interpolating sets for A^∞, *J. Math. Anal. Appl.* **36**, 556–566.

Allen, H. A., and Belna, C. L.
  [1972] Singular inner functions with derivatives in B^p, *Michigan Math. J.* **19**, 185–
    188.

Amar, E.
  [1973] Sur un théorème de Mooney relatif aux fonctions analytiques bornées, *Pacific
    J. Math.* **16**, 191–199.
  [1977a] Thèse, Faculté des Sciences d'Orsay, Université de Paris XI.
  [1977b] Suites d'interpolation harmoniques, *J. Analyse Math.* **32**, 197–211.

Amar, E., and Lederer, A.
  [1971] Points exposés de la boule unité de $H^\infty(D)$, *C. R. Acad. Sci. Paris. Sér. A*
    272, 1449–1552.

Anderson, J. M.
  [1975] A note on a basis problem, *Proc. Amer. Math. Soc.* **51**, 330–334.
  [1979] On division by inner factors, *Comment. Math. Helv.* **54**, 309–317.

Anderson, J. M., Clunie, J., and Pommerenke, Ch.
[1974]   On Bloch functions and normal functions, *J. Reine Angew. Math.* **270**, 12–37.

Ando, T.
[1977]   Uniqueness of predual of H^∞, preprint.

Axler, S.
[1977]   Factorization of L^∞ functions, *Ann. of Math.* **106**, 567–572.

Axler, S., Berg, I. D., Jewell, N. P., and Shields, A.
[1979]   Approximation by compact operators and the space $H^\infty + C$, *Ann. of Math.* **109**, 601–612.

Axler, S., Chang, S.-Y., and Sarason, D.
[1978]   Products of Toeplitz operators, *Integral Equations Operator Theory* **1/3**, 285–309.

Baernstein, A.
[1974]   Integral means, univalent functions and circular symmetrization, *Acta Math.* **133**, 139–169.
[1976]   Univalence and bounded mean oscillation, *Michigan Math. J.* **23**, 217–223.
[1979]   Some sharp inequalities for conjugate functions, *Proc. Symp. Pure Math.* **35** (1), 409–416.
[1980]   Analytic functions of bounded mean oscillation in "Aspects of Contemporary Complex Analysis. Academic Press, New York.

Bagemihl, F., and Seidel, W.
[1954]   Some boundary properties of analytic functions, *Math. Z.* **61**, 186–189.

Barbey, K.
[1975]   Ein Sätz über abstrakte analytische Funktionen, *Arch. Math. (Basel)* **26**, 521–527.

Bear, H. S.
[1970]   "Lecture on Gleason Parts" (Lecture Notes in Mathematics, Vol. 121). Springer-Verlag, Berlin and New York.

Behrens, M.
[1971]   The maximal ideal space of algebras of bounded analytic functions on infinitely connected domains, *Trans. Amer. Math. Soc.* **161**, 358–380.
[1981]   Interpolation and Gleason parts in *L*-domains, *Trans. Amer. Math. Soc.* **286** (1984), 203–225.

Belna, C., and Piranian, C.
[1981]   A Blaschke product with a level-set of infinite length, to be published.

Bernard, A.
[1971]   Algèbres quotients d'algèbres uniformes, *C. R. Acad. Sci. Paris, Sér. A* **272**, 1101–1104.

Bernard, A., Garnett, J. B., and Marshall, D. E.
[1977]   Algebras generated by inner functions, *J. Funct. Anal.* **25**, 275–285.

Besicovitch, A.
[1923]   Sur la nature des fonctions à carré sommable measurables, *Fund. Math.* **4**, 172–195.

Bessaga, C., and Pelczynski, A.
[1960]   Spaces of continuous functions (IV), *Studia Math.* **19**, 53–62.

Beurling, A.

[1949] On two problems concerning linear transformations in Hilbert space, *Acta Math.* **81**, 239–255.

Bishop, E.

[1959] A minimal boundary for function algebras, *Pacific J. Math.* **9**, 629–642.

[1962] A general Rudin–Carleson theorem, *Proc. Amer. Math. Soc.* **13**, 140–143.

[1963] Holomorphic completions, analytic continuations, and the interpolation of seminorms, *Ann. of Math.* **78**, 468–500.

[1964] Representing measures for points in a uniform algebra, *Bull. Amer. Math. Soc.* **70**, 121–122.

[1965] Abstract dual extremal problems, *Notices Amer. Math. Soc.* **12**, 123.

Bishop, E., and Phelps, R. R.

[1961] A proof that every Banach space is subreflexive, *Bull. Amer. Math. Soc.* **67**, 97–98.

Blaschke, W.

[1915] Eine Erweiterung des Satzes von Vitali über Folgen analytischer Funktionen, *S.-B. Säcks Akad. Wiss. Leipzig Math.-Natur. Kl.* **67**, 194–200.

Bočkarev, V.

[1974] Existence of a basis in the space of analytic functions, and some properties of the Franklin system. *Math. USSR-Sb.* **24**, 1–16.

Boyd, D.

[1979] Schur's algorithm for bounded holomorphic functions, *Bull. London Math. Soc.* **11**, 145–150.

Browder, A.

[1969] "Introduction to Function Algebras." Benjamin, New York.

Brown, L.

[1970] Subalgebras of L_∞ of the circle group, *Proc. Amer. Math. Soc.* **25**, 585–587.

Brown, L., Shields, A., and Zeller, K.

[1960] On absolutely convergent exponential sums, *Trans. Amer. Math. Soc.* **96**, 162–183.

Burkholder, D. L.

[1973] Distribution function inequalities for martingales, *Ann. Probability* **1**, 19–42.

[1976] Harmonic analysis and probability, in "Studies in Harmonic Analysis" (J. M. Ash, ed.), Studies in Math. Vol. 13, **13**, pp 136–149.

[1979] Martingale theory and harmonic analysis in Euclidean spaces, *Proc. Symposia in Pure Math.* **35** (2), 283–302.

Burkholder, D. L., and Gundy, R. F.

[1972] Distribution function inequalities for the area integral, *Studia Math.* **44**, 527–544.

Burkholder, D. L., Gundy, R. F., and Silverstein, M. L.

[1971] A maximal function characterization of the class H^p, *Trans. Amer. Math. Soc.* **157**, 137–153.

Calderón, A. P.

[1950a] On theorems of M. Riesz and Zygmund, *Proc. Amer. Math. Soc.* **1**, 533–535.

[1950b]   On the behavior of harmonic functions near the boundary, *Trans. Amer. Math. Soc.* **68**, 47–54.

Calderón, A. P., and Zygmund, A.
[1952]   On the existence of certain singular integrals, *Acta Math.* **88**, 85–139.
[1956]   On singular integrals, *Amer. J. Math.* **78**, 249–271.

Campanato, S.
[1963]   Propreità de hölderianità di alcane classi di funzioni, *Ann. Scuola Norm. Sup. Pisa* (3) **17**, 175–188.

Cantor, D. G.
[1981]   Oral communication.

Carathéodory, C.
[1911]   Über den Variabilitätsbereich der Fourier'schen Konstanten von positiven harmonischen Funktionen, *Rend. Circ. Mat. Palermo* **32**, 193–217.
[1929]   Über die Winkelderivierten von beschränkten analytischen Funktionen, *Sitzungsber Preuss. Akad. Phys.-Math.* **4**, 1–18.
[1954]   "Theory of Functions of a Complex Variable," Vols. I and II. Chelsea, New York.

Cargo, G. T.
[1962]   Angular and tangential limits of Blaschke products and their successive derivatives, *Canad. J. Math.* **14**, 334–348.

Carleson, L.
[1952]   Sets of uniqueness for functions analytic in the unit disc, *Acta. Math.* **87**, 325–345.
[1957]   Representations of continuous functions, *Math. Z.* **66**, 447–451.
[1958]   An interpolation problem for bounded analytic functions, *Amer. J. Math.* **80**, 921–930.
[1962a]   Interpolations by bounded analytic functions and the corona problem, *Ann. of Math.* **76**, 547–559.
[1962b]   Interpolations by bounded analytic functions and the corona problem, *Proc. Internat. Congr. Math. Stockholm.* 314–316.
[1967]   "Selected Problems on Exceptional Sets." Van Nostrand-Reinhold, Princeton, New Jersey.
[1970]   The corona theorem, *Proc. Scand. Congr., 15th, Oslo* (Lecture Notes in Mathematics, Vol. 118). Springer-Verlag, Berlin and New York.
[1972]   A moment problem and harmonic interpolation, preprint, Inst. Mittag-Leffler, Djursholm.
[1976]   Two remarks on H^1 and BMO, *Advances in Math*, **22**, 269–277.

Carleson, L., and Garnett, J.
[1975]   Interpolating sequences and separation properties, *J. Analyse Math.* **28**, 273–299.

Carleson, L., and Jacobs, S.
[1972]   Best approximation by analytic functions, *Ark. Mat.* **10**, 219–229.

Caughran, J. G., and Shields, A. L.
[1969]   Singular inner factors of analytic functions, *Michigan Math. J.* **16** (1969), 409–410.

438   BIBLIOGRAPHY

Chang, S.-Y.

[1976]   A characterization of Douglas subalgebras, *Acta Math.* **137**, 81–89.

[1977a]   Structure of subalgebras between L^∞ and H^∞, *Trans. Amer. Math. Soc.* **227**, 319–332.

[1977b]   On the structure and characterization of some Douglas subalgebras, *Amer. J. Math.* **99**, 530–578.

Chang, S.-Y., and Garnett, J. B.

[1978]   Analytically of functions and subalgebras of L^∞ containing H^∞, *Proc. Amer. Math. Soc.* **72**, 41–46.

Chang, S.-Y., and Marshall, D. E.

[1977]   Some algebras of bounded analytic functions containing the disc algebra, "Banach Spaces of Analytic Functions" (Lecture Notes in Math., Vol. 604, pp. 12–20). Springer-Verlag, Berlin and New York.

Chaumat, J.

[1978]   Quelques proprietes du couple d'espace vectoriels $(L^1(m)/H^{\infty\perp}, H^\infty)$, preprint. Univ. de Paris-Sud, Orsay.

Cima, J. A., and Petersen, K. E.

[1976]   Some analytic functions whose boundary values have bounded mean oscillation, *Math. Z.* **147**, 237–247.

Cima, J. A., and Schober, G.

[1976]   Analytic functions with bounded mean oscillation and logarithms of H^p-functions, *Math. Z.* **151**, 295–300.

Clark, D. N.

[1968a]   Hankel forms, Toeplitz forms and meromorphic functions, *Trans. Amer. Math. Soc.* **134**, 109–116.

[1968b]   On the spectra of bounded, hermitian, Hankel matrices, *Amer. J. Math.* **90**, 627–656.

[1968c]   On matrices associated with generalized interpolation problems, *Pacific J. Math.* **27**, 241–253.

[1970]   On interpolating sequences and the theory of Hankel and Toeplitz matrices, *J. Func. Anal.* **5**, 247–258.

Cohen, P. J.

[1961]   A note on constructive methods in Banach algebras, *Proc. Amer. Math. Soc.* **12**, 159–163.

Coifman, R. R.

[1974]   A real variable characterization of H^p, *Studia Math.* **51**, 269–274.

Coifman, R. R., and Fefferman, C.

[1974]   Weighted norm inequalities for maximal functions and singular integrals, *Studia Math.* **51**, 241–250.

Coifman, R. R., and Meyer, Y.

[1975]   On commutators of singular integrals and bilinear singular integrals, *Trans. Amer. Math. Soc.* **212**, 315–331.

Coifman, R. R., and Rochberg, R.

[1980]   Another characterization of BMO, *Proc. Amer. Math. Soc.* **79**, 249–254.

Coifman, R. R., Rochberg, R., and Weiss, G.

[1976] Factorization theorems for Hardy spaces in several complex variables, *Ann. of Math.* **103**, 611–635.

Coifman, R. R., and Weiss, G.

[1971] "Analysis harmonique non-commutative sur certains espaces homogènes" (Lecture Notes in Mathematics, Vol. 242). Springer-Verlag, Berlin and New York.

[1977] Extensions of Hardy spaces and their uses in analysis, *Bull. Amer. Math. Soc.* **83**, 569–646.

Collingwood, E. F., and Lohwater, A. J.

[1966] "The Theory of Cluster Sets." Cambridge Univ. Press, London and New York.

Córdoba, A., and Fefferman, C.

[1976] A weighted norm inequality for singular integrals, *Studia Math.* **57**, 97–101.

Dahlberg, B. J. E.

[1980] Approximation of harmonic functions, *Ann. Inst. Fourier (Grenoble)* **30**, 2, 97–107.

Davie, A. M.

[1972] Linear extension operators for spaces and algebras of functions, *Amer. J. Math.* **94**, 156–172.

Davie, A. M., Gamelin, T. W., and Garnett, J. B.

[1973] Distance estimates and pointwise bounded density, *Trans. Amer. Math. Soc.* **175**, 37–68.

Davis, B. J.

[1974] On the weak type $(1, 1)$ inequality for conjugate functions, *Proc. Amer. Math. Soc.* **44**, 307–311.

[1976] On Kolmogorov's inequalities $\|\hat{f}\|_p \le c_p\|f\|_1, 0 < p < 1$, *Trans. Amer. Math. Soc.* **222**, 179–192.

Dawson, D. W.

[1975] Subalgebras of H^∞, Thesis, Indiana Univ., Bloomington, Indiana.

deLeeuw, K., and Rudin, W.

[1958] Extreme points and extreme problems in H_1, *Pacific J. Math.* **8**, 467–485.

Denjoy, A.

[1929] Sur une classe de fonctions analytiques, *C.R. Acad. Sci. Paris, Sér. A.-B* **188**, 140, 1084.

Diximier, J.

[1951] Sur certains espaces considérés par M. H. Stone, *Summa Brasil Math.* **2**, 151–182.

Donoghue, W. F.

[1974] "Monotone Matrix Functions and Analytic Continuation." Springer-Verlag, Berlin and New York.

Douglas, R. G.

[1969] On the spectrum of Toeplitz and Wiener–Hopf operators, in "Abstract Spaces and Approximation Theory," pp. 53–66. Birkhauser, Basel.

[1972] "Banach Algebra Techniques in Operator Theory." Academic Press, New York.

Douglas, R. G., and Rudin, W.
    [1969]   Approximation by inner functions, *Pacific J. Math.* **31**, 313–320.

Douglas, R. G., Shapiro, H. S., and Shields, A. L.
    [1970]   Cyclic vectors and invariant subspaces for the backward shift operator, *Ann. Inst. Fourier (Grenoble)* **20**, 1, 37–76.

Drury, S.
    [1970]   Sur les ensembles de Sidon, *C.R. Acad. Sci. Paris, Sér. A*, **271**, 162–163.

Dunford, N., and Schwartz, J.
    [1958]   "Linear Operators," Part I, Wiley (Interscience), New York.

Duren, P. L.
    [1970]   "Theory of H^p Spaces." Academic Press, New York.

Duren, P. L., Romberg, B. W., and Shields, A. L.
    [1969]   Linear functionals on H^p spaces with $0 < p < 1$, *J. Reine Angew. Math.* **238**, 32–60.

Duren, P. L., Shapiro, H. S., and Shields, A. L.
    [1966]   Singular measures and domains not of Smirnov type, *Duke Math. J.* **33**, 247–254.

Dym, H., and McKean, H. P.
    [1976]   "Gaussian Processes, Function Theory, and the Inverse Spectral Problem." Academic Press, New York.

Earl, J. P.
    [1970]   On the interpolation of bounded sequences by bounded functions, *J. London Math. Soc.* (2) **2**, 544–548.
    [1976]   A note on bounded interpolation in the unit disc, *J. London Math. Soc.* (2) **13**, 419–423.

Fatou, P.
    [1906]   Séries trigonometriques et séries de Taylor, *Acta Math.* **30**, 335–400.

Fefferman, C.
    [1971]   Characterizations of bounded mean oscillation, *Bull. Amer. Math. Soc.* **77**, 587–588.
    [1974]   Harmonic analysis and H^p spaces, in "Studies in Harmonic Analysis" (J. M. Ash, ed.), Studies in Math. Vol. 13, pp. 38–75.

Fefferman, C., and Stein, E. M.
    [1972] H^p spaces of several variables, *Acta Math.* **129**, 137–193.

Fisher, S.
    [1968]   The convex hull of the finite Blaschke products, *Bull. Amer. Math. Soc.* **74**, 1128–1129.
    [1969a]   Another theorem on convex combinations of unimodular functions, *Bull. Amer. Math, Soc.* **75**, 1037–1039.
    [1969b]   Exposed points in spaces of bounded analytic functions, *Duke Math. J.* **36**, 479–484.
    [1971]   Approximation by unimodular functions, *Canad. J. Math.* **23**, 257–269.

Forelli, F.
    [1963a]   The Marcel Riesz theorem on conjugate functions, *Trans. Amer. Math. Soc.* **106**, 369–390.

[1963b] Analytic measures, *Pacific J. Math.* **13**, 571–578.

[1964] The isometries of H^P, *Can. J. Math.* **16**, 721–728.

Fournier, J.

[1974] An interpolation problem for coefficients of H^∞ functions, *Proc. Amer. Math. Soc.* **42**, 402–408.

Frostman, O.

[1935] Potential d'equilibre et capacité des ensembles avec quelques applications á la théorie des fonctions. *Medd. Lunds Univ. Mat. Sem.* **3**, 1–118.

Gamelin, T. W.

[1964] Restrictions of subspaces of $C(X)$, *Trans. Amer. Math. Soc.* **112**, 278–286.

[1969] "Uniform Algebras." Prentice Hall, Englewood Cliffs, New Jersey.

[1970] Localization of the corona problem, *Pacific J. Math.* **34**, 73–81.

[1972] "Lectures on $H^\infty(D)$." Univ. Nacional de la Plata, Argentine.

[1973] Extremal problems in arbitrary domains, *Michigan Math. J.* **20**, 3–11.

[1974] The Silov boundary of $H^\infty(U)$, *Amer. J. Math.* **94**, 79–103.

[1979] "Uniform Algebras and Jensen Measures." London Math. Society Lecture Notes Series, 32. Cambridge Univ. Press, London and New York.

[1981] Wolff's proof of the corona theorem, *Israel J. Math.*, in press.

Gamelin, T. W., and Garnett, J.

[1970] Distinguished homomorphisms and fiber algebras, *Amer. J. Math.* **92**, 455–474.

Gamelin, T. W., Garnett, J. B., Rubel, L. A., and Shields, A. L.

[1976] On badly approximable functions, *J. Approx. Theory* **17**, 280–296.

Garnett, J.

[1971a] Vitushkin's localization operator, *Indiana Univ. Math. J.* **20**, 905–907.

[1971b] Interpolating sequences for bounded harmonic functions, *Indiana Univ. Math. J.* **21**, 187–192.

[1977] Two remarks on interpolation by bounded analytic functions, "Banach Spaces of Analytic Functions" (Baker *et al.*, eds.) (Lecture Notes in Math. Vol. 604). Springer-Verlag, Berlin.

[1978] Harmonic interpolating sequences, L^p and BMO, *Ann. Inst. Fourier (Grenoble)* **28**, 215–228.

Garnett, J. B., and Jones, P. W.

[1978] The distance in BMO to L^∞, *Ann. of Math.* **108**, 373–393.

Garnett, J. B., and Latter, R. H.

[1978] The atomic decomposition for Hardy spaces in several complex variables, *Duke Math. J.* **45**, 815–845.

Garsia, A.

[1970] "Topics in Almost Everywhere Convergence." Markham, Chicago, Illinois.

[1971] A presentation of Fefferman's theorem, unpublished notes.

[1973] "Martingale Inequalities." Benjamin, New York.

Gehring, F. W.

[1973] The L^p-integrability of the partial derivatives of a quasiconformal mapping, *Acta Math.* **130**, 265–277.

Glicksberg, I.

[1962]   Measures orthogonal to algebras and sets of antisymmetry, *Trans. Amer. Math. Soc.* **105**, 415–435.

[1967]   The abstract F. Riesz and M. Riesz Theorem, *J. Funct. Anal.* **1**, 109–122.

Goluzin, G. M.

[1952]   "Geometric Theory of Functions of a Complex Variable, Moscow, 1952, (in Russian) [*English Transl.*: Amer. Math. Soc., Providence, Rhode Island, 1969].

Grenander, U., and Rosenblatt, M.

[1957]   "Statistical Analysis of Stationary Time Series." Wiley, New York.

Grothendieck, A.

[1955]   Une caracterization vectoriellemetrique des espaces L^1, *Canad. J. Math.* **7**, 552–561.

Gundy, R. F., and Wheeden, R. L.

[1974]   Weighted integral inequalities for the non-tangential maximal function, Lusin area integral, and Walsh–Paley series, *Studia Math.* **49**, 107–124.

Hall, T.

[1937]   Sur la mesure harmonique de certains ensembles, *Ark. Mat. Astr. Fys.* **25A**, No. 28.

Hardy, G. H., and Littlewood, J. E.

[1927]   Some new properties of Fourier constants, *Math. Ann.* **97**, 159–209.

[1930]   A maximal theorem with function-theoretic applications, *Acta Math.* **54**, 81–116.

[1932]   Some properties of conjugate functions, *J. Reine Angew. Math.* **167**, 405–423.

[1936]   Some more theorems concerning Fourier series and Fourier power series, *Duke Math. J.* **2**, 354–382.

Hartman, P.

[1958]   On completely continuous Hankel matrices, *Proc. Amer. Math. Soc.* **9**, 862–866.

Havin, V. P.

[1973]   Weak completeness of the space L^1/H_0^1, *Vestnik Leningrad Univ.* **13**, 77–81 (in Russian).

[1974]   The spaces H^∞ and L^1/H_0^1, *Issl. po Lin. Operat. Tieor. Funk. IV, Zap. Naučn Sem. Leningrad. Otdel. Mat. Inst. Steklov (LOMI)* **29**, 120–148 (in Russian).

Havin, V. P., and Vinogradov, S. A.

[1974]   Free interpolation in H^∞ and some other function classes, 1, *Zap. Naučn. Sem. Leningrad. Otdel. Mat. Inst. Steklov (LOMI)* **47**, 15–54 (in Russian).

Havinson, S. Ja.

[1949]   On an extremal problem in the theory of analytic functions, *Usp. Mat. Nauk.* **4**, No. 4 (32), 158–159 (in Russian).

[1951]   On some external problems in the theory of analytic functions, *Moskov. Gos. Univ. Ucen. Zap.* **148**, Matem. 4, 133–143 (in Russian); *Amer. Math. Soc. Transl.* (2) **32**, 139–154.

Hayman, W. K.

[1974]   On a theorem of Tord Hall, *Duke Math. J.* **41**, 25–26.

Hayman, W. K., and Kennedy, P. L.

[1976]   "Subharmonic Functions." Academic Press, New York.

Hayman, W. K., and Pommerenke, Ch.
[1978]   On analytic functions of bounded mean oscillation, *Bull. London Math. Soc* **10**, 219–224.

Heard, E. A., and Wells, J. H.
[1969]   An interpolation problem for subalgebras of H^∞, *Pacific J. Math.* **28**, 543–553.

Helson, H.
[1964]   "Lectures on Invariant Subspaces." Academic Press, New York.

Helson, H., and Sarason, D.
[1967]   Past and future, *Math. Scand.* **21**, 5–16.

Helson, H., and Szegö, G.
[1960]   A problem in prediction theory, *Ann. Mat. Pura Appl.* **51**, 107–138.

Hörmander, L.
[1966]   "An Introduction to Complex Analysis in Several Variables." Van Nostrand-Reinhold, Princeton, New Jersey.
[1967a]   Generators for some rings of analytic functions, *Bull. Amer. Math. Soc.* **73**, 943–949.
[1967b] L^p estimates for (pluri-) subharmonic functions, *Math. Scand.* **20**, 65–78.

Hoffman, K.
[1962a]   "Banach Spaces of Analytic Functions." Prentice Hall, Englewood Cliffs, New Jersey.
[1962b]   Analytic functions and logmodular Banach algebras, *Acta Math.* **108**, 271–317.
[1967]   Bounded analytic functions and Gleason parts, *Ann. of Math.* **86**, 74–111.

Hoffman, K., and Ramsey, A.
[1965]   Algebras of bounded sequences, *Pacific J. Math.* **5**, 1239–1248.

Hoffman, K., and Rossi, H.
[1967]   Extensions of positive weak*-continuous functionals, *Duke Math. J.* **34**, 453–466.

Hoffman, K., and Singer, I. M.
[1957]   Maximal subalgebras of $C(\Gamma)$, *Amer. J. Math.* **79**, 295–305.
[1960]   Maximal algebras of continuous functions, *Acta Math.* **103**, 217–241.

Hunt, R. A., Muckenhoupt, B., and Wheeden, R. L.
[1973]   Weighted norm inequalities for the conjugate function and Hilbert transform, *Trans. Amer. Math. Soc.* **176**, 227–251.

Jensen, J. L. W. V.
[1899]   Sur un nouvel et important théorème de la théorie des fonctions, *Acta Math.* **22**, 219–251.
[1906]   Sur les fonctions convexes et les inégalités entre les valeurs moyennes, *Acta Math.* **30**, 175–193.

Jerison, D.
[1976]   Sur un théorème d'interpolation de R. Nevanlinna, *C.R. Acad. Sci. Paris, Sèr. A* **282**, 1291–1293.

Jewell, N.
[1976] Continuity of Derivations and Uniform Algebras on Odd Spheres. Thesis, Univ. of Edinburgh.

John, F., and Nirenberg, L.
[1961] On functions of bounded mean oscillation, *Comm. Pure Appl. Math.* **14**, 415–426.

Jones, P.
[1979a] A complete bounded complex submanifold of C^3, *Proc. Amer. Math. Soc.* **76**, 305–306.
[1979b] Constructions for BMO ($\mathbb{R}^n$) and $A_p(\mathbb{R}^n)$, *Proc. Symposia in Pure Math.* **35** (1), 409–416.
[1979c] Extension theorems for BMO, *Indiana Univ. Math. J.* **29**, 41–66.
[1980a] Bounded holomorphic functions with all level sets of infinite length, *Michigan Math. J.* **27**, 75–79.
[1980b] Carleson measures and the Fefferman–Stein decomposition of BMO($\mathbb{R}$), *Ann. of Math.* **111**, 197–208.
[1980c] Factorization of A_p weights, *Ann. of Math.* **111**, 511–530.
[1980d] Estimates for the corona problem, *J. Funct. Anal.* **39**, 162–181.
[1981] Ratios of interpolating Blaschke products, *Pacific J. Math.*, in press.

Julia, G.
[1920] Extension d'un lemma de Schwarz, *Acta Math.* **42**, 349–355.

Kahane, J. P.
[1967] Another theorem on bounded analytic functions, *Proc. Amer. Math. Soc.* **18**, 818–826.
[1969] Trois notes sur les ensembles parfaits linéaires, *Enseignement Math.* **15**, 185–192.
[1974] Best approximation in $L^1(T)$, *Bull. Amer. Math. Soc.* **80**, 788–804.

Katznelson, Y.
[1968] "An Introduction to Harmonic Analysis." Wiley, New York.

Kelleher, J. J., and Taylor, B. A.
[1971] Finitely generated ideals in rings of analytic functions, *Math. Ann.* **193**, 225–237.

Kelley, J. L.
[1955] "General Topology." Van Nostrand-Reinhold, Princeton, New Jersey.

Kerr-Lawson, A.
[1965] A filter description of the homomorphisms of H^∞, *Canad. J. Math.* **17**, 734–757.
[1969] Some lemmas on interpolating Blaschke products and a correction, *Canad. J. Math.* **21**, 531–534.

König, H.
[1969] On the Gleason and Harnack metrics for uniform algebras, *Proc. Amer. Math. Soc.* **22**, 100–101.

Kolmogoroff, A. N.
[1925] Sur les fonctions harmoniques conjuquées et les séries de Fourier, *Fund. Math.* **7**, 24–29.

[1941] Stationary sequences in Hilbert space, *Bull. Math. Univ. Moscow* **2**, No. 6 (in Russian).

Koosis, P.
[1971] Weighted quadratic means of Hilbert transforms, *Duke Math. J.* **38**, 609–634.
[1973] Moyennes quadratiques de transformées de Hilbert et fonctions de type exponential, *C.R. Acad. Sci. Paris, Sér. A* **276**, 1201–1204.
[1978] Sommabilité de la fonction maximale et appartenance á H_1, *C.R. Acad. Sci. Paris, Sér. A* **286**, 1041–1043.
[1979] Sommabilité de la fonction maximale et appartenance á H_1, Cas de plusieurs variables, *C.R. Acad. Sci. Paris, Sér. A* **288**, 489–492.
[1980] "Lectures on H_p Spaces," London Math. Society Lecture Notes Series, 40. Cambridge Univ. Press, London and New York.

Landau, E.
[1913] Abschätzung der Koeffizientensumme einer Potenzreihe, *Arch. Math. Phys.* **21**, 42–50, 250–255.
[1916] "Darstellung und Begründung einiger neuerer Ergebnisse der Funktionentheorie." Berlin; [1946] reprinted by Chelsea, New York.

Latter, R. H.
[1978] A decomposition of $H^p(\mathbb{R}^n)$ in terms of atoms, *Studia Math.* **62**, 92–101.

Lindelöf, E.
[1915] Sur un principle générale de l'analyse et ces applications a la théorie de la représentation conforme, *Acta Soc. Sci. Fenn.* **46**, No. 4.

Littlewood, J. E.
[1929] On functions subharmonic in a circle, II. *Proc. London Math. Soc.* **28**, 383–394.

Littlewood, J. E., and Paley, R. E. A. C.
[1931] Theorems on Fourier series and power series, (I), *J. London Math. Soc.* **6**, 230–233; (II) *ibid.* **42**, 52–89; (III) *ibid.* **43**, 105–126.

Loomis, L. H.
[1946] A note on the Hilbert transform, *Bull. Amer. Math. Soc.* **52**, 1082–1086.

Luecking, D.
[1980] The compact Hankel operators form an M-ideal in the space of Hankel operators, *Proc. Amer. Math. Soc.* **79**, 222–224.

Lumer, G.
[1965] Analytic functions and the Dirichlet problem, *Bull. Amer. Math. Soc.* **70**, 98–104.
[1969] "Algèbres de fonctions et espaces de Hardy" (Lecture Notes in Math., Vol. 75). Springer-Verlag, Berlin and New York.

Macintyre, A. J., and Rogosinski, W. W.
[1950] Extremum problems in the theory of analytic functions, *Acta Math.* **82**, 275–325.

Marden, M.
[1966] Geometry of polynomials, *Amer. Math. Soc. Math. Surveys*, No. 3, 2nd ed.

Marshall, D. E.

[1974] An elementary proof of the Pick–Nevanlinna interpolation theorem, *Michigan Math. J.* **21**, 219–233.

[1976a] Blaschke products generate H^∞, *Bull. Amer. Math. Soc.* **82**, 494–496.

[1976b] Subalgebras of L^∞ containing H^∞, *Acta Math.* **137**, 91–98.

[1976c] Approximation and interpolation by inner functions. Thesis, Univ. of California, Los Angeles, California.

Meyers, N. G.

[1964] Mean oscillation over cubes and Hölder continuity, *Proc. Amer. Math. Soc.* **15**, 717–721.

Mooney, M. C.

[1973] A theorem on bounded analytic functions, *Pacific J. Math.* **43**, 457–463.

Muckenhoupt, B.

[1972] Weighted norm inequalities for the Hardy maximal function, *Trans. Amer. Math. Soc.* **165**, 207–226.

[1974] The equivalence of two conditions for weighted functions, *Studia Math.* **49**, 101–106.

[1979] Weighted norm inequalities for classical operators, *Proc. Symposia in Pure Math.* **35** (1), 69–84.

Muckenhoupt, B., and Wheeden, R. L.

[1974] Weighted norm inequalities for fractional integrals, *Trans. Amer. Math. Soc.* **192**, 261–274.

Natzitz, B.

[1970] A note on interpolation, *Proc. Amer. Math. Soc.* **25**, 918.

Negrepontis, S.

[1967] On a theorem of Hoffman and Ramsey, *Pacific J. Math.* **20**, 281–282.

Neri, U.

[1977] Some properties of functions with bounded mean oscillation, *Studia Math.* **61**, 63–75.

Neuwirth, J., and Newman, D. J.

[1967] Positive $H^{1/2}$ functions are constant, *Proc. Amer. Math. Soc.* **18**, 958.

Nevanlinna, F., and Nevanlinna, R.

[1922] Eigenschaften analyscher Funktionen in der Umgebung einer singulären Stelle oder Linie, *Acta Soc. Sci. Fenn.* **50**, No. 5.

Nevanlinna, R.

[1919] Über beschränkte Funktionen, die in gegebenen Punkten vorgeschrieben Werte annehmen, *Ann. Acad. Sci. Fenn. Ser. A*, **13**, No. 1.

[1929] Über beschränkte analytische Funktionen, *Ann. Acad. Sci. Fenn.* **32**, No. 7.

[1953] "Eindeutige analytische Funktionen," Vol. 2, Auflage. Springer-Verlag, Berlin and New York.

Newman, D. J.

[1959a] Interpolation in H^∞, *Trans. Amer. Math. Soc.* **92**, 501–507.

[1959b] Some remarks on the maximal ideal space structure of H^∞, *Ann. of Math.* **70**, 438–445.

Nikolski, N. K., Havin, V. P., and Kruschev, S. V.

[1978] "Investigations on Linear Operators and the Theory of Functions. 99 Unsolved Problems in Linear and Complex Analysis," *Zap. Naucn. Sem. Leningrad. Otdel. Mat. Inst. Steklor.* (LOMI), **81**.

Noshiro, K.

[1960] "Cluster Sets." Springer-Verlag, Berlin and New York.

Ohtsuka, M.

[1954] Note on functions bounded and analytic in the unit circle, *Proc. Amer. Math. Soc.* **5**, 533–535.

Øksendal, B. K.

[1971] A short proof of the F. and M. Riesz theorem, *Proc. Amer. Math. Soc.* **30**, 204.

Øyma, K.

[1977] Extremal interpolatory functions in H^∞, *Proc. Amer. Math. Soc.* **64**, 272–276.

Paley, R. E. A. C.

[1933] On the lacunary coefficients of power series, *Ann. of Math.* **34**, 615–616.

Parreau, M.

[1951] Sur les moyennes des fonctions harmoniques et analytiques et la classification des surfaces de Riemann, *Ann. Inst. Fourier (Grenoble)* **3**, 103–197.

Pelczynski, A.

[1977] "Banach Spaces of Analytic Functions and Absolutely Summing Operators." Americal Mathematical Society, Providence, Rhode Island.

Petersen, K. E.

[1977] "Brownian Motion, Hardy Spaces and Bounded Mean Oscillation." London Math. Society Lecture Notes Series, 28, Cambridge Univ. Press, London and New York.

Phelps, R. R.

[1965] Extreme points in function algebras, *Duke Math. J.* **32**, 267–277.

Pichorides, S. K.

[1972] On the best values of the constants in the theorems of M. Riesz, Zygmund and Kolmogorov, *Studia Math.* **44**, 165–179.

Pick, G.

[1916] Uber die Beschränkungen analytische Funktionen, welche durch vorgegebene Funktionswerte bewirkt werden, *Math. Ann.* **77**, 7–23.

Piranian, G.

[1966] Two monotonic, singular, uniformly almost smooth functions, *Duke Math. J.* **33**, 255–262.

Piranian, G., Shields, A., and Wells, J. H.

[1967] Bounded analytic functions and absolutely continuous measures, *Proc. Amer. Math. Soc.* **18**, 818–826.

Piranian, G., and Weitsman, A.

[1978] Level sets of infinite length, *Comment. Math. Helv.* **53**, 161–164.

Plessner, A. I.

[1927] Über das Verhalten analytischer Funktionen am Rande ihres Definitionsbereichs, *J. Reine Angew. Math.* **158**, 219–227.

Pommerenke, C.

[1970]   On Bloch functions, *J. London Math. Soc.* (2) **2**, 689–695.

[1977]   Schlichte Funktionen und analytische Funktionen beschränkter mittler Oszillation, *Comment. Math. Helv.* **52**, 591–602.

[1978]   On univalent functions, Bloch functions and VMOA, *Math. Ann.* **236**, 199–208.

Poreda, S. J.

[1972]   A characterization of badly approximable functions, *Trans. Amer. Math. Soc.* **169**, 249–256.

Privalov, I. I.

[1916]   Sur les fonctions conjugées, *Bull. Soc. Math. France* **44**, 100–103.

[1919]   Intégrale de Cauchy, *Bull. Univ. Saratov* 1–104.

[1956]   "Randeigenschaften Analytischer Funktionen." Deutscher V. der Wiss., Berlin.

Protas, D.

[1973]   Blaschke products with derivative in H^p and B^p, *Michigan Math. J.* **20**, 393–396.

Rao, K. V. R.

[1967]   On a generalized corona problem, *J. Analyse Math.* **18**, 277–278.

Riesz, F.

[1920]   Über Potenzreihen mit vorgeschrieben Anfangsglieder, *Acta Math.* **42**, 147–171.

[1923]   Über die Randwerte einer analytischen Funktionen, *Math. Z.* **18**, 87–95.

[1926]   Sur les fonctions subharmoniques et leur rapport à la théorie du potentiel, I, *Acta Math.* **48**, 329–343.

[1930]   *Ibid.*, II, *Acta Math.* **54**, 321–360.

Riesz, F., and Riesz, M.

[1916]   Über Randwerte einer analytischen Funktionen, *Quatrième Congrès des Math. Scand. Stockholm*, pp. 27–44.

Riesz, M.

[1924]   Les fonctions conjugées et les séries de Fourier, *C. R. Acad. Sci. Paris, Sér. A-B* **178**, 1464–1467.

[1927]   Sur les fonctions conjugées, *Math. Z.* **27**, 218–244.

Rogosinski, W. W., and Shapiro, H. S.

[1953]   On certain extremum problems for analytic functions, *Acta Math.* **90**, 287–318.

Rosay, J. P.

[1968]   Sur un problème posé par W. Rudin, *C. R. Acad. Sci. Paris, Sér. A.* **267**, 922–925.

Roydan, H.

[1962]   The boundary values of analytic and harmonic functions, *Math. Z.* **78**, 1–24.

[1965]   Algebras of bounded analytic functions on Riemann surfaces, *Acta Math.* **144**, 113–141.

Rudin, W.

[1955a]   Analytic functions of class H^p, *Trans. Amer. Math. Soc.* **78**, 46–56.

[1955b]   The radial variation of analytic functions, *Duke Math. J.* **22**, 235–242.

[1956]   Boundary values of continuous analytic functions, *Proc. Amer. Math. Soc.* **7**, 808–811.

[1967]   A generalization of a theorem of Frostman, *Math. Scand.* **21**, 136–143.

[1969]   Convex combinations of unimodular functions, *Bull. Amer. Math. Soc.* **75**, 795–797.

[1974]   "Real and Complex Analysis," 2nd ed. McGraw-Hill, New York.

Sarason, D.

[1967]   Generalized interpolation in H^∞, *Trans. Amer. Math. Soc.* **127**, 179–203.

[1972]   Approximation of piecewise continuous functions by quotients of bounded analytic functions, *Canad. J. Math.* **24**, 642–657.

[1973]   Algebras of functions on the unit circle, *Bull. Amer. Math. Soc.* **79**, 286–299.

[1975]   Functions of vanishing mean oscillation, *Trans. Amer. Math. Soc.* **207**, 391–405.

[1976]   Algebras between L^∞ and H^∞, "Spaces of Analytic Functions" (Lecture Notes in Math. Vol. 512, pp. 117–129). Springer-Verlag, Berlin and New York.

[1979]   "Function Theory on the Unit Circle." Virginia Poly. Inst. and State Univ., Blacksburg, Virginia.

Schark, I. J.

[1961]   The maximal ideals in an algebra of bounded analytic functions, *J. Math. Mech.* **10**, 735–746.

Schur, I.

[1917]   Über Potenzreihen die im Innern des Einheitskreises beschränkt sind, *J. Reine Angew. Math.* **147**, 205–232.

[1918]   *Ibid., J. Reine Angew. Math.* **148**, 122–145.

Seidel, W.

[1934]   On the distribution of values of bounded analytic functions, *Trans. Amer. Math. Soc.* **36**, 201–226.

Shapiro, H. S.

[1968]   Generalized analytic continuation, "Symposia in Theoretical Physics and Mathematics," Vol. 8. Plenum Press, New York.

Shapiro, H. S., and Shields, A.

[1961]   On some interpolation problems for analytic functions, *Amer. J. Math.* **83**, 513–532.

Sidney, S. J., and Stout, E. L.

[1968]   A note on interpolation, *Proc. Amer. Math. Soc.* **19**, 380–382.

Sine, R.

[1967]   On a paper of Phelps, *Proc. Amer. Math. Soc.* **18**, 484–486.

Smirnov, V. I.

[1929]   Sur les valeurs limits des fonctions, regulière à l'intérieur d'un cercle, *J. Soc. Phys. Math. Léningrade* **2**, No. 2, 22–37.

[1933]   Über die Ränderzuordnung bei konformer Abbildung, *Math. Ann.* **107**, 313–323.

Somadasa, H.
[1966] Blaschke products with zero tangential limits, *J. London Math. Soc.* **41**, 293–303.

Spanne, S.
[1966] Sur l'interpolation entre les espaces L $_k^{p,\Phi}$, *Ann. Scuola Norm Sup. Pisa.* (3) **20**, 625–648.

Srinivasan, T., and Wang, J.-K.
[1965] On closed ideals of analytic functions, *Proc. Amer. Math. Soc.* **16**, 49–52.

Stegenga, D. A.
[1976] Bounded Toeplitz operators on H^1 and applications of the duality between H^1 and the functions of bounded mean oscillation, *Amer. J. Math.* **98**, 573–589.
[1979] A geometric condition which implies BMOA, *Proc. Symp. Pure Math.* **35** (1), 427–430.

Stein, E. M.
[1967] Singular integrals, harmonic functions and differentiability properties of functions of several variables, *Proc. Symp. in Pure Math.* **10**, 316–335.
[1969] Note on the class $L \log L$, *Studia Math.* **31**, 305–310.
[1970] "Singular Integrals and Differentiability Properties of Functions." Princeton Univ. Press, Princeton, New Jersey.

Stein, E. M., and Weiss, G.
[1959] An extension of a theorem of Marcinkiewicz and some of its applications, *J. Math. Mech.* **8**, 263–284.
[1960] On the theory of harmonic functions of several variables, *Acta Math.* **103**, 26–62.
[1971] "Introduction to Fourier Analysis on Euclidean Spaces." Princeton Univ. Press, Princeton, New Jersey.

Stein, P.
[1933] On a theorem of M. Riesz, *J. London Math. Soc.* **8**, 242–247.

Stout, E. L.
[1971] "The Theory of Uniform Algebras." Bogden and Quigley, Belmont, California.

Stromberg, J.-O.
[1976] Bounded mean oscillation with Orlicz norms and duality of Hardy spaces, *Bull. Amer. Math. Soc.* **82**, 953–955.
[1979] Bounded mean oscillation with Orlicz norms and duality of Hardy spaces, *Indiana Univ. Math. J.* **28**, 511–544.

Szegö, G.
[1920] Beiträge zur Theorie der Toeplitzschen Formen, I, *Math. Z.* **6**, 167–202.
[1921] Über die Randwerte einer analytischen Funktionen, *Math. Ann.* **84**, 232–244.

Sz.-Nagy, B., and Korányi, A.
[1956] Relations d'un problem de Nevanlinna et Pick avec la théorie des opérateurs de l'espace Hilbertien, *Acta Math. Acad. Sci. Hungar.* **7**, 295–303.

Taylor, B. A., and Williams, D. L.
[1970] The peak sets of A^m, *Proc. Amer. Math. Soc.* **24**, 604–606.

[1971]   Zeros of Lipschitz functions analytic in the unit disc, *Michigan Math. J.* **18**, 129–139.

Toeplitz, O.
[1911]   Über die Fouriersche Entwicklung positiver Funktionen, *Rend. Circ. Mat. Palermo.* **32**, 191–192.

Tsuji, M.
[1959]   "Potential Theory in Modern Function Theory." Maruzen, Tokyo.

Uchiyama, A.
[1981]   The construction of certain BMO functions and the corona problem, *Pacific J. Math.*, in press.

Varopoulos, N. Th.
[1970]   Groups of continuous functions in harmonic analysis, *Acta Math.* **125**, 109–154.
[1971a] Ensembles pics et ensembles d'interpolation pour les algebras uniformes, *C.R. Acad. Sci. Paris, Sér. A.* **272**, 866–867.
[1971b] Sur la reunion de deux ensembles d'interpolation d'une algèbre uniforme, *C.R. Acad. Sci. Paris, Sér. A.* **272**, 950–952.
[1971c] Un problème d'extension linéare dans les algèbres uniforms, *Ann. Inst. Fourier (Grenoble)* **21**, 263–270.
[1972]   Sur un problème d'interpolation, *C.R. Acad. Sci. Paris, Sér. A.* **274**, 1539–1542.
[1977a] BMO functions and the $\bar{\partial}$ equation, *Pacific J. Math.* **71**, 221–273.
[1977b] A remark on BMO and bounded harmonic functions, *Pacific J. Math.* **73**, 257–259.
[1980]   A probabilistic proof of the Garnett–Jones Theorem on BMO, *Pacific J. Math.* **90**, 201–221.

Walsh, J. L.
[1960]   "Interpolation and Approximation by Rational Functions in the Complex Domain." Americal Mathematical Society, Providence, Rhode Island.

Weiss, G.
[1979]   Some problems in the theory of Hardy spaces, *Proc. Symp. Pure Math.* **35** (1), 189–200.

Wermer, J.
[1953]   On algebras of continuous functions, *Proc. Amer. Math. Soc.* **4**, 866–869.
[1960]   Dirichlet algebras, *Duke Math. J.* **27**, 373–382.
[1964]   "Seminar über Funktionen-Algebren" (Lecture Notes in Math. Vol. 1). Springer-Verlag, Berlin and New York.
[1971]   "Banach Algebras and Several Complex Variables." Markham, Chicago, Illinois.

Wolff, T.
[1979]   Some theorems on vanishing mean oscillation, Thesis, Univ. of California, Berkeley, California.
[1980]   Oral communication.

Yang, P.
[1977]   Curvature of complex manifolds, *Proc. Symposia in Pure Math.* **30** (2), 135–137. Amer. Math. Soc., Providence, Rhode Island.

Ziskind, S.
    [1976]   Interpolating sequences and the Shilov boundary of $H^\infty(\Delta)$, *J. Functional Anal.* **21**, 380–388.

Zygmund, A.
    [1929]   Sur les fonctions conjugées, *Fund. Math.* **13**, 284–303.
    [1955]   "Trigonometrical Series." Dover, New York.
    [1968]   "Trigonometric Series," 2nd ed. Cambridge Univ. Press, London and New York.

Index